State-of-the-Art Wastewater Treatment Techniques

State-of-the-Art Wastewater Treatment Techniques

Guest Editors

Yi Yang
Ying Mei
Fuqiang Fan
Shangwei Zhang

MDPI

Basel • Beijing • Wuhan • Barcelona • Belgrade • Novi Sad • Cluj • Manchester

Guest Editors
Yi Yang
Faculty of Arts and Sciences
Beijing Normal University
Zhuhai
China

Ying Mei
Advanced Institute of Natural Sciences
Beijing Normal University
Zhuhai
China

Fuqiang Fan
Advanced Institute of Natural Sciences
Beijing Normal University
Zhuhai
China

Shangwei Zhang
Advanced Interdisciplinary Institute of Environment and Ecology
Beijing Normal University
Zhuhai
China

Editorial Office
MDPI AG
Grosspeteranlage 5
4052 Basel, Switzerland

This is a reprint of the Special Issue, published open access by the journal *Processes* (ISSN 2227-9717), freely accessible at: https://www.mdpi.com/si/processes/744O88PC32.

For citation purposes, cite each article independently as indicated on the article page online and as indicated below:

Lastname, A.A.; Lastname, B.B. Article Title. *Journal Name* **Year**, *Volume Number*, Page Range.

ISBN 978-3-7258-5579-7 (Hbk)
ISBN 978-3-7258-5580-3 (PDF)
https://doi.org/10.3390/books978-3-7258-5580-3

Cover image courtesy of Yi Yang

Contents

About the Editors

Yi Yang

Yi Yang, Ph.D., is an Associate Professor and Ph.D. supervisor of Beijing Normal University. He has been honored with the 2024 IChemE Global Award—Young Researcher Award by the Institution of Chemical Engineers (IChemE) and the 2024 Processes Young Investigator Award by MDPI, and was listed in World's Top 2% Scientists by Stanford–Elsevier in 2025. He has long-term engagement in wastewater treatment and the design of novel porous materials. Dr. Yi Yang serves as an editorial board member and guest editor for renowned academic publishers such as Elsevier, MDPI, Frontiers, Bon View, and Bentham Science, and also acts as a reviewer for prestigious academic journals. In recent years, he has published over 40 papers in famous academic journals, including several ESI highly cited papers, hot papers, and cover articles, and holds more than 10 international patents, four of which have been commercially applied. His accolades include the Gold Medal and Special Award at the 2024 London International Invention and Trade Fair (one of the world's most famous invention exhibitions), the Gold Medal at the Moscow Archimedes International Invention and Innovation Exhibition, the Prix Eiffel International Invention and Innovation Contest, the China International Invention Innovation Exhibition, among other domestic and international awards.

Ying Mei

Ying Mei, Ph.D., is currently an Associate Professor and Ph.D. supervisor at Beijing Normal University, Zhuhai Campus. Her research focuses on the development of novel membrane materials and innovative membrane processes for unconventional water treatment. She has long been engaged in advancing membrane technologies for emerging contaminant removal, valuable resource recovery, and clean energy generation. Her work aims to establish high-efficiency, low-energy membrane fabrication and process design strategies by integrating fundamental studies on membrane separation mechanisms, formation processes of selective layers, and system-level process optimization through combined experimental and numerical simulation approaches. She has published 16 papers as first or corresponding author in leading SCI journals in environmental and materials science, including *Environmental Science & Technology*, *Water Research*, *ACS Energy Letters*, *Journal of Membrane Science*, and *Desalination*, with one paper recognized as an ESI Highly Cited Paper. She has served as Principal Investigator for projects funded by the National Natural Science Foundation of China (NSFC) and the Key Research Program of the Department of Education of Guangdong Province, and has actively contributed to major collaborative research projects supported by the Research Grants Council of Hong Kong (HKD 6.65 million), the Innovation and Technology Fund of Hong Kong, and The University of Hong Kong's Excellence Research Initiative.

Fuqiang Fan

Fuqiang Fan is an Associate Professor of hydraulic engineering at the Advanced Institute of Natural Sciences, Beijing Normal University at Zhuhai. He has been the deputy director of the Center for Water Research since 2020. His main research interests include water pollution control, odor treatment, and ecological restoration of oil reservoir/soil environment. He demonstrates a profound research accumulation in the development of soil and water pollution control and remediation technologies, as well as environmental microbial research and application, especially in the development of high-tech phosphorus removal materials and algae prevention and control. He has published over 60 academic papers in journals such as *Water Research*, *Chemical Engineering*

Journal, *Journal of Hazardous Materials*, etc., and obtained more than ten patent authorizations. He shows expertise in a multitude of chemical and microbial environmental techniques.

Shangwei Zhang

Shangwei Zhang is an Associate Professor at Beijing Normal University. He seeks to understand how microorganisms transform persistent organohalides in order to address the longstanding environmental challenges raised by organohalides. His primary research interest is decoding the mechanisms of microbial organohalide respiration through laboratory experiments and quantum chemical computations. So far, he has unraveled a series of mechanisms underlying microbial organohalide respiration, including electron transfer, interaction mode, and microscopic reaction processes in this microbial-mediated event. It furthers the understanding of organohalide respiration.

processes

MDPI

Editorial

Advancing the Frontiers of Wastewater Treatment—Synthesis and Future Perspectives in State-of-the-Art Techniques

Yi Yang [1,*], Ying Mei [2,*], Fuqiang Fan [3,*] and Shangwei Zhang [4,*]

1 Faculty of Arts and Sciences, Beijing Normal University, Zhuhai 519087, China
2 Research and Development Center for Watershed Environmental Eco-Engineering, Advanced Institute of Natural Sciences, Beijing Normal University, Zhuhai 519087, China
3 Center for Water Research, Advanced Institute of Natural Sciences, Beijing Normal University, Zhuhai 519087, China
4 Guangdong Provincial Key Laboratory of Wastewater Information Analysis and Early Warning, Advanced Interdisciplinary Institute of Environment and Ecology, School of Technology for Sustainability, Beijing Normal University, Zhuhai 519087, China
* Correspondence: yiyang@bnu.edu.cn (Y.Y.); meiying@bnu.edu.cn (Y.M.); ffan02@bnu.edu.cn (F.F.); zhangshangwei@bnu.edu.cn (S.Z.)

Water scarcity, exacerbated by population growth, industrialization, and climate change, demands innovative and sustainable solutions [1–3]. Wastewater treatment stands as a cornerstone in addressing this global crisis, transforming polluted effluents into valuable water resources [4,5]. This Special Issue presents cutting-edge research addressing the complex challenges of modern wastewater treatment, covering novel materials, process optimization, mechanistic insights, predictive modeling, and more. The field of wastewater treatment is undergoing a dynamic transformation, shaped by several key trends.

(1) Advanced material development. Significant research efforts have focused on engineering functional materials, such as modified biochars, nanocomposites, and specialized adsorbents, tailored for pollutant removal (e.g., heavy metals, nutrients, pharmaceuticals, etc.) [6–10]. Surface functionalization and hybrid material design enhance selectivity and capacity. Beyond mere pollutant removal, technologies are increasingly targeting the recovery of valuable resources like nutrients (N, P), energy, and embedded materials (e.g., metals from waste streams), promoting circular economy principles [11–14].

(2) Sustainable wastewater treatment technology. Understanding the complex microbial consortia in treatment systems (e.g., aerobic granular sludge, biofilms in wetlands) and manipulating community structure for enhanced nutrient removal and stability is becoming a major research thrust [15–17]. Persistent and potentially hazardous pollutants, including microplastics, pharmaceutical residues, antibiotic resistance genes (ARGs), and industrial solvents, remain a research priority [18–21]. Engineered natural systems, such as constructed wetlands, continue to be optimized for their cost-effectiveness and ecological benefits, particularly in decentralized treatment and non-point source pollution control [22–25].

(3) Process intensification and hybridization. The integration of physical, chemical, and biological processes (e.g., coupling adsorption with biological degradation or combining membranes with advanced oxidation) enhances the efficiency and resilience of the water treatment systems, particularly for recalcitrant compounds [26–31]. The application of sophisticated machine learning (ML) algorithms and statistical models for predicting treatment performance, pollutant fate, material behavior, and system optimization is rapidly growing, enabling smarter design and operation [32–35].

Despite significant progress in the field, several critical challenges remain.

(1) Translating innovation into practical applications. Scaling up the synthesis of novel materials cost-effectively and ensuring their long-term stability, regenerability, and safe disposal, and translating laboratory-scale innovations into reliable, large-scale implementation remain key barriers. Furthermore, many promising technologies lack rigorous, long-term field validation under realistic, variable operating conditions, hindering confidence and adoption. Complex processes like Advanced Oxidation Processes (AOPs) face issues such as catalyst deactivation and high resource demands, limiting real-world adoption.

(2) Complexity for effective prediction and control. The inherent complexity of wastewater systems and pollutant behavior presents fundamental scientific and engineering challenges. Elucidating precise mechanisms of pollutant removal within heterogeneous matrices containing contaminant mixtures requires deeper investigation. For example, the full lifecycle impact of microfibers and microplastics within treatment plants—including their fragmentation, interactions with other pollutants, and effects on microbial communities and sludge properties—is still poorly understood. Developing robust and generalizable predictive models capable of accurately forecasting treatment performance, pollutant fate, and system response under dynamic conditions (e.g., shock loads, seasonal changes) is hampered by insufficient data and modeling limitations.

(3) Ensuring holistic sustainability and impact management. Truly sustainable water treatment extends beyond simple removal efficiency, assessing and optimizing the broader environmental and economic footprint. Comprehensive techno-economic analyses (TEA) and lifecycle assessments (LCA) comparing novel technologies (especially those focused on resource recovery) to conventional methods are often incomplete, making holistic sustainability claims difficult. Ultimately, achieving sustainability requires integrating solutions that minimize negative environmental consequences while maximizing resource recovery and economic viability.

As a result, this Special Issue addresses these gaps through contributions in three core areas:

(1) Engineered adsorbents and functional materials. Feng et al. provide a review focusing on improving persulfate (PS) activation strategies, emphasizing the role of chelating/reducing agents in overcoming inherent limitations like Fe(III) accumulation/sluggish reaction rates for efficient and sustainable contaminant destruction. Struhs et al. and Mirkouei et al. explore nutrient recovery (N, P) from aquaculture effluent using modified biochar and contribute to closing the loop in fish farming. The focus on a specific real-world application (Idaho aquaculture) provides valuable field-relevant data. Geng et al. identify acid-modified rice husk biochar as an optimal adsorbent for polycyclic aromatic hydrocarbons (PAHs) and a microbial carrier for Rhodococcus sp. DG1. Their study details adsorption interactions (π-π/n-π EDA, H-bonding, hydrophobicity) and demonstrates field applicability potential with significantly enhanced phenanthrene degradation. They also highlight the crucial impact of immobilization methods (adsorption vs. adsorption–embedding) on microbial activity and nutrient access, a key consideration for hybrid technologies. Niu et al. present a multifunctional, sustainable adsorbent derived from iron-rich microalgae, capable of high Cu^{2+} adsorption, efficient oil/water emulsion separation, and photo-Fenton antifouling activity, offering a solution for complex wastewaters. The integrated functionality (adsorption, separation, self-cleaning) addresses the need for versatile materials.

(2) Microbial processes, community dynamics, and biological treatment. Onyedibe et al. investigate the impact of polyester/denim microfibers (MFs) on aerobic granular sludge (AGS) systems. They provide vital mechanistic insights, linking decline to nutrient transport blockage and toxic leachates, and strongly advocate for upstream MF control. Zheng et al. explore algae-assisted nitrogen fixation and nitrification to diversify the sources of irrigation water. Yuan et al. focus on field-scale performance, examining nitrogen removal in lakeshore multicell constructed wetlands (MCWs) under seasonal nutrient shocks, emphasizing the role of forebay storage ponds in mitigating shock loads and identifying hydraulic factors as primary drivers of microbial community variation. They reveal how shock loads and periodic saturation create complex micro-environments that foster diverse nitrogen-cycling microbes, enhancing resilience and performance. They also expand the view to receiving environments, linking water level fluctuations (WLF) and vegetation cover to sediment microbial communities and N-cycling genes in Lake Erhai, demonstrating that WLF, by creating oxic–anoxic interfaces, filters for adaptable microbes and increases abundance of key functional genes. Mai et al. investigate cyanobacteria interactions and the management of blue-green algae outbreaks in reservoirs. Csutak et al. explore a novel application of yeast (C. parapsilosis CMGB-YT) to produce biosurfactants that enhance heavy metal removal (Pb, Cd) by Rhodotorula mucilaginosa from wastewater and inhibit pathogenic biofilm. They highlight the potential of microbial-derived surfactants for multi-contaminant management in complex waste streams.

(3) Modeling, risk assessment, and remediation strategies. Jung et al. provide a comprehensive review of the pollution issues associated with shipyards (heavy metals, petroleum hydrocarbons, and organic solvents) and explore effective strategies for soil and groundwater remediation. Cheng et al. apply a sophisticated Copula Bayesian Network model to Qilu Lake. It quantifies spatial and temporal pollution risks, identifies key drivers (agricultural runoff, water scarcity), and establishes early warning thresholds (total nitrogen TN, total phosphorus TP, COD) for "deteriorated" status. This provides a powerful tool in proactive lake management and data-driven decision-making, moving beyond simple descriptive monitoring. Hou et al. quantify the disproportionate impact of extreme rainfall (ER) events on surface runoff and diffuse nitrogen loss in a subtropical monsoon watershed. They reveal the limited effectiveness of natural systems (forests, pastures) in controlling losses during ER and emphasize the dominance of water-soluble ON-N and the amplifying role of antecedent soil moisture. Rodziewicz et al. investigate the effect of carbon dosing on nutrient removal in a rotating electro-biological disk contactor (REBDC) treating hydroponic tomato wastewater and evaluate the impact on effluent COD, finding that the rational solution enables high substrate utilization and nitrogen removal. Wang et al. bridge the gap between lab-scale remediation concepts and field implementation for complex industrial contaminants, providing critical validated field data and a practical framework for remediating groundwater contaminated by challenging organic solvents. Shi et al. introduce the novel IDISO metaheuristic algorithm, enhancing the DISO algorithm with nonlinear shrinking and Cauchy mutation, to optimize XGBoost model hyperparameters. Demonstrating superior performance on hydrochar element prediction, they provide a valuable computational tool for accelerating material design and characterization in the biochar/hydrochar domain. Zhao et al. highlight the detection of diverse ARGs in a reclaimed water river system, linking their distribution to environmental factors. This study contributes to understanding the drivers of antibiotic resistance in water reuse scenarios, essential for managing public health risks associated with water reclamation.

Building on the insights from this Special Issue and recognizing the persisting challenges, several future directions emerge. Regenerable and scalable functional materials with enhanced stability and specificity are still important. Promising avenues include improved modification techniques (e.g., co-doping, nanohybrids), biocomposite design, and stimuli-responsive materials (responsive to pH, light, or specific triggers). Correspondingly, advanced process understanding and hybridization promote the application of novel materials. For example, advanced in situ techniques should be applied to investigate real-time adsorption, degradation, and microbial interaction mechanisms. Systematic design of hybrid systems, combining physical (e.g., membranes, adsorption), chemical (e.g., tailored AOPs), and biological processes (e.g., AGS, CWs), holds potential for tailored treatment solutions. Enhancing system robustness against variability is another point to be considered, like integrating climate projection into the design and management of treatment systems and developing models predicting performance under extreme hydrology. Leveraging AI or ML and sensor networks for real-time system monitoring and control can help to improve the accuracy and convenience of the operation. Based on the above, accelerating translation to practice is the final target, and long-term pilot- and full-scale demonstrations of promising technologies across diverse geographies and wastewater types are essential. Collaborative platforms for sharing comprehensive datasets on treatment performance, material properties, and environmental conditions will support model generalizability and accelerate practical adoption. Overall, we hope this Special Issue inspires continued innovation and collaboration, paving the way for state-of-the-art wastewater treatment technologies to become standard practice in the pursuit of sustainable water management.

Author Contributions: Conceptualization, Y.Y.; resources, Y.Y., Y.M., F.F., S.Z.; data curation, Y.Y., Y.M., F.F., S.Z.; writing—original draft preparation, Y.Y., Y.M., F.F., S.Z.; writing—review and editing, Y.Y., Y.M., F.F., S.Z.; project administration, Y.Y. All authors have read and agreed to the published version of the manuscript.

Conflicts of Interest: The authors declare no conflict of interest.

List of Contributions:

1 Feng, L.; Zeng, Y.; Wang, P.; Duan, N.; Ji, H.; Zhao, X. A mini-review on the use of chelating or reducing agents to improve fe(ii)-fe(iii) cycles in persulfate/fe(ii) systems. *Processes* **2024**, *12*, 2361.

2 Struhs, E.; Bare, W.F.R.; Mirkouei, A. Overturf, Magnesium-modified biochar for removing phosphorus from aquaculture facilities: A case study in idaho, usa. *Processes* **2025**, *13*, 1021.

3 Bare, W.F.R.; Struhs, E.; Mirkouei, A.; Overturf, K.; Chacón-Patiño, M.L.; McKenna, A.M.; Chen, H.; Raja, K.S. Controlling eutrophication of aquaculture production water using biochar: Correlation of molecular composition with adsorption characteristics as revealed by ft-icr mass spectrometry. *Processes* **2023**, *11*, 2883.

4 Geng, S.; Mao, S.; Xu, G.; Ding, A.; Chen, F.; Dou, J.; Fan, F. Performance evaluation of modified biochar as a polycyclic aromatic hydrocarbon adsorbent and microbial-immobilized carrier. *Processes* **2024**, *12*, 2939.

5 Niu, X.; Si, J.; Chen, B.; Wang, Q.; Zeng, S.; Cui, Z. Preparation of bioaerogel from iron-rich microalgae for the removal of water pollutants. *Processes* **2024**, *12*, 1313.

6 Onyedibe, V.O.; Waseem, H.; Aqeel, H.; Liss, S.N.; Gilbride, K.A.; Sühring, R.; Hamza, R. Influence of polyester and denim microfibers on the treatment and formation of aerobic granules in sequencing batch reactors. *Processes* **2025**, *13*, 2272.

7 Zheng, H.; Wang, X.; Huang, C.; Bao, Z.; Zhao, X.; Tan, Z.; Xie, E. Effects of irrigation with slightly algae-contaminated water on soil moisture, nutrient redistribution, and microbial community. *Processes* **2024**, *12*, 1639.

8 Yuan, J.; Wang, B.; Hou, Z.; Peng, J.; Li, D.; Chu, Z. Response of nitrogen removal performance and microbial distribution to seasonal shock nutrients load in a lakeshore multicell constructed wetland. *Processes* **2023**, *11*, 2781.
9 Yuan, J.; Cao, J.; Liao, W.; Zhu, F.; Hou, Z.; Chu, Z. Effects of vegetation cover varying along the hydrological gradient on microbial community and n-cycling gene abundance in a plateau lake littoral zone. *Processes* **2024**, *12*, 1276.
10 Mai, Y.; Hong, C.; Liu, D.; Yang, F.; Xiao, G.; Zhang, Z.; Liu, S. Dynamics of bacterial communities and identification of microbial indicators in a cylindrospermopsis-bloom reservoir in western Guangdong province, China. *Processes* **2025**, *13*, 2129.
11 Csutak, O.E.; Nicula, N.; Lungulescu, E.; Marinescu, V.E.; Gifu, I.C.; Corbu, V.M. Candida parapsilosis cmgb-yt biosurfactant for treatment of heavy metal- and microbial-contaminated wastewater. *Processes* **2024**, *12*, 1471.
12 Jung, J.H.; Khirul, M.A.; Kang, D.; Jee, H.; Park, C.; Jung, Y.; Song, S.; Yang, E. Cutting-edge solutions for soil and sediment remediation in shipyard environments. *Processes* **2025**, *13*, 2010.
13 Cheng, X.; Wang, S.; Dong, Y.; Ni, Z.; Hong, Y. Spatiotemporal analysis and risk prediction of water quality using copula Bayesian networks: A case in Qilu Lake, China. *Processes* **2024**, *12*, 2922.
14 Hou, C.; Yang, Z. Ouyang, Surface runoff and diffuse nitrogen loss dynamics in a mixed land use watershed with a subtropical monsoon climate. *Processes* **2023**, *11*, 1910.
15 Wang, Z.; Lao, K.; Chen, C.; Zhu, H.; Yang, Y.; Chen, H.; Pang, H. Field study on washing of 4-methoxy-2-nitroaniline from contaminated site by dye intermediates. *Processes* **2024**, *12*, 2801.
16 Shi, J.; Zhang, D.; Sui, Z.; Wu, J.; Zhang, Z.; Hu, W.; Huo, Z.; Wu, Y. Improved dujiangyan irrigation system optimization (idiso): A novel metaheuristic algorithm for hydrochar characteristics. *Processes* **2024**, *12*, 1321.
17 Zhao, X.; Wang, X.; Lang, H.; Zhang, P.; Ni, J.; Wu, W. Effects of reclaimed water supplementation on the occurrence and distribution characteristics of antibiotic resistance genes in a recipient river. *Processes* **2024**, *12*, 1717.
18 Rodziewicz, J.; Mielcarek, A.; Bryszewski, K.; Kwietniewski, M.; Janczukowicz, W. Limited Utilization of an External Carbon Source in a Rotating Electro-Biological Disc Contactor (REBDC). *Processes* **2025**, *13*, 3115.

References

1. Schmitz, B.W.; Polanco, J.A.; Chen, H.; Manaktala, A.; Gu, X.; Goh, S.G.; Gin, K.Y. Virus surrogates throughout a full-scale advanced water reuse system. *Water Res.* **2024**, *256*, 121556. [CrossRef]
2. Liu, M.; You, X.; Li, Y.; Yang, Y. Critical review on development of methylene blue degradation by wet catalytic methods. *Rev. Chem. Eng.* **2025**, *41*, 179–195. [CrossRef]
3. Yang, Y.; You, X.; Shu, W.; Qu, W.; Liu, M.; Li, Y.; Chen, Z.; Chen, J.P. Coupling mechanisms for the simultaneous treatment of phosphate- and hexavalent-chromium-contaminated wastewater. *Acs EsT Water* **2025**, *5*, 4461–4471. [CrossRef]
4. Molinos-Senante, M.; Poch, M.; Rosso, D.; Garrido-Baserba, M. From wastewater treatment plants to decentralized resource factories. *NPJ Clean Water* **2024**, *7*, 46. [CrossRef]
5. Yang, Y.; Liu, D.; Chen, Y.; He, J.; Li, Q. Mechanistic study of highly effective phosphate removal from aqueous solutions over a new lanthanum carbonate fabricated carbon nanotube film. *J. Environ. Manag.* **2024**, *359*, 120938. [CrossRef]
6. Sable, H.; Kumar, V.; Singh, V.; Rustagi, S.; Chahal, S.; Chaudhary, V. Strategically engineering advanced nanomaterials for heavy-metal remediation from wastewater. *Coord. Chem. Rev.* **2024**, *518*, 216079. [CrossRef]
7. Wang, S.; Zhuang, H.; Shen, X.; Zhao, L.; Pan, Z.; Liu, L.; Lv, S.; Wang, G. Copper removal and recovery from electroplating effluent with wide ph ranges through hybrid capacitive deionization using cuse electrode. *J. Hazard. Mater.* **2023**, *457*, 131785. [CrossRef]
8. Bhandari, G.; Chaudhary, P.; Gangola, S.; Gupta, S.; Gupta, A.; Rafatullah, M.; Chen, S. A review on hospital wastewater treatment technologies: Current management practices and future prospects. *J. Water Process. Eng.* **2023**, *56*, 104516. [CrossRef]
9. Tran, D.; Pham, T.; Dang, V.; Pham, T.; Nguyen, M.; Dang, N.; Ha, M.; Nguyen, V.; Nghiem, L.D. A facile technique to prepare mgo-biochar nanocomposites for cationic and anionic nutrient removal. *J. Water Process. Eng.* **2022**, *47*, 102702. [CrossRef]
10. Yang, Y.; Wang, Y.; Zheng, C.; Lin, H.; Xu, R.; Zhu, H.; Bao, L.; Xu, X. Lanthanum carbonate grafted zsm-5 for superior phosphate uptake: Investigation of the growth and adsorption mechanism. *Chem. Eng. J.* **2022**, *430*, 133166. [CrossRef]

11. Tasic, T.; Milankovic, V.; Potkonjak, N.; Unterweger, C.; Pasti, I.; Lazarevic-Pasti, T. Valorization of viscose textile waste for the adsorptive removal of organophosphate pesticides from water. *J. Water Process. Eng.* **2025**, *69*, 106793. [CrossRef]
12. Sheikh, M.; Harami, H.R.; Rezakazemi, M.; Cortina, J.L.; Aminabhavi, T.M.; Valderrama, C. Towards a sustainable transformation of municipal wastewater treatment plants into biofactories using advanced nh3-n recovery technologies: A review. *Sci. Total Environ.* **2023**, *904*, 166077. [CrossRef]
13. Xiang, Y.; Liu, Y.; Cong, B.; Cai, Z.; Wang, N.; Zhang, H.; He, C.; Lai, B. Rapid and efficient removal of organophosphorus pollutant and recovery of valuable elements: A boosted strategy for eliminating organophosphorus from wastewater. *Water Res.* **2024**, *267*, 122494. [CrossRef]
14. Vareda, J.P.; Valente, A.J.M.; Duraes, L. Ligands as copper and nickel ionophores: Applications and implications on wastewater treatment. *Adv. Colloid. Interface Sci.* **2021**, *289*, 102364. [CrossRef]
15. Marazzi, F.; Fornaroli, R.; Clagnan, E.; Brusetti, L.; Ficara, E.; Bellucci, M.; Mezzanotte, V. Wastewater from textile digital printing as a substrate for microalgal growth and valorization. *Bioresour. Technol.* **2023**, *375*, 128828. [CrossRef]
16. Zhou, M.; Wang, J.; Wang, H.; Ran, X.; Xue, H.; Liu, C.; Wang, Y. Revealing the comprehensive impact of organic compounds on the partial nitrification-anammox system during incineration leachate treatment: Metabolic hierarchy and adaptation. *Water Res.* **2024**, *255*, 121534. [CrossRef]
17. Zhang, X.; Zhang, X.; Chen, J.; Wu, P.; Yang, Z.; Zhou, L.; Zhu, Z.; Wu, Z.; Zhang, K.; Wang, Y.; et al. A critical review of improving mainstream anammox systems: Based on macroscopic process regulation and microscopic enhancement mechanisms. *Environ. Res.* **2023**, *236*, 116770. [CrossRef]
18. Lin, Z.; Shapiro, E.F.; Barajas-Rodriguez, F.J.; Gaisin, A.; Ateia, M.; Currie, J.; Helbling, D.E.; Gwinn, R.; Packman, A.I.; Dichtel, W.R. Trace organic contaminant removal from municipal wastewater by styrenic β-cyclodextrin polymers. *Environ. Sci. Technol.* **2023**, *57*, 19624–19636. [CrossRef]
19. Brown, C.L.; Maile-Moskowitz, A.; Lopatkin, A.J.; Xia, K.; Logan, L.K.; Davis, B.C.; Zhang, L.; Vikesland, P.J.; Pruden, A. Selection and horizontal gene transfer underlie microdiversity-level heterogeneity in resistance gene fate during wastewater treatment. *Nat. Commun.* **2024**, *15*, 5412. [CrossRef]
20. Liu, S.; Zhao, Y.; Li, T.; Hu, T.; Zheng, K.; Shen, M.; Long, H. Removal of micro/nanoplastics in constructed wetland: Efficiency, limitations and perspectives. *Chem. Eng. J.* **2023**, *475*, 146033. [CrossRef]
21. Khan, A.H.; Aziz, H.A.; Palaniandy, P.; Naushad, M.; Cevik, E.; Zahmatkesh, S. Pharmaceutical residues in the ecosystem: Antibiotic resistance, health impacts, and removal techniques. *Chemosphere* **2023**, *339*, 139647. [CrossRef]
22. Bibi, A.; Bibi, S.; Abu-Dieyeh, M.; Al-Ghouti, M.A. Towards sustainable physiochemical and biological techniques for the remediation of phenol from wastewater: A review on current applications and removal mechanisms. *J. Clean. Prod.* **2023**, *417*, 137810. [CrossRef]
23. Zhao, L.; Fu, D.; Wu, X.; Liu, C.; Yuan, X.; Wang, S.; Duan, C. Opposite response of constructed wetland performance in nitrogen and phosphorus removal to short and long terms of operation. *J. Environ. Manag.* **2024**, *351*, 120002. [CrossRef]
24. Kang, Y.; Wang, J.; Li, Z. Enhancing pollutants removal in hospital wastewater: Comparative analysis of pac coagulation vs. Bio-contact oxidation, highlighting the impact of outdated treatment plants. *J. Hazard. Mater.* **2024**, *471*, 134340. [CrossRef]
25. Zeng, L.; Liu, X.; Ma, J.; Yang, J.; Yang, J.; Zhou, Y. Current progress on manganese in constructed wetlands: Bibliometrics, effects on wastewater treatment, and plant uptake. *Environ. Res.* **2024**, *249*, 118382. [CrossRef]
26. Peng, X.; Li, Y.; Jiang, Z.; Zhu, K.; An, Q.; Xiao, Z.; Dong, X.; Zhai, S. Photothermal-synergistic peroxymonosulfate activation promoting carbamazepine degradation by porphyra-derived porous biochar composites: Performance, mechanism, transformation pathway and practical application. *Chem. Eng. J.* **2024**, *489*, 151263. [CrossRef]
27. Sharma, B.; Tiwari, S.; Kumar, R.; Kumar, M.; Tewari, L. Eco-friendly detoxification of hazardous congo red dye using novel fungal strain trametes flavida wtfp2: Deduced enzymatic biomineralization process through combinatorial in-silico and in-vitro studies. *J. Hazard. Mater.* **2023**, *455*, 131503. [CrossRef]
28. Kizza, R.; Eskicioglu, C. Ultrafiltration fractionation of potentially inhibitory substances of hydrothermal liquefaction aqueous phase derived from municipal sludge. *Water Res.* **2024**, *257*, 121703. [CrossRef]
29. Trommetter, G.; Mendret, J.; Khaska, S.; La Salle, C.L.; Brosillon, S.; Goetz, V.; Plantard, G. Removal efficiencies for 52 pesticides and pharmaceuticals from wastewater effluent by coupling solar heterogeneous photo-oxidation with tio2 and infiltration in saturated soil column. *Sci. Total Environ.* **2024**, *956*, 177278. [CrossRef]
30. Yang, Y.; Liu, M.; You, X.; Li, Y.; Lin, H.; Chen, J.P. A novel bimetallic fe-cu-cnt catalyst for effective catalytic wet peroxide oxidation: Reaction optimization and mechanism investigation. *Chem. Eng. J.* **2024**, *479*, 147320. [CrossRef]
31. Yang, Y.; Lin, H.; Long, Y.; Mei, Y.; Chen, J.P. Development of catalytic zero-valent iron incorporated pan catalytic film for efficient degradation of organic matters. *NPJ Clean Water* **2024**, *7*, 38. [CrossRef]
32. Dansawad, P.; Li, Y.; Li, Y.; Zhang, J.; You, S.; Li, W.; Yi, S. Machine learning toward improving the performance of membrane-based wastewater treatment: A review. *Adv. Membr.* **2023**, *3*, 100072. [CrossRef]

33. Ganthavee, V.; Trzcinski, A.P. Artificial intelligence and machine learning for the optimization of pharmaceutical wastewater treatment systems: A review. *Environ. Chem. Lett.* **2024**, *22*, 2293–2318. [CrossRef]
34. Chen, J.; Wan, J.; Ye, G.; Wang, Y. Prediction and optimization of wastewater treatment process effluent chemical oxygen demand and energy consumption based on typical ensemble learning models. *Bioresour. Technol.* **2024**, *411*, 131362. [CrossRef] [PubMed]
35. Jeon, J.; Choi, M.; Park, S.; Bae, H. Management strategy of granular sludge settleability in saline denitrification: Insights from machine learning. *Chem. Eng. J.* **2024**, *493*, 152747. [CrossRef]

processes

MDPI

Article

Limited Utilization of an External Carbon Source in a Rotating Electro-Biological Disc Contactor (REBDC)

Joanna Rodziewicz [1], Artur Mielcarek [1], Kamil Bryszewski [2], Marian Kwietniewski [3] and Wojciech Janczukowicz [1,*]

1 Department of Environment Engineering, University of Warmia and Mazury in Olsztyn, Warszawska St. 117a, 10-719 Olsztyn, Poland; joanna.rodziewicz@uwm.edu.pl (J.R.); artur.mielcarek@uwm.edu.pl (A.M.)
2 Hanna Instruments Sp. z.o.o., Al. J. Piłsudskiego St. 73, 10-449 Olsztyn, Poland; r1@hanna-polska.com
3 Faculty of Environmental Engineering, Warsaw University of Technology, Nowowiejska St. 20, 00-653 Warsaw, Poland; marian.kwietniewski@is.pw.edu.pl
* Correspondence: jawoj@uwm.edu.pl

Abstract: Overdosing an external carbon source can lead to its incomplete utilization. With this in mind, this study aimed to investigate the effect of carbon dosing on nutrient removal in a rotating electro-biological disc contactor (REBDC) treating hydroponic tomato wastewater and to evaluate its impact on effluent COD under a constant electric current (2.5 A/m^2). In REBDC, discs served as the cathode, and an aluminum sheet placed in the tank served as the anode. Sodium acetate was added to provide carbon to nitrogen (C/N) ratios of 0.5, 1.0, 2.0, and 3.0. The HRT was 24 h. The phosphorus removal efficiency in the REBDC exceeded 90% at C/N ratios of 0.5, 1.0, and 2.0. A several-fold increase was observed in nitrogen removal efficiency at C/N = 2.0 and C/N = 3.0 compared to C/N = 0.5 (9 and 11 times higher, respectively). At C/N ratios of 2.0 and 3.0, the efficiency was 56.0% and 65%, respectively. Considerable amounts of unutilized organic carbon were found in the REBDC-treated wastewater. The rational solution would be to extend the HRT, which would enable greater substrate consumption and nitrogen removal.

Keywords: tomato soilless cultivation wastewater; nutrient removal; rotating electro-biological disc contactor (REBDC); external carbon source; COD residuals

1. Introduction

The growing demand for vegetables, fruits, herbs, and other crops has significantly increased the use of protected cultivation methods, including vertical farms. These are highly efficient systems that enable year-round production, like soilless cultivation, which isolates plants from soil and entails fertigation (a combination of fertilization and irrigation) [1]. This approach allows for the precise optimization of crop yields by adjusting the nutrient medium's composition and dosage to meet the plants' demands for nutrients under specified environmental conditions.

To prevent salt accumulation in the rhizosphere and ensure optimal nutrient supply, crops cultivated in this system are irrigated with an excess of a nutrient solution, typically ranging from 25% to 45%, which generates approximately 4.5 L/m^2 of nutrient-dense greenhouse wastewater (GW) [2].

Our previous study addressing greenhouse wastewater from soilless tomato cultivation revealed that the wastewater had low levels of organic matter but high concentrations of total nitrogen (TN) and total phosphorus (TP), and had increased salinity (expressed as electrical conductivity, EC). Notably, more than 99% of TN was present in the form of nitrates [1]. The analyzed wastewater originated from two greenhouses for soilless tomato cultivation under shelter, located in the north-eastern Poland. The greenhouses differed in the production scale, i.e., the area of Greenhouse 1 (located at the University of Warmia and Mazury in Olsztyn) was approximately 100 m^2, whereas that of Greenhouse 2 was 200,000 m^2 (private enterprise).

The composition of GW varied as follows:

- COD: 50.07–75.82 mg O_2/L (Greenhouse 1) and 37.35–78.12 mg O_2/L (Greenhouse)
- TN: 403.59–614.89 mg N/L (Greenhouse 1) and 270.00–577.40 mg N/L (Greenhouse 2)
- TP: 35.44–78.00 mg P/L (Greenhouse 1) and 54.10–104.00 mg P/L (Greenhouse 2)
- EC: 3.53–6.93 mS/cm (Greenhouse 1) and 4.94–6.94 mS/cm (Greenhouse 2).

According to Polish and EU regulations [3], wastewater with such high concentrations of biogenic compounds cannot be discharged untreated into surface waters or soil. For biodegradable industrial wastewater originating from the production and processing of fruit and vegetables (including tomato processing), the biogene concentrations have been stipulated below 30.0 mg N/L for total nitrogen (TN) and below 3.0 mg P/L for total phosphorus (TP). These quality requirements underscore the necessity of treating or pre-treating GW prior to discharging it into surface waters or sewage systems. The pre-treatment is also essential if the wastewater is intended for recirculation within the cultivation system [1]. Other studies have supported these observations. For instance, Du et al. [4] reported nitrate concentrations ranging from 150 to 600 mg N/L in effluents from soilless tomato cultivation, whereas Kwon et al. [5] determined values of up to 494 mg N/L (TN), 96.9 mg P/L (TP), and an EC of 37.5 μmho/cm in wastewater from tomato and capsicum cultivation.

The treatment of wastewater having a low C/N ratio requires supplementing it with an external carbon source, which acts as an electron donor in the denitrification process [6,7]. Various solid (SCS), liquid, or gaseous substances can serve as external carbon sources. According to Henze et al. [8], the reduction of 1 mg N–NO_3^- to nitrogen requires 2–5 mg COD of organic material. For methanol, this ratio is approximately 2.5 mg methanol per mg N–NO_3^-, whereas for acetic acid it ranges from 3.0 to 4.0 mg COD, when biomass growth is taken into account. The theoretical carbon demand for complete hydrogenotrophic denitrification has been estimated at 0.20 mg C (as bicarbonate) per 1 mg NO_3^- reduced to gaseous nitrogen [9], and at 0.12–0.21 mg C (as CO_2 gas) per 1 mg NO_3^- [10]. Liquid carbon sources include methanol [11], acetic acid [12], sodium acetate [13], citric acid [14], sugars [15], industrial wastewater [16], and liquid fermentation products derived from sewage sludge and food waste [17,18].

Biodegradable polymers (BDPs), composed of polyesters and polyolefins, have been increasingly recognized as viable solid carbon sources. They ensure controlled carbon release, while minimizing the risk of exceeding COD discharge limits, thereby ensuring safe and reliable wastewater treatment [19–21]. Furthermore, solid carbon sources, which are generally insoluble in water, not only serve as stable electron donors for denitrification but also act as a growth medium for denitrifying microorganisms. This helps prevent the overdosage of carbon sources, which could otherwise adversely affect effluent quality and process control [17,22]. Despite these advantages, their use is often constrained by reactor design, often requiring the supplementation of liquid carbon sources. One such case involves bio-electro reactors (BERs), which, when combined with an external carbon source, create optimal conditions for nitrate and phosphorus removal from GW.

One of the technologically prospective BERs is the rotating electro-biological disc contactor (REBDC). It consists of flat discs mounted on a horizontal rotating shaft [23]. The electric current applied in REBDC facilitates key processes, including water electrolysis, hydrogenotrophic denitrification, electrochemical nitrate reduction, and efficient phosphorus electrocoagulation.

Hydrogenotrophic denitrification, driven by the electric current applied, may reduce the demand for organic substrates required for heterotrophic denitrification. Sodium acetate serves as a carbon source for heterotrophic denitrifying bacteria, both within the biofilm on the discs and the suspended biomass, influencing dissolved organic carbon (DOC) levels in the effluent. Additionally, hydrogen and carbon dioxide produced during water electrolysis are consumed by autotrophic denitrifying microorganisms [23].

Rodziewicz et al. [13] examined the use of sodium acetate as a carbon source in REBDC systems at a C:N ratio of 0.5. Their research showed that maximum nitrogen and phosphorus removal efficiencies were achieved at a hydraulic retention time (HRT) of 24 h. Key findings from their study were as follows:

- At current densities above 2.5 A/m^2, hydrogenotrophic denitrification became the dominant nitrogen removal mechanism, while heterotrophic denitrification and electrochemical nitrate reduction were suppressed.
- At a 24-h HRT and 2.5 A/m^2 current density, phosphorus removal efficiency exceeded 99%, while nitrogen removal efficiency reached 38.7%.
- Higher current densities (>2.5 A/m^2) increased the wastewater temperature above 33.5 °C.

During the experiment, the COD of synthetic wastewater was 52.5 ± 4.5 mg O_2/L. After adding sodium acetate, it increased to 259.5 ± 39.7 mg O_2/L, and ultimately reached 24.5 ± 2.9 mg O_2/L in the treated wastewater. These findings confirm the complete utilization of the external carbon source during heterotrophic denitrification.

However, the nitrogen removal efficiency recorded in these experiments (<40%) suggests that increasing the external carbon dose could further enhance heterotrophic denitrification and overall nitrogen removal. Nevertheless, reactor design must correspond to the specific characteristics of the treated wastewater to minimize carbon and nitrogen footprints.

In addition to direct emissions (e.g., greenhouse gases) and aquatic pollutants (e.g., reactive nitrogen species and phosphorus), also energy and raw material consumption during wastewater treatment must be considered. The priority should be to optimize energy and resource efficiency by leveraging the full potential of the selected technology.

While the addition of an external carbon source enhances heterotrophic denitrification in REBDC systems, precise dosing is critical. Insufficient carbon supply leads to incomplete denitrification, rendering the treated wastewater non-compliant with discharge standards, whereas excess carbon increases COD levels in the effluent, thereby posing potential environmental risks.

The objective of this study was to evaluate the effect of external carbon dosing on nitrogen and phosphorus removal efficiency in a rotating electro-biological disc contactor treating wastewater from soilless tomato cultivation, and to assess its impact on COD levels in the effluent. The results of the tests conducted in REBDC were compared with those obtained in a biological reactor without electric current flow (rotating biological disc contactor, RBDC) fed with the same doses of the organic substrate as REBDC.

2. Materials and Methods

2.1. Wastewater Characteristics

The study was conducted with wastewater collected from a greenhouse specializing in soilless tomato cultivation, located in Legajny, Poland. The wastewater was collected at the outflow of the fertigation system, ensuring a representative sample of drainage water from the cultivation process.

Prior to the experiments, composite samples were analyzed for:

- Chemical oxygen demand (COD),
- Total nitrogen (TN),
- Nitrate nitrogen (N-NO_3),
- Total phosphorus (TP),
- Electrical conductivity (EC).

The initial wastewater parameters are presented in Table 1.

Table 1. Wastewater characteristics.

Parameters	Value Mean
COD [mg O_2/L]	51.45 ± 4.45
TOC [mg C/L]	15.64 ± 1.41
Total nitrogen [mg N/L]	610.73 ± 14.89
Nitrate [mg N/L]	609.98 ± 11.52
Ammonium nitrogen [mg N/L]	0.69 ± 0.25
Nitrite [mg N/L]	0.06 ± 0.01
Total phosphorus [mg P/L]	57.75 ± 3.08
pH	5.75 ± 0.7
Electrolytic conductivity [mS/cm]	4.89 ± 0.03

The wastewater had a low C/N ratio, with over 99% of nitrogen present in the form of nitrates.

2.2. Reactor Setups—Rotating Biological Disc Contactor (RBDC) and Rotating Electro-Biological Disc Contactor (REBDC)

The treatment experiments were conducted using a laboratory-scale RBDC and REBDC, designed to simulate the operation of a full-scale reactor (Table 2).

Table 2. Rotating disc contactor parameters.

Parameter	Values
Number of discs [pcs.]	8
Disc diameter [cm]	22
The total area of the discs [m^2]	0.56
Submergence of the discs [%]	40
Flow chamber volume [L]	2.0
Rotation speed [rpm]	10

The system consisted of (Figure 1):

- A cylindrical biological tank with a working volume of 2.0 L,
- A horizontal shaft equipped with flat PVC discs (diameter: 22 cm),
- Aluminum plate electrodes mounted on the tank walls, acting as an anode,
- A DC power supply (HANTEK PPS2116A, Qingdao, China) providing adjustable constant current densities ranging from 0 to 10 A/m^2.

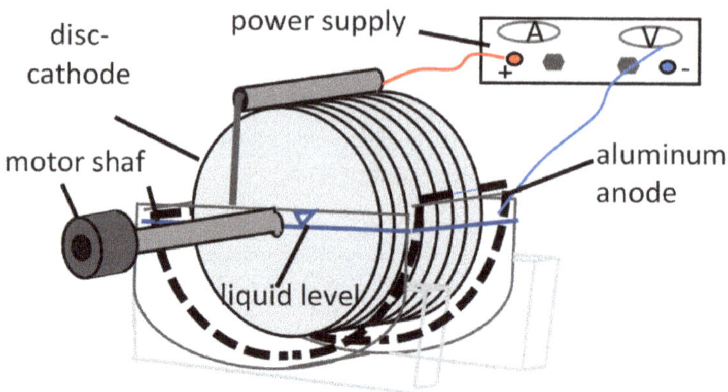

Figure 1. Scheme of a single-stage REBDC.

The rotational speed of the discs was maintained at 10 rpm, ensuring periodic submergence and exposure to air. The hydraulic retention time (HRT) was set at 24 h throughout the study.

2.3. External Carbon Source Dosing

Sodium acetate (CH_3COONa, analytical purity, powder, VWR) was selected as the external carbon source because it is readily biodegradable, safe and effective, and provides predictable results in denitrification and other biological processes. It has been used and tested in previous studies by the authors' [10]. The dosing aimed to achieve C:N ratios of 0.5, 1.0, 2.0, and 3.0 in the influent, was calculated based on TN concentration. The lowest value (0.5), used in the study by Rodziewicz et al. [13], was found to be insufficient for effective nitrate removal in an electro-biological disc reactor. The sodium acetate solution was prepared by dissolving analytical-grade CH_3COONa in deionized water and afterwards was introduced into the reactor's influent using a peristaltic pump which ensured its continuous and stable dosing.

2.4. Operational Parameters

The experiments were conducted at ambient laboratory temperature (22 ± 2 °C). The current density applied to the electrodes was set at 2.5 A/m^2, selected based on findings from a previous study [9] to balance nitrogen removal efficiency and energy consumption. The aeration was provided by natural rotation of the discs; no additional air supply was used.

The study was divided into two phases (Table 3):

1. Phase I—biological phase; operation of RBDC (without electric current flow), with external carbon source dosing (C:N = 0.5, 1.0, 2.0, and 3.0)
2. Phase II—electro-biological phase; operation of REBDC with external carbon source dosing (C:N = 0.5, 1.0, 2.0, and 3.0).

The two-phase study employing RBDC and REBDC under fixed technological parameters enabled determining contribution of individual biological processes (mainly heterotrophic denitrification) in the overall performance of the electro-biological contactor.

Table 3. Technological parameters used in the study.

Parameter	Phase I—Biological Reactor: Rotating Biological Disc Contactor (RBDC)	Phase II—Electro-Biological Reactor: Rotating Electro-Biological Disc Contactor (REBDC)
HRT [h]	24	24
C/N [gC/gN]	0.5, 1.0, 2.0, 3.0	0.5, 1.0, 2.0, 3.0
J [A/m^2]	-	2.5

Biomass from the denitrification tank of the Olsztyn municipal wastewater treatment plant (20°5876′ E 54°128′ N) was used as the inoculum during the biological phase. After completing analyses in the rotating biological disc contactors (RBDC), which lasted 3.5 months (including biofilm cultivation), the experiments in the electro-biological reactor were initiated. The biofilm cultivation and adaptation period in the electro-biological reactor lasted over 2 months.

The analytical monitoring of the processes, both in the biological and electro-biological phases, commenced after the adaptation period of the reactors. Samples were collected at 24-h intervals during both phases of the study. The results of physicochemical analyses of 25 samples were used to calculate the average values for each phase.

All experiments were conducted under laboratory conditions at a temperature of 20.0 ± 1 °C.

2.5. Analytical Methods

Samples of the influent and the effluent were collected daily and analyzed for the following parameters:

- COD: dichromate method [24],
- TN: persulfate digestion followed by UV-spectrophotometric detection [25],
- N-NO_3: ion chromatography [26],
- TP: molybdenum blue spectrophotometric method [27],
- EC: conductometric method [28].

Content of total nitrogen was determined using a Total Organic Carbon Analyzer TOC-L CPH/CPN with TNM-L device (Shimadzu Corporation, Kyoto, Japan), with a method of oxidative combustion—chemiluminescence, content of total phosphorus using a UV-VIS 5000 DR spectrophotometer (HACH Lange, Duesseldorf, Germany), with an HACH Lange LCK 348–350 method, nitrate nitrogen using a VWR UV-3100PC Spectrophotometer (VWR International, Shanghai, China), with a colorimetric method and chemical oxygen demand (COD; titrimetric method), using a Gerhardt KI 16 (Königswinter, Germany) laboratory heater—with a dichromate method. Analytical control of the sewage treatment process also included measurements of pH value, electrolytic conductivity, and temperature using an HQ4300 multimeter (HACH Company, Loveland, CO, USA).

Ammonium nitrogen (N-NH4; HACH Lange LCK303- 305) and nitrite (N-NO2; HACH Lange LCK341-342) levels were determined using a DR5000 HACH Lange Spectrophotometer (Malente, Germany).

All analyses were performed in accordance with Polish Standards (PN) or ISO methods [24–28]. Each measurement was conducted in triplicate to ensure analytical accuracy.

The research results obtained were subjected to statistical analysis in order to organize the data and assess the nature of their distribution. In the first phase, the arithmetic mean was calculated as a measure of the tendency, which allows the average level of the analyzed characteristic to be determined. The standard deviation was then calculated to assess the degree of variability of the results and their dispersion around the mean.

3. Results and Discussion

3.1. pH

The low pH of the raw wastewater (5.75) was due to the pH of the nutrient solution supplied to the cultivated crops, which in turn resulted from the fact that the highest availability of most nutrients in hydroponic cultivation occurs at a pH range of 5.5–5.8 [1].

During the study, the pH of the effluent from both reactors increased as a result of the treatment processes (Figure 2). A similar upward trend in pH was noted in both reactors as the doses of sodium acetate and the C/N ratio increased. Simultaneously, in the electro-biological reactor, the pH values were lower (ranging from 7.59 to 9.39) than those recorded in the biological reactor (ranging from 8.50 to 9.48).

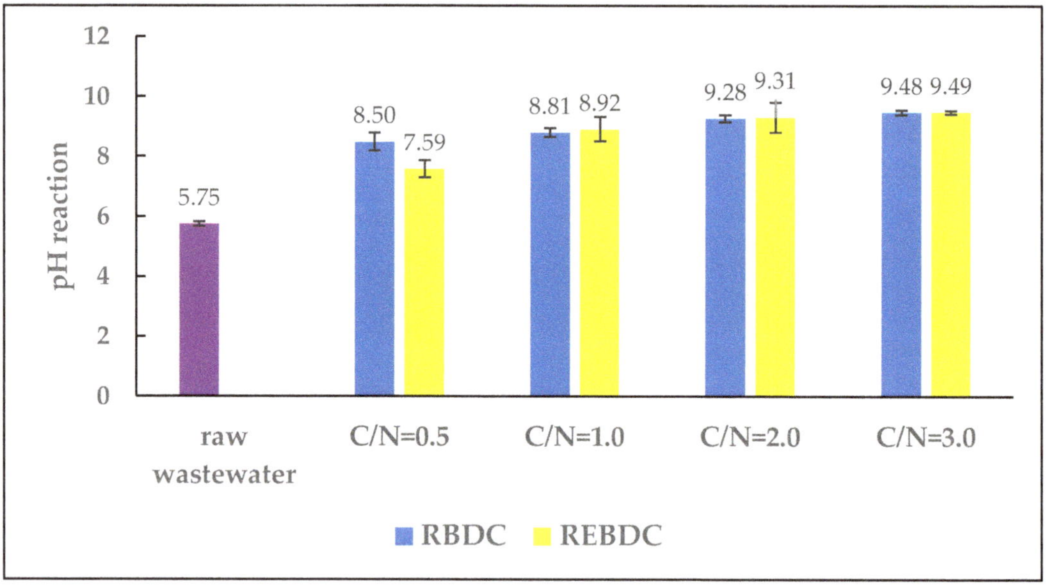

Figure 2. The effect of the C/N ratio on effluent pH.

The alkaline pH of the effluent from RBDC and REBDC was primarily due to the pH of the sodium acetate (strongly alkaline) added to the influent fed to both types of reactors. The differences in pH levels between the reactors stemmed from the processes occurring therein. The pH increase in the biological reactor was a consequence of heterotrophic denitrification, as the removal of one mole of NO_3^- results in the formation of one mole of OH^-, which increases pH and can simultaneously cause nitrite accumulation and reduce the nitrate removal rate. Additionally, the pH increase during denitrification is due to the reaction converting nitrite to nitrogen gas [29–31].

In the electro-biological reactor, the pH increase results from concurrent processes: water electrolysis (producing OH^- ions), hydrogenotrophic denitrification, and heterotrophic denitrification (further increasing OH^- concentration). In contrast, lower pH values in this reactor are due to hydrogen gas evolution on the cathode surface during water electrolysis, accompanied by the release of H^+ ions, leading to a pH decrease. In turn, the pH increase from hydrogen evolution is more or less compensated by the reactions releasing H^+ ions [32,33].

pH is one of the main factors influencing the denitrification process. Controlling pH values can effectively enhance nitrate (V) removal efficiency. Uncontrolled or fluctuating pH can lead to partial denitrification and accumulation of nitrite (III) [34]. However, different denitrifying bacterial cultures have various optimal pH ranges. Rezania et al. [35] studied the effect of pH in the range of 7.5–9.5 on auto-hydrogenotrophic denitrification of nitrogen-contaminated water and achieved maximum removal rates at pH = 9.5. Conversely, Kurt et al. [36] found the highest denitrification rates at pH = 7.5.

It should be noted that, in addition to the denitrification process which contributes to pH increase, the use of BERs for denitrification also raises alkalinity since water electrolysis and pH control become more complex. The production of 1 mole of H_2 results in the formation of 2 moles of OH^-. Moreover, during denitrification, the removal of one mole of NO_3^- forms one mole of OH^-, increasing pH, which leads to nitrite (III) accumulation and reduces nitrate (V) removal rates. The pH increase during denitrification is also due to the half-reaction converting nitrite (III) to nitrogen gas [29,30].

3.2. The Electrical Conductivity (EC)

The electrical conductivity, expressed in mS/cm, is an indicator of nutrient concentration in the nutrient solution. Average EC values for hydroponic solutions range from 1.5 to 4.5 mS/cm [1]. During the study, the EC of the raw wastewater was slightly higher, reaching 4.89 mS/cm (Figure 3).

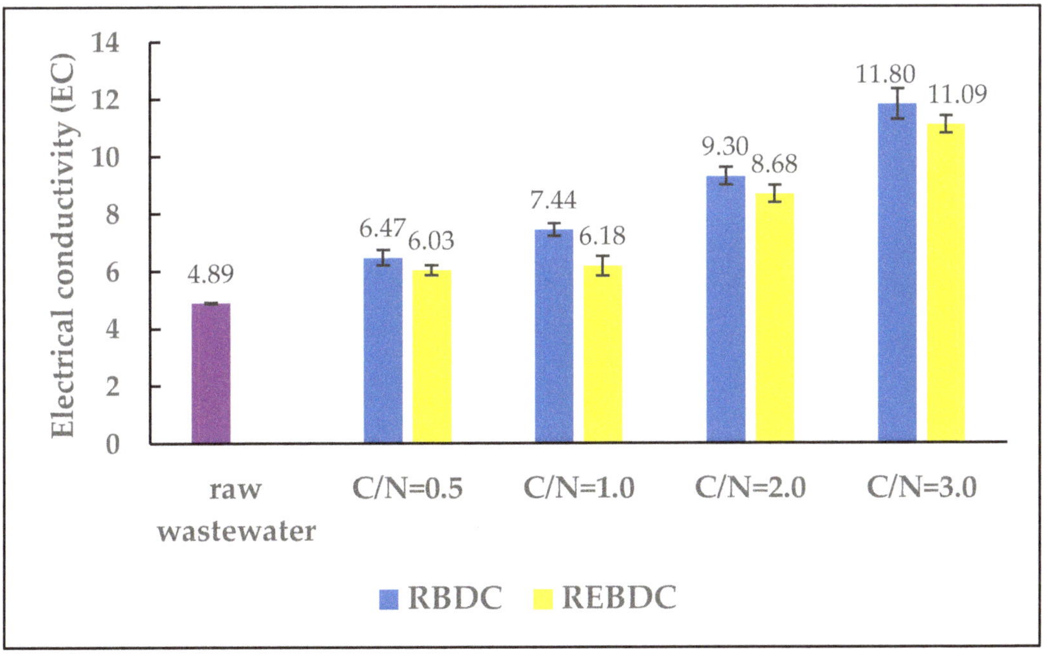

Figure 3. The effect of the C/N ratio on the electrical conductivity (EC) of the effluent.

In reactors of both types, the EC values increased with the amount of carbon supplied. In the RBDC reactor, EC increased from 4.89 to 6.47, 7.44, 9.3, and 11.80 mS/cm for C/N ratios of 0.5, 1.0, 2.0, and 3.0, respectively. In the REBDC reactor, the EC values were lower than in the biological reactor, ranging from 6.03 to 11.09 mS/cm.

The observed increase in EC values in the effluents from both reactors was primarily due to the addition of sodium acetate, a strong electrolyte with considerable conductivity, and the simultaneous removal of nitrates via heterotrophic denitrification, hydrogenotrophic denitrification, and electrochemical nitrate reduction. This finding corroborates results from a previous research by Huang et al. [37,38], who found that nitrogen removal decreased solution conductivity. The presence of carbon and nitrates in the treated wastewater leads to an intense electric current flow through the bio-electrochemical reactor (BER) [37]. Organic compounds in BER are consumed by microorganisms during heterotrophic degradation or for electron production at the anode, thereby increasing current intensity [37]. Nitrogen removal reduces solution's conductivity, whereas the breakdown of organic substances by microorganisms at the anode generates electrons, increasing the current flow in the reactor [39]. Lower EC values in the effluent of electro-biological reactors are most likely due to higher nitrogen removal efficiency in these reactors and the lower concentration of nitrogen in the wastewater.

The magnitude of EC increase was a consequence of the C/N ratio, with the highest EC values in the reactors observed at C/N = 3.0. Lower EC values were noted in the electro-biological reactor, which resulted from higher nitrogen removal efficiency in this reactor (Figure 4). These findings align with observations made by Huang et al. [37,38], who demonstrated that nitrogen removal decreased solution's conductivity. The presence of carbon and nitrates in the effluent leads to a significant increase in electric current flow through the electro-biological reactor, which could be caused by the addition of electrolytes that increase solution's conductivity. Nitrogen removal reduces conductivity, and vice versa—the degradation of organic substances by microorganisms at the anode generates electrons, increasing the current flow in the reactor [37,38].

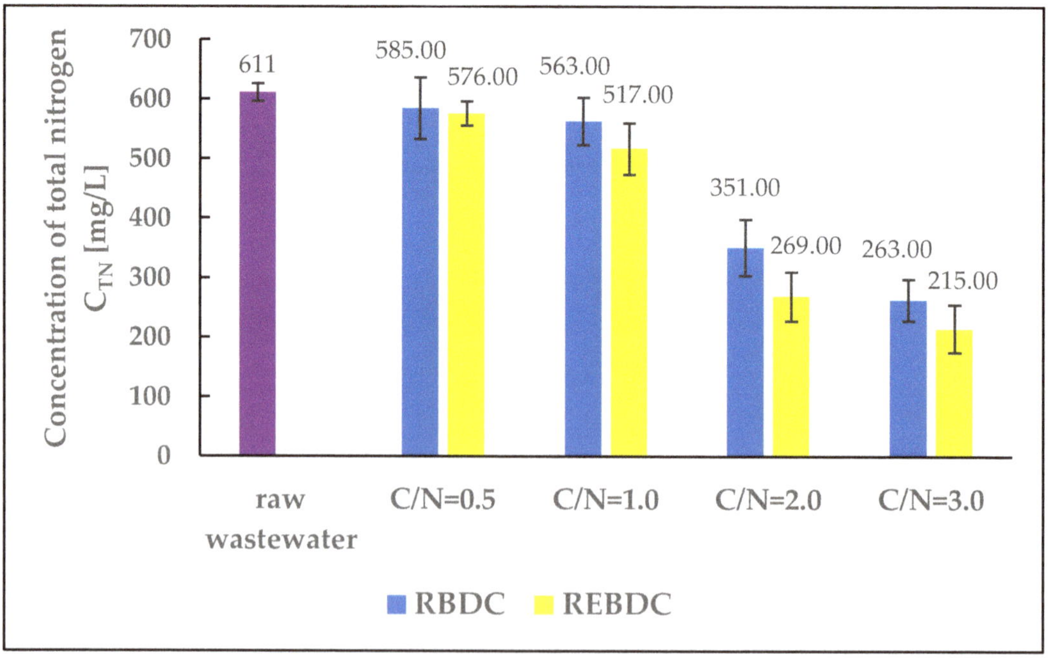

Figure 4. The effect of the C/N ratio on the concentration of total nitrogen (TN) in the effluent.

3.3. Total Phosphorus (TP)

The total phosphorus concentration in the influent wastewater fed to the reactors was 57.75 mg P/L. The research results showed a distinct difference in phosphorus removal efficiency between the biological and electro-biological reactors. Phosphorus concentrations in the effluent from the REBDC reactor ranged from 0.19 to 9.69 mg P/L, whereas in the biological reactor, the respective values were significantly higher, ranging from 16.61 to 45.25 mg P/L (Figure 5). In the biological reactor, phosphorus removal was primarily associated with biofilm development. Achieving high phosphorus removal efficiencies in biological rotating contactors remains a challenge for engineers and operators due to the difficulty in controlling the sequential aerobic and anaerobic conditions required for the growth and accumulation of polyphosphate-accumulating organisms (PAOs). However, it cannot be definitively stated that phosphate binding and release processes did not occur within the biofilm, given the high share of anaerobic biofilm in the biomass colonizing the discs, especially under high organic compound loads, which was the case at C/N ratios of 2.0 and 3.0. The thickness of the biofilm influences phosphorus removal efficiency, with the maximum efficiency of up to 70% achieved at biofilm thickness lesser than 1.8 mm [40–42].

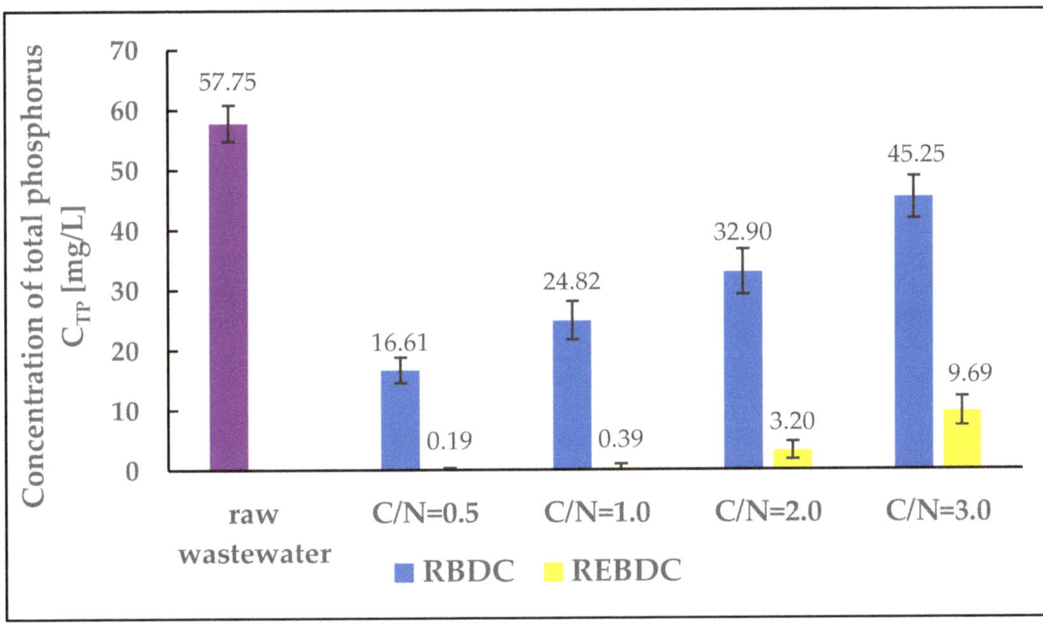

Figure 5. The effect of the C/N ratio on the concentration of total phosphorus (TP) in the effluent.

The decrease in phosphorus removal efficiency noted in the reactors at high C/N values results from the fact that high concentrations of organic substrates in the biofilm promote the colonization by glycogen-accumulation microorganisms (GAOs, which do not accumulate phosphorus) that may outcompete PAOs, leading to reduced phosphorus removal efficiency [41].

In the electro-biological reactor, phosphorus removal efficiencies exceeding 90% were achieved at C/N ratios of 0.5, 1.0, and 2.0, whereas over 83% efficiency was noted at C/N = 3.0. These results are due to two processes: phosphorus uptake during biofilm biomass growth and, primarily, electrocoagulation. Similarly high phosphorus removal efficiencies in electro-biological reactors have been reported by Rodziewicz et al. [43], Lagum [44], and Hackbarth et al. [45]. All cited researchers highlighted the significantly higher phosphorus removal efficiencies in electrochemical and electro-biological reactors compared to the biological reactors, where phosphorus removal primarily resulted from phosphorus incorporation into biofilm biomass, as well as phosphate release and accumulation processes.

At the same time, the observed trend indicated that increasing substrate availability in the REBDC reactor led to a decrease in phosphorus removal efficiency and an increase in phosphorus concentration in the effluent (with the highest efficiency determined at C/N = 0.5 and the lowest one at C/N = 3.0). This is because the increase in available organic substrate can upset the electrochemical balance in the reactor (Figure 5). Higher amounts of substrate mean a larger pool of electron donors, which are primarily utilized by microorganisms, thus diminishing the role of the electrode. Additionally, excess carbon sources promote faster and more intense biofilm growth. Biofilm thickness affects oxygen and electron diffusion, which can suppress PAOs bacterial activity and diminish phosphorus accumulation in the biofilm biomass. High concentrations of organic substrates may also promote the growth of GAO bacteria (not accumulating phosphorus), which compete with PAOs for organic substrates [46].

3.4. Total Nitrogen (TN)

The efficiency of total nitrogen removal increased with the dose of the organic substrate supplied to reactors of both types (Figure 4). At the lowest substrate dose (C/N = 0.5), TN concentration decreased from 611.0 mg N/L in raw wastewater to 585.0 mg N/L in the effluent from the biological reactor and 576.0 mg N/L in that from the electro-biological reactor, indicating denitrification efficiency below 6.0% (4.1% and 5.7%, respectively). At the highest substrate dose (C/N = 3.0), the TN concentration was 263.0 mg N/L in the effluent from RBDC (56.7% efficiency) and 215.0 mg N/L in that from REBDC (64.8% efficiency). The research results show a significant, multi-fold increase in nitrogen removal efficiency at C/N = 2.0 and C/N = 3.0 compared to C/N = 0.5. In the electro-biological reactor, the efficiency at C/N = 2.0 was more than 9 times higher, and at C/N = 3.0 over 11 times higher. This proves that increasing the amount of organic substrate supplied to both biological and electro-biological reactors to achieve higher nitrogen removal efficiencies (in the case of wastewater from soilless tomato cultivation, primarily nitrates) turned out to be a viable technological solution.

In the biological reactor, nitrogen removal efficiency resulted from heterotrophic denitrification and nitrogen incorporation into biomass. In the electro-biological reactor, nitrate removal efficiency was also influenced by hydrogenotrophic denitrification and electrochemical nitrate reduction due to electric current flow. This is why higher nitrogen removal efficiencies were achieved in the electro-biological reactor, corroborating earlier findings by Rodziewicz et al. [13] and Chen et al. [47].

The effectiveness of all the mentioned processes depends on the pH of the wastewater in the reactor. pH is one of the main factors influencing the denitrification process. For heterotrophic denitrification, the optimal pH is considered to be between 7.0 and 8.0. At pH values below 7.0, N_2O and NO concentrations increase, while pH values below 4.0 promote the accumulation of large amounts of nitrite (III), which are toxic to most enzymes. Denitrification halts in strongly acidic and highly alkaline environments, respectively, below pH 4.0 and above pH 9.5 [48,49]. Salem et al. [50] considered the optimal pH range for heterotrophic denitrifiers to be 5.95–7.9. In biological reactors, where pH ranged from 8.5 to 9.5 (Figure 2), the conditions were far from optimal; however, they would rather not inhibit denitrification.

Hydrogenotrophic denitrification is also strongly dependent on pH. The optimal values are considered to be between 7.6 and 8.6 [9,47]. Nevertheless, different denitrifying bacteria cultures have various optimal pH ranges. Rezania et al. [35] studied the influence of pH in the range of 7.5–9.5 on auto-hydrogenotrophic denitrification of nitrogen-contaminated water and achieved maximum removal rates at pH = 9.5, whereas Kurt et al. [36] found the highest denitrification rates at pH = 7.5. It should be noted that in addition to denitrification, which causes pH to rise, a characteristic feature of electro-biological reactors is an increase in alkalinity, leading to nitrite accumulation and a decrease in the nitrate removal rate [29,30]. Nitrites are also by-products of electrochemical nitrate reduction [39,51].

3.5. Nitrites and Ammonium Nitrogen

The research results confirm nitrite accumulation in the treated wastewater (Figure 6). As the amount of organic substrate supplied increased, and with associated changes such as wastewater pH shifts in both reactors, nitrite concentrations in the effluent increased as well. Nitrite concentrations were higher in the effluent from the electro-biological reactor because, generally, the pH value and energy source concentration (H_2) depend on the electric current intensity. Water electrolysis in the cathode zone generates H_2 and OH^-. The OH^- concentration increases proportionally to the electric current intensity, which can raise the pH and consequently decrease reactor efficiency. However, the activity of hydrogenotrophic bacteria depends on the concentration of electron donors as denitrification-controlling factors. Complete denitrification was achieved at various H_2 concentrations [30]. Lower H_2 concentrations were reported to lead to partial denitrification and nitrite accumulation as well as to inhibit biomass growth rates. Due to the low solubility of H_2, ensuring an appropriate contact time increases the denitrification rate [32,52]. Such phenomena induced by electric current flow did not occur in the biological reactor; hence, the lower nitrite concentrations in the RBDC effluent.

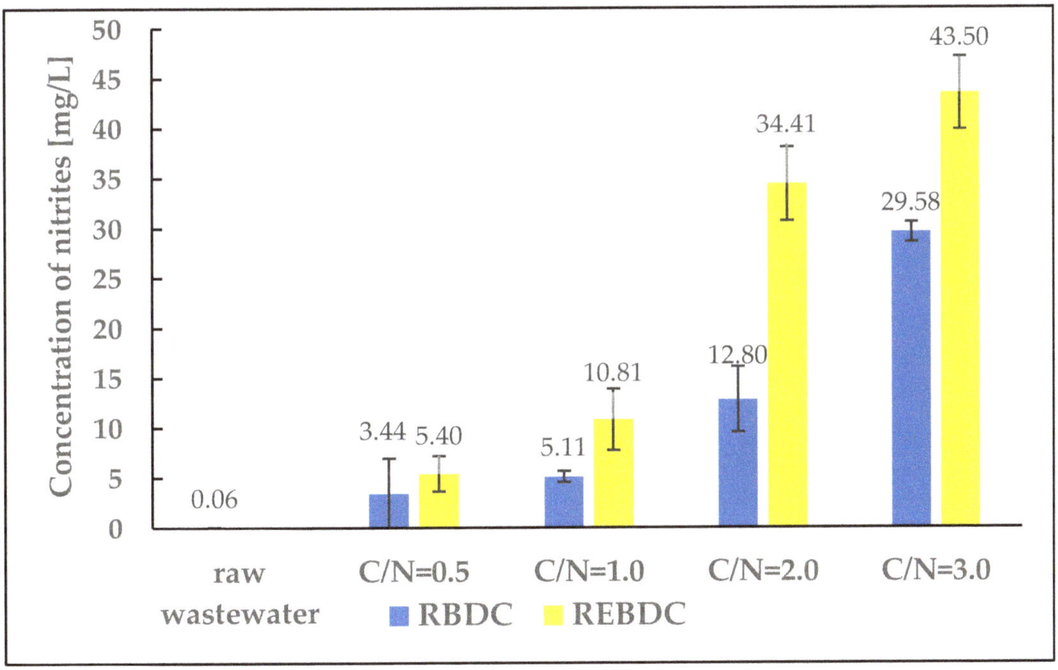

Figure 6. The effect of the C/N ratio on the concentration of nitrites in the effluent.

Ammonium nitrogen was also present in the effluent from both RBDC and REBDC, but in very low concentrations, ranging from 1.34 to 3.84 mg N-NH_4/L. This is most likely the result of one of the two phenomena: electrochemical nitrate reduction at the cathode [53] or dissimilatory nitrate reduction to ammonium nitrogen [54].

3.6. COD

The COD concentration of raw wastewater was 51.0 mg O_2/L (Figure 7). It increased after sodium acetate addition in the amounts corresponding to the C/N ratio. The content of organic matter decreased along with the progress of processes in the reactors. The COD of the effluent from the biological reactor ranged from 237.0 to 566.0 mg O_2/L at C/N = 0.5 and C/N = 3.0, respectively.

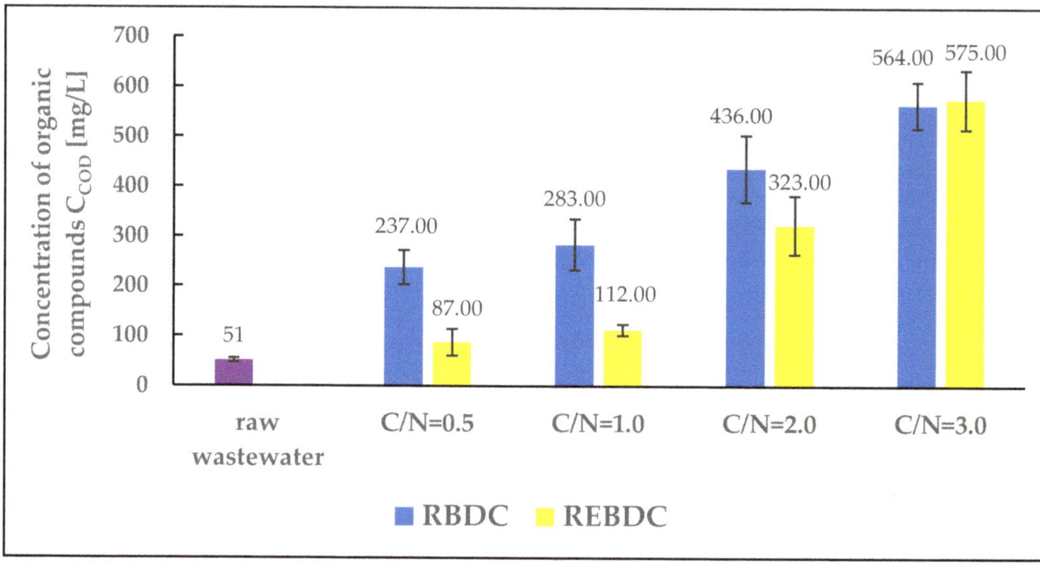

Figure 7. The effect of the C/N ratio on COD concentration in the effluent.

An increase in the amount of an external carbon source supplied to the RBDC reactor resulted in an increase in the COD of the treated wastewater compared to the COD of the raw influent. These values were more than 4 times higher (at C/N = 0.5) and 11 times higher (at C/N = 3.0) than the COD concentration in the influent. This indicates the inefficient utilization of the supplied organic substrate by heterotrophic microorganisms within the biofilm during its biodegradation and during the heterotrophic denitrification occurring in the deeper layers of the biofilm. For this type of reactor and a hydraulic retention time (HRT) of 24 h, the demand of microorganisms (responsible for biodegradation and heterotrophic denitrification) for a carbon and energy source, such as sodium acetate, was significantly lower than its amount supplied. As a result of denitrification, biomass growth, and biodegradation processes, only a portion of the organic substances was consumed [55].

In the electro-biological reactor, the supplied organic compounds were removed via ongoing physicochemical, biological, and electrochemical processes. Biological degradation occurred as aerobic and anaerobic microorganisms metabolized the supplied organic substrate as a source of carbon and energy, leading to a reduction in COD within the reactor and its increase in the biomass. The flow of electric current through the REBDC reactor enhanced the activity of microorganisms, decomposed organic compounds on the electrodes via electro-oxidation, produced hydroxyl radicals oxidizing organic compounds, and induced electrocoagulation of organic matter. Simultaneously, part of the organic substrate was consumed during heterotrophic denitrification as an electron donor for denitrifying bacteria [56].

The extent of organic compound utilization in the electro-biological reactor was greater than in the biological reactor. At C/N ratios of 0.5, 1.0, and 2.0, the COD concentration in the effluent from this reactor was significantly lower than in that from the biological reactor. At the highest C/N ratio of 3.0, the COD concentrations in the effluent from both reactors were similar, but slightly higher in the effluent from the REBDC reactor. According to Hao et al. [57], the presence of both heterotrophic and autotrophic bacteria responsible for nitrate removal in the electro-biological reactor depends on the C/N ratio. Similar conclusions were drawn by Lin et al. [58]. Meanwhile, Tong et al. [59] observed an increase in organic matter concentration, expressed as COD, in the effluent from the electro-biological reactor at higher C/N ratios. This increase could be due to autotrophic denitrifying bacteria consuming part of the nitrates during electric current flow, resulting in lower organic carbon consumption [60].

In the REBDC reactor with C/N = 0.5, the COD concentration in the treated wastewater was 36 mg O_2/L higher than in the raw influent (87 and 51 mg O_2/L, respectively), indicating a significant, though not complete, utilization of the external carbon source. Its complete utilization was observed in the study by Rodziewicz et al. [13]. To achieve complete utilization of the external carbon source at current density applied, a rational solution would be to extend the wastewater retention time, which would additionally enable higher nitrogen removal efficiency and lower pollutant concentrations in the effluent.

However, it should be noted that for both C/N values (0.5 and 1.0), the nitrogen removal efficiency was low and did not exceed a few or several percent, respectively, in the first and second case (Figure 5). Only at C/N = 2.0 did the REBDC reactor achieve an efficiency above 50% (56.0%), and at C/N = 3.0 over 65%. Nevertheless, significant amounts of unused organic carbon were determined in the effluent from the reactors. The COD to TN ratio (COD/N), calculated based on final concentrations of organic compounds and nitrogen, was 0.23, 0.50, 1.24, and 2.07 for the RBDC reactor, and 0.15, 0.21, 1.20, and 2.67 for the REBDC reactor, respectively. In the case of the first two C/N values (0.5 and 1.0), we can speak of significant and much more efficient carbon utilization in the electro-biological reactor. For the remaining sodium acetate doses (C/N = 2.0 and C/N = 3.0), it is clearly evident that the external carbon source was utilized inefficiently in both reactors, and that the discharged wastewater had high concentrations of not only nitrogen compounds but also carbon compounds, indicating the need for further treatment.

It may be undoubtedly concluded that achieving higher nitrogen removal efficiency combined with more effective utilization of the external carbon source in the reactors discussed in this publication is feasible. However, a more reasonable approach would be to conduct research using an electro-biological reactor that ensures higher efficiency of nutrient removal. Taking into account the results from a previous study [13], the primary factor to be considered should be the extension of the wastewater retention time [55], rather than increasing the current density (due to operating and investment costs) [43]. The future research should focus on C/N ratios of 2.0 and 3.0 because they are expected to ensure the highest nitrogen removal efficiency and, simultaneously, a high phosphorus removal efficiency. This is most likely to be the case with REBDC at a C/N value of 2.0. The aim of the future studies would be to determine the impact of extending the HRT compared to 24 h on the efficiency of nitrogen removal and the utilization of the external carbon source.

4. Conclusions

1. Phosphorus removal efficiency in the biological reactor was significantly lower than in the electrobiological reactor because phosphorus removal in the biological reactor was primarily associated with biofilm formation, a process of low efficiency.
2. Phosphorus removal efficiencies above 90% were obtained in the electro-biological reactor at C/N ratios of 0.5, 1.0, and 2.0. At C/N = 3.0, the efficiency exceeded 83%. This reactor performance resulted from two mechanisms: phosphorus uptake linked to biofilm biomass growth and, more importantly, electrocoagulation.
3. High C/N ratios reduced phosphorus removal efficiency because elevated organic substrate levels in the biofilm promoted the growth of GAOs, which outcompeted PAOs.
4. The efficiency of total nitrogen (TN) removal increased with the dose of organic substrate fed to reactors of both types. In the electrobiological reactor, the TN removal efficiency was more than 9 times higher at C/N = 2.0 and more than 11 times higher at C/N = 3.0 compared to the values obtained at C/N = 0.5.
5. At the highest substrate dose (C/N = 3.0), the TN concentration was 263.0 mg N/L in the effluent from RBDC (56.7% TN removal efficiency) and 215.0 mg N/L in that from REBDC (64.8% efficiency).
6. The higher nitrogen removal rates were achieved in the electrobiological reactor, as nitrogen removal in the biological reactor was achieved upon heterotrophic denitrification and nitrogen incorporation into the biomass. In the electrobiological reactor, the efficiency of nitrogen removal was also influenced by hydrogenotrophic denitrification and electrochemical nitrate reduction due to the electric current flow.
7. The extent of organic compound utilization in the electro-biological reactor was higher than in the biological reactor.
8. In the reactor operated at a C/N = 0.5, the COD concentration in the effluent (87 mg O_2/L) exceeded that in the influent (51 mg O_2/L), indicating substantial, though incomplete, utilization of the external carbon source.
9. At C/N ratios of 0.5 and 1.0, nitrogen removal efficiencies remained very low, not exceeding dozen percent. A tangible improvement was observed only at C/N = 2.0, where the REBDC reactor reached 56.0%, and at C/N = 3.0, where the efficiency exceeded 65%. Despite this enhancement, considerable residual organic carbon persisted in the effluent.
10. To enhance nitrogen removal in the electrobiological reactor while ensuring more efficient utilization of the external carbon source, the technological approach should involve either extending the hydraulic retention time or implementing a multi-stage REBDC system. Further research should be carried out in this respect.

Author Contributions: Conceptualization, J.R. and W.J.; methodology, K.B. and A.M.; validation, A.M. and M.K.; formal analysis, J.R. and K.B.; data curation, K.B. and J.R.; writing—original draft preparation, W.J.; writing—review and editing, A.M. and M.K.; supervision, W.J. All authors have read and agreed to the published version of the manuscript.

Funding: This research was funded in the framework of a Project No. 29.610.023-300 of the University of Warmia and Mazury in Olsztyn, Poland.

Data Availability Statement: The data presented in this study are available in the article.

Conflicts of Interest: Author Kamil Bryszewski was employed by the company Hanna Instruments Sp. z.o.o., Al. J. Piłsudskiego St. 73, 10-449 Olsztyn, Poland. The remaining authors declare that the research was conducted in the absence of any commercial or financial relationships that could be construed as a potential conflict of interest.

Abbreviations

The following abbreviations are used in this manuscript:

BER	Bio-electrochemical reactor
REBDC	Rotating electro-biological disc contactor
RBDC	Rotating biological disc contactor
BDPs	Biodegradable polymers
HRT	Hydraulic retention time
TN	Total nitrogen
TP	Total phosphorus
DOC	Dissolved organic carbon
EC	Electrical conductivity
PAOs	Polyphosphate-accumulating organisms
GAOs	Glycogen-accumulating organisms

References

1. Mielcarek, A.; Rodziewicz, J.; Janczukowicz, W.; Dobrowolski, A. Analysis of wastewater generated in greenhouse soilless tomato cultivation in central Europe. *Water* **2019**, *11*, 2538. [CrossRef]
2. van der Salm, C.; Voogt, W.; Beerling, E.; van Ruijven, J.; van Os, E. Minimising emissions to water bodies from NW European greenhouses; with focus on Dutch vegetable cultivation. *Agric. Water Manag.* **2020**, *242*, 106398. [CrossRef]
3. Regulation of the Minister of Maritime Economy and Inland Navigation of July 12, 2019 on Substances Particularly Harmful to the Aquatic Environment and Conditions to Be Met When Introducing Sewage into Waters or into the Ground, as Well as When Discharging Rainwater or Snowmelt into Waters or into Water Equipment. Dz. U. (J. Laws) 2019, 1311. Available online: https://isap.sejm.gov.pl/isap.nsf/DocDetails.xsp?id=WDU20190001311 (accessed on 23 September 2025). (In Polish)
4. Du, J.; Xu, B.; Ma, L.; Jiao, H.; Tian, B.; Li, K.; Lian, J. Study on denitrification of hydroponic wastewater reverse osmosis concentrate using sulfur-autotrophic denitrification. *J. Environ. Chem. Eng.* **2023**, *11*, 111195. [CrossRef]
5. Kwon, M.J.; Hwang, Y.; Lee, J.; Ham, B.; Rahman, A.; Azam, H.; Yang, J.-S. Waste nutrient solutions from full-scale open hydroponic cultivation: Dynamics of effluent quality and removal of nitrogen and phosphorus using a pilot-scale sequencing batch reactor. *J. Environ. Manag.* **2021**, *281*, 111893. [CrossRef] [PubMed]
6. Gu, X.; Leng, J.; Zhu, J.; Zhang, K.; Zhao, J.; Wu, P.; Xing, Q.; Tang, K.; Li, X.; Hu, B. Influence mechanism of C/N ratio on heterotrophic nitrification- aerobic denitrification process. *Bioresour. Technol.* **2022**, *343*, 126116. [CrossRef]
7. Zhao, B.; Sun, Z.; Liu, Y. An overview of in-situ remediation for nitrate in groundwater. *Sci. Total Environ.* **2022**, *804*, 149981. [CrossRef]
8. Henze, M.; Harremoës, P.; Jes La Cour, J.; Arvin, E. *Wastewater Treatment. Biological and Chemical Processes*; Kielce University of Technology Publishing House: Kielce, Poland, 2002. (In Polish)
9. Lee, K.C.; Rittmann, B.E. Effects of pH and precipitation on autohydrogenotrophic denitrification using the hollow-fiber membrane-biofilm reactor. *Water Res.* **2003**, *37*, 1551–1556. [CrossRef]
10. Mansell, B.O.; Schroeder, E.D. Hydrogenotrophic denitrification in a microporous membrane bioreactor. *Water Res.* **2002**, *36*, 4683–4690. [CrossRef]
11. Peng, Y.-Z.; Ma, Y.; Wang, S.-Y. Denitrification potential enhancement by addition of external carbon sources in a pre-denitrification process. *J. Environ. Sci.* **2007**, *19*, 284–289. [CrossRef] [PubMed]
12. Chen, H.B.; Wang, D.B.; Li, X.M.; Yang, Q.; Zeng, G.M. Enhancement of post-anoxic denitrificationfor biological nutrient removal: Effect of different carbon sources. *Environ. Sci. Pollut. Res.* **2015**, *22*, 5887–5894. [CrossRef]
13. Rodziewicz, J.; Mielcarek, A.; Janczukowicz, W.; Jóźwiak, T.; Struk–Sokołowska, J.; Bryszewski, K. The share of electrochemical reduction, hydrogenotrophic and heterotrophic denitrification in nitrogen removal in rotating electrobiological contactor (REBC) treating wastewater from soilless cultivation systems. *Sci. Total Environ.* **2019**, *683*, 21–28. [CrossRef] [PubMed]
14. Mielcarek, A.; Rodziewicz, J.; Janczukowicz, W.; Dąbrowska, D.; Ciesielski, S.; Thornton, A.; Struk-Sokołowska, J. Citric acid application for denitrification process support in biofilm reactor. *Chemosphere* **2017**, *171*, 512–519. [CrossRef] [PubMed]
15. Lee, D.-U.; Lee, I.-S.; Choi, Y.-D.; Bae, J.-H. Effects of external carbon source and empty bed contact time on simultaneous heterotrophic and sulfur-utilizing autotrophic denitrification. *Process Biochem.* **2001**, *36*, 1215–1224. [CrossRef]
16. Fernández, F.J.; Castro, M.C.; Villasenor, J.; Rodríguez, L. Agro-food wastewaters as external carbon source to enhance biological phosphorus removal. *Chem. Eng. J.* **2011**, *166*, 559–567. [CrossRef]
17. Fu, X.; Hou, R.; Yang, P.; Qian, S.; Feng, Z.; Chen, Z.; Wang, F.; Yuan, R.; Chen, H.; Zhou, B. Application of external carbon source in heterotrophic denitrification of domestic sewage: A review. *Sci. Total Environ.* **2022**, *817*, 153061. [CrossRef]

18. Zhang, Y.; Wang, X.C.; Cheng, Z.; Li, Y.; Tang, J. Effect of fermentation liquid from food waste as a carbon source for enhancing denitrification in wastewater treatment. *Chemosphere* **2016**, *144*, 689–696. [CrossRef]
19. Cui, H.; Feng, Y.; Yin, Z.; Qu, K.; Wang, L.; Li, J.; Jin, T.; Bai, T.; Cui, Z. Organic carbon release, denitrification performance and microbial community of solid-phase denitrification reactors using the blends of agricultural wastes and artificial polymers for the treatment of mariculture wastewater. *Ecotoxicol. Environ. Saf.* **2023**, *255*, 114791. [CrossRef]
20. Liu, D.; Li, J.; Li, C.; Deng, Y.; Zhang, Z.; Ye, Z.; Zhu, S. Poly(butylene succinate)/bamboo powder blends as solid-phase carbon source and biofilm carrier for denitrifying biofilters treating wastewater from recirculating aquaculture system. *Sci. Rep.* **2018**, *8*, 3289. [CrossRef]
21. Yang, X.L.; Jiang, Q.; Song, H.L.; Gu, T.T.; Xia, M.Q. Selection and application of agricultural wastes as solid carbon sources and biofilm carriers in MBR. *J. Hazard. Mater.* **2015**, *283*, 186–192. [CrossRef]
22. Kłobukowska, K.; Rodziewicz, J.; Mielcarek, A.; Bryszewski, K.; Janczukowicz, W.; Bugajski, P.; Jóźwiakowski, K.; Operacz, A. Novel materials as exogenous carbon sources for denitrifying biofilters. *Appl. Sci.* **2024**, *14*, 176. [CrossRef]
23. Krzemieniewski, M.; Rodziewicz, J. Nitrogen compounds removal in a rotating electrobiological contactor. *Environ. Eng. Sci.* **2005**, *22*, 816–822. [CrossRef]
24. *PN-ISO 6060:2006*; Water Quality—Determination of the Chemical Oxygen Demand. Polish Committee for Standardization: Warsaw, Poland, 2006.
25. *PN-EN 12260:2005*; Water Quality—Determination of Nitrogen—Determination of Bound Nitrogen (TN), Following Oxidation To Nitrogen Oxides. Polish Committee for Standardization: Warsaw, Poland, 2005.
26. *PN-EN ISO 10304-1:2009*; Water Quality—Determination of Dissolved Anions by Liquid Chromatography of Ions—Part 1: Determination of Bromide, Chloride, Fluoride, Nitrate, Nitrite, Phosphate and Sulfate. Polish Committee for Standardization: Warsaw, Poland, 2009.
27. *PN-EN ISO 6878:2006*; Water Quality—Determination of Phosphorus—Ammonium Molybdate Spectrometric Method. Polish Committee for Standardization: Warsaw, Poland, 2006.
28. *PN-EN 27888:1999*; Water Quality—Determination of Electrical Conductivity. Polish Committee for Standardization: Warsaw, Poland, 1999.
29. Ghafari, S.; Hasan, M.; Aroua, M.K. Improvement of autohydrogenotrophic nitrite reduction rate through optimization of pH and bicarbonate dose in a batch experiments. *J. Biosci. Bioeng.* **2009**, *107*, 275–280. [CrossRef] [PubMed]
30. Mousavi, S.A.R.; Ibrahim, S.; Aroua, M.K.; Ghafari, S. Development of nitrate elimination by autohydrogenotrophic bacteria in bio-electrochemical reactors—A review. *Biochem. Eng. J.* **2012**, *67*, 251–264. [CrossRef]
31. Di Capua, F.; Pirozzi, F.; Lens, P.N.L.; Esposito, G. Electron donors for autotrophic denitrification. *Chem. Eng. J.* **2019**, *362*, 922–937. [CrossRef]
32. Prosnansky, M.; Sakakibara, Y.; Kuroda, M. High-rate denitrification and SS rejection by biofilm electrode reactor (BER) combined with microfiltration. *Water Res.* **2002**, *36*, 4801–4810. [CrossRef] [PubMed]
33. Rahimi, S.; Modin, O.; Mijakovic, I. Technologies for biological removal and recovery of nitrogen from wastewater. *Biotechnol. Adv.* **2020**, *43*, 107570. [CrossRef]
34. Glass, C.; Silverstein, J. Denitrification kinetics of high nitrate concentration water: pH effect on inhibition and nitrite accumulation. *Water Res.* **1998**, *32*, 831–839. [CrossRef]
35. Rezania, B.; Cicek, N.; Oleszkiewicz, J.A. Kinetics of hydrogen-dependent denitrification under varying pH and temperature conditions. *Biotechnol. Bioeng.* **2005**, *92*, 900–906. [CrossRef]
36. Kurt, M.; Dunn, J.; Bourne, J.R. Biological denitrification of drinking water using autotrophic organisms with H2 in a fluidized-bed biofilm reactor. *Biotechnol. Bioeng.* **1987**, *29*, 493–501. [CrossRef]
37. Huang, B.; Feng, H.; Wang, M.; Li, N.; Cong, Y.; Shen, D. The effect of C/N ratio on nitrogen removal in a bioelectrochemical system. *Bioresour. Technol.* **2013**, *132*, 91–98. [CrossRef]
38. Huang, W.; Gong, B.; Wang, Y.; Lin, Z.; He, L.; Zhou, J.; He, Q. Metagenomic analysis reveals enhanced nutrients removal from low C/N municipal wastewater in a pilot-scale modified AAO system coupling electrolysis. *Water Res.* **2020**, *173*, 115530. [CrossRef]
39. Zhang, L.; Hao, S.; Wang, Y.; Lan, S.; Dou, Q.; Peng, Y. Rapid start-up strategy of partial denitrification and microbially driven mechanism of nitrite accumulation mediated by dissolved organic matter. *Bioresour. Technol.* **2021**, *340*, 125663. [CrossRef]
40. Hassard, F.; Biddle, J.; Cartmell, E.; Jefferson, B.; Tyrrel, S.; Stephenson, T. Rotating biological contactors for wastewater treatment—A review. *Process Saf. Environ.* **2015**, *94*, 285–306. [CrossRef]
41. Mahto, K.U.; Das, S. Bacterial biofilm and extracellular polymeric substances in the moving bed biofilm reactor for wastewater treatment: A review. *Bioresour. Technol.* **2022**, *345*, 126476. [CrossRef]
42. Gaspar, E.; Munteanu, I.; Sintea, S. Removal of N and P in a Rotating Biological Contactor Plant: Case Study Agnita, Romania. *Water* **2022**, *14*, 3670. [CrossRef]

43. Rodziewicz, J.; Mielcarek, A.; Bryszewski, K.; Janczukowicz, W.; Kłobukowska, K. Energy consumption for nutrient removal from high-nitrate and high-phosphorus wastewater in aerobic and anaerobic bioelectrochemical reactors. *Energies* **2022**, *15*, 7251. [CrossRef]
44. Lagum, A.A. Integrating electrochemical and biological phosphorus removal processes via electrokinetic-based technology. *J. Environ. Chem. Eng.* **2021**, *9*, 106609. [CrossRef]
45. Hackbarth, M.; Gescher, J.; Horn, H.; Reiner, J.E. A scalable, rotating disc bioelectrochemical reactor (RDBER) suitable for the cultivation of both cathodic and anodic biofilms. *Bioresour. Technol. Rep.* **2023**, *21*, 101357. [CrossRef]
46. Rittmann, B.E.; McCarty, P.L. *Environmental Biotechnology: Principles and Applications*; McGraw-Hill Education: New York, NY, USA, 2020.
47. Chen, Y.Z.; Zhang, L.J.; Ding, L.Y.; Zhang, Y.Y.; Wang, X.S.; Qiao, X.J.; Pan, B.Z.; Wang, Z.W.; Xu, N.; Tao, H.C. Sustainable treatment of nitrate-containing wastewater by an autotrophic hydrogen-oxidizing bacterium. *Environ. Sci. Ecotechnol.* **2022**, *9*, 100146. [CrossRef] [PubMed]
48. Pan, Y.; She, D.; Shi, Z.; Cao, T.; Xia, Y.; Shan, J. Salinity and high pH reduce denitrification rates by inhibiting denitrifying gene abundance in a saline-alkali soil. *Sci. Rep.* **2023**, *13*, 2155. [CrossRef]
49. González, V.L.; Dombrowski, P.M.; Lee, M.D.; Ramsburg, C.A. The Influence of pH on Subsurface Denitrification Stimulated with Emulsified Vegetable Oil. *Water* **2023**, *15*, 883. [CrossRef]
50. Salem, Z.; Lebik, H.; Cherafa, W.K.; Allia, K. Valorisation of olive pits using biological denitrification. *Desalination* **2007**, *204*, 72–78. [CrossRef]
51. Dash, B.P.; Chaudhari, S. Electrochemical denitrification of simulated ground water. *Water Res.* **2005**, *39*, 4065–4072. [CrossRef]
52. Ding, A.; Zheng, P.; Zhang, M.; Zhang, Q. Impacts of electron donor and acceptor on the performance of electrotrophic denitrification. *Environ. Sci. Pollut. Res.* **2017**, *24*, 19693–19702. [CrossRef]
53. Li, M.; Feng, C.P.; Zhang, Z.N.; Lei, X.H.; Chen, R.Z.; Yang, Y.N.; Sugiura, N. Simultaneous reduction of nitrate and oxidation by-products using electrochemical method. *J. Hazard. Mater.* **2009**, *171*, 724–730. [CrossRef] [PubMed]
54. Zhang, J.M.; Feng, C.P.; Hong, S.Q.; Hao, H.L.; Yang, Y.N. Behavior of solid carbon sources for biological denitrification in groundwater remediation. *Water Sci. Technol.* **2012**, *65*, 1696–1704. [CrossRef] [PubMed]
55. Metcalf & Eddy, Inc. *Wastewater Engineering: Treatment and Resource Recovery*, 5th ed.; McGraw Hill: New York, NY, USA, 2014.
56. Sanjrani Manzoor Ahmed, B.Z.; Rind, S.; Rani, K. Systematic review: External carbon source for biological denitrification for wastewater. *Biotechnol. Bioeng.* **2023**, *120*, 642–658. [CrossRef] [PubMed]
57. Hao, R.X.; Li, S.M.; Li, J.B.; Meng, C.C. Denitrification of simulated municipal wastewater treatment plant effluent using a three-dimensional biofilm-electrode reactor: Operating performance and bacterial community. *Bioresour. Technol.* **2013**, *143*, 178–186. [CrossRef]
58. Lin, X.; Yin, H.; Wang, L.; Chen, Y.; Zhao, F.; Pu, Y.; Tang, X. Study of a three-dimensional biofilm-electrode reactor (3D-BER) that combined heterotrophic and autotrophic denitrification (HAD) to remove nitrate from water. *RSC Adv.* **2023**, *21*, 14675–14684. [CrossRef]
59. Tong, S.; Zhang, B.; Feng, C.; Zhao, Y.; Chen, N.; Hao, C.; Pu, J.; Zhao, L. Characteristics of heterotrophic/biofilm-electrode autotrophic denitrification for nitrate removal from groundwater. *Bioresour. Technol.* **2013**, *148*, 121–127. [CrossRef]
60. Wu, L.; Wei, W.; Xu, J.; Chen, X.; Liu, Y.; Peng, L.; Wang, D.; Ni, B.-J. Denitrifying biofilm processes for wastewater treatment: Developments and perspective. *Environ. Sci. Water Res. Technol.* **2021**, *7*, 40–67. [CrossRef]

processes

MDPI

Article

Influence of Polyester and Denim Microfibers on the Treatment and Formation of Aerobic Granules in Sequencing Batch Reactors

Victoria Okhade Onyedibe [1], Hassan Waseem [2], Hussain Aqeel [2], Steven N. Liss [2], Kimberley A. Gilbride [2], Roxana Sühring [2] and Rania Hamza [1,*]

[1] Department of Civil Engineering, Toronto Metropolitan University, Toronto, ON M5B 2K3, Canada; oonyedibe@torontomu.ca

[2] Department of Chemistry & Biology, Toronto Metropolitan University, Toronto, ON M5B 2K3, Canada; hassan.waseem@torontomu.ca (H.W.); hussain.aqeel@torontomu.ca (H.A.); steven.liss@torontomu.ca (S.N.L.); gilbride@torontomu.ca (K.A.G.); roxana.suehring@torontomu.ca (R.S.)

* Correspondence: rhamza@torontomu.ca

Abstract: This study examines the effects of polyester and denim microfibers (MFs) on aerobic granular sludge (AGS) over a 42-day period. Treatment performance, granulation, and microbial community changes were assessed at 0, 10, 70, 210, and 1500 MFs/L. Reactors with 70 MFs/L achieved rapid granulation and showed improved settling by day 9, while 0 and 10 MFs/L reactors showed delayed granule formation, which was likely due to limited nucleation and weaker shear conditions. Severe clogging and frequent maintenance occurred at 1500 MFs/L. Despite > 98% MF removal in all reactors, treatment performance declined at higher MF loads. Nitrogen removal dropped from 93% to 68%. Phosphate removal slightly increased in reactors with no or low microfiber loads (96–99%), declined in reactors with 70 or 210 MFs/L (92–91%, 89–88%), and dropped significantly in the reactor with1500 MFs/L (86–70%, $p < 0.05$). COD removal declined with increasing MF load. *Paracoccus* (denitrifiers) dominated low-MF reactors; *Acinetobacter* (associated with complex organic degradation) and *Nitrospira* (nitrite-oxidizing genus) were enriched at 1500 MFs/L. Performance decline likely stemmed from nutrient transport blockage and toxic leachates, highlighting the potential threat of MFs to wastewater treatment and the need for upstream MF control.

Keywords: polyester and denim fibers; microfiber removal; wastewater treatment; aerobic granular sludge; microbial community dynamics; sequencing batch reactors

1. Introduction

Anthropogenic microfibers (MFs), are prevalent and persistent contaminants in wastewater treatment systems. They are long, slender, and threadlike strands with a consistent width throughout their entire length, typically measuring less than 5 mm [1–3]. They originate largely from synthetic and natural textiles that are released into wastewater systems through activities like household laundry. Due to their small sizes, MFs can bypass pretreatment barriers and enter the downstream treatment processes [4–8], leading to a significant enrichment in treated wastewater and sewage sludge [9–11]. Recent findings show that more than 50 million microplastic (MP) particles may routinely be discharged in wastewater treatment plant (WWTP) effluents, with fibers being the most dominant form, accounting for approximately 59% of the particles analyzed [12]. Their prevalence is influenced by factors such as wastewater sources (municipal vs. industrial) and the treatment technologies employed. For instance, textile industry WWTPs have higher influent concentrations of MFs compared to municipal WWTPs [11,13].

Microfibers pose ecological and health risks by containing, adsorbing, transporting, and releasing harmful contaminants such as endocrine disruptors and bacterial pathogens [14,15], yet their impact on wastewater treatment systems themselves remain poorly understood. Among the various MFs, denim and polyester fibers are the most prevalent particles in wastewater [16,17].

Denim jeans made from cotton, polyester, lyocell, and polyurethane are widely worn globally [18,19], with 46–76% of Canadians and Americans wearing them daily and 450 million pairs being sold annually in the U.S. [20,21]. Washing jeans releases MFs into wastewater, with a single pair of jeans shedding up to 560,000 MFs per wash [22]. In Canada alone, laundering jeans releases 130 million MFs annually, while Canadian and American households discharge 533 million MFs annually, equivalent to 22 kilotonnes of MFs entering wastewater [22,23]. Polyester, which accounts for 49% of global clothing production [24–26], is expected to reach a production volume of 84 million metric tonnes by 2025 [24,27]. A single garment washed domestically can release over 1900 fibers, with polyester-based clothing being the primary source of microfiber release [28]. Congruently, polyester MFs are the dominant MF found in effluents from commercial washing machines [17].

Conventional and industrial WWTPs are not designed to remove microplastics, making them both for and sources of environmental microplastic contamination [8,29]. Although some microplastics are removed during pretreatment, up to 50% can reach the biological treatment stages, where they can interfere with microbial processes [30]. Even in systems achieving high removal rates (e.g., membrane bioreactors (MBRs) with efficiencies up to 99.9%, dissolved air flotation (DAF) at ~97%, and rapid sand filtration (RSF) at ~95%), microplastics, including MFs, can still persist in effluents [13]. In addition, a significant portion, often over 80% of the influent concentration, is retained in sludge and potentially redistributed through land-applied biosolids [13,30–32]. Within wastewater treatment systems, growing evidence shows that microplastics can disrupt floc formation and inhibit key microbial groups involved in nitrification and denitrification, ultimately compromising treatment performance and stability [8,30].

The potential effects of microplastics on wastewater treatment depend on their type, concentration, and the reactor system involved. For example, previous research has shown that low concentrations of polyethylene terephthalate (PET) microplastics had no impact on chemical oxygen demand (COD) removal and only slightly inhibited total nitrogen (TN) removal in an MBR (membrane bioreactor), while higher PET concentrations (75–300 particles/L) led to COD inhibition and reduced methane production in an upflow anaerobic sludge blanket (UASB) reactor [33,34]. Similarly, polyamides (PA66) at concentrations of 0.1–0.5 g/L was reported to inhibit ammonia removal and to slightly affect COD and TN removals in an aerobic granular sludge (AGS) system [35]. Interestingly, the microbial community in the AGS system adapted over time, resulting in a recovery and eventual improvement of both ammonia and TN removal efficiency in the system [35]. These results indicated, that AGS might be resilient to potential toxic effects from microplastic contamination in wastewater.

The reported resilience of AGS to microplastic impacts on wastewater treatment effectiveness is consistent with previous reports on AGS performance that reported high settling performance, simultaneous nitrogen and phosphorus removal, and resistance to a variety of toxins [36,37]. The treatment efficiency and resilience of AGS in comparison to conventional activated sludge systems, is rooted in its compact granule structure that supports high microbial activity and diversity [38,39]. Despite these encouraging findings for AGS application in the presence of ubiquitous microplastic contamination, the extent–and limits–of this resilience are not well understood.

For example, previous research on polyethylene (PE) microplastics in an AGS system reduced the relative abundance of *Proteobacteria*, which are key denitrifying organisms in AGS [8]. Similarly, the number of *Bacteroidetes*, which are associated with microbial immunity, also declined, suggesting a potential reduction in the system's resistance to environmental toxicity [40]. Furthermore, prolonged exposure to bisphenol A (BPA) weakened the activity of heterotrophic and nitrifying bacteria, resulting in the reduced removal of COD and ammonia in AGS systems [8,41]. These findings highlight a critical knowledge gap in understanding how microplastics and their associated additives interact with AGS and affect its long-term treatment performance.

Another important knowledge gap lies in the focus of previous research on the potential impact of spherical microplastic particles [33] on AGS systems. Conversely, MFs remain a highly abundant yet underexplored form of microplastics. Given their prevalence in wastewater and wastewater treatment systems, understanding their interactions with biological systems is critical for advanced technologies like AGS. In this research, we investigated the impacts and fate of polyester and denim MFs on AGS-based treatment systems. The objectives of this work were to (1) assess the effects of increasing MF concentrations on granule formation and settling properties; (2) evaluate their impact on the removal of organic matter and nutrients, particularly with respect to key nitrifying and denitrifying organisms; and (3) examine changes in microbial community structure and dynamics in response to MF exposure. The results enable us to assess AGS performance under realistic loading scenarios and identify its potential vulnerabilities and adaptive responses to MF pollution.

2. Materials and Methods

2.1. Sources and Fiber Concentrations

The denim fibers, averaging 4 mm in size, were procured from levis.com in Canada, while the polyester fibers were obtained from the Fabric Land Home Decor Centre in Ontario, Canada. These fibers were precision-cut to 4 mm in length using needles, Fiskars scissors, and fabric cutters. Influent concentrations of 10, 70, 210, and 1500 microfibers/ liter (MFs/L) were chosen, as they represented the minimum, median, triple of the median, and 95th percentile concentration of microplastics in the influents of WWTPs globally from an extensive literature review covering 31 peer-reviewed journal articles (Supplementary Table S1).

2.2. Reactor Configuration and Experimental Setup

The experimental study employed five polyvinyl chloride bubble column sequencing batch reactors (SBR), one for each of the tested MF concentrations (SBR-0, SBR-10, SBR-70, SBR-210, and SBR-1500), with a uniform configuration: 70 cm in height, 1.38 L total volume, and 5 cm internal diameter. The reactors operated at a height-to-diameter ratio (H/D) of 12.2 and a working volume of 1.2 L. Aeration was provided by fine bubble diffusers at the reactor bottom, with airflow regulated at 3 L min, resulting in a superficial air velocity of 2.55 cm/s. Each reactor ran for 42 days. The experiments were conducted at ambient temperatures (22–25 °C) in the laboratory, with no external pH control and with dissolved oxygen maintained above 2 mg/L.

Feeding influents into and discharging treated effluents out of the reactors was achieved with the aid of peristaltic pumps. The feeding port of the reactors was located at the bottom of each reactor, while the decant ports were located at the midpoint of each reactor, leading to a volumetric exchange ratio (VER) of 50%. The reactors were operated sequentially on a 4-h cycle. The sequential operation included influent filling (40 min), aeration (140 min), settling (30 min), effluent decanting (20 min), and idle time (10 min). A schematic diagram of the AGS reactor setup is shown in Figure 1. Throughout the duration of the experimental study, influent filling, aeration, treated effluent discharge, and idle values remained constant.

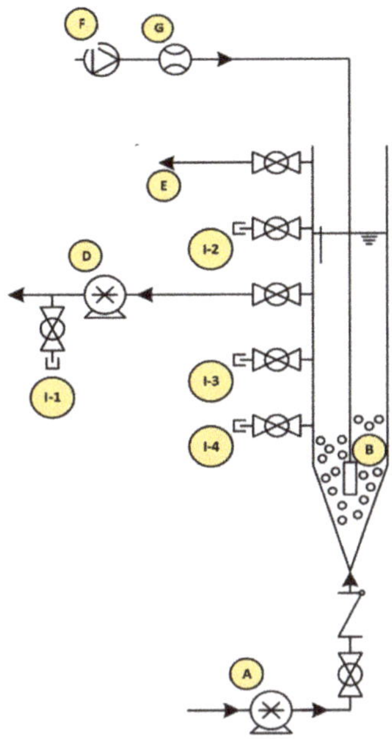

Figure 1. Schematic of the AGS reactor setup: (A) feed pump, (B) air diffuser, (C) pH probe, (D) decant pump, (G) air pump, (H) flow meter, (I-1) effluent sample port, (I-2-3) spare sample ports, and (1-4) sludge sample port.

2.3. *Seed Sludge and Feed Wastewater*

Return-activated sludge (RAS) from the Ashbridge's Bay WWTP in Toronto, Ontario, Canada was used to inoculate the SBRs. Synthetic wastewater as the feed was prepared based on a study by Tay et al. (2003) [42] with slight modifications. In brief, sodium acetate was used as the carbon source, NH_4Cl as the nitrogen source, and monobasic and dibasic phosphorus (K_2HPO_4 and KH_2PO_4) as the phosphorus sources. A chemical oxygen demand: nitrogen: phosphorus (COD:N:P) ratio of 100:4.5:0.45 was used, with a sodium acetate anhydrous concentration of 1.92 g/L (COD of 1500 g/L), a NH_4Cl concentration of 0.26 g/L, and K_2HPO_4 and KH_2PO_4 concentrations of 0.023 g/L and 0.015 g/L, respectively. The pH of the wastewater ranged from 7 to 8.05 without any external manipulation.

2.4. *Wastewater Characterization and Microfiber Analysis*

The biomass and wastewater characteristics of the influents and effluents, such as mixed-liquor suspended solids (MLSS), mixed-liquor volatile suspended solids (MLVSS), total suspended solids (TSS), volatile suspended solids (VSS), and sludge volume index (SVI), were determined according to the *Standard Methods for the Examination of Water and Wastewater* (APHA/AWWA/WEF, 2012) [43].

MLSS and MLVSS were measured according to Standard Methods 2540 D and 2540 E [43]. A well-mixed sample was filtered through a pre-weighed 47 mm glass fiber filter (Whatman GF/C, ~1.2 μm). The filter and retained solids were dried at 103–105 °C to determine the MLSS, then they were ignited at 550 °C to determine the MLVSS as the volatile fraction. All effluent samples were filtered using 0.45 μm Acrodisc® sterile syringe filters with Supor ® polyethersulfone membranes (Pall Laboratory, Port Washington, NY, USA). Soluble chemical demand (sCOD) was determined using the USEPA reactor digestion method with a COD HR HACH kit (HACH Company, Loveland, CO, USA), also supplied by VWR International (Mississauga, ON, Canada). Digestion was conducted using a DRB 200 reactor (HACH Company, Loveland, CO, USA). A DR 3900 laboratory spectrophotometer (HACH Company, Loveland, CO, USA) was used to analyze sCOD, TN, reactive phosphorus (P and PO_4^{3-}) and ammonia concentrations using HACH reagent kits, as follows: the salicylate method (Method 10031) for NH_3, the persulfate digestion method (Method 10071) for TN, and the molybdovanadate method (Method 8114) for reactive phosphorus.

MFs with an average size of 4 mm were dosed daily into the influent wastewater of the SBRs for the duration of the experiment at concentrations of 0 MFs/L, 10 MFs/L, 70 MFs/L, 210 MFs/L, and 1500 MFs/L with equal amounts of polyester MFs and denim MFs. Fiber amounts were determined based on a five-point calibration curve, including 0, 50, 500, 2000, and 8000 dry and wet fibers (Supplementary Figure S1). The MFs were analyzed visually and with the aid of a DM1000 LED microscope (Leica Microsystems, Wetzlar, Hesse, Germany). A subset of fibers was further characterized chemically to confirm polymer type using an 8700 Laser Direct Infrared (LDIR) chemical imaging system (Agilent Technologies, Santa Clara, CA, USA). This is in line with previous research, which encourages chemical analysis after visual identification [44]. Biomass samples were collected weekly from the lowest sample port of each reactor and the granule structure and morphology were analyzed using the Leica DM1000 LED microscope.

2.5. Microbial Community Analysis

The granules and biofilms (biomass attached to MFs) were processed for genomic DNA extraction using a previously published method [45]. Briefly, the MFs were picked from the granular sludge with forceps. The MFs were rinsed in deionized water to remove loosely bound granular sludge from the biofilms. The biomass samples were centrifuged at 8000 rpm for five minutes, after careful separation of the MFs and granular sludge. The supernatant was discarded, and the pellet was stored at −80 °C until further processing. DNA extractions were conducted with the DNeasy PowerSoil Pro Kit (QIAGEN, Germantown, MD, USA) in conjunction with Qiagen's QIAcube instrument, adhering rigorously to the manufacturer's provided protocols [46]. An MBI nanophotometer was used to accurately quantify genomic DNA. Microbial community structure was determined by 16S rRNA high-throughput sequencing (Illumina MiSeq-PE250, San Diego, CA, USA), conducted at Génome Québec (Montreal, QC, Canada). The V3–V4 region of the 16S rRNA gene was amplified using the primers 347F (GGAGGCAGCAGTRRGGAAT) and 803R (CTACCRGGGTATCTAATCC). Equimolar concentrations of genomic DNA were used for sequencing. The forward and reverse sequences were merged, and low-quality sequences were removed using the DADA2 software (v1.26) [47]. The sequences were analyzed using Microbiome Analyst (v2.0) to estimate the relative abundance of microbial communities and alpha diversity analyses [48]. The 16S rRNA gene sequencing data can be accessed from the National Library of Medicine, in the National Center for Biotechnology Information.

The Shannon diversity index value, indicating the richness and abundance of species in the AGS samples, was calculated by taking an average of the relative abundance of 3 technical replicates for each of the 5 reactors [49].

2.6. Quality Assurance

Numerous steps were taken to avoid contamination of the samples throughout the sample processing and analysis. All laboratory surfaces were thoroughly wiped with 70% ethanol three times before use. During sampling, cotton laboratory coats and nitrile gloves were worn to decrease contamination. All bottles, buckets, covers, and containers used were covered whenever possible. All the containers used were thoroughly washed with bleach and then rinsed with hot water and distilled water three times. Any filter used in the experiments was microscopically examined to ensure that there were no fibers on it before use. Before performing MLSS and TSS tests, aluminum dishes and filters were heated in the incinerator for 15 min to incinerate any impurities that were present and immediately stored in a tightly closed desiccator to prevent atmospheric contamination. Brightly colored fibers were used for ease of identification (red and yellow) and to prevent the fibers from getting lost within the biomass. All tests were performed in duplicate. The fibers were stored in tightly closed glass bottles with aluminum covers to prevent atmospheric contamination. Three blank control samples were tested to evaluate potential atmospheric laboratory contamination with MFs. These blanks contained none of the tested MFs. Direct filtration was applied where possible to minimize transfer and to reduce the risk of MF loss. Closed or semi-closed systems prevented MF loss through air or evaporation. Equipment and surfaces were rinsed with deionized water to recover any adhering MFs. Sampling occurred in a controlled environment with minimal airflow. Finally, the polymer type of the tested fibers was confirmed using an Agilent Technology 8700 LDIR (Agilent Technology, Santa Clara, CA, USA). For microbial analysis, low-quality sequences and chimera sequences were removed from the downstream analyses.

2.7. Data Analysis

Statistical analyses were conducted using Microsoft Excel. To assess the significant differences between reactors, Student's *t*-tests were performed. Pearson correlation was employed to examine the correlations between fiber concentration and removal efficiencies, as well as between runtime and removal efficiencies. The significance level was set at $p < 0.05$. Normality was checked using a quantile plot.

3. Results and Discussion

3.1. Inhibitory Effect of Microfibers on COD, TN, NH_3, and PO_4^{3-} Removal Efficiency

The COD removal efficiencies of the control reactor SBR-0 and the reactor with the lowest amount of MFs, (SBR-10), were high and remained consistent throughout the experiment (Figure 2), with an average removal rate of 96% $\pm$ 0.3 and 96% $\pm$ 0.2, respectively. The remaining three reactors showed slightly decreasing COD removal rates, with SBR-70 decreasing from 94–93%, SBR-210 from 92.3–92%, and SBR-1500 from 92–89% (Figure 2). Despite the overall low decrease in COD removal, the consistent temporal and fiber concentration-dependent effect led us to infer that a high concentration of denim and polyester MFs had a certain inhibitory effect on COD removal. These results were consistent with the inhibitory effects on COD removal reported by previous research on the effect of up to 0.5 g/L polyamide microplastics on AGS, wherein COD removal declined from 89% to 85% [35]. Likewise, the exposure of AGS to 75–300 MP/L of PET microplastics was previously reported to decrease the COD removal efficiency from 89.9% to 60.5% [33]. The reduction of COD removal could be due to the obstruction of nutrient transport channels within AGS, at high microplastic concentrations. Such a blockage could limit the diffusion of oxygen and the nutrients necessary for microbial activity [50].

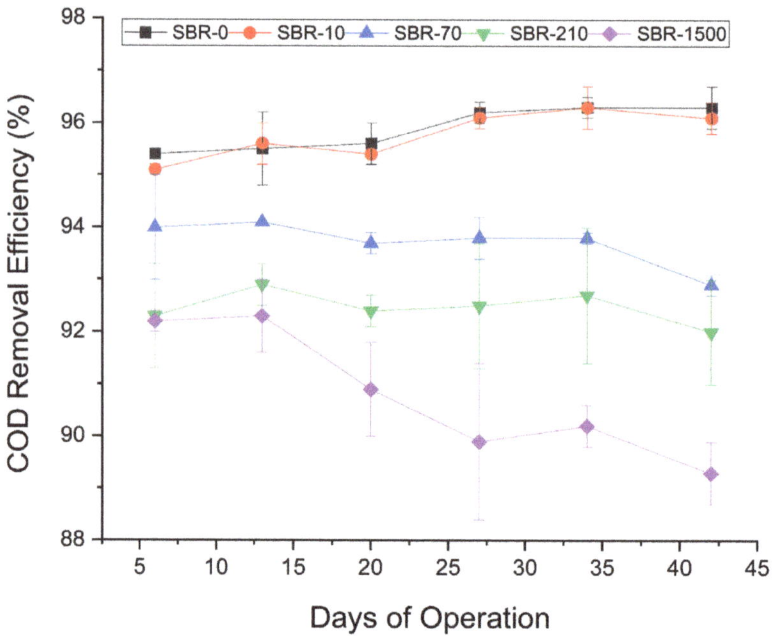

Figure 2. Profile of COD removal efficiencies for SBR-0, SBR-10, SBR-70, SBR-210, and SBR-1500.

TN removal efficiency fluctuated in all five reactors during the early stages of operation but later stabilized (Figure 3). The highest average TN removal was observed in SBR-0 and SBR-10 (94% and 90%, respectively). After day 20, TN removal in SBR-10 began to decline to 91%, while SBR-0 remained stable at 94%. Conversely, Average TN removal decreased substantially in the reactors with higher MF concentrations, with average TN removal efficiencies across increasing MF concentrations (SBR-0 to SBR-1500) of 93%, 93%, 81%, 75%, and 68%, respectively. The statistically significant decline in TN removal (Student's *t*-test, $p < 0.05$) in reactors with MF concentrations above 70 MF/L compared to the control, as well as the negative correlation between increasing MF concentrations and TN removal (Pearson $R = -0.79$), highlighting the adverse impact of elevated MF loads. A previous AGS study reported similar trends, wherein 200 MP/L of polyethylene microplastics reduced TN removal from 83% to 73% [37]. Interestingly, the ammonia (NH_3) removal remained consistently high (>95%) in all SBRs, suggesting that nitrification was unaffected despite the impact on TN removal. This decoupling from a decline in TN removal suggested the selective disruption of denitrification at higher MF concentrations. Likely causes include an altered biofilm structure, limited carbon availability, or potential inhibitory effects of leached plastic additives and dyes on denitrifiers. Although nitrate and nitrite were not measured—a limitation of this study—microbial analysis confirmed the presence of denitrifiers (e.g., *Paracoccus*), suggesting that denitrification occurred at least to some degree. Without intermediate nitrogen species data the extent and completeness of this process remain uncertain. Future work should include nitrate/nitrite profiling to clarify the impact of MFs on denitrification.

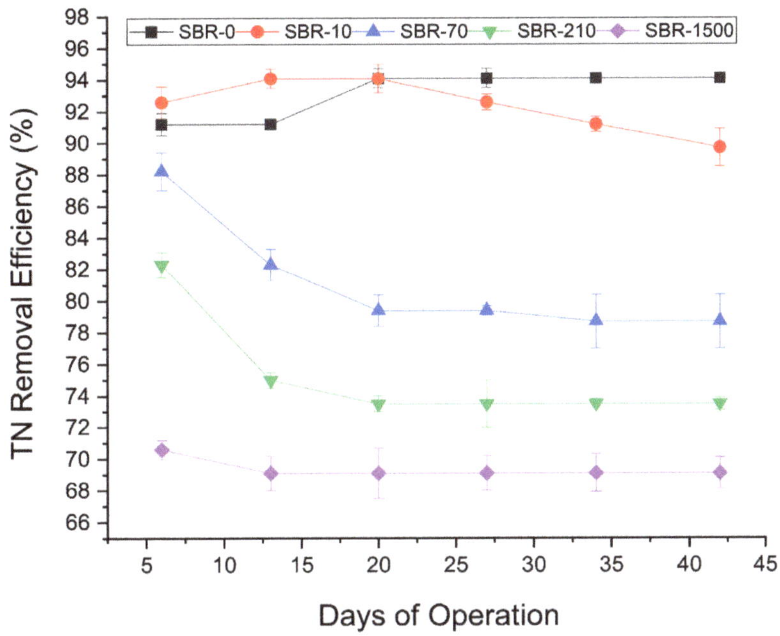

Figure 3. Profile of TN removal efficiencies for SBR-0, SBR-10, SBR-70, SBR-210, and SBR-1500.

Nitrogen assimilation into biomass may have contributed to removal as well, particularly in the early stages. However, given the stable COD loading (Figure 2) and consistent nitrification performance (Figure 3), assimilation alone is unlikely to explain the removal of TN at low MF concentrations or the continued removal of ammonia even at high MF concentrations. These observations suggest that reduced TN removal efficiency is primarily due to disruptions in denitrification or other dissimilatory nitrogen pathways. This effect has not previously been reported and is an important indication of the potential toxic impacts of MFs on WWTPs, warranting further studies.

The phosphate removal profiles (Figure 4) across the five reactors showed three discernible trends: (1) the reactor with no fibers and that with low influent MF concentrations, namely, SBR-0 and SBR-10, exhibited a slight increase in PO_4^{3-} removal, increasing from 96% to 99%; (2) the reactors with MF concentrations of 70 MFs/L and 210 MFs/L showed a marginal decline in PO_4^{3-} removal, decreasing from 92% to 91% and 89% to 88%, respectively; and (3) most notably, the reactor dosed with the highest MF concentration of 1500 fibers/L exhibited a significant reduction in PO_4^{3-} removal, declining from 86% to 70% (Student's t-test, $p < 0.05$) showing the negative impact of MFs on the removal of phosphate from WWTPs (Figure 4). This was consistent with a previous study on PE microplastics in AGS, which likewise reported a significant inhibitory effect on the removal of total phosphate (74% to 62%) [40]. Similar to the decreased nitrogen removal, the decrease in phosphate removal rates might have been caused by MF toxicity, changes in biofilm structures, or due to physical barriers for the nutrient transport channels in AGS, making it difficult for nutrients like nitrogen and phosphate to be absorbed and processed by microbial communities [51]. The potential toxic effects of MFs could be explained by the release of plastic additives and dyes that are toxic to the microbes involved in nutrient removal. Previous research has shown that many toxic water-soluble plastic additives are released from microplastics once they are in contact with water [52]. Likewise, it has been reported that synthetic textile dyes can reduce dissolved oxygen in aquatic environments, which leads to anoxic conditions that are not suitable for many microbes [53].

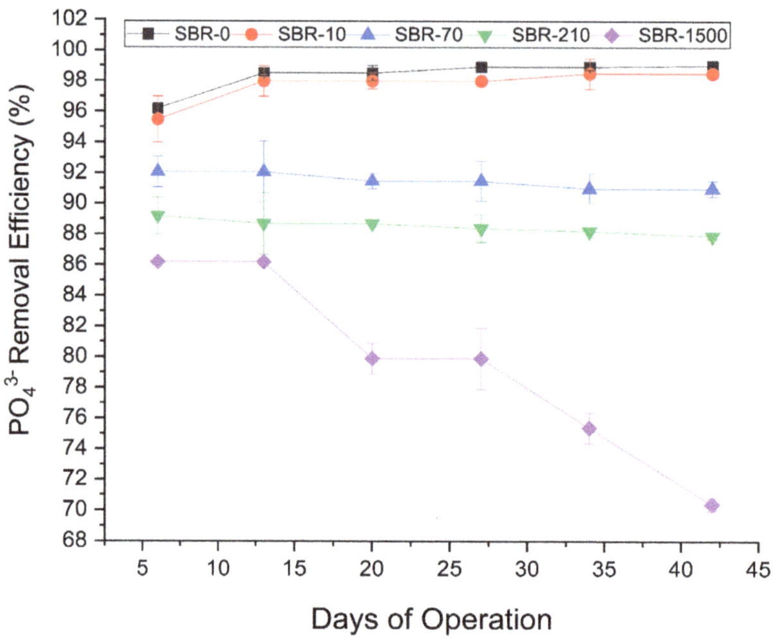

Figure 4. Profile of PO_4^{3-} removal efficiencies for SBR-0, SBR-10, SBR-70, SBR-210, and SBR-1500.

3.2. Impact of MFs on the Granulation Process

In addition to the nutrient removal, MFs significantly influenced the formation of granules in this study (Student's t-test, $p < 0.05$). The reactors that operated at 0 or 10 MFs/L required more time for the formation of visible granular structures (see Figure 5, SBR-0 and SBR-10). By the sixth day, we noted the development of small aggregates, approximately 30 μm in size, which subsequently evolved into well-defined granules ranging from 48 to 140 μm in diameter by the ninth day (Figure 5) but required until day 28 to form fully matured granules. Reactors with elevated fiber concentrations, i.e., SBR-70, SBR-210, and SBR-1500, on the other hand, already showed mature granules after 10 days. By the 28th day, all reactors displayed stable and mature granules of different sizes, ranging from 210 μm in SBR-10 to 523 μm in SBR-1500. The microscopic observation highlighted that two types of granules were present in the reactors, one fluffy and irregular without definite cores, while the others had smooth margins and well-defined cores. On day 28, the reactors with high concentrations of MFs (70, 210, and 1500 MFs/L) displayed granules that mostly showed well-defined cores, while the control and SBR-10 reactors still showed some fluffy and irregular granules (Figure 5). The accelerated pace and improved structure of the granules in the presence of MFs can be attributed to the increase in shear force due to the aeration and immobilization of biomass and attachment to the MFs [54]. Microplastics can promote AGS granulation through various mechanisms, including enhancing extracellular polymeric substance (EPS) production [55], modifying microbial communities by enriching the functional bacteria that support granulation, acting as nucleation sites for microbial aggregation [51], and improving sludge settling properties and compactness, due to enhanced hydrophobicity.

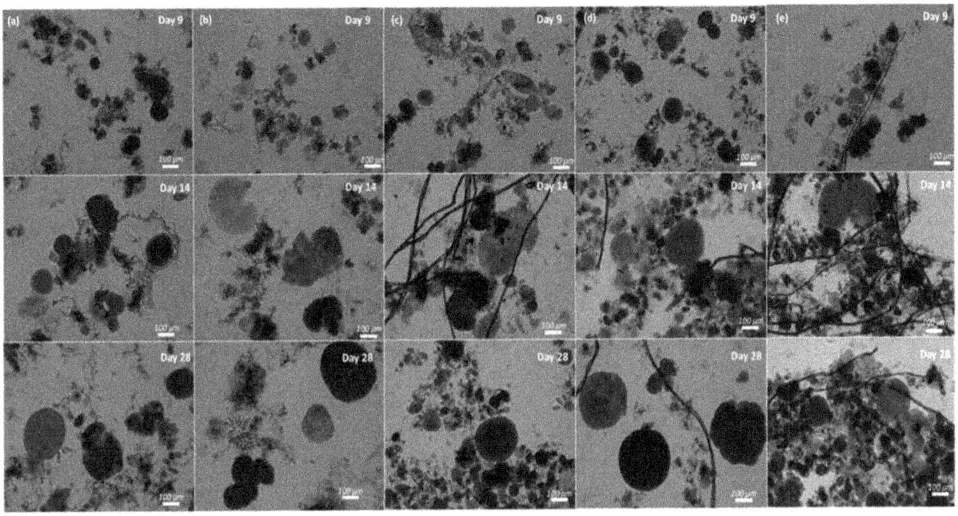

Figure 5. Granulation process from cultivation to maturation in SBR-0 (**a**), SBR-10 (**b**), SBR-70 (**c**), SBR-210 (**d**), and SBR-1500 (**e**) on days 9, 14, and 28.

To fully explore the impact of MFs on the aerobic granulation process, granulation indices other than size, such as MLSS, SVI_{30}, and SVI_5, were investigated (Figure 6). Biomass settleability showed continuous improvement, where the SVI profiles of reactors SBR-0, SBR-10, SBR-70, SBR-210, and SBR-1500 exhibited a consistently declining trend over the experimental period (Figure 6). The SVI_{30} and SVI_5 values across all reactors (SBR-0, SBR-10, SBR-70, SBR-210, and SBR-1500) demonstrated a reduction over time, ranging from 60 to 160 mL/g and 78 to 360 mL/g, respectively, throughout the 42-day duration of the study. Furthermore, the SVI_5 values (after 5 min of settling time) gradually approached proximity with the SVI_{30} values (after 30 min of settling time) in all reactors on the 42nd day, which is indicative of stable granulation. A difference between SVI_5 and SVI_{30} of within 10% is typically seen as a measure for full granulation [45], while lower SVI_{30}/SVI_5 ratios indicate the coexistence of granules and dense flocs within an AGS system [56,57].

The SVI_{30}/SVI_5 ratios for SBR-0, SBR-10, SBR-70, SBR-210, and SBR-1500 started at 0.44 and increased progressively to 0.61, 0.63, 0.71, 0.73, and 0.77, respectively (Figure 6), corresponding to granulation percentages of 61%, 63%, 71%, 73%, and 77%. Our findings demonstrate a clear positive relationship between granulation speed and overall granulation percentage with MF concentration (Figure 6). This suggests that MFs may have a positive impact on the physical aspects of the AGS system, such as promoting microbial aggregation and enhancing sludge settleability.

At the same time, MF accumulation led to a progressive increase in MLSS levels, particularly in SBR-1500 (Supplementary Figure S2). MLSS concentrations increased in all reactors during the experiment, with substantial fluctuations in the first 14 days, ranging from 4180 to 8010 mg/L. The initial MLSS concentration of 4000 mg/L rose to 6500 mg/L in SBR-0, 6900 mg/L in SBR-10, 7980 mg/L in SBR-70, and 9600 mg/L in SBR-210 (Supplementary Figure S2). The highest increase occurred in SBR-1500, reaching 13,770 mg/L. The observed buildup was due not only to improved biomass retention from granulation but also to the physical entrapment of synthetic fibers. While moderate increases in MLSS may support reactor stability, excessive accumulation can disrupt internal hydrodynamics, reduce effective reactor volume, and impair mass transfer [58]. In SBR-1500, these effects culminated in physical clogging and the need for frequent maintenance, highlighting the negative operational impacts of high MF loads.

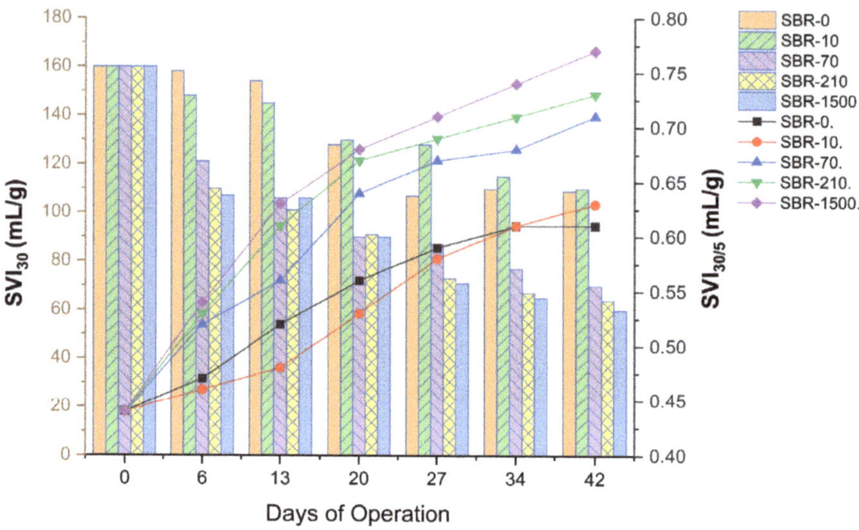

Figure 6. SVI_{30} (bars) and $SVI_{30}/_{5}$ (lines) profiles for SBR-0, SBR-10, SBR-70, SBR-210, and SBR-1500.

Notably, the physical benefits of MF presence did not translate into improved biological performance, as evidenced by the significant reduction in TN removal at higher MF concentrations. This apparent contradiction may be explained by a decoupling of structural and functional performance in the AGS system. While MFs likely acted as physical scaffolds that enhanced microbial aggregation and were stabilized within the EPS matrix, they may have simultaneously disrupted dissimilatory nitrogen removal pathways and potential toxic effects for key microbes. Increased granule density and altered porosity could have impaired internal substrate diffusion, limiting the penetration of nitrate and carbon sources into the anoxic zones required for effective denitrification. In addition, the adsorption of bioavailable COD onto synthetic fibers may have further reduced its accessibility to heterotrophic denitrifiers. These effects were compounded by the microbial community shifts observed at elevated MF concentrations, suggesting the potential inhibition or displacement of key denitrifying taxa. Thus, although MFs enhanced granule formation and settling properties, they negatively impacted the microenvironmental and microbial conditions essential for complete nitrogen removal and effective wastewater treatment.

3.3. Microbial Community Structure in AGS

To investigate the potential origins of the observed reduction in TN and phosphate removal, changes in the microbial communities within the reactors were investigated. The composition of microorganisms in the SBRs was found to be distinctive and dynamic at the family level (Figure 7). However, certain trends were observed. For instance, the community of microbes in reactors with lower concentrations of MFs appeared to be similar.

In SBR-0, Rhodocyclaceae dominated from day 7 to day 21. In SBR-10 and SBR-70, Rhodocyclaceae dominated on day 7 but shifted to Rhodobacteraceae by day 21 (Figure 7). SBR-210 was dominated by Rhodobacteraceae, with Moraxellaceae becoming dominant on day 21. In contrast, SBR-1500 showed Rhodocyclaceae dominance on day 7, shifting to Moraxellaceae dominance by day 21 (Figure 7). The abundance of denitrifying bacteria related to the family Rhodocyclaceae has been observed when readily available dissolved organic substrates, such as sodium acetate, are fed into the wastewater treatment system [59]. Rhodobacteraceae are capable of aerobic denitrification and degrade complex organic substrates [60,61]. Similarly, bacteria related to the Moraxellaceae family are also capable of degrading complex organic substrates and are linked to the removal of nitrogen and phosphorus [62]. Therefore, the metabolic activity of Rhodobacteraceae and Moraxellaceae may have benefited from the presence of MFs in the bioreactors.

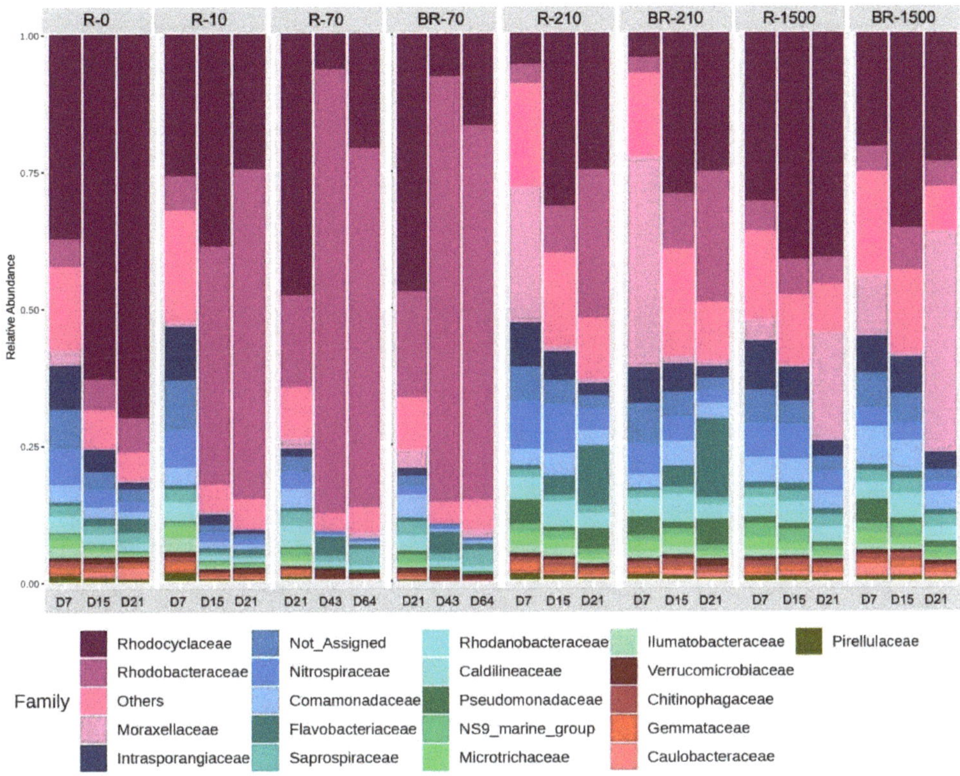

Figure 7. Relative abundance of the most abundant bacterial families in the studied samples (R-0 = SBR-0, R-10 = SBR-10, R-70 = SBR-70, BR-70 = biofilm from reactor SBR-70, R-210 = SBR-210, R-1500 = SBR-1500, and BR-1500 = biofilm from reactor SBR-1500).

The initial dominance of Rhodocyclaceae in reactors with lower MF concentrations likely reflects their metabolic versatility in terms of degrading various organic compounds [63]. The shift to Rhodobacteraceae in reactors with higher MF concentrations suggests an adaptation to diverse carbon sources under these conditions [64]. This transition may also result from the adverse environmental effects of MFs on certain microbial families, as microplastic pollution can reduce bacterial diversity and alter taxonomic abundance [65]. The dominance of Moraxellaceae by day 21 in the R-1500 and BR-1500 reactors suggests their role in late-stage degradation or resilience to prolonged MF exposure, particularly as they were significantly more abundant in reactors with higher MF concentrations (R-210 and R-1500) (Student's t-test, $p < 0.05$) [66]. These findings highlight the dynamic, adaptive nature of microbial communities in aerobic SBRs, emphasizing the need for further research into their specific roles in MF degradation, microbial interactions, and reactor performance.

Microbial distribution at the genus level in the SBRs was dynamic over 21 days. Denitrifiers such as *Paracoccus, Thauera*, and *Zoogloea* were the most prevalent, alongside others like *Dechloromonas*, *Fusibacter*, and *Flavobacterium*, which was consistent with other AGS studies [67] (Figure 8). *Paracoccus* increased in relative abundance across most reactors, except in SBR-1500, where it remained below 5% throughout the experiment. Despite SBR-1500 having the highest fiber content and microbial diversity, *Paracoccus* was likely outcompeted by other genera due to niche partitioning and resource competition in this environment [68]. *Thauera* (related to the family Rhodocyclaceae) and *Zoogloea* (related to the family Zoogloeaceae) denitrifiers are predominant in wastewater treatment systems that are fed with readily available dissolved organic substrates, such as sodium acetate [59]. *Thauera* and *Zoogloea* were also observed to be predominant in the control bioreactor, which was operating without MFs. *Paracoccus* (related to the family Rhodobacteraceae) are capable of aerobic denitrification and also of degrading complex organic substrates [68], which could explain their predominance in the bioreactors with MFs that leached e.g. additives and dyes.

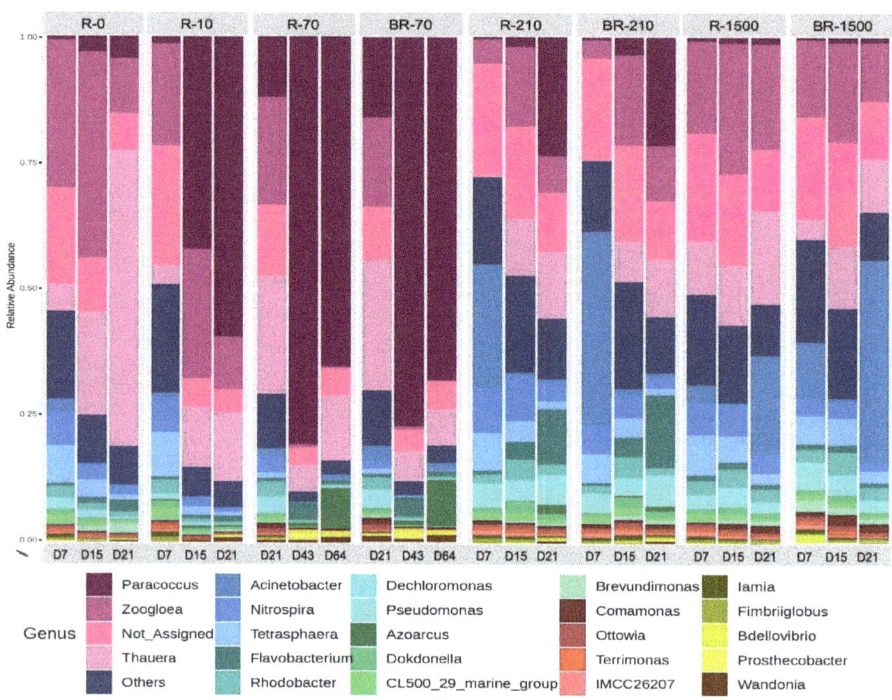

Figure 8. Relative abundance of the most abundant bacterial genera in the studied samples (R-0 = SBR-0, R-10 = SBR-10, R-70 = SBR-70, BR-70 = biofilm from reactor SBR-70, R-210 = SBR-210, R-1500 = SBR-1500, and BR-1500 = biofilm from reactor SBR-1500).

Nitrospira and *Acinetobacter* were most abundant during the early days of reactor operation, with high MF concentrations (Figure 8). *Nitrospira*, which is capable of both ammonia and nitrite oxidation, is crucial for nitrification and nitrogen reduction in wastewater. *Acinetobacter* abundance declined over time in all reactors except SBR-1500, where it persisted (Figure 8). Known for heterotrophic nitrification, aerobic denitrification, and phosphorus removal, *Acinetobacter* thrives under low-nutrient conditions [69,70]. This presence of heterotrophic nitrifiers was consistent with the observed ammonia removal even at high MF concentrations. *Pseudomonas* was most abundant in SBR-1500 and in reactors with high fiber concentrations compared to SBR-0 and SBR-10.

Thauera and *Zoogloea* showed significant fluctuations (Student's t-test, $p < 0.05$) in reactors with lower MF concentrations ($\leq$SBR-70) but remained stable in SBR-1500 with the highest MF concentration (Figure 8). This likely explains the higher TN removal efficiency seen in SBR-0 compared to reactors with higher MF concentrations. *Thauera* plays a key role in simultaneous carbon and nitrogen removal, contributing to the high COD and TN removal efficiencies observed in SBR-0 [71]. *Zoogloea*, which is capable of utilizing diverse carbon sources and enhancing biological phosphorus removal, was more abundant in SBR-0, especially early on, likely explaining the high observed PO_4^{3-} removal [71,72]. The observed changes in the microbial community structure may have resulted from toxic substances leaching from MFs [73,74] and the adsorption of MFs on granular sludge surfaces [41].

Interestingly, among the five SBRs tested (SBR-0, SBR-10, SBR-70, SBR-210, and SBR-1500), SBR-1500 and its biofilm exhibited the highest microbial diversity, with 657 and 660 observed species, respectively (Supplementary Figure S3). The diversity levels in the other reactors were: SBR-0 (580), SBR-10 (606), SBR-70 (615), SBR-210 (600), BR3 (637), and BR5 (615). The increased diversity in AGS reactors with fibers can be attributed to the layered granule structures, creating distinct aerobic, anaerobic, and anoxic zones, which was consistent with findings from polyethylene microplastic studies [75,76] and the observations of enhanced granulation in this study. The fibers provided stratified structures that supported diverse microbial communities [77], as evidenced by SBR-0's lower diversity (580 species) compared to fiber-containing reactors. However, as discussed previously, this enhanced structure, and diversity did not seem to protect the system against the negative impacts of MFs on nutrient removal.

3.4. Microfibre Removal Efficiencies of AGS-SBR

Contrary to the nutrient removal, all reactors showed high MF removal efficiency. SBR-10 achieved 100% removal, likely due to low influent fiber concentrations and efficient settling processes in the sludge (Supplementary Figure S4). SBR-70, SBR-210, and SBR-1500 achieved removal efficiencies of 99 $\pm$ 0.6%, 99 $\pm$ 0.3%, and 98 $\pm$ 0.1%, respectively, primarily through biomass entrapment. These results align with previous studies reporting 98–99% MF removal in SBRs and WWTPs [15,78]. The persistently high effluent concentrations reported in the literature [78,79] despite these high removal efficiencies highlight the exposure of WWTP to substantial MFs loads and the need for source control to reduce influent concentrations, especially given their potential impact on treatment performance. Such measures should address both textile manufacturing practices and consumer behavior to reduce MF release at the origin point. Key strategies include the development of more environmentally sustainable textile designs, modifications to fiber manufacturing processes, behavioral shifts in laundering practices, and the promotion of washing machine filters.

These strategies must be supported by robust regulatory frameworks, including mandatory standards for textile shedding and appliance-based filtration. In parallel, research into the chronic ecotoxicological impacts of MFs and associated additives on aquatic ecosystems is urgently needed to inform evidence-based policy. In particular, long-term studies on low-dose exposure, trophic transfer, and impacts on microbial processes in engineered systems are essential. Additionally, further investigation into the resilience of biological treatment systems under chronic MF stress would help guide design and operational adaptations for future-proofing wastewater infrastructure.

When implemented collectively, these source control and policy initiatives—coupled with targeted research—have the potential to significantly reduce MF emissions and their downstream risks as well as increase the resilience of wastewater treatment to impacts of microplastics. Given the limitations of end-of-pipe removal technologies, upstream interventions remain essential for minimizing the ecological and operational burden posed by microfibers [80–82].

3.5. Future Directions and Policy Implications

To build on these findings and strengthen AGS resilience, future work should address key limitations. Specifically, full nitrogen profiling, including nitrate and nitrite measurements, is needed to clarify the transformation pathways. EPS composition analysis would offer insight into structural microbial responses under MF stress. Additionally, longer exposure periods and the evaluation of plastic additives and textile dyes are essential to understanding chronic toxicity and system adaptability.

These findings confirm that MF pollution poses a process-level threat to biological nutrient removal, even in advanced systems like AGS. Mitigating this risk will require upstream source control (e.g., filtration at the laundry stage and at industrial discharge points), regulatory support, and continued investigation into the fate of MFs, microbial interactions, and system-level resilience.

4. Conclusions

The addition of MFs significantly affected nutrient removal in AGS SBRs. Higher MF concentrations (70, 210, and 1500 MFs/L) inhibited COD, PO_4^{3-}, and TN removal efficiencies, while NH_3 removal remained unaffected. MLSS levels increased across all reactors, with the steepest rise seen in SBR-1500, posing risks of clogging and reduced WWTP performance.

The results of our study highlight that polyester and denim MFs, although efficiently captured in AGS systems (>98% removal), impose concentration-dependent functional and microbial burdens, even in resilient biological processes. As MF loading increased, TN removal declined from 93% to 68%, and phosphate removal dropped significantly from 86% to 70% ($p < 0.05$). COD removal, while remaining high, decreased slightly from 96% to 89%. Nitrification (NH_3 removal) was unaffected, but MLSS levels increased with MF concentration, with SBR-1500 showing the steepest rise and operational instability. Microbial shifts accompanied these changes: *Paracoccus* dominated low-MF reactors, while *Acinetobacter* and *Pseudomonas* increased under higher MF loads. Diversity was greater in reactors exposed to $\geq$70 MFs/L, indicating ecological restructuring. Dominant families such as Rhodocyclaceae, Rhodobacteraceae, and Moraxellaceae shifted with MF exposure, pointing to a functional reorganization of the microbial community.

Overall, the results highlighted the risk that MF pose for biological wastewater treatment at concentrations currently seen in global facilities. This information is crucial for the assessment of potential secondary effects of MFs on water quality and the true impacts of MF contamination in our wastewater systems.

Supplementary Materials: The following supporting information can be downloaded at: https://www.mdpi.com/article/10.3390/pr13072272/s1, Table S1: Influent and effluent concentrations of MPs in WWTPs globally [4,5,15,83–109]; Table S2: SVI30 and SVI30/5 profiles for SBR-0, SBR-10, SBR-70, SBR-210, and SBR-1500; Figure S1: The five-point calibration curve for dosing; Figure S2: MLSS profiles for SBR-0, SBR-10, SBR-70, SBR-210, and SBR-1500. Figure S3: Shannon diversity values of the microbial communities within the five studied reactors. Figure S4: Fibers enmeshed in and around biomass.

Author Contributions: Conceptualization, R.S. and R.H.; Methodology, V.O.O. and R.H.; Software, H.A.; Validation, H.A. and R.H.; Formal analysis, R.H. and R.S.; Investigation, V.O.O., H.A. and K.A.G.; Resources, S.N.L., K.A.G., R.S. and R.H.; Data curation, V.O.O. and H.W.; Writing—original draft, V.O.O., H.W., R.S. and R.H.; Writing—review and editing, H.W., S.N.L., K.A.G., R.S. and R.H.; Supervision, R.S. and R.H.; Funding acquisition, R.S. and R.H. All authors have read and agreed to the published version of the manuscript.

Funding: This research was funded by the Natural Sciences and Engineering Council of Canada (NSERC) Discovery grant (RGPIN-2021-03770) and the Faculty of Engineering and Architectural Science (FEAS) at Toronto Metropolitan University, Toronto, Canada.

Data Availability Statement: Data will be made available on request.

Conflicts of Interest: The authors declare no conflicts of interest.

References

1. Roblin, B.; Ryan, M.; Vreugdenhil, A.; Aherne, J. Ambient atmospheric deposition of anthropogenic microfibers and microplastics in the remote region of Newfoundland, Canada. *Environ. Sci. Technol.* **2020**, *54*, 11100–11108. [CrossRef]
2. Welsh, B.; Aherne, J.; Paterson, A.; Yao, H.; McConnell, C. Atmospheric deposition of anthropogenic particles and microplastics in south-central Ontario, Canada. *Sci. Total Environ.* **2022**, *835*, 155426. [CrossRef] [PubMed]
3. Henry, B.; Laitala, K.; Klepp, I.G. Microfibres from apparel and home textiles: Prospects for including microplastics in environmental sustainability assessment. *Sci. Total Environ.* **2019**, *652*, 483–494. [CrossRef] [PubMed]
4. Carr, S.A.; Liu, J.; Tesoro, A.G. Transport and fate of microplastic particles in wastewater treatment plants. *Water Res.* **2016**, *91*, 174–182. [CrossRef] [PubMed]
5. Dris, R.; Imhof, H.; Sanchez, W.; Gasperi, J.; Galgani, F.; Tassin, B.; Laforsch, C. Beyond the ocean: Contamination of freshwater ecosystems with micro-plastic particles. *Environ. Chem.* **2015**, *12*, 539–550. [CrossRef]
6. Mintenig, S.M.; Int-Veen, I.; Löder, M.G.J.; Primpke, S.; Gerdts, G. Identification of microplastic in effluents of wastewater treatment plants using focal plane array-based micro-Fourier-transform infrared imaging. *Water Res.* **2017**, *108*, 365–372. [CrossRef]
7. Raju, S.; Carbery, M.; Kuttykattil, A.; Senathirajah, K.; Subashchandrabose, S.R.; Evans, G.; Thavamani, P. Transport and fate of microplastics in wastewater treatment plants: Implications to environmental health. *Rev. Environ. Sci. Biotechnol.* **2018**, *17*, 637–653. [CrossRef]
8. Zhou, S.; Wang, L.; Liu, J.; Zhang, C.; Liu, X. Microplastics' toxic effects and influencing factors on microorganisms in biological wastewater treatment units. *Water Sci. Technol.* **2024**, *89*, 1539–1553. [CrossRef]
9. Yaseen, A.; Assad i Sofi, M.; Hashmi, M.; Bhat, S. A global review of microplastics in wastewater treatment plants: Understanding their occurrence, fate and impact. *Environ. Res.* **2022**, *212 Pt B*, 113258. [CrossRef]
10. Sol, D.; Laca, A.; Laca, A.; Díaz, M. Microplastics in Wastewater and Drinking Water Treatment Plants: Occurrence and Removal of Microfibres. *Appl. Sci.* **2021**, *11*, 10109. [CrossRef]
11. Liu, J.; Liu, Q.; An, L.; Wang, M.; Yang, Q.; Zhu, B.; Ding, J.; Ye, C.; Xu, Y. Microfiber Pollution in the Earth System. *Rev. Environ. Contam. Toxicol.* **2022**, *260*, 13. [CrossRef]
12. Priya, A.; Anusha, G.; Thanigaivel, S.; Karthick, A.; Mohanavel, V.; Velmurugan, P.; Balasubramanian, B.; Ravichandran, M.; Kamyab, H.; Kirpichnikova, I.M. Removing microplastics from wastewater using leading-edge treatment technologies: A solution to microplastic pollution—A review. *Bioprocess Biosyst. Eng.* **2023**, *46*, 309–321. [CrossRef]
13. Xu, X.; Hou, Q.; Xue, Y.; Jian, Y.; Wang, L. Pollution characteristics and fate of microfibers in the wastewater from textile dyeing wastewater treatment plant. *Water Sci. Technol.* **2018**, *78*, 2046–2054. [CrossRef]
14. Bayo, J.; Olmos, S.; López-Castellanos, J.; Alcolea, A. Microplastics and microfibers in the sludge of a municipal wastewater treatment plant. *Int. J. Sustain. Dev. Plan.* **2016**, *11*, 812–821. [CrossRef]
15. Lares, M.; Ncibi, M.C.; Sillanpää, M.; Sillanpää, M. Occurrence, identification and removal of microplastic particles and fibers in conventional activated sludge process and advanced MBR technology. *Water Res.* **2018**, *133*, 236–246. [CrossRef] [PubMed]
16. Grbić, J.; Helm, P.; Athey, S.; Rochman, C.M. Microplastics entering northwestern Lake Ontario are diverse and linked to urban sources. *Water Res.* **2020**, *174*, 115623. [CrossRef] [PubMed]
17. Pedrotti, M.L.; Petit, S.; Eyheraguibel, B.; Kerros, M.E.; Elineau, A.; Ghiglione, J.F.; Loret, J.F.; Rostan, A.; Gorsky, G. Pollution by anthropogenic microfibers in North-West Mediterranean Sea and efficiency of microfiber removal by a wastewater treatment plant. *Sci. Total Environ.* **2021**, *758*, 144195. [CrossRef] [PubMed]
18. McQueen, R.H.; Batcheller, J.C.; Moran, L.J.; Zhang, H.; Hooper, P.M. Reducing laundering frequency to prolong the life of denim jeans. *Int. J. Consum. Stud.* **2017**, *41*, 36–45. [CrossRef]
19. Periyasamy, A.P.; Duraisamy, G. Carbon footprint on denim manufacturing. *Handb. Ecomater.* **2019**, *3*, 1581–1598. [CrossRef]
20. Memon, N.A. Cotton denim fabrics: Pakistan key supplier in Global market. *Pak. Text. J.* **2017**, *66*, 44–45.
21. Miller, D.; Woodward, S. Manifesto for a study of denim. *Soc. Anthropol.* **2007**, *15*, 335–351. [CrossRef]

22. Athey, S.N.; Adams, J.K.; Erdle, L.M.; Jantunen, L.M.; Helm, P.A.; Finkelstein, S.A.; Diamond, M.L. The Widespread Environmental Footprint of Indigo Denim Microfibers from Blue Jeans. *Environ. Sci. Technol. Lett.* **2020**, *7*, 840–847. [CrossRef]
23. Vassilenko, E.; Watkins, M.; Chastain, S.; Mertens, J.; Posacka, A.M.; Patankar, S.; Ross, P.S. Domestic laundry and microfiber pollution: Exploring fiber shedding from consumer apparel textiles. *PLoS ONE* **2021**, *16*, e0250346. [CrossRef] [PubMed]
24. Acharya, S.; Rumi, S.S.; Hu, Y.; Abidi, N. Microfibers from synthetic textiles as a major source of microplastics in the environment: A review. *Text. Res. J.* **2021**, *91*, 2136–2156. [CrossRef]
25. Cesa, F.S.; Turra, A.; Baruque-Ramos, J. Synthetic fibers as microplastics in the marine environment: A review from textile perspective with a focus on domestic washings. *Sci. Total Environ.* **2017**, *598*, 1116–1129. [CrossRef]
26. Corami, F.; Rosso, B.; Bravo, B.; Gambaro, A.; Barbante, C. A novel method for purification, quantitative analysis and characterization of microplastic fibers using Micro-FTIR. *Chemosphere* **2020**, *238*, 124564. [CrossRef]
27. Hartline, N.L.; Bruce, N.J.; Karba, S.N.; Ruff, E.O.; Sonar, S.U.; Holden, P.A. Microfiber Masses Recovered from Conventional Machine Washing of New or Aged Garments. *Environ. Sci. Technol.* **2016**, *50*, 11532–11538. [CrossRef]
28. Carney Almroth, B.M.; Åström, L.; Roslund, S.; Petersson, H.; Johansson, M.; Persson, N.K. Quantifying shedding of synthetic fibers from textiles; a source of microplastics released into the environment. *Environ. Sci. Pollut. Res.* **2018**, *25*, 1191–1199. [CrossRef]
29. Fries, E.; Grewal, T.; Sühring, R. Persistent, mobile, and toxic plastic additives in Canada: Properties and prioritization. *Environ. Sci. Process. Impacts* **2022**, *24*, 1945–1956. [CrossRef]
30. Li, L.; Song, K.; Yeerken, S.; Geng, S.; Liu, D.; Dai, Z.; Xie, F.; Zhou, X.; Wang, Q. Effect evaluation of microplastics on activated sludge nitrification and denitrification. *Sci. Total Environ.* **2020**, *707*, 135953. [CrossRef]
31. Fuschi, C.; Pu, H.; MacDonell, M.; Picel, K.; Negri, M.; Chen, J. Microplastics in the Great Lakes: Environmental, health, and socioeconomic implications and future directions. *ACS Sustain. Chem. Eng.* **2022**, *10*, 14074–14091. [CrossRef]
32. Munno, K.; Helm, P.A.; Rochman, C.; George, T.; Jackson, D.A. Microplastic contamination in Great Lakes fish. *Conserv. Biol.* **2022**, *36*, e13794. [CrossRef]
33. Zhang, Y.T.; Wei, W.; Huang, Q.S.; Wang, C.; Wang, Y.; Ni, B.J. Insights into the microbial response of anaerobic granular sludge during long-term exposure to polyethylene terephthalate microplastics. *Water Res.* **2020**, *179*, 115898. [CrossRef]
34. Yi, K.; Huang, J.; Li, X.; Li, S.; Pang, H.; Liu, Z.; Zhang, W.; Liu, S.; Liu, C.; Shu, W. Long-term impacts of polyethylene terephthalate (PET) microplastics in membrane bioreactor. *J. Environ. Manag.* **2022**, *323*, 116234. [CrossRef]
35. Zhao, L.; Su, C.; Liu, W.; Qin, R.; Tang, L.; Deng, X.; Wu, S.; Chen, M. Exposure to polyamide 66 microplastic leads to effects performance and microbial community structure of aerobic granular sludge. *Ecotoxicol. Environ. Saf.* **2019**, *190*, 110070. [CrossRef] [PubMed]
36. Rosa-Masegosa, A.; Perez-Bou, L.; Muñoz-Palazon, B.; Monteoliva-García, A.; Gonzalez-Martinez, A.; Gonzalez-Lopez, J.; Correa-Galeote, D. Effects of sulphur amino acids on the size and structure of microbial communities of aerobic granular sludge bioreactors. *Amino Acids* **2022**, *54*, 1403–1419. [CrossRef] [PubMed]
37. Zhang, B.; Huang, S.; Wu, L.; Guo, Y.; Shi, W.; Lens, P.N.L. Micro(nano)plastic size and concentration co-differentiate the treatment performance and toxicity mechanism in aerobic granular sludge systems. *Chem. Eng. J.* **2023**, *457*, 141212. [CrossRef]
38. Aqeel, H.; Weissbrodt, D.G.; Cerruti, M.; Wolfaardt, G.M.; Wilén, B.M.; Liss, S.N. Drivers of bioaggregation from flocs to biofilms and granular sludge. *Environ. Sci. Water Res. Technol.* **2019**, *5*, 2072–2089. [CrossRef]
39. Dai, M.; Wu, Y.; Wang, J.; Lv, Z.; Li, F.; Zhang, Y.; Kong, Q. Constructed wetland-microbial fuel cells enhanced with iron carbon fillers for ciprofloxacin wastewater treatment and power generation. *Chemosphere* **2022**, *305*, 105048. [CrossRef]
40. Zheng, X.; Han, Z.; Shao, X.; Zhao, Z.; Zhang, H.; Lin, T.; Yang, S.; Zhou, C. Response of aerobic granular sludge under polyethylene microplastics stress: Physicochemical properties, decontamination performance, and microbial community. *J. Environ. Manag.* **2022**, *323*, 116215. [CrossRef]
41. Lin, X.; Su, C.; Deng, X.; Wu, S.; Tang, L.; Li, X.; Liu, J.; Huang, X. Influence of polyether sulfone microplastics and bisphenol A on anaerobic granular sludge: Performance evaluation and microbial community characterization. *Ecotoxicol. Environ. Saf.* **2020**, *205*, 111318. [CrossRef]
42. Tay, J.H.; Tay, S.T.L.; Ivanov, V.; Pan, S.; Jiang, H.L.; Liu, Q.S. Biomass and porosity profiles in microbial granules used for aerobic wastewater treatment. *Lett. Appl. Microbiol.* **2003**, *36*, 297–301. [CrossRef] [PubMed]
43. APHA (American Public Health Association); American Water Works Association; Water Environment Federation. *Standard Methods for the Examination of Water and Wastewater*, 22nd ed.; APHA Press: Washington, DC, USA, 2012.
44. Hidalgo-Ruz, V.; Gutow, L.; Thompson, R.C.; Thiel, M. Microplastics in the marine environment: A review of the methods used for identification and quantification. *Environ. Sci. Technol.* **2012**, *46*, 3060–3075. [CrossRef] [PubMed]
45. Aqeel, H.; Liss, S.N. Fate of sloughed biomass in integrated fixed-film systems. *PLoS ONE* **2022**, *17*, e0262603. [CrossRef] [PubMed]

46. Oliva, G.; Pahunang, R.R.; Vigliotta, G.; Zarra, T.; Ballesteros, F.C.; Mariniello, A.; Buonerba, A.; Belgiorno, V.; Naddeo, V. Advanced treatment of toluene emissions with a cutting-edge algal bacterial photo-bioreactor: Performance assessment in a circular economy perspective. *Sci. Total Environ.* **2023**, *878*, 163005. [CrossRef]
47. Callahan, B.J.; McMurdie, P.J.; Rosen, M.J.; Han, A.W.; Johnson, A.J.A.; Holmes, S.P. DADA2: High-resolution sample inference from Illumina amplicon data. *Nat. Methods* **2016**, *13*, 581–583. [CrossRef]
48. Dhariwal, A.; Chong, J.; Habib, S.; King, I.L.; Agellon, L.B.; Xia, J. Microbiome Analyst: A web-based tool for comprehensive statistical, visual and meta-analysis of microbiome data. *Nucleic Acids Res.* **2017**, *45*, W180–W188. [CrossRef]
49. Feng, L.; Xu, J.; Xu Xiang Zhu, L.; Xu, J.; Ding, W.; Luan, J. Enhanced Biological Nitrogen Removal via Dissolved Oxygen Partitioning and Step Feeding in a Simulated River Bioreactor for Contaminated Source Water Remediation. *Int. Biodeterior. Biodegrad.* **2012**, *71*, 72–79. [CrossRef]
50. Guo, X.; Ma, X.; Niu, X.; Li, Z.; Wang, Q.; Ma, Y.; Cai, S.; Li, P.; Li, H. The impacts of biodegradable and non-biodegradable microplastic on the performance and microbial community characterization of aerobic granular sludge. *Front. Microbioogyl.* **2024**, *15*, 1389046. [CrossRef]
51. Fries, E.; Sühring, R. The unusual suspects: Screening for persistent, mobile, and toxic plastic additives in plastic leachates. *Environ. Poll.* **2023**, *335*, 122263. [CrossRef]
52. Jamee, R.; Siddique, R. Biodegradation of synthetic dyes of textile effluent by microorganisms: An environmentally and economically sustainable approach. *Eur. J. Microbiol. Immunol.* **2019**, *9*, 114–118. [CrossRef] [PubMed]
53. Xu, J.; Xu, D.; Wang, X.; Long, S.; Yang, J. Improved interfacial shear strength of carbon fiber/polyphenylene sulfide composites by graphene. *High Perform. Polym.* **2017**, *29*, 913–921. [CrossRef]
54. Tan, C.; Wang, Q.; Li, F. *Effects of Microplastics on Granular Sludge: A Review*; Institute of Coastal Environmental Pollution Control, Key Laboratory of Marine Environment and Ecology, Ministry of Education, College of Environmental Science and Engineering, Ocean University of China: Qingdao, China, 2024.
55. Hamza, R.A.; Iorhemen, O.T.; Zaghloul, M.S.; Tay, J.H. Rapid formation and characterization of aerobic granules in pilot-scale sequential batch reactor for high-strength organic wastewater treatment. *J. Water Process Eng.* **2018**, *22*, 27–33. [CrossRef]
56. Aqeel, H.; Basuvaraj, M.; Hall, M.; Neufeld, J.D.; Liss, S.N. Microbial dynamics and properties of aerobic granules developed in a laboratory-scale sequencing batch reactor with an intermediate filamentous bulking stage. *Appl. Microbiol. Biotechnol.* **2016**, *100*, 447–460. [CrossRef]
57. Campbell, K.; Wang, J.; Daigger, G.T. Filamentous organisms degrade oxygen transfer efficiency by increasing mixed liquor apparent viscosity: Mechanistic understanding and experimental verification. *Water Res.* **2020**, *173*, 115570. [CrossRef]
58. Aqeel, H.; Asefa, B.; Liss, S.N. Nitrospira dominant pin-point flocs with granule-like settleability in stirred tank reactors with oxic/hypoxic/oxic zones. *Front. Microbiol.* **2023**, *14*, 1307727. [CrossRef]
59. Rajeev, M.; Cho, J.C. Rhodobacteraceae are Prevalent and Ecologically Crucial Bacterial Members in Marine Biofloc Aquaculture. *J. Microbiol.* **2024**, *62*, 985–997. [CrossRef]
60. Gui, X.; Wang, Z.; Li, K.; Li, Z.; Mao, X.; Geng, J.; Pan, Y. Enhanced nitrogen removal in sewage treatment is achieved by using kitchen waste hydrolysate without a significant increase in nitrous oxide emissions. *Sci. Total Environ.* **2024**, *906*, 167108. [CrossRef]
61. Wang, B.Y.; Xin, Y.; Lin, G.F.; Gao, D.W.; Liu, L. Nitrogen removal from decentralized domestic wastewater in gravel-bed coupled with wood chip under variable hydraulic loading rate. *Desalination Water Treat.* **2025**, *322*, 101219. [CrossRef]
62. Fang, D.; Wu, A.; Huang, L.; Shen, Q.; Zhang, Q.; Jiang, L.; Ji, F. Polymer substrate reshapes the microbial assemblage and metabolic patterns within a biofilm denitrification system. *Chem. Eng. J.* **2020**, *387*, 124128. [CrossRef]
63. Chen, Y.J.; Leung, P.M.; Wood, J.L.; Bay, S.K.; Hugenholtz, P.; Kessler, A.J.; Shelley, G.; Waite, D.W.; Franks, A.E.; Cook, P.L.M.; et al. Metabolic flexibility allows bacterial habitat generalists to become dominant in a frequently disturbed ecosystem. *ISME J.* **2021**, *15*, 2986–3004. [CrossRef]
64. Jain, R.; Gaur, A.; Suravajhala, R.; Chauhan, U.; Pant, M.; Tripathi, V.; Pant, G. Microplastic pollution: Understanding microbial degradation and strategies for pollutant reduction. *Sci. Total Environ.* **2023**, *905*, 167098. [CrossRef] [PubMed]
65. Sun, H.; Yu, P.; Li, Q.; Ren, H.; Liu, B.; Ye, L.; Zhang, X.X. Transformation of anaerobic granules into aerobic granules and the succession of bacterial community. *Appl. Microbiol. Biotechnol.* **2017**, *101*, 7703–7713. [CrossRef] [PubMed]
66. Fall, C.; Barrón-Hernández, L.M.; Gonzaga-Galeana, V.E.; Olguín, M.T. Ordinary heterotrophic organisms with aerobic storage capacity provide stable aerobic granular sludge for C and N removal. *J. Environ. Manag.* **2022**, *308*, 114662. [CrossRef] [PubMed]
67. Chang, F.; Liu, W.; Fan, X.; Yu, C.; Liu, H.; Li, Q. Characteristics of aerobic denitrification and ammonia assimilation for nitrogen removal by *Paracoccus binzhouensis* wg1. *J. Environ. Chem. Eng.* **2025**, *13*, 116529. [CrossRef]
68. Su, J.F.; Zhang, K.; Huang, T.L.; Wen, G.; Guo, L.; Yang, S.F. Heterotrophic nitrification and aerobic denitrification at low nutrient conditions by a newly isolated bacterium, Acinetobacter sp. SYF26. *Microbiology* **2015**, *161*, 829–837. [CrossRef]
69. Hamza, R.A.; Zaghloul, M.S.; Iorhemen, O.T.; Sheng, Z.; Tay, J.H. Optimization of organics to nutrients (COD:N:P) ratio for aerobic granular sludge treating high-strength organic wastewater. *Sci. Total Environ.* **2019**, *650*, 3168–3179. [CrossRef]

70. Zhao, Y.; Huang, J.; Zhao, H.; Yang, H. Microbial community and N removal of aerobic granular sludge at high COD and N loading rates. *Bioresour. Technol.* **2013**, *143*, 439–446. [CrossRef]
71. Huang, T.-L.; Zhou, S.-L.; Zhang, H.-H.; Bai, S.-Y.; He, X.-X.; Yang, X. Nitrogen removal characteristics of a newly isolated indigenous aerobic denitrifier from oligotrophic drinking water reservoir, Zoogloea sp. *N299*. *Int. J. Mol. Sci.* **2015**, *16*, 10038–10060. [CrossRef]
72. Zhang, B.; Wu, L.; Shi, W.; Zhang, Z.; Lens, P.N.L. A novel strategy for rapid development of a self-sustaining symbiotic algal-bacterial granular sludge: Applying algal-mycelial pellets as nuclei. *Water Res.* **2022**, *214*, 118210. [CrossRef]
73. Zhang, Q.; Zhao, L.; Zhang, J.; Liu, W.; Cai, S.; Chen, L.; Cai, T.; Ji, X.M. Nitrogen contribution and microbial community of size-fractionated anammox sludge in continuous stirred-tank reactors. *Bioresour. Technol.* **2022**, *362*, 127857. [CrossRef]
74. Campo, R.; Lubello, C.; Lotti, T.; Di Bella, G. Aerobic granular sludge–membrane bioreactor (Ags–mbr) as a novel configuration for wastewater treatment and fouling mitigation: A mini-review. *Membranes* **2021**, *11*, 261. [CrossRef]
75. Ali, I.; Ding, T.; Peng, C.; Naz, I.; Sun, H.; Li, J.; Liu, J. Micro- and nanoplastics in wastewater treatment plants: Occurrence, removal, fate, impacts and remediation technologies—A critical review. *Chem. Eng. J.* **2021**, *423*, 130205. [CrossRef]
76. Kelly, J.J.; London, M.G.; McCormick, A.R.; Rojas, M.; Scott, J.W.; Hoellein, T.J. Wastewater treatment alters microbial colonization of microplastics. *PLoS ONE* **2021**, *16*, e0244443. [CrossRef] [PubMed]
77. Lee, S.; Kim, J. Microplastic removal efficiency in sequencing batch reactors and wastewater treatment plants. *Environ. Sci. Technol.* **2018**, *52*, 12345–12356.
78. Cristaldi, A.; Fiore, M.; Zuccarello, P.; Oliveri Conti, G.; Grasso, A.; Nicolosi, I.; Copat, C.; Ferrante, M. Efficiency of Wastewater Treatment Plants (WWTPs) for Microplastic Removal: A Systematic Review. *Int. J. Environ. Res. Public Health* **2020**, *17*, 8014. [CrossRef] [PubMed]
79. Haque, M.; Kabir, A.; Latifi, E.; Mahmud, S.; Hossain, M.; Himu, H.; Fatema, U.; Tareq, S. Microfiber prevalence and removal efficiency of textile effluent treatment plants in Bangladesh. *J. Hazard. Mater. Adv.* **2024**, *14*, 100436. [CrossRef]
80. Weis, J.S.; De Falco, F. Microfibers: Environmental Problems and Textile Solutions. *Microplastics* **2022**, *1*, 626–639. [CrossRef]
81. Acarer, S. Microplastics in wastewater treatment plants: Sources, properties, removal efficiency, removal mechanisms, and interactions with pollutants. *Water Sci. Technol.* **2023**, *87*, 685–710. [CrossRef]
82. Allen, E.; Henninger, C.E.; Garforth, A.; Asuquo, E. Microfiber Pollution: A Systematic Literature Review to Overcome the Complexities in Knit Design to Create Solutions for Knit Fabrics. *Environ. Sci. Technol.* **2024**, *58*, 4031–4045. [CrossRef]
83. Lv, X.; Dong, Q.; Zuo, Z.; Liu, Y.; Huang, X.; Wu, W.-M. Microplastics in a Municipal Wastewater Treatment Plant: Fate, Dynamic Distribution, Removal Efficiencies, and Control Strategies. *J Clean. Prod.* **2019**, *225*, 579–586. [CrossRef]
84. Talvitie, J.; Heinonen, M.; Pääkkönen, J.P.; Vahtera, E.; Mikola, A.; Setälä, O.; Vahala, R. Do Wastewater Treatment Plants Act as a Potential Point Source of Microplastics? Preliminary Study in the Coastal Gulf of Finland, Baltic Sea. *Water Sci. Technol.* **2015**, *72*, 1495–1504. [CrossRef]
85. Conley, K.; Allan, C.; Jestine, D.; Haven, L.; Barbara, B. Wastewater Treatment Plants as a Source of Microplastics to an Urban Estuary: Removal Efficiencies and Loading per Capita over One Year. *Water Res. X* **2019**, *3*, 100030. [CrossRef]
86. Akarsu, C.; Halil, K.; Kerem, G.; Ahmet, K.; Anna, S.-V. Microplastics Composition and Load from Three Wastewater Treatment Plants Discharging into Mersin Bay, North Eastern Mediterranean Sea. *Mar. Pollut. Bull.* **2020**, *150*, 110776. [CrossRef]
87. Ziajahromi, S.; Neale, P.A.; Rintoul, L.; Leusch, F.D.L. Wastewater Treatment Plants as a Pathway for Microplastics: Development of a New Approach to Sample Wastewater-Based Microplastics. *Water Res.* **2017**, *112*, 93–99. [CrossRef]
88. Magni, S.; Andrea, B.; Lucia, P.; Carlo, G.A.; Camilla, D.T.; Camilla, C.P.; Stefania, G.; Francesco, R. The Fate of Microplastics in an Italian Wastewater Treatment Plant. *Sci. Total Environ.* **2019**, *652*, 602–610. [CrossRef] [PubMed]
89. Bayo, J.; Sonia, O.; Joaquín, L.-C. Microplastics in an Urban Wastewater Treatment Plant: The Influence of Physicochemical Parameters and Environmental Factors. *Chemosphere* **2020**, *238*, 124593. [CrossRef] [PubMed]
90. Long, Z.; Pan, Z.; Wang, W.; Ren, J.; Yu, X.; Lin, L.; Lin, H.; Chen, H.; Jin, X. Microplastic Abundance, Characteristics, and Removal in Wastewater Treatment Plants in a Coastal City of China. *Water Res.* **2019**, *155*, 255–265. [CrossRef] [PubMed]
91. Talvitie, J.; Mikola, A.; Setälä, O.; Heinonen, M.; Koistinen, A. How Well Is Microlitter Purified from Wastewater? A Detailed Study on the Stepwise Removal of Microlitter in a Tertiary Level Wastewater Treatment Plant. *Water Res.* **2017**, *109*, 164–172. [CrossRef]
92. Raju, S.; Carbery, M.; Kuttykattil, A.; Senthirajah, K.; Lundmark, A.; Rogers, Z.; Suresh, S.; Evans, G.; Palanisami, T. Improved Methodology to Determine the Fate and Transport of Microplastics in a Secondary Wastewater Treatment Plant. *Water Res.* **2020**, *173*, 115549. [CrossRef]
93. Lee, H.; Kim, Y. Treatment Characteristics of Microplastics at Biological Sewage Treatment Facilities in Korea. *Mar. Pollut. Bull.* **2018**, *137*, 1–8. [CrossRef] [PubMed]
94. Magnusson, K.; Norén, F. *Screening of Microplastic Particles in and Down-Stream a Wastewater Treatment Plant*; Report C55; IVL Swedish Environmental Research Institute: Stockholm, Sweden, 2014.

95. Murphy, F.; Ewins, C.; Carbonnier, F.; Quinn, B. Wastewater Treatment Works (WWTW) as a Source of Microplastics in the Aquatic Environment. *Environ. Sci. and Technol.* **2016**, *50*, 5800–5808. [CrossRef] [PubMed]
96. Ren, P.J.; Dou, M.; Wang, C.; Li, G.Q.; Jia, R. Abundance and Removal Characteristics of Microplastics at a Wastewater 020-09611-5. Treatment Plant in Zhengzhou. *Environ. Sci. Pol. Res.* **2020**, *27*, 36295–36305. [CrossRef] [PubMed]
97. Gündoğdu, S.; Cem, Ç.; Evşen, G.; Serdar, K. Microplastics in Municipal Wastewater Treatment Plants in Turkey: A Comparison of the Influent and Secondary Effluent Concentrations. *Environ. Mon Assess.* **2018**, *190*, 626. [CrossRef]
98. Bilgin, M.; Meral, Y.; Fatih, K. Microplastic Removal by Aerated Grit Chambers versus Settling Tanks of a Municipal Wastewater Treatment Plant. *J. Water Process Eng.* **2020**, *38*, 101604. [CrossRef]
99. Gies, E.A.; Jessica, L.L.; Marie, N.; Anahita, E.; Farida, B.; Eric, R.H.; Peter, S.R. Retention of Microplastics in a Major Secondary Wastewater Treatment Plant in Vancouver, Canada. *Mar. Pollut. Bull.* **2018**, *133*, 553–561. [CrossRef]
100. Ziajahromi, S.; Neale, P.A.; Silveira, I.T.; Chua, A.; Leusch, F.D.L. An Audit of Microplastic Abundance throughout Three Australian Wastewater Treatment Plants. *Chemosphere* **2021**, *263*, 128294. [CrossRef]
101. Leslie, H.A.; Brandsma, S.H.; Van-Velzen, M.J.M.; Vethaak, A.D. Microplastics En Route: Field Measurements in the Dutch River Delta and Amsterdam Canals, Wastewater Treatment Plants, North Sea Sediments and Biota. *Environ. Int.* **2017**, *101*, 133–142. [CrossRef]
102. Xu, X.; Jian, Y.; Xue, Y.; Hou, Q.; Wang, L.P. Microplastics in the Wastewater Treatment Plants (WWTPs): Occurrence and Removal. *Chemosphere* **2019**, *235*, 1089–1096. [CrossRef]
103. Liu, X.; Yuan, W.; Di, M.; Li, Z.; Wang, J. Transfer and Fate of Microplastics during the Conventional Activated Sludge Process in One Wastewater Treatment Plant of China. *Chem. Eng. J.* **2019**, *362*, 176–182. [CrossRef]
104. Michielssen, M.R.; Michielssen, E.R.; Ni, J.; Duhaime, M.B. Fate of Microplastics and Other Small Anthropogenic Litter (SAL) in Wastewater Treatment Plants Depends on Unit Processes Employed. *Environ. Sci. Water Res. Technol.* **2016**, *2*, 1064–1073. [CrossRef]
105. Edo, C.; Miguel, G.-P.; Francisco, L.; Francisca, F.-P.; Rosal, R. Fate of Microplastics in Wastewater Treatment Plants and Their Environmental Dispersion with Effluent and Sludge. *Environ Poll.* **2020**, 259. [CrossRef] [PubMed]
106. Franco, A.A.; Arellano, J.M.; Albendín, G.; Rodríguez-Barroso, R.; Quiroga, J.M.; Coello, M.D. Microplastic Pollution in Wastewater Treatment Plants in the City of Cádiz: Abundance, Removal Efficiency and Presence in Receiving Water Body. *Sci Tot Environ.* **2021**, *776*, 145795. [CrossRef]
107. Talvitie, J.; Heinonen, M. *Preliminary Study on Synthetic Microfibers and Particles at a Municipal Wastewater Treatment Plant*; HELCOM BASE Project—Implementation of the Baltic Sea Action Plan in Russia; Baltic Marine Environment Protection Commission; Helsinki Commission: Helsinki, Finland, 2014.
108. Hidayaturrahman, H.; Tae-Gwan, L. A Study on Characteristics of Microplastic in Wastewater of South Korea: Identification, Quantification, and Fate of Microplastics during Treatment Process." Mar. *Pollut. Bull.* **2019**, *146*, 696–702. [CrossRef]
109. Simon, M.; Alst, N.V.; Vollertsen, J. Quantification of Microplastic Mass and Removal Rates at Wastewater Treatment Plants Applying Focal Plane Array (FPA)-Based Fourier Transform Infrared (FT-IR) Imaging. *Water Res.* **2018**, *142*, 1–9. [CrossRef]

processes MDPI

Article

Dynamics of Bacterial Communities and Identification of Microbial Indicators in a *Cylindrospermopsis*-Bloom Reservoir in Western Guangdong Province, China

Yingwen Mai [1,2,*], Changhong Hong [1,2,3], Da Liu [1,2,3], Fengjuan Yang [1,2], Gengfeng Xiao [1,2], Zhilin Zhang [1,2] and Shuai Liu [1,2]

1 Guangdong Research Institute of Water Resources and Hydropower, Guangzhou 510610, China; changjianghong@126.com (C.H.); gdsky_liuda@foxmail.com (D.L.); yangjuan0609@126.com (F.Y.); xgf05191995@gmail.com (G.X.); scutzzl@163.com (Z.Z.); liushuai0822@mail.bnu.edu.cn (S.L.)
2 State and Local Joint Engineering Laboratory of Estuarine Hydraulic Technology, Guangzhou 510610, China
3 Guangdong Key Laboratory of Hydrodynamic Research, Guangzhou 510610, China
* Correspondence: yingwenmaiii@gmail.com

Abstract: Blue-green algae blooms present persistent environmental challenges in freshwater ecosystems, yet ecological interactions within the bacterial communities of *Cylindrospermopsis*-bloom reservoirs remain poorly understood. In this study, water samples were collected from February to May 2024 from 11 sampling sites in a *Cylindrospermopsis*-bloom reservoir in western Guangdong province, China. At each sampling point, a water sample was collected every month. High-throughput sequencing was applied to analyze the interaction between *Cylindrospermopsis* and other bacteria. As shown in our results, the phyla Actinobacteriota, Proteobacteria, Bacteroidota, Verrucomicrobiota, and Cyanobacteria were revealed as dominant phyla. Bacterial communities exhibited significant seasonal differences between flood and non-flood periods (ANOSIM: R = 0.472, $p = 0.001$). *Cylindrospermopsis* (dominance index Y = 0.53) acted as the keystone in the co-occurrence network ($Z_i < 2.5$, $P_i > 0.62$) and closely interacted with other bacteria. For better management of the blue-green algae bloom reservoir, the phyla of Actinobacteriota, Dependentiae, Acidobacteriota, Armatimonadota, Gemmatimonadota, and Desulfobacterota were proposed as microbial indicators for the eutrophic process. This study provides a new insight into the interactions of Cyanobacteria with other bacteria and the management of blue-green algae outbreaks in reservoirs.

Keywords: eutrophication; *Cylindrospermopsis*; interaction; high-throughput sequencing; indicator species

1. Introduction

With the rapid development of industrialization and agriculture, excessive inputs of nitrogen and phosphorus have exacerbated eutrophication in aquatic environments. Eutrophication could lead to serious negative ecological effects: for instance, increased microbial biomass, decreased biodiversity, lower efficiency of nutrient use, and accelerated cycles [1,2]. Recently, lakes in China have experienced various degrees of eutrophic pollution during the rapid anthropogenic development period, leading to a series of problems. Le et al.'s study [3] proposed that lake water eutrophication in China has disrupted the drinking water supply. Jin et al. [4] demonstrated that the water quality and lake ecosystems are declining, as many lakes in China commonly undergo eutrophication. Wang et al. [5]

 https://doi.org/10.3390/pr13072129

indicated that Lake Wuli in Jiangsu Province, China, has experienced an evidently deteriorating trend due to overexploitation. Reservoirs are characterized by slow flow rates and long hydraulic retention times, usually corresponding to eutrophic status and further leading to water quality deterioration, ecosystem instability, and a high frequency of blue-green algal outbreaks. Qin et al.'s study [6] considered that shallow lakes, which are mixed systems without thermal stratification and sediment resuspension, similarly to reservoirs, are commonly related to the loss of control of eutrophication. Jiang et al. [7] found that new reservoirs in China covered by villages and farmlands were easily eutrophic with low biodiversity and self-purification capacity. Rocha et al. [8] assumed that the reservoirs with low water volume corresponded with a higher nutrient level. Except for abnormal ecological responses, Cyanobacterial blooms are of particular concern, with their potential production of toxins, such as cylindrospermopsin, saxitoxin [9], and microcystin [10] produced by *Cylindrospermopsis* and *Microcystis*, posing severe threats to aquatic ecosystems and human health [11,12]. With its concentrated distribution in subtropical and tropical areas [13,14] and sensitivity to growth environment factors [15,16], there is relatively less research focusing on *Cylindrospermopsis* compared with *Microcystis* and other Cyanobacterial algae, and the mechanism of *Cylindrospermopsis* bloom needs to be further investigated [17].

The reasons for the outbreak of *Cylindrospermopsis* are comprehensive and are related not only to abiotic factors but also to biotic factors [18]. There is a complex interaction between Cyanobacteria and other bacteria in aquatic microbial communities, which plays an important role in their respective growth [19,20]. Among the microbes, some bacteria could promote the growth of *Cylindrospermopsis* by providing some essential compounds [21,22], such as nitrates and phosphates, or enhancing their stress tolerance, and some algicidal bacteria inhibit blue-green algae growth by competing for nutrients or lysing algal cells [23]. The interactions between them are important in shaping aquatic microbial communities' structure and the development of harmful blue-green algal blooms. Exploring the changes in microbial communities in algal outbreaks is of great significance for the control of *Cylindrospermopsis* blooms.

However, the understanding of microbial communities' dynamics in a blue-green algal bloom aquatic environment and the interaction of *Cylindrospermopsis* and other bacteria is still unclear. For better prevention and the management of algal outbreaks, current studies are mainly focused on bacteria inhibiting the growth of Cyanobacteria, such as *Bacilluscereus L7* and *Lysobacter* sp., which can lyse algal cells [24,25]. Some studies have focused on identifying the microbial indicators in eutrophic water, which were reported to be closely related to eutrophic environmental factors, yet lack a discussion on the interactions with Cyanobacteria. For instance, *Thermoanaerobaculum* (Acidobacteria), *Synechococcus* (Cyanobacteria), and *Blastocatella* (Chlorobi) were mentioned to positively relate with the concentrations of Chla ($p < 0.05$) in the eutrophic Gehu lake, and it was suggested that the microbial taxa could thrive in nutrient-rich conditions and could be used for monitoring eutrophication progress [26]. Ji et al. [27] applied the ratio of the abundance of Cyanobacteria and Actinobacteria to indicate the early warning of eutrophication in water, which is also based on environmental parameters. In general, taking the relationships between Cyanobacteria and other bacteria into consideration is more comprehensive when identifying microbial indicators. For the better management of blue-green algae blooms, the research of interactions between Cyanobacteria and other bacteria requires further investigation, which can improve the criterion for identifying other bacterial indicators.

A reservoir in western Guangdong province, China, serves multiple functions, including urban and industrial water supply, agricultural irrigation, flood prevention, and ecological environment maintenance. However, due to rapid socioeconomic development and climate change, the reservoir has become severely eutrophic, particularly during flood seasons. In recent years, Cyanobacterial blooms dominated by *Cylindrospermopsis* have frequently occurred in the reservoir. Similarly, Han et al. [28] reported that most drinking water reservoirs downstream in Guangdong Province are approaching a eutrophic state, as upstream rivers suffer from deteriorating water quality under the pressure of societal development. Lei et al. [29] confirmed the widespread distribution of *Cylindrospermopsis* in Dongguan city, southern China, observing it in 21 out of 25 reservoirs. Lei et al.'s study [30] indicated that *Cylindrospermopsis* was present in 19 out of 20 reservoirs, with biomass ranging from 0.0001 to 39.740 mg·L^{-1}. Yang et al. [31] also reported a high abundance of filamentous Cyanobacteria, including *Cylindrospermopsis*, in tropical urban reservoirs in southern China. To improve the management of Cyanobacterial blooms, it is essential to investigate the dynamics of microbial communities, examine the interactions between *Cylindrospermopsis* and other bacteria, and identify potential bacterial indicators. In this study, we conducted a four-month monitoring campaign to assess environmental factors and employed 16S rRNA high-throughput sequencing to analyze microbial community shifts across different eutrophic levels. Additionally, we aimed to identify bacterial indicators associated with *Cylindrospermopsis* blooms. This research provides new insights into the interactions between *Cylindrospermopsis* and other bacteria, offering potential strategies for controlling Cyanobacterial blooms in eutrophic reservoirs.

2. Materials and Methods

2.1. Study Area

The study reservoir is located in Lianjiang, Zhanjiang, Guangdong Province (21°39′ N, 110°17′ E), in China. It has a total storage capacity of 1.144 billion m^3, a maximum water surface area of 122 km^2, and a catchment area of 1495 km^2. Over the past 60 years, its average annual water yield has been approximately 1.48 billion m^3. To ensure representative spatial coverage and sampling efficiency, 11 sampling points were strategically selected across the reservoir. Their distribution is illustrated in Figure 1.

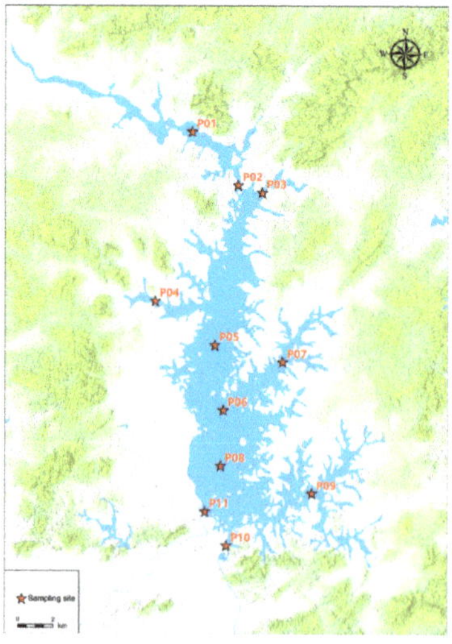

Figure 1. Sample points' distribution in the reservoir.

2.2. Sampling and Analysis

Water samples were collected from February to May 2024, covering both non-flood (February–March) and flood (April–May) seasons. To minimize variability, all sampling operations were consistently conducted between 10:00 a.m. and 12:00 noon each month.

Samples were collected monthly at each of the 11 sampling points. A total of 44 samples were collected. At each sampling location, 2 L of surface water samples was collected from 0.5 m below the water surface. Samples were immediately transported to the laboratory and stored at 4 °C until analysis. From each sample, 200 mL of water sample was used for water quality analysis. Field measurements of temperature (T), dissolved oxygen (DO), and pH were obtained using a YSI 650MDS multiparameter water quality meter (YSI Incorporated., Yellow Springs, OH, USA). Laboratory analyses included the total nitrogen (TN), total phosphorus (TP), ammonia nitrogen (NH_4^+-N), and permanganate index (COD_{Mn}), performed according to standard methods outlined in the "Water and Wastewater Monitoring and Analysis Method (China)". Chlorophyll-a (Chla) concentrations were determined using a portable PAM fluorometer AquaPen-CAP-C 100 (PSI, Czech Republic). All physicochemical parameters were analyzed in triplicate to ensure data quality.

Genomic DNA was extracted from filter membranes using Soil DNA Extraction CZ Kit (China) according to the manufacturer's protocol. DNA purity and concentration were determined using a NanoDrop One spectrophotometer (Thermo Fisher Scientific, MA, USA). For bacterial community analysis, the V3-V4 regions of the 16S rRNA gene were amplified using the universal primer pair 338F (5′-ACTCCTACGGGAGGCAGCCAG-3′) and 806R (5′-GACTACHVGGGTWTCTAAT-3′). PCR amplification was performed using barcoded specific primers and Premix Taq DNA polymerase. The amplification products were verified by 1% agarose gel electrophoresis [32], and only samples exhibiting distinct target bands were selected for subsequent analysis. All high-throughput sequencing procedures, including library construction, quality control, and sequencing, were conducted by Magigene Sequencing (Magigene, Shanghai, China). Data was analyzed on an IlluminaMiseq platform (Guangdong Magigene Biotechnology Co., Ltd., Guangzhou, China). For the absolute quantitation of Cyanobacteria, parallel water samples (50 mL) were collected and immediately preserved with 5 mL of 10% Lugol's iodine solution. Following 24–48 h of sedimentation for fixation, samples were concentrated and washed with distilled water prior to microscopic examination. Cyanobacterial identification was based on comprehensive morphological characterization, including colony morphology and dimensions, filamentous structure and arrangement, cellular morphology and size, trichome characteristics, and cell wall thickness. Quantitative analysis was performed using an Olympus BX53 light microscope (Olympus Co., Ltd., Japan) at 40× magnification. Cell densities were enumerated and expressed as cells per liter and calculated according to the following standard counting equation [33]:

$$\text{Algal density (cells/L)} = \text{C}/(\text{A} \times \text{D} \times \text{F})$$

where C = the number of cells counted, A = the area of the field (mm^2), D = the depth of field (mm), and F = the number of fields counted

The dominance Y of blue-green algae was determined by the following equation:

$$Y = \frac{n_x}{N} f_x$$

where n_x is the number of x-th species, N is the total number of all species, and f_x is the frequency of the x-th species appearing in each sample.

2.3. Data Processing

The DNA sequencing data were optimized and processed through the Magigene Cloud Platform (http://cloud.magigene.com/ (accessed on 10 February 2025)). After removing low-quality and chimeric sequences, paired-end reads were merged and clustered into operational taxonomic units (OTUs) at a 97% similarity threshold using UPARSE. The most abundant sequence within each OUT was selected as the representative sequence for downstream analysis. We conducted comprehensive microbial community analyses through the Magigene cloud platform, including the following: α-diversity indices (Chao1 richness and Shannon diversity), community richness index, redundancy analysis (RDA), principal coordinate analysis (PCoA), community composition analysis (relative abundance), significant difference testing (one-way ANOVA), and environmental factor correlation analysis (Spearman's rank correlation). PCoA was employed to visualize sample dissimilarity through a dimensionality reduction in the distance matrix. The distance algorithms of PCoA in this study were Bray–Curtis and Euclidean. Spatial proximity in PCoA plots indicates greater similarity in microbial community composition. Co-occurrence network analysis was performed using the igraph package in R (v 4.42) [34], with network topology determined by the following: number of connections between nodes, correlation coefficients, and betweenness centrality (identifying keystone taxa).

The trophic-level index (TLI, derived from China's Environmental Monitoring Station) was calculated using five parameters: chlorophyll-a (Chla), total phosphorus (TP), total nitrogen (TN), Secchi depth (SD), and permanganate index (COD_{Mn}). The trophic state classification was as follows: TLI < 30, referred to as "Oligotropher"; 30 ≤ TLI ≤ 50, referred to as "Mesotropher"; TLI > 50, referred to as "Eutropher". All geospatial visualizations were generated using ArcGIS 10.3 software (ESRI, Redlands, CA, USA).

3. Results

3.1. Water Quality and Trophic Status

Environmental parameters (Table 1) differed between flood and non-flood seasons (detailed data in Table S1). Chlorophyll-a (Chla), COD_{Mn}, total phosphorus (TP), dissolved oxygen (DO), and NH_4^+-N increased during the flood season, while transparency (Secchi depth, SD) and total nitrogen (TN) decreased. These changes corresponded to an elevation in the trophic-level index (TLI) from 40.74, indicating progression toward eutrophic conditions during flood periods. Cyanobacterial abundance demonstrated pronounced seasonal variation (Figure 2), with cell densities increasing from 3.87×10^7 cells/L in the non-flood season to 1.09×10^8 cells/L during the flood season. The dominance *Y* index of Cyanobacteria is shown in Figure 3a, and the main species are *Cylindrospermopsis* sp. (dominance *Y* = 0.53), *Pseudanabaena* sp. (dominance *Y* = 0.13), and *Microcystis* sp. (dominance *Y* = 0.04). Notably, *Cylindrospermopsis* sp. exhibited increased dominance during the flood season, establishing itself as the reservoir's primary Cyanobacterial species (complete dominance data in Table S2).

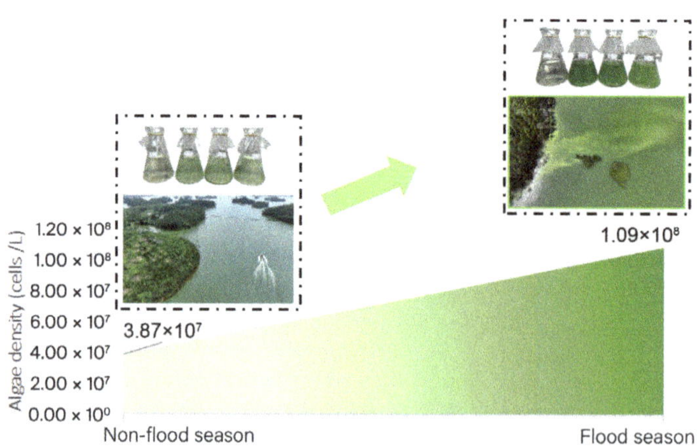

Figure 2. Average of Cyanobacteria's density (cells/L) in two seasons.

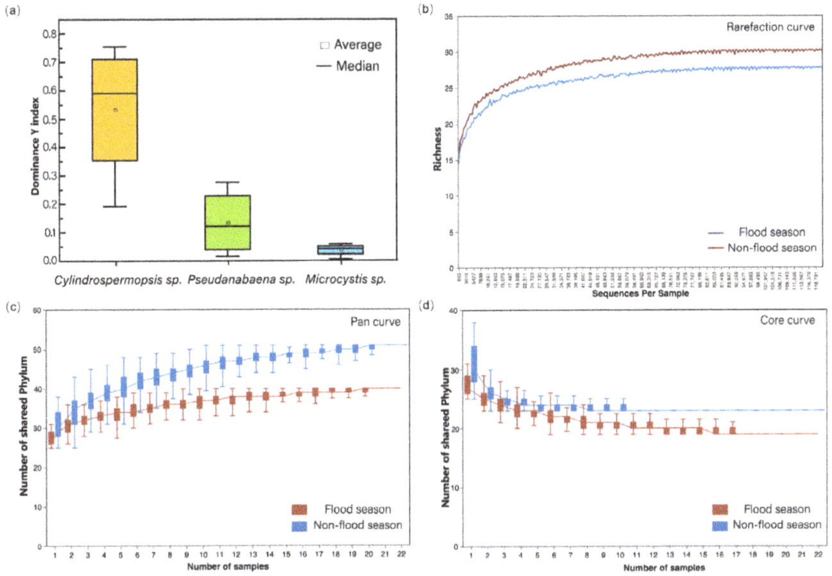

Figure 3. Dominance *Y* index of 3 species of Cyanobacteria in the reservoir (**a**), rarefaction curve (**b**), pan curve (**c**), and core curve (**d**) of OTUs.

Table 1. Average of water parameters in the flood and non-flood seasons.

Sampling Time	DO (mg/L)	T (°C)	SD (cm)	pH	Chla (μg/L)	NH_4^+-N (mg/L)	TP (mg/L)	TN (mg/L)	COD_{Mn} (mg/L)	TLI
Non-flood season	7.85	17.15	81.41	8.02	17.64	0.03	0.14	2.00	2.14	35.32
Flood season	9.17	29.09	47.91	9.19	25.88	0.06	0.17	1.58	3.37	40.74

3.2. Bacterial Diversity and Composition

A total of 4,688,216 sequences were obtained through high-throughput sequencing, with an average of 106,550 per sample and an average length of 469 bp. High-quality sequences were clustered at 97% similarity, yielding an average of 7384 OTUs per sample. The analysis of OTU diversity revealed that the rarefaction curves, pan curves, and core curves (Figure 3b–d) all approached plateaus. This indicates that the sampling depth was sufficient, and the data were representative.

Bacterial diversity across the 44 samples was compared using Shannon and Chao1 indices. The Shannon index reflects community diversity, while the Chao1 index indicates community richness. As shown in Table 2 (detailed results in Table S3), both the average Chao1 and Shannon indices during the flood season were similar to those in the non-flood season, indicating analogous bacterial diversity and richness between seasons. In addition, ANOSIM similarity analysis revealed significant differences in microbial composition between flood season and non-flood seasons (R = 0.472, $p = 0.001$).

Table 2. Average of the alpha-diversity index of Shannon and Chao1.

Index	Chao1	Shannon
Flood season	29.09	2.68
Non-flood season	30.85	2.43

The relative abundance of bacterial communities at the phylum level during the flood season and the non-flood season is shown in Figure 4a. The bacterial phyla with the highest average relative abundance were Actinobacteriota (41.08%), Proteobacteria (16.92%), Bacteroidota (10.63%), Verrucomicrobiota (9.64%), Cyanobacteria (7.48%), Planctomycetota (3.47%), Chloroflexi (1.45%), and Patescibacteria (1.09%). During the flood season, the dominant phyla were Actinobacteriota and Proteobacteria, accounting for 35.24% and 16.69% of the total sample sequences, respectively. During the non-flood season, the dominant phylum was Actinobacteriota, accounting for 46.92% of the total sample sequences. The proportions of Actinobacteriota and Proteobacteria were higher during the non-flood season than during the flood season, while Bacteroidetes showed higher relative abundance during the flood season.

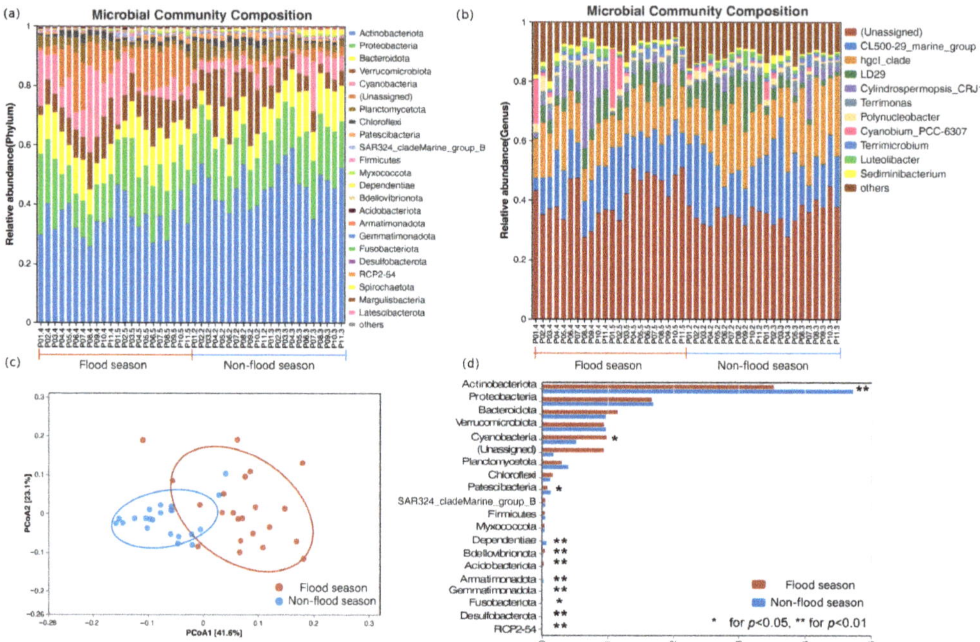

Figure 4. Relative abundance of bacterial community at the phylum (**a**) and genus (**b**) level, principal coordinate analysis (PCoA) (**c**), and significance test with between-group variation (**d**) (* for $p < 0.05$; ** for $p < 0.01$).

After excluding unclassified genera, 545 genera were annotated (Figure 4b). *CL500_29_marine_group* (17%), *hgcl_clade* (16.64%), *LD29* (5.73%), *Cylindrospermopsis_CRJ1* (3.59%), *Terrimonas* (1.83%), *Cyanobium_PCC-6307* (1.63%), and *Polynucleatobacter* (1.58%) were the dominant bacterial genera. The dominant bacterial genera during the flood season included *CL500-29_marine_group* (15.26%), *hgcI2 clade* (14.17%), *LD29* (5.41%), and *Cylindrospermopsis-CRJ1* (5.26%). The dominant bacterial genera in the non-flood season included *hgcI2 clade* (19.12%), *CL500-29_marine_group* (18.74%), and *LD29* (6.06%).

From the PCoA (Figure 4c), the explained variance of principal component 1 (PCoA1) and principal component 2 (PCoA2) was 23.1% and 41.6%, respectively. Non-flood season samples were mainly grouped on the negative axis of PCoA1, while flood season samples were grouped on the positive axis side. The microbial community composition during the flood season was distinct from that during the non-flood season. Further intergroup difference significance tests were conducted on the bacterial communities of reservoirs during flood and non-flood seasons (Figure 4d). Actinobacteriota, Patescibacteria, Dependentiae, Armatimonadota, Fusobacteriota, Desulfobacterota, and RCP2-54 showed significantly higher abundances in non-flood seasons than in flood seasons ($p < 0.05$), while Cyanobacteria, Bdellovibrionota, Acidobacteriota, and Gemmatimonadota showed significantly higher abundances in flood seasons than in non-flood seasons ($p < 0.05$).

3.3. Microbial–Environmental Correlations

The relationship between bacterial communities and environmental factors was revealed by redundancy analysis (RDA), as shown in Figure 5a. The arrows, dots, and their angles represent environmental factors, samples, and their correlations, respectively. An acute angle indicates a positive correlation between environmental factors and samples, while a blunt angle indicates a negative correlation. Longer arrow lengths denote stronger influences on microbial community composition. The results showed that algal density, permanganate index (COD_{Mn}), pH, temperature (T), dissolved oxygen (DO), trophic-level index (TLI), chlorophyll-a (Chla), total phosphorus (TP), and ammonia nitrogen ($NH_4{}^+$-N) were all positively correlated with bacterial communities during flood seasons, while transparency (SD) and total nitrogen (TN) were positively correlated with bacterial communities during non-flood seasons.

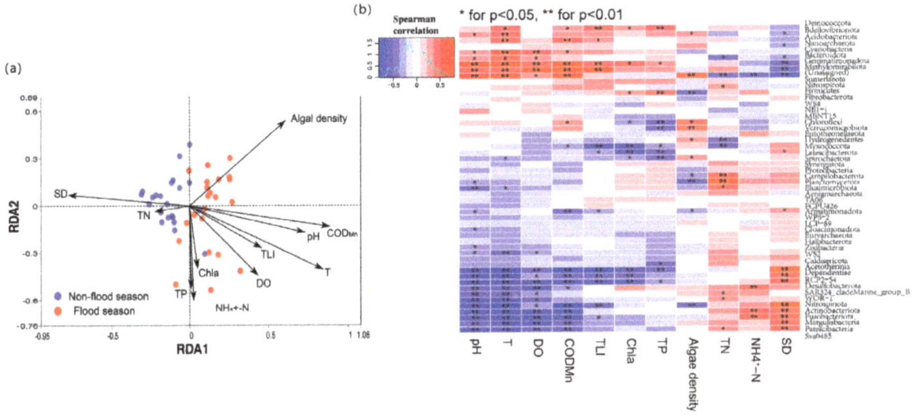

Figure 5. Redundancy analysis (RDA) (**a**) and heatmap of the correlation analysis of bacterial community and environmental parameters (**b**).

The correlation heatmap (Figure 5b) shows the correlation between bacterial and environmental factors. The colors of red and blue represent positive and negative correlations, respectively, and darker colors indicate stronger correlations. Cyanobacteria, Bacteroidota, and Gemmatimonadota show significant positive correlations with dissolved oxygen, while Acidobacteriota, Cyanobacteria, and Gemmatimonadota correlate positively with the permanganate index. Deinococcota, Acidobacteriota, Bdellovibrionota, Cyanobacteria, and Gemmatimonadota exhibit positive correlations with the trophic-level index, while Gemmatimonadota, Deinococcota, and Firmicutes demonstrate positive correlations with the total phosphorus and Chla. Campilobacterota, Planctomycetota, Nitrospirota, and Elusimicrobiota are positively correlated with the total nitrogen, while Desulfobacterota, Fusobacteriota, and Margulisbacteria show positive correlations with ammonia nitrogen.

3.4. Co-Occurrence Network Analysis

Microbial co-occurrence network analyses (Figure 6a,b) serve as a key method for investigating microbial community structure and interactions [35]. By comparing the overall composition, node characteristics, and topological structure of different networks during the non-flood season (node: 369; edge: 2914) and flood season (node: 336; edge: 1824), it was found that the microbial network during the flood season exhibited lower complexity, connectivity, and modularity [36]. In addition, the microbial networks in both seasons were mainly composed of the phyla of Proteobacteria (29.36%), Bacteroidota (20.57%), Actinobacteriota (14.99%), Verrucomicrobiota (6.53%), Cyanobacteria (4.38%), and Planctomycetota (3.83%). Cyanobacteria demonstrated interactions with multiple bacterial groups, including Actinobacteriota, Proteobacteria, Verrucomicrobiota, Bacteroidota, Planctomycetota, Patescibacteria, Firmicutes, Myxococcota, and Chloroflexi. Detailed node and edge information for other bacteria is provided in Tables S4 and S5.

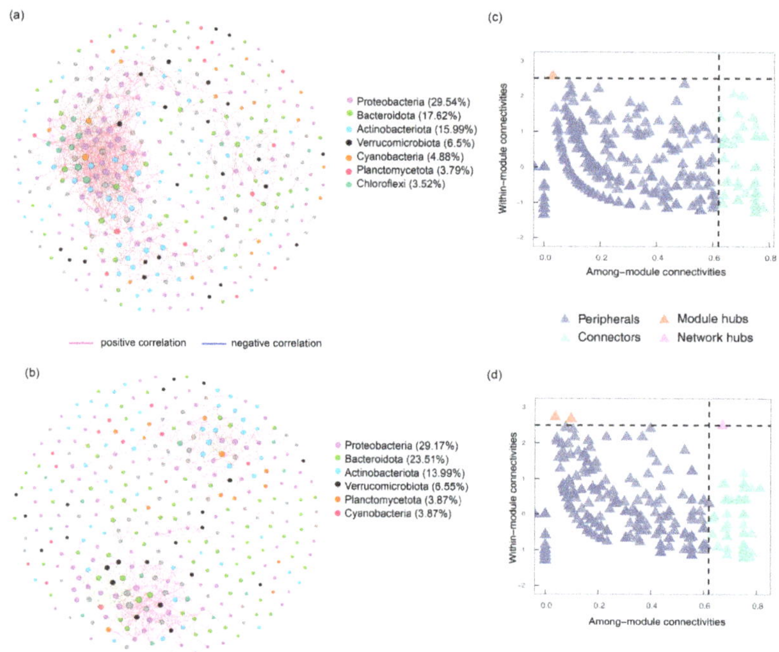

Figure 6. Co-occurrence network analysis: non-flood season (**a**), flood season (**b**), and identification of keystones in the non-flood season (**c**) and flood season (**d**).

In co-occurrence networks, those with high node degrees and high betweenness centrality are identified as keystones (Figure 6c,d), and they are categorized as peripheral hubs, connector hubs, module hubs, and network hubs. During the non-flood season, a total of 383 peripheral hubs (Zi < 2.5, Pi < 0.62), 63 connector hubs (Zi < 2.5, Pi > 0.62), and 4 module hubs (Zi > 2.5, Pi < 0.62) were observed in the co-occurrence network. During the flood season, a total of 398 peripherals hubs (Zi < 2.5, Pi < 0.62), 66 connectors hubs (Zi < 2.5, Pi > 0.62), 4 module hubs (Zi > 2.5, Pi < 0.62), and 1 network hub (Zi > 2.5, Pi > 0.62) were identified. The microbial communities' details of the keystones are presented in Tables S6 and S7.

4. Discussion

4.1. Characteristics of Reservoir Bacterial Communities

The nutritional level of the reservoir gradually increases during flood seasons, coinciding with *Cylindrospermopsis* blooms (dominance $Y = 0.53$) and consistent with a previous study [37]. Although nutrient levels increase from non-flood seasons to flood seasons, bacterial community richness and diversity remain similar, which is probably attributed to non-significant differences in trophic indices between periods. Some studies indicated a significant decrease in the richness and diversity of bacterial communities in higher-level eutrophic lakes, where excessive nutrients promote algal proliferation that competitively excludes other bacteria through spatial and resource limitation [38]. In addition, elevated trophic conditions may stimulate toxin production by bloom-forming species, such as *Microcystis aeruginosa* and *Cylindrospermopsis*, directly leading to the death of bacteria.

The phyla Actinobacteriota (41.08%), Proteobacteria (16.92%), Bacteroidota (10.63%), Verrucomicrobiota (9.64%), and Cyanobacteria (7.48%) were among the dominant microbial communities in the reservoir, which is consistent with many previous studies, including the study of 21 plateau lakes in China [39], Qionghai Lake in Sichuan [40], and Keruk Lake and Puma Lake in the Qinghai Tibet Plateau [41]. Actinobacteriota and Proteobacteria predominated across tropic states, representing their keystone role in maintaining ecosystem stability [19,42–44]. During flood seasons, Cyanobacteria abundance increased significantly ($p < 0.05$), likely driven by elevated temperature, nutrients, and light availability, intensifying the competition for nutrients and compounds in flood seasons, as reported in [45,46]. The increased abundance of Cyanobacteria exacerbates eutrophication and water quality deterioration [47] while potentially suppressing Actinobacteriota through competitive exclusion under high-nutrient conditions [48–50]. This was consistent with the results in this study, where the abundance of Actinobacteriota significantly decreased in the flood season ($p < 0.05$). Bacteroidota maintained stable abundance throughout sampling. As primary degraders of algal polysaccharides [51], Bacteroidota likely mediate carbohydrate metabolism in this reservoir. Similarly, Verrucomicrobiota also accounted for a stable proportion during periods, and studies indicated that it was a specialist consumer of microalgal sugars during algal bloom [52]. The phylum Verrucomicrobiota possibly helped decompose organic compounds and maintain the ecological balance in the reservoir.

4.2. Investigation of Microbial Interactions and Eutrophic Indicators

By comparing network topology between non-flood (node: 369, edge: 2914) and flood seasons (node: 336; edge: 1824), our results revealed reduced complexity, connectivity, and modularity during the flood season. This suggests that progressive eutrophication may render microbial communities more unstable and susceptible to environmental disturbances [53]. The phyla of Proteobacteria, Bacteroidota, Actinobacteriota, Verrucomicrobiota, and Cyanobacteria were dominant components in the co-occurrence network, indicating their extensive interactions within the microbial community. Notably, Cyanobacteria maintained stable proportions in both seasonal networks, with *Cylindrospermopsis* identified as a keystone genus. This implies that bloom-forming Cyanobacteria establish complex associations with other bacteria and significantly influence microbial community composition [54,55], consistent with prior studies [49]. Li et al. [56] demonstrated that blue-green algae maintain stable interactions through nutrient exchange, dissolved oxygen regulation, and allelopathic substance secretion, thereby shaping bacterial compositions. Meanwhile, co-occurring bacteria might exert either positive or negative effects on *Cylindrospermopsis* growth during blooms.

Other keystones in the network also play essential roles in maintaining interspecies interactions. Among the keystone taxa in the network, Bdellovibrionota, Acidobacteriota, Gemmatimonadota, Actinobacteriota, Patescibacteria, Dependentiae, and Armatimonadota were observed with significant abundance changes ($p < 0.05$) from the non-flood season to the flood season according to the significance test's between-group variation, consistent with previous studies [57,58]. Notably, Actinobacteriota show direct correlations with *Cylindrospermopsis*, indicating tight Cyanobacterial associations [59]. These prominent keystone taxa might efficiently respond to nutrient fluctuations by modulating metabolic functions and gene expression preferences [60]. Studies showed that keystone bacteria played important roles in maintaining community stability by driving nutrient cycling through enhanced organic matter decomposition, nutrient conversion, and energy transfer in an eutrophic environment [61]. Zhou et al. [62] demonstrated metabolic shifts during algal blooms: an increase in the gene expression level of carbon fixation, degradation, denitrification, phosphorus acquisition, and sulfur metabolism genes during bloom formation; and transitions to oxidative stress response and protease synthesis during bloom decline. Significantly, these keystone taxa correlated strongly with eutrophic factors ($p < 0.05$), including the trophic-level index and Chla, indicating high sensitivity to trophic status changes. Considering the high sensitivity to trophic status changes and the pivotal roles of the communities, these bacteria can serve as microbial indicators for monitoring shifts in reservoir nutrient statuses [63]. When the abundance of Actinobacteriota, Dependentiae, Armatimonadota, and Desulfobacterota in the reservoir significantly decreases and the abundance of Acidobacteriota and Gemmatimonadota shows a significant upward trend, there is a risk of eutrophic-level improvement in the reservoir. These bioindicators offer novel insights for monitoring approaches [64–66], enabling early warning systems for blue-green algal bloom prevention.

5. Conclusions

This study reveals key dynamics and interactions within the microbial communities of *Cylindrospermopsis*-bloom reservoirs: Microbial community structures differed significantly between flood and non-flood seasons (ANOSIM: $R = 0.472$, $p = 0.001$) alongside shifting trophic conditions. Notably, Cyanobacteria, Bdellovibrionota, Acidobacteriota, and Gemmatimonadota increased significantly during flood seasons ($p < 0.05$), while Actinobacteriota, Patescibacteria, Patescibacteria, Armatimonadota, and Desulfobacterota decreased ($p < 0.05$). Despite seasonal variations, Cyanobacteria maintained stable network proportions, with *Cylindrospermopsis* identified as a keystone genus. This suggests complex associations with other bacteria and a pivotal role in shaping microbial community composition. Other keystone taxa, including Bdellovibrionota, Acidobacteriota, Gemmatimonadota, Actinobacteriota, Patescibacteria, and Armatimonadota, showed significant abundance shifts under eutrophic conditions ($p < 0.05$) and correlated strongly with eutrophication indicators ($p < 0.05$), such as the trophic-level index and Chla. These sensitive bacteria serve as effective bioindicators for monitoring eutrophication progression and Cyanobacterial blooms. Their interactions with Cyanobacteria provide a comprehensive framework for identifying indicator species in eutrophic aquatic systems. Future research should elucidate *Cylindrospermopsis*–bacteria interactions to improve bloom management strategies in global subtropical/tropical reservoirs.

Supplementary Materials: The following supporting information can be downloaded at https://www.mdpi.com/article/10.3390/pr13072129/s1. Table S1: Water parameters of 44 samplings. Table S2. Average of dominance Y index of various blue-green algae in four months. Table S3. Alpha-diversity index of Shannon and Chao1 of each sampling. Table S4. The information of co-occurrence network' s node and edge in flood season. Table S5. The information of co-occurrence network' s node and edge in non-flood season. Table S6. The information of co-occurrence network' s keystones in non-flood season. Table S7. The information of co-occurrence network' s keystones in flood season.

Author Contributions: Conceptualization, Y.M. and F.Y.; formal analysis, Y.M. and G.X.; funding acquisition, D.L. and C.H.; methodology, Y.M. and S.L.; project administration, Y.M. and C.H.; resources, Y.M. and F.Y.; supervision, D.L. and C.H.; investigation, Y.M., G.X. and F.Y.; writing—original draft, Y.M. and G.X.; visualization, Y.M. and Z.Z.; writing—review and editing, Y.M. and S.L. All authors have read and agreed to the published version of the manuscript.

Funding: This research was funded by the Science and Technology Innovation Program from Water Resources of Guangdong Province, grant numbers 2024-03 and 2025-07, and the Special Fund for Stability Support of Guangdong Provincial Scientific Research Institutions (221016).

Data Availability Statement: The original contributions presented in this study are included in the article/Supplementary Material. Further inquiries can be directed to the corresponding author.

Conflicts of Interest: The authors declare no conflict of interest.

References

1. Qin, B.; Gao, G.; Zhu, G.; Zhang, Y.; Song, Y.; Tang, X.; Xu, H.; Deng, J. Lake Eutrophication and Its Ecosystem Response. *Chin. Sci. Bull.* **2013**, *58*, 961–970. [CrossRef]
2. Qin, B.; Yang, L.; Chen, F.; Zhu, G.; Zhang, L.; Chen, Y. Mechanism and Control of Lake Eutrophication. *Chin. Chin. Chin. Sci. Bull.* **2006**, *51*, 2401–2412. [CrossRef]
3. Le, C.; Zha, Y.; Li, Y.; Sun, D.; Lu, H.; Yin, B. Eutrophication of Lake Waters in China: Cost, Causes, and Control. *Environ. Manag.* **2010**, *45*, 662–668. [CrossRef]
4. Jin, X.; Xu, Q.; Huang, C. Current Status and Future Tendency of Lake Eutrophication in China. *Sci. China Ser. C Life Sci.* **2005**, *48*, 948–954. [CrossRef]
5. Wang, J.; Fu, Z.; Qiao, H.; Liu, F. Assessment of Eutrophication and Water Quality in the Estuarine Area of Lake Wuli, Lake Taihu, China. *Sci. Total Environ.* **2019**, *650*, 1392–1402. [CrossRef]
6. Qin, B.; Zhang, Y.; Zhu, G.; Gao, G. Eutrophication Control of Large Shallow Lakes in China. *Sci. Total Environ.* **2023**, *881*, 163494. [CrossRef]
7. Jiang, C.; Zhu, L.; Hu, X.; Cheng, J.; Xie, M. Reasons and Control of Eutrophication in New Reservoirs. In *Eutrophication: Causes, Consequences and Control*; Ansari, A.A., Singh Gill, S., Lanza, G.R., Rast, W., Eds.; Springer: Dordrecht, The Netherlands, 2011; pp. 325–340, ISBN 978-90-481-9625-8.
8. da Rocha, C.A.N.; da Costa, M.R.A.; Menezes, R.F.; Attayde, J.L.; Becker, V. Water Volume Reduction Increases Eutrophication Risk in Tropical Semi-Arid Reservoirs. *Acta Limnol. Bras.* **2018**, *30*, e106.
9. Moreira, C.; Azevedo, J.; Antunes, A.; Vasconcelos, V. Cylindrospermopsin: Occurrence, Methods of Detection and Toxicology. *J. Appl. Microbiol.* **2013**, *114*, 605–620. [CrossRef]
10. De Figueiredo, D.R.; Azeiteiro, U.M.; Esteves, S.M.; Gonçalves, F.J.; Pereira, M.J. Microcystin-Producing Blooms—A Serious Global Public Health Issue. *Ecotoxicol. Environ. Saf.* **2004**, *59*, 151–163. [CrossRef]
11. Zhang, Y.; Deng, J.; Zhou, Y.; Zhang, Y.; Qin, B.; Song, C.; Shi, K.; Zhu, G.; Hou, X.; Zhang, Y. Drinking Water Safety Improvement and Future Challenge of Lakes and Reservoirs. *Sci. Bull.* **2024**, *69*, 3558–3570. [CrossRef]
12. Igwaran, A.; Kayode, A.J.; Moloantoa, K.M.; Khetsha, Z.P.; Unuofin, J.O. Cyanobacteria Harmful Algae Blooms: Causes, Impacts, and Risk Management. *Water Air Soil Pollut.* **2024**, *235*, 71. [CrossRef]
13. McGregor, G.B.; Fabbro, L.D. Dominance of Cylindrospermopsis Raciborskii (Nostocales, Cyanoprokaryota) in Queensland Tropical and Subtropical Reservoirs: Implications for Monitoring and Management. *Lakes Reserv. Res. Manag.* **2000**, *5*, 195–205. [CrossRef]
14. Yatigammana, S.; Perera, M. Distribution of Cylindrospermopsis Raciborskii (Cyanobacteria) in Sri Lanka. *Ceylon J. Sci.* **2017**, *46*, 65–80. [CrossRef]
15. Chislock, M.F.; Sharp, K.L.; Wilson, A.E. Cylindrospermopsis Raciborskii Dominates under Very Low and High Nitrogen-to-Phosphorus Ratios. *Water Res.* **2014**, *49*, 207–214. [CrossRef]
16. Bonilla, S.; González-Piana, M.; Soares, M.; Huszar, V.L.; Becker, V.; Somma, A.; Marinho, M.M.; Kokociński, M.; Dokulil, M.; Antoniades, D. The Success of the Cyanobacterium Cylindrospermopsis Raciborskii in Freshwaters Is Enhanced by the Combined Effects of Light Intensity and Temperature. *J. Limnol.* **2016**, *75*, 606–617. [CrossRef]
17. Bhagowati, B.; Ahamad, K.U. A Review on Lake Eutrophication Dynamics and Recent Developments in Lake Modeling. *Ecohydrol. Hydrobiol.* **2019**, *19*, 155–166. [CrossRef]
18. Lan, J.; Liu, P.; Hu, X.; Zhu, S. Harmful Algal Blooms in Eutrophic Marine Environments: Causes, Monitoring, and Treatment. *Water* **2024**, *16*, 2525. [CrossRef]
19. Wang, Y.; Guo, M.; Li, X.; Liu, G.; Hua, Y.; Zhao, J.; Huguet, A.; Li, S. Shifts in Microbial Communities in Shallow Lakes Depending on Trophic States: Feasibility as an Evaluation Index for Eutrophication. *Ecol. Indic.* **2022**, *136*, 108691. [CrossRef]
20. Shao, K.; Yao, X.; Wu, Z.; Jiang, X.; Hu, Y.; Tang, X.; Xu, Q.; Gao, G. The Bacterial Community Composition and Its Environmental Drivers in the Rivers around Eutrophic Chaohu Lake, China. *BMC Microbiol.* **2021**, *21*, 179. [CrossRef]
21. Palacios, O.A.; López, B.R.; de-Bashan, L.E. Microalga Growth-Promoting Bacteria (MGPB): A Formal Term Proposed for Beneficial Bacteria Involved in Microalgal–Bacterial Interactions. *Algal Res.* **2022**, *61*, 102585. [CrossRef]

22. Xiong, X.; Li, Y.; Zhang, C. Cable Bacteria: Living Electrical Conduits for Biogeochemical Cycling and Water Environment Restoration. *Water Res.* **2024**, *253*, 121345. [CrossRef]
23. Jin, Y.; Ren, S.; Wu, Y.; Zhang, X.; Chen, Z.; Xie, B. Microbial Community Structures and Bacteria-Cylindrospermopsis Raciborskii Interactions in Yilong Lake. *FEMS Microbiol. Ecol.* **2024**, *100*, fiae048. [CrossRef]
24. Meyer, N.; Bigalke, A.; Kaulfuß, A.; Pohnert, G. Strategies and Ecological Roles of Algicidal Bacteria. *FEMS Microbiol. Rev.* **2017**, *41*, 880–899. [CrossRef]
25. Coyne, K.J.; Wang, Y.; Johnson, G. Algicidal Bacteria: A Review of Current Knowledge and Applications to Control Harmful Algal Blooms. *Front. Microbiol.* **2022**, *13*, 871177. [CrossRef]
26. Chen, Q.; Liu, Y.; Zhang, M.; Lin, K.; Wang, Z.; Liu, L. Seasonal Responses of Microbial Communities to Water Quality Variations and Interaction of Eutrophication Risk in Gehu Lake. *Sci. Total Environ.* **2024**, *955*, 177199. [CrossRef]
27. Ji, B.; Liang, J.; Ma, Y.; Zhu, L.; Liu, Y. Bacterial Community and Eutrophic Index Analysis of the East Lake. *Environ. Pollut.* **2019**, *252*, 682–688. [CrossRef]
28. Han, B.-P.; Liu, Z.; Dumont, H.J. Water Supply and Eutrophication of Reservoirs in Guangdong Province, South China. In *Tropical and Sub-Tropical Reservoir Limnology in China: Theory and Practice*; Han, B.-P., Liu, Z., Eds.; Springer: Dordrecht, The Netherlands, 2012; pp. 279–292, ISBN 978-94-007-2007-7.
29. Lei, L.; Peng, L.; Huang, X.; Han, B.-P. Occurrence and Dominance of Cylindrospermopsis Raciborskii and Dissolved Cylindrospermopsin in Urban Reservoirs Used for Drinking Water Supply, South China. *Environ. Monit. Assess.* **2014**, *186*, 3079–3090. [CrossRef]
30. Lei, M.-T.; Peng, L.; Han, B.-P.; Lei, L.-M. Distribution and Factors Affecting *Cylindrospermopsis Raciborskii* in Guangdong Reservoirs. *Huan Jing Ke Xue* **2018**, *39*, 5523–5531.
31. Yang, Y.; Zheng, X.; Tang, Q.; Gu, J.; Lei, L.; Han, B.-P. Species Diversity and Seasonal Dynamics of Filamentous Cyanobacteria in Urban Reservoirs for Drinking Water Supply in Tropical China. *Ecotoxicology* **2020**, *29*, 780–789. [CrossRef]
32. Knebelsberger, T.; Stöger, I. DNA Extraction, Preservation, and Amplification. In *DNA Barcodes: Methods and Protocols*; Humana Press: Totowa, NJ, USA, 2012; pp. 311–338.
33. Harper, D.A.; Owen, A.W. *Quantitative and Morphometric Methods in Taxonomy*; Wiley: Hoboken, NJ, USA, 1999.
34. Csardi, G.; Nepusz, T. The Igraph Software. *Complex. Syst.* **2006**, *1695*, 1–9.
35. Paver, S.F.; Hayek, K.R.; Gano, K.A.; Fagen, J.R.; Brown, C.T.; Davis-Richardson, A.G.; Crabb, D.B.; Rosario-Passapera, R.; Giongo, A.; Triplett, E.W. Interactions between Specific Phytoplankton and Bacteria Affect Lake Bacterial Community Succession. *Environ. Microbiol.* **2013**, *15*, 2489–2504. [CrossRef]
36. Zhang, H.; Yang, L.; Li, Y.; Wang, C.; Zhang, W.; Wang, L.; Niu, L. Pollution Gradients Shape the Co-Occurrence Networks and Interactions of Sedimentary Bacterial Communities in Taihu Lake, a Shallow Eutrophic Lake. *J. Environ. Manag.* **2022**, *305*, 114380. [CrossRef]
37. Hou, P.; Chang, F.; Duan, L.; Zhang, Y.; Zhang, H. Seasonal Variation and Spatial Heterogeneity of Water Quality Parameters in Lake Chenghai in Southwestern China. *Water* **2022**, *14*, 1640. [CrossRef]
38. Feng, C.; Jia, J.; Wang, C.; Han, M.; Dong, C.; Huo, B.; Li, D.; Liu, X. Phytoplankton and Bacterial Community Structure in Two Chinese Lakes of Different Trophic Status. *Microorganisms* **2019**, *7*, 621. [CrossRef]
39. Liao, J.; Cao, X.; Zhao, L.; Wang, J.; Gao, Z.; Wang, M.C.; Huang, Y. The Importance of Neutral and Niche Processes for Bacterial Community Assembly Differs between Habitat Generalists and Specialists. *FEMS Microbiol. Ecol.* **2016**, *92*, fiw174. [CrossRef]
40. Liu, Q.; Zhang, H.; Chang, F.; Xie, P.; Zhang, Y.; Wu, H.; Zhang, X.; Peng, W.; Liu, F. eDNA Revealed in Situ Microbial Community Changes in Response to Trapa Japonica in Lake Qionghai and Lake Erhai, Southwestern China. *Chemosphere* **2022**, *288*, 132605. [CrossRef]
41. Long, Q.; Feng, X.; Liu, J.; Zhang, X.; Shen, G.; Zhu, D. Microbial Diversity of Keluke-Tuosu Lake Wetland Reserve in Qinghai-Tibet Plateau. *Earth Environ.* **2017**, *4*, 399–407.
42. Liu, S.; Yu, H.; Yu, Y.; Huang, J.; Zhou, Z.; Zeng, J.; Chen, P.; Xiao, F.; He, Z.; Yan, Q. Ecological Stability of Microbial Communities in Lake Donghu Regulated by Keystone Taxa. *Ecol. Indic.* **2022**, *136*, 108695. [CrossRef]
43. Dong, X.; Greening, C.; Rattray, J.; Chakraborty, A.; Chuvochina, M.; Mayumi, D.; Dolfing, J.; Li, C.; Brooks, J.; Bernard, B. Metabolic Potential of Uncultured Bacteria and Archaea Associated with Petroleum Seepage in Deep-Sea Sediments. *Nat. Commun.* **2019**, *10*, 1816. [CrossRef]
44. Zhou, Z.; Tran, P.Q.; Kieft, K.; Anantharaman, K. Genome Diversification in Globally Distributed Novel Marine Proteobacteria Is Linked to Environmental Adaptation. *ISME J.* **2020**, *14*, 2060–2077. [CrossRef]
45. McMahon, K.D.; Read, E.K. Microbial Contributions to Phosphorus Cycling in Eutrophic Lakes and Wastewater. *Annu. Rev. Microbiol.* **2013**, *67*, 199–219. [CrossRef]
46. Chen, Z.; Zhang, J.; Li, R.; Tian, F.; Shen, Y.; Xie, X.; Ge, Q.; Lu, Z. Metatranscriptomics Analysis of Cyanobacterial Aggregates during Cyanobacterial Bloom Period in Lake Taihu, China. *Environ. Sci. Pollut. Res.* **2018**, *25*, 4811–4825. [CrossRef]

47. Cao, X.; Wang, J.; Liao, J.; Gao, Z.; Jiang, D.; Sun, J.; Zhao, L.; Huang, Y.; Luan, S. Bacterioplankton Community Responses to Key Environmental Variables in Plateau Freshwater Lake Ecosystems: A Structural Equation Modeling and Change Point Analysis. *Sci. Total Environ.* **2017**, *580*, 457–467. [CrossRef]
48. Haukka, K.; Kolmonen, E.; Hyder, R.; Hietala, J.; Vakkilainen, K.; Kairesalo, T.; Haario, H.; Sivonen, K. Effect of Nutrient Loading on Bacterioplankton Community Composition in Lake Mesocosms. *Microb. Ecol.* **2006**, *51*, 137–146. [CrossRef]
49. Lan, C.; Sun, L.; Hu, Y.; Zhang, Y.; Xu, J.; Ding, H.; Tang, R.; Hou, J.; Li, Y.; Wu, X. Diversity and Their Response to Environmental Factors of Prokaryotic Ultraplankton in Spring and Summer of Cihu Lake and Xiandao Lake in China. *Sustainability* **2023**, *15*, 11532. [CrossRef]
50. Shen, M.; Li, Q.; Ren, M.; Lin, Y.; Wang, J.; Chen, L.; Li, T.; Zhao, J. Trophic Status Is Associated with Community Structure and Metabolic Potential of Planktonic Microbiota in Plateau Lakes. *Front. Microbiol.* **2019**, *10*, 2560. [CrossRef]
51. Chen, J.; Robb, C.S.; Unfried, F.; Kappelmann, L.; Markert, S.; Song, T.; Harder, J.; Avcı, B.; Becher, D.; Xie, P. Alpha-and Beta-mannan Utilization by Marine Bacteroidetes. *Environ. Microbiol.* **2018**, *20*, 4127–4140. [CrossRef]
52. Orellana, L.H.; Francis, T.B.; Ferraro, M.; Hehemann, J.-H.; Fuchs, B.M.; Amann, R.I. Verrucomicrobiota Are Specialist Consumers of Sulfated Methyl Pentoses during Diatom Blooms. *ISME J.* **2022**, *16*, 630–641. [CrossRef]
53. Hu, A.; Ju, F.; Hou, L.; Li, J.; Yang, X.; Wang, H.; Mulla, S.I.; Sun, Q.; Bürgmann, H.; Yu, C. Strong Impact of Anthropogenic Contamination on the Co-occurrence Patterns of a Riverine Microbial Community. *Environ. Microbiol.* **2017**, *19*, 4993–5009. [CrossRef]
54. Faust, K.; Raes, J. Microbial Interactions: From Networks to Models. *Nat. Rev. Microbiol.* **2012**, *10*, 538–550. [CrossRef]
55. Zhao, D.; Shen, F.; Zeng, J.; Huang, R.; Yu, Z.; Wu, Q.L. Network Analysis Reveals Seasonal Variation of Co-Occurrence Correlations between Cyanobacteria and Other Bacterioplankton. *Sci. Total Environ.* **2016**, *573*, 817–825. [CrossRef]
56. Li, S.-N.; Zhang, C.; Li, F.; Ren, N.-Q.; Ho, S.-H. Recent Advances of Algae-Bacteria Consortia in Aquatic Remediation. *Crit. Rev. Environ. Sci. Technol.* **2023**, *53*, 315–339. [CrossRef]
57. Yue, Y.; Yang, Z.; Wang, F.; Chen, X.; Huang, Y.; Ma, J.; Cai, L.; Yang, M. Effects of Cascade Reservoirs on Spatiotemporal Dynamics of the Sedimentary Bacterial Community: Co-Occurrence Patterns, Assembly Mechanisms, and Potential Functions. *Microb. Ecol.* **2024**, *87*, 18. [CrossRef]
58. Wang, H.; Wang, Z.; Yu, J.; Ma, C.; Liu, L.; Xu, D.; Zhang, J. The Function and Keystone Microbiota in Typical Habitats under the Influence of Anthropogenic Activities in Baiyangdian Lake. *Environ. Res.* **2024**, *247*, 118196. [CrossRef]
59. Zhang, Z.; Lu, J.; Zhang, S.; Tian, Z.; Feng, C.; Liu, Y. Analysis of Bacterial Community Structure, Functional Variation, and Assembly Mechanisms in Multi-Media Habitats of Lakes during the Frozen Period. *Ecotoxicol. Environ. Saf.* **2024**, *284*, 116903. [CrossRef]
60. Xie, G.; Zhang, Y.; Gong, Y.; Luo, W.; Tang, X. Extreme Trophic Tales: Deciphering Bacterial Diversity and Potential Functions in Oligotrophic and Hypereutrophic Lakes. *BMC Microbiol.* **2024**, *24*, 348. [CrossRef]
61. Mikhailov, I.S.; Zakharova, Y.R.; Bukin, Y.S.; Galachyants, Y.P.; Petrova, D.P.; Sakirko, M.V.; Likhoshway, Y.V. Co-Occurrence Networks among Bacteria and Microbial Eukaryotes of Lake Baikal during a Spring Phytoplankton Bloom. *Microb. Ecol.* **2019**, *77*, 96–109. [CrossRef]
62. Zhou, J.; Lao, Y.; Song, J.; Jin, H.; Zhu, J.; Cai, Z. Temporal Heterogeneity of Microbial Communities and Metabolic Activities during a Natural Algal Bloom. *Water Res.* **2020**, *183*, 116020. [CrossRef]
63. Ghai, R.; Mizuno, C.M.; Picazo, A.; Camacho, A.; Rodriguez-Valera, F. Key Roles for Freshwater A Ctinobacteria Revealed by Deep Metagenomic Sequencing. *Mol. Ecol.* **2014**, *23*, 6073–6090. [CrossRef]
64. Carlson, R.E. A Trophic State Index for Lakes 1. *Limnol. Oceanogr.* **1977**, *22*, 361–369. [CrossRef]
65. Liang, D.; Wang, Q.; Wei, N.; Tang, C.; Sun, X.; Yang, Y. Biological Indicators of Ecological Quality in Typical Urban River-Lake Ecosystems: The Planktonic Rotifer Community and Its Response to Environmental Factors. *Ecol. Indic.* **2020**, *112*, 106127. [CrossRef]
66. Adamovich, B.; Medvinsky, A.; Nikitina, L.; Radchikova, N.; Mikheyeva, T.; Kovalevskaya, R.; Veres, Y.K.; Chakraborty, A.; Rusakov, A.; Nurieva, N. Relations between Variations in the Lake Bacterioplankton Abundance and the Lake Trophic State: Evidence from the 20-Year Monitoring. *Ecol. Indic.* **2019**, *97*, 120–129. [CrossRef]

processes

MDPI

Article

Magnesium-Modified Biochar for Removing Phosphorus from Aquaculture Facilities: A Case Study in Idaho, USA

Ethan Struhs [1], William F. Rance Bare [2], Amin Mirkouei [1,2,*] and Kenneth Overturf [3]

[1] Department of Nuclear Engineering and Industrial Management, University of Idaho, Idaho Falls, ID 83402, USA; estruhs@gmail.com
[2] Department of Biological Engineering, University of Idaho, Idaho Falls, ID 83402, USA
[3] Agricultural Research Service, United States Department of Agriculture, Hagerman, ID 83332, USA
* Correspondence: amirkouei@uidaho.edu

Abstract: This study aims to improve water quality and reduce eutrophication downstream when receiving water from aquaculture facilities that can support the aquaculture industry and increase fish production capacity. The primary objective is to investigate a novel approach, using magnesium (Mg)-modified biochar water treatment systems from pinewood, to remove the main eutrophication agents (i.e., phosphorous and nitrogen) from the effluents of aquaculture facilities in Magic Valley, Idaho. The downstream water contains approximately 0.14 mg/L of phosphorous and 2.25 mg/L of nitrogen. The results show that the initial P_2O_5 concentration between Mg-modified and non-modified biochar is comparable. After exposure to aquaculture production water, the modified biochar is shown to have a significant increase in phosphorous and nitrogen adsorption. Non-modified biochar started with noticeably higher concentration levels of nitrogen than modified biochar. Over time in the treatment water, the modified biochar showed a significant increase in nitrogen concentration. Mg bonded to the modified biochar is shown to decrease drastically after exposure to the effluent. This could be due to the insufficient bonding of the magnesium to biomass feedstocks during pre-processing and biochar production. The amount of biochar near the end of experimentation is almost comparable to the non-modified char. We concluded that the proposed approach, using a Mg-modified biochar water treatment system, could sequester more nitrogen and phosphorous over time.

Keywords: aquaculture production; magnesium-modified biochar; eutrophication; water treatment

1. Introduction

Idaho is considered the most prominent commercial producer of rainbow trout in the United States [1]. Products of Idaho's aquaculture industry also include cold-water facilities that raise trout, sturgeon, salmon, and steelhead, as well as warm-water facilities that raise tropical fish, tilapia, and catfish [1]. Approximately 70% of the 115 permitted aquaculture facilities in Idaho operate in the Magic Valley. Effluent from these facilities is discharged directly to the Snake River or its tributaries [1]. The effluent contains solid and liquid pollutants, such as fish feces, parasites, uneaten fish food, algae, pathogens, drugs, and excess nutrients, which detrimentally affect downstream surface and groundwater, posing potential environmental problems [1].

Currently, all the fish farms operate under a discharge permit issued by the State of Idaho, Department of Water Resources, including 49 in operation and 12 non-producing. A permit issued by the Federal Government to the State of Idaho identifies a limit to the pollutants that can be locally discharged from the Hagerman area into the Snake River. The individual farm permits identify discharge limits through a series of regularly monitored water quality parameters based upon the requirements established by the federal permit. This monitoring has determined that phosphorous is the chief pollutant discharged in farm effluents and has the greatest impact on freshwater aquatic ecosystems [2]. The recommended limit for total phosphorous in streams was set at 0.05 mg/L, and larger flowing waters at 0.1 mg/L [3]. These limits are identified in the federal permit, which was last renewed in 2019 but is subject to change for future renewals. If the federal permit does alter the area's phosphorous limit, it will most likely be lowered, as phosphorus is a chief eutrophication element. Current monitoring shows that aquaculture effluent struggles to meet the requirements, with some reports showing phosphorus levels are 35% higher than allowed. The same monitoring also revealed that 65% of the farms had exceeded their permit limits for phosphorus at some point in the last five years. Due to these challenges, fish farm operators are trying several methods to meet compliance. These methods include limiting production, diverting effluents to agriculture irrigation for part of the year, obtaining permission to use the permit rights from producers who are not currently operating, and developing bio-filtration ponds.

Mismanagement of these pollutants results in an increase in pH levels, phosphorus, nitrogen, and sediment. Other water conditions (e.g., temperature, oxygen supply, and nutrient levels) can also be negatively impacted, resulting in harm to aquatic organisms. Therefore, sustainable water treatment and aquaculture waste management practices are vital to the longevity and quality of ground and surface water in Idaho. The motivation behind this study lies in improving water quality and sustainability benefits across Idaho's aquaculture industry by developing and improving a biochar-based water treatment system. The main objective of this study is to explore a magnesium (Mg)-modified biochar water treatment approach for reducing excess phosphorus and nitrogen levels that can subsequently reduce the overall eutrophication of the Snake River.

Previous studies have examined modified biochar contaminant adsorption capacities and effects on water quality [4–7]. Table 1 summarizes the key parameters from these studies, four of these studies have considered Mg-modified biochar, of which three different biomass feedstocks were tested (e.g., soybean straw, poplar chips, and sawdust/sediment). All these studies, however, examined contaminant adsorption in controlled lab settings with prepared eutrophic solutions. This study aims to develop Mg-modified biochar and test adsorption capacity as a part of field experiments at fish farms in Hagerman, Idaho. We have tested non-modified biochar and examined its potential in prior studies. Other goals are to expand on the preceding research, compare modified and non-modified biochar, and create a more efficient pathway for fish farm eutrophication reduction.

Table 1. Similarities and dissimilarities of the prior conducted studies with this project.

Authors	Biochar Modification	Eutrophic Target	Biomass Type	Ref.
Cai et al. (2017)	Fe-oxide	PO_4^{3-}	Water hyacinth	[8]
Ngatia et al. (2017)	-	PO_4^{3-}	Switchgrass, kudzu, and Chinese tallow	[9]
Yin et al. (2018)	Al and Mg	NH_4^+, NO_3^-, PO_4^{3-}	Soybean straw	[10]
Novais et al. (2018)	Al	PO_4^{3-}	Poultry manure and sugarcane straw	[11]
Yin et al. (2019)	Mg		Poplar chips	[12]

Table 1. *Cont.*

Authors	Biochar Modification	Eutrophic Target	Biomass Type	Ref.
Liu et al. (2019)	Fe-Mn-Ce	As(III)	-	[13]
Zhu et al. (2019)	-	NH_4^+,	Phyllostachys pubescens	[14]
Ren et al. (2021)	Fe	PO_4^{3-}	Reed straw	[15]
Yang et al. (2021)	Fe and Mg	PO_4^{3-}	Sawdust and sediment	[16]
Cheng et al. (2022)	-	NH_4^+, NO_3^-	-	[17]
Konczak and Huber (2022)	-	NH_4^+, PO_4^{3-}	Sewage sludge	[18]
Bare et al. (2023)	-	NH_4^+, NO_3^-, PO_4^{3-}	Pine	[19]
Bare et al. (2023)	-	NH_4^+, NO_3^-, PO_4^{3-}	Pine	[20]
This project *	Mg	NH_4^+, NO_3^-, PO_4^{3-}	Pine	-

* Results from the proposed approach in this study.

2. Materials and Methods

This study aims to improve water quality through a reduction in pollutants through the application of modified biochar and quantitatively determine the characteristics that support sustainable water treatment, employing emerging technologies and biomass-based products. In this study, we measured the nitrogen, phosphorous, pH, and temperature of the water using various sensors and testers in aquaculture facilities located near the Snake River in the Hagerman and Buhl, southern Idaho area, USA. Samples are taken before and after biochar applications to determine impact. Each experiment is repeated three times to promote repeatability and identify critical factors.

The Mg-modified biochar samples were produced using a built-in-house, custom ball-bearing mixer and slow pyrolysis reactor. Magnesium chloride was mixed with the pinewood and dried prior to undergoing pyrolysis. The Mg–pine mixing ratio was 1:1, 100 g of pinewood mixed with 100 g of magnesium chloride hexahydrate ($MgCl_2{\cdot}6H_2O$). Magnesium chloride hexahydrate is a technical-grade chemical used for a variety of different applications. The primary active component (magnesium chloride) is a natural mineral produced by the evaporation of seawater. Because it is an exothermic reaction with water, it is commonly used for the low-temperature de-icing of highways, parking lots, and walkways. Biochars were produced from pinewood, which was shown previously to have desirable pollutant adsorption capabilities [19]. The pyrolysis process was performed under inert conditions, purging the reactor with nitrogen under a flow rate of 0.5 L/min. The reactor was heated using internal heating cartridges to 450 °C and ran for 1 h and 30 min. The pressure was considered to be ambient.

This study explored Mg-modified biochar production through biomass conversion in the presence of magnesium chloride to create an efficient water pollution mitigation and abatement candidate. The collected data can generate knowledge to assess the feasibility of this water treatment pathway moving forward. Table 2 shows the detailed experimental tasks in this study. The required materials for the experiments are as follows:

- A total of 24 water filter bags (remove particle size down to 100 mesh);
- A total of 12 non-modified biochar bags and 12 Mg-modified biochar bags (each has 500 g of biochar);
- In total, 48 ft. of 2-inch flexible hose (2 ft per tank);
- pH meter and thermometer;
- A total of 25 zip ties and 25 Ziploc bags for collecting the samples;
- A total of 3 sample bottles for collecting water samples.

Table 2. Detailed experimental tasks in this study.

Day	Experimental Tasks
0	Collect 3 untreated water samples and check water pH, temp, and Magic Valley Lab for N, P, and C (3 times). Set up the water filter bags with 500 g biochar—lay them flat for better absorption. Duration: 1, 3, 5, and 7 days (24 water filter bags).
1	Collect 3 modified biochar bags and 3 non-modified biochar bags.
3	Collect 3 modified biochar bags and 3 non-modified biochar bags.
5	Collect 3 modified biochar bags and 3 non-modified biochar bags.
7	Collect 3 modified biochar bags and 3 non-modified biochar bags.

2.1. Techno-Economic Assessment

In this study, a life cycle costing model was developed to assess the feasibility of a proposed portable pyrolysis unit for water nutrient adsorption from fish farms. Both the capital (fixed) and operational (variable and labor) costs were incorporated into the total cost estimation, encompassing activities, such as collection, pretreatment (grinding and dewatering), magnesium chloride extraction/production, portable refining, storage, transportation, and water treatment. The total cost, as defined by Equation (1), includes the cost components for pine collection (C1), grinding (C2), drying (C3), magnesium production (C4), conversion (C5), storage (C6), distribution (C7), and water treatment (C8). Detailed nomenclature for these parameters is provided in the Supplementary Materials.

$$\text{Min } Z = \sum_{i=1}^{n} C_i = C_1 + \ldots + C_8 \quad (1)$$

$$C_1 = \sum_a \sum_b \sum_t C_{C-col} \times Z_{abt} + (C_{V-col}) \times \frac{PN_{abt}}{U_{col}}$$
$$C_2 = \sum_a \sum_b \sum_t (C_{C-gr} + C_{V-gr}) \times \frac{PN_{abt}}{U_{gr}}$$
$$C_3 = \sum_a \sum_b \sum_t (C_{C-dr} + C_{V-dr}) \times \frac{PN_{abt}}{U_{dr}}$$
$$C_4 = \sum_a \sum_b \sum_t (C_{C-mg} + C_{V-mg}) \times \frac{MGN_{abt}}{U_{mg}}$$
$$C_5 = \sum_a \sum_b \sum_t (C_{C-pyr} + C_{V-pyr}) \times \frac{PN_{abt}}{U_{pyr}}$$
$$C_6 = \sum_b \sum_c \sum_t (C_{C-bs} + C_{V-bs}) \times \frac{B_{bct}}{U_{bs}} + \sum_b \sum_c \sum_t (C_{C-os} + C_{V-os}) \times \frac{O_{bct}}{U_{os}}$$
$$C_7 = \sum_c \sum_e \sum_t (C_{C-di} + C_{V-di}) \times \frac{B_{cet}}{U_d}$$
$$C_8 = \sum_c \sum_e \sum_t (C_{C-wt} + C_{V-wt}) \times \frac{B_{cet}}{U_{wt}}$$

2.2. Environmental Impact Assessment

A life cycle assessment (LCA) study was performed using the open-source OpenLCA v2.1.2 software in conjunction with data from previous studies to evaluate the environmental impacts of the proposed water treatment method [19,21,22]. The assessment was structured into four distinct phases: (1) definition of the goal and scope, (2) life cycle inventory analysis, (3) life cycle impact assessment, and (4) interpretation.

Goal and scope definition: The objective of this study is to assess the environmental impacts of producing modified biochars for nutrient adsorption in fish farms. The LCA study evaluates both the global warming potential (GWP) and eutrophication effects throughout the conversion of residues into biomaterials for treating downstream waters. GWP is quantified using greenhouse gas (GHG) emission factors expressed in kg CO_2 equivalents, with 28 kg CO_2 eq./kg CH_4 and 265 kg CO_2 eq./kg N_2O as provided by the Intergovernmental Panel on Climate Change (IPCC) for a 100-year time horizon. Eutrophication is measured in PO_4 equivalents based on factors of 3.07 kg CO_2 eq./kg P and 0.24 kg CO_2 eq./kg N_2O. The study's scope covers upstream and midstream processes, including residue collection, on-site pre-processing (size reduction and drying), conversion into biomaterials, utilization of intermediate products (pyrolysis oil and gas) for process energy (heat and electricity), and distribution. The system boundary is cradle-to-gate, with one kilogram of Mg-modified biochar for phosphorus removal as the functional unit.

Life cycle inventory: Parameter data for both inputs and outputs were sourced from prior studies and the AGRIBALYSE and OpenLCA v2.1.2 databases [19,23,24]. The machinery required for collecting pine residues includes a forwarder and a loader, and key inputs in this stage are the pine residues and diesel fuel, while diesel combustion emissions constitute the primary output. For magnesium chloride production, the process examined involves brine processing and the use of hydrochloric acid. GHG emission factor for the upstream stage of the product system accounts for both the collection and hauling of forest residues and the production of magnesium chloride. After collection, the residues are processed using an on-site grinding machine to generate suitably sized pine particles, followed by dewatering in a rotary dryer. During this pretreatment stage, pine residues and diesel fuel (for the grinding and drying operations) are consumed, producing pretreated residues alongside emissions from fuel combustion and water vapor from drying. Notably, enhancing residue quality for the pyrolysis process requires additional energy input, thereby increasing emissions and the overall GWP. Subsequently, a portable pyrolysis unit converts the pretreated residues into biomaterials. This process employs nitrogen as an inert carrier gas, utilizes energy from the combustion of pyrolysis oil and gas, and may require supplementary diesel fuel. The inputs for the pyrolysis process include the pretreated residues, nitrogen, energy derived from byproduct combustion, and cooling water for condensing the pyrolysis oil, while the outputs comprise the target biomaterials and all associated process emissions. Emissions from the combustion of pyrolysis byproducts are treated as biogenic GHGs. Finally, the produced biomaterials are loaded and transported by truck to fish farms for water treatment, with diesel fuel consumption and environmental impacts primarily influenced by the distance between the collection site and the fish farms. An LCA study was performed using data from a previous case study in Hagerman, ID, and the process was modeled as a production system in OpenLCA. This comprehensive assessment of pathway impacts and emissions offers valuable insight into the biomaterial-based water treatment approach.

Life cycle impacts assessment: The impact assessment was conducted using the CML-IA baseline method within OpenLCA. GWP100 was calculated based on Equations (2)–(9) and encompasses both upstream and midstream processes. These processes include residue collection, on-site pre-processing (i.e., size reduction and drying), conversion of residues into biomaterials, on-site reuse of intermediate products (pyrolysis oil and gas) for process energy (heat and electricity), and biomaterial distribution. Upstream emission factors and GWP for biomaterial production are calculated using Equations (2) and (3). Midstream emission factors and GWP for biomaterial production are calculated using Equations (4) and (5). Emission factors and GWP of biomaterial transportation are calculated using Equations (6) and (7). Emission factors and eutrophication potential of modified biomaterial production and application are calculated using Equations (8) and (9).

$$US_{EF} = ER_{CO2} \times US_{EFCO2} + ER_{CH4} \times US_{EFCH4} + ER_{N2O} \times US_{EFN2O} \quad (2)$$

$$US_{GWP} = M_P \times US_{EF} \quad (3)$$

$$MS_{EF} = ER_{CO2} \times MS_{EFCO2} + ER_{CH4} \times MS_{EFCH4} + ER_{N2O} \times MS_{EFN2O} \quad (4)$$

$$MS_{GWP} = M_{bc} \times MS_{EF} \quad (5)$$

$$TR_{EF} = ER_{CO2} \times TR_{EFCO2} + ER_{CH4} \times TR_{EFCH4} + ER_{N2O} \times TR_{EFN20} \quad (6)$$

$$TR_{GWP} = M_{bc} \times TR_{EF} \times D \quad (7)$$

$$EP_{EF} = EP_{P} \times EP_{EFP} + EP_{N2O} \times EP_{EFN2O} \quad (8)$$

$$E_{P} = M_{bc} \times EP_{EF} \quad (9)$$

Interpretation: The interpretation step, which integrates the findings from the life cycle inventory and the impact assessment, is detailed in Section 4.

3. Case Study

Hagerman is located within the Snake River Canyon, approximately 40 miles west of Twin Falls, Idaho. It is the home of over 60 fish farms, which produce up to 75% (around 41 million pounds per year) of the nation's commercial rainbow trout depending upon how many farms are operating each year. The desirability of operating in this location depends upon two factors: (1) a sustainable freshwater supply and (2) a temperature range favorable to the rearing of rainbow trout (54–65 °F) [25]. These conditions are met by the outflow of the Snake Plain aquifer, also known as "Thousand Springs", which serves as the water source for the majority of the farms. These farms have a significant financial impact on the community and the entire Southern Idaho region, currently estimated at more than 150 million dollars annually (Figure 1).

Figure 1. Field experiments using USDA-ARS testing tanks (dotted orange line) at Riverence fish farm in southern Idaho, USA.

4. Results and Discussion

The study investigated the potential for the Mg-modified biochar to remove excess nutrients (e.g., phosphorus- and nitrogen-based compounds) efficiently and economically from the effluent water of fish farms. The detailed analytical techniques for laboratory and field sample analyses are provided in an earlier published study by Bare et al. (2023) [19]. Table 3 presents the water sample results.

The initial phosphorus concentrations of the Mg-modified and non-modified biochar are comparable. Both show a linear increase in phosphorus adsorption through day 5. A noticeable deviation between the two is observed by day 7, with the Mg-modified biochar exhibiting a substantial increase in phosphorus adsorption, but with increased variation. The initial drop in phosphorus adsorption on day 3 for Mg-modified biochar may be attributed to competing ions, clogging the biochar surface, or microbial biofilm formation. The later surge in phosphorus adsorption by day 7 suggests that the biochar underwent physicochemical changes, improving its adsorption capacity. Further data on competing ions and biofilm activity could help confirm the exact mechanism.

Table 3. Field experiment results of water samples.

Parameter	Sample			Method
	1	2	3	
Total organic carbon (mg/L)	1.15	1.22	1.08	SM 5310B
Total P (mg/L)	0.09	0.09	0.09	EPA 36.1
Nitrate/N (mg/L)	2.58	2.61	2.59	EPA 300
Nitrite/N (mg/L)	<0.40	<0.40	<0.40	EPA 300

A similar trend is seen for the adsorption of nitrogen. Non-modified char started with noticeably higher concentration levels than the modified char. Over time, particularly after day 5, the modified biochar is shown to greatly increase nitrogen concentration (Figure 2).

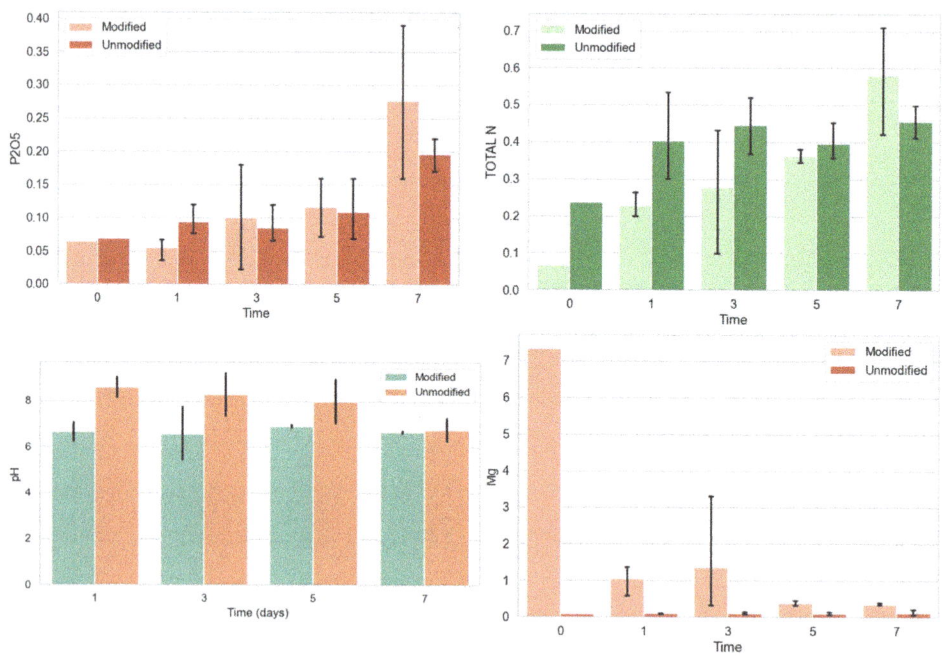

Figure 2. Results comparison from field experiments.

While pH is relatively constant for the Mg-modified biochar, the unmodified biochar shows a decrease with time. This decrease is likely due to a combination of nutrient adsorption, microbial activity, and chemical changes in the biochar. Initially, biochar's alkaline functional groups provide some buffering, but as phosphorus and nitrogen species adsorb, proton release lowers the pH. Microbial processes, particularly nitrification, further contribute to acidification by generating additional H^+. Over time, the gradual leaching of carbon compounds and the depletion of alkaline minerals (e.g., K_2O and CaO) reduce the biochar's buffering capacity, leading to further pH decline. The Mg-modified biochar could also be slightly more acidic in nature due to the slight hydrolysis of magnesium species and the increase in acidic functional groups formed during the pyrolysis step.

The modified biochar shows a gradual increase in total nitrogen with a p-value of 0.0028, indicating the significance of the rise in nitrogen over time. On the other hand, the unmodified biochar remains relatively flat, showing less nitrogen adsorption. Both the modified and unmodified biochar water treatment systems show a significant increase in phosphorus adsorption as well, with p-values of 0.0059 and 0.0082, respectively. The modified biochar significantly improves nitrogen, nitrate, and phosphorus adsorption over time. The unmodified biochar shows weaker trends and decreases pH. These findings suggest that the modified biochar is a more effective long-term solution for nutrient adsorption.

Magnesium levels stay constant in the unmodified char as there is no significant amount of Mg in the pre-processed biochar. Mg bonded to the modified char is shown to drastically decrease shortly after exposure to the effluent. This could be due to the insufficient bonding of the magnesium to the biomass/biochar during processing. The amount of biochar near the end of experimentation is almost comparable to the unmodified biochar. Further work should be put into optimizing Mg/C binding mechanisms to increase nutrient adsorption capacity.

For both the modified and unmodified biochars, similar correlations are observed in the heatmaps (Figure 3), although the magnitudes differ slightly. The correlation study used for generating the heatmaps was based on Pearson's correlation coefficient. This is the most common method for measuring linear relationships between numerical variables. A strong positive correlation between nitrate-nitrogen and phosphorus adsorption suggests that these nutrients tend to be retained simultaneously, likely due to the inherent surface properties of the biochar. Similarly, a positive correlation between pH and potassium oxide (K_2O) indicates that higher pH levels enhance potassium retention, possibly through improved cation exchange capacity.

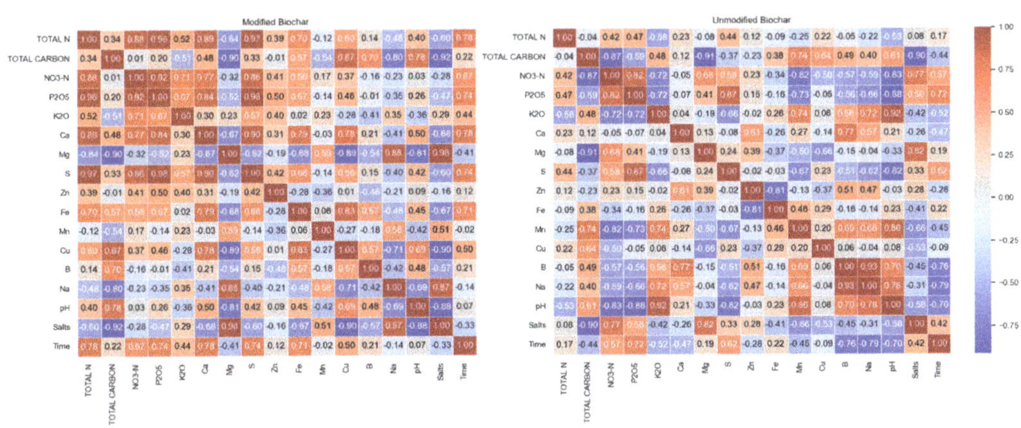

Figure 3. Correlation heatmaps for biochar parameters.

Conversely, a strong negative correlation between nitrate-nitrogen adsorption and total carbon content implies that surface chemical modifications occurred, altering nitrogen adsorption behavior. Exposure to wastewater may have induced oxidation or protonation of functional groups, leading to shifts in surface charge distribution and affecting nitrate binding. Additionally, the adsorption of anions like nitrate could have influenced electrostatic interactions, potentially reducing the availability of active sites for further nitrogen retention. These findings suggest that the observed changes in nitrogen adsorption are more likely due to biochar surface chemistry. Additionally, the data indicate that higher nitrate retention is associated with a decrease in pH, potentially due to the release of H^+ ions during the adsorption process. Table S1 presents the results of field experiments.

Techno-economic results: In the proposed techno-economic analysis, 29,400 metric tons of pinewood are processed annually to produce 11,760 metric tons of modified biomaterials.

Under the base-case scenario—with a processing capacity of 10 metric tons of ground, dry forest residue per day—the total cost, which encompasses both capital and variable expenditures, is estimated at USD 51,709 per year over a 10-year period. This results in a unit cost of USD 436 per metric ton of biomaterials. Detailed breakdowns of the capital and operational costs for each pathway are provided in Table 4, with approximately 67% of the total cost attributed to capital expenses, primarily due to the high initial investment required for the mobile pyrolysis refinery. The use of magnesium chloride for biochar modification had a significant impact on operational costs compared to the previous study.

Table 4. Capital and operational expenditures of each process.

Process	Capital (USD/yr)	Variable (USD/yr)
Collection (Wood, $MgCl_2$)	4445	4361
Grinding	10,912	11,653
Drying	13,976	17,254
Portable refinery	59,976	4954
Biomaterials storage	6076	2116
Transportation	9076	6914
Fish farm	-	4457
Total	104,461	51,709

Environmental impact results: The largest contributing factor to the global warming potential in this process is the upstream production of the magnesium chloride additive (Table 5). Although it represents only about 5% of the feed by weight, its synthesis is highly energy-intensive and generates a disproportionate amount of CO_2-eq emissions (~75%). This chemical production stage drives a significant portion of the overall GWP because the emissions per unit mass of magnesium chloride are much higher than those associated with the biomass itself.

Table 5. Life cycle impact assessment data using CML baseline.

Impact Category	Per Metric Ton Biochar	Unit
Global Warming Potential (GWP100)	692	kg CO_2-eq
Acidification Potential	2.4	kg SO_2-eq
Ecotoxicity Potential	0.18	kg 1,4-DCB-eq
Human Toxicity Potential	0.36	kg 1,4-DCB-eq
Photochemical Oxidation Potential	1.44	kg C_2H_4-eq
Eutrophication	−1124	kg PO_4-eq

In addition, the energy required to initially heat the pyrolysis reactor—primarily via natural gas combustion—also contributes notably to the process's carbon footprint. However, when comparing the two, the emissions from the chemical production of the additive remain the dominant source of GWP. This means that efforts to reduce the overall environmental impacts of the process should focus on either improving the efficiency of the additive's production or reducing the required dosage while maintaining biochar performance.

5. Conclusions

The primary objectives of this study are as follows: (a) improving water quality and pollution reduction in downstream water from aquaculture facilities using raw and magnesium (Mg)-modified biochar water treatment systems from pinewood, (b) estimating water quality using data from laboratory and field experiments, as well as molecular and spectroscopy analysis, and (c) conducting characterization and statistical analysis to identify the primary causes that drive the fish production away from being sustainable. Particularly, this study involves the investigation for developing a sustainable methodology to remove the main eutrophication agents, i.e., phosphorous and nitrogen, from the effluents of aquaculture facilities. The key results of this study include the following: (i) modified biochar removed 56.3% more NO_3-N and 13.3% more P_2O_5 compared to unmodified biochar, making it significantly more effective for nutrient removal from fish farm effluent; (ii) the total global warming potential (GWP) for 1 ton of modified biochar was estimated at 1750 kg CO_2-eq, with 400 kg CO_2-eq sequestration in the soil phase, reducing its overall footprint; (iii) pH and K_2O adsorption ($r \approx 0.92$) showed a high positive correlation, suggesting that increased phosphorous and nitrogen retention enhances pH stabilization; and (iv) NO_3-N and P_2O_5 adsorption ($r \approx 0.82$) were strongly correlated, indicating that biochar modification enhances the simultaneous retention of these nutrients. This study provided the following outcomes:

- Understanding the complexities of modified biochar production, reaction mechanisms, and multi-functional performance in water treatment.
- Generating valuable data and a base of knowledge to effectively assess water quality in the Thousand Springs and Hagerman areas of Idaho.
- Developing an environmentally friendly and economical method to remove the eutrophic contaminants from the farm effluents to sustain and improve the productivity of fish farming in Idaho and the nation.
- Reducing water pollution generated from aquaculture production facilities to enhance sustainability benefits to the surrounding area and aquaculture industry using engineered biochar water treatment systems.
- Increasing fish production within the EPA permit limits, and consequently stabilizing employment and increasing profitability.
- The collected nutrient-rich biochar after water treatment has the potential as a commercial slow-release fertilizer to improve soil fertility and crop health. The possible supplementary benefits could result in the recycling of excess nutrients from downstream water through repurposing as soil amendments.

Supplementary Materials: The following supporting information can be downloaded at https://www.mdpi.com/article/10.3390/pr13041021/s1. Table S1: Field experiment results of Mg-modified and non-modified biochar samples.

Author Contributions: E.S., W.F.R.B., A.M., and K.O. contributed to the conceptualization, methodology, data collection, analyses, and writing the manuscript. E.S. produced Idaho biochar and K.O. performed the field experiments. All the authors contributed to writing the manuscript. All authors have read and agreed to the published version of the manuscript.

Funding: This research was funded by the United States Geological Survey, 104b grant.

Data Availability Statement: The original contributions presented in this study are included in the article/Supplementary Materials. Further inquiries can be directed to the corresponding author.

Acknowledgments: The authors wish to acknowledge inputs, resources, and support from the United States Department of Agriculture, Agricultural Research Service (USDA-ARS), University of Idaho Aquaculture Research Institute (UI-ARI), Riverence Provisions LLC, and Idaho Water Resources Research Institute (IWRRI).

Conflicts of Interest: The authors declare no conflicts of interest.

References

1. Idaho Department of Environmental Aquaculture in Idaho. Available online: https://www.deq.idaho.gov/water-quality/wastewater/ (accessed on 15 April 2024).
2. Low-Phosphorous Feeds Development. Available online: https://wracuw.org/products-and-publications/low-phosphorous-feeds-development (accessed on 18 March 2024).
3. Litke, D. *Review of Phosphorus Control Measures in the United States and Their Effects on Water Quality*; U.S. Geological Survey: Reston, VA, USA, 1999.
4. Silvani, L.; Vrchotova, B.; Kastanek, P.; Demnerova, K.; Pettiti, I.; Papini, M.P. Characterizing Biochar as Alternative Sorbent for Oil Spill Remediation. *Sci. Rep.* **2017**, *7*, 43912. [CrossRef]
5. Silvani, L.; Di Palma, P.R.; Riccardi, C.; Eek, E.; Hale, S.E.; Viotti, P.; Papini, M.P. Use of Biochar as Alternative Sorbent for the Active Capping of Oil Contaminated Sediments. *J. Environ. Chem. Eng.* **2017**, *5*, 5241–5249.
6. Moller, G.; Strawn, D. Biochar Water Treatment. U.S. Patent 10,351,455, 16 July 2019.
7. DeMessie, B.; Sahle-Demessie, E.; Sorial, G.A. Cleaning Water Contaminated with Heavy Metal Ions Using Pyrolyzed Biochar Adsorbents. *Sep. Sci. Technol.* **2015**, *50*, 2448–2457. [CrossRef]
8. Cai, R.; Wang, X.; Ji, X.; Peng, B.; Tan, C.; Huang, X. Phosphate Reclaim from Simulated and Real Eutrophic Water by Magnetic Biochar Derived from Water Hyacinth. *J. Environ. Manag.* **2017**, *187*, 212–219. [CrossRef] [PubMed]
9. Ngatia, L.W.; Hsieh, Y.P.; Nemours, D.; Fu, R.; Taylor, R.W. Potential Phosphorus Eutrophication Mitigation Strategy: Biochar Carbon Composition, Thermal Stability and pH Influence Phosphorus Sorption. *Chemosphere* **2017**, *180*, 201–211. [CrossRef] [PubMed]
10. Yin, Q.; Wang, R.; Zhao, Z. Application of Mg–Al-Modified Biochar for Simultaneous Removal of Ammonium, Nitrate, and Phosphate from Eutrophic Water. *J. Clean. Prod.* **2018**, *176*, 230–240.
11. Novais, S.V.; Zenero, M.D.O.; Barreto, M.S.C.; Montes, C.R.; Cerri, C.E.P. Phosphorus Removal from Eutrophic Water Using Modified Biochar. *Sci. Total Environ.* **2018**, *633*, 825–835. [CrossRef] [PubMed]
12. Yin, Q.; Liu, M.; Ren, H. Removal of Ammonium and Phosphate from Water by Mg-Modified Biochar: Influence of Mg Pretreatment and Pyrolysis Temperature. *BioResources* **2019**, *14*, 6203–6218. [CrossRef]
13. Liu, X.; Gao, M.; Qiu, W.; Khan, Z.H.; Liu, N.; Lin, L.; Song, Z. Fe–Mn–Ce Oxide-Modified Biochar Composites as Efficient Adsorbents for Removing As (III) from Water: Adsorption Performance and Mechanisms. *Environ. Sci. Pollut. Res.* **2019**, *26*, 17373–17382. [CrossRef] [PubMed]
14. Zhu, Y.; Tang, W.; Jin, X.; Shan, B. Using Biochar Capping to Reduce Nitrogen Release from Sediments in Eutrophic Lakes. *Sci. Total Environ.* **2019**, *646*, 93–104. [CrossRef] [PubMed]
15. Ren, L.; Li, Y.; Wang, K.; Ding, K.; Sha, M.; Cao, Y.; Kong, F.; Wang, S. Recovery of Phosphorus from Eutrophic Water Using Nano Zero-Valent Iron-Modified Biochar and Its Utilization. *Chemosphere* **2021**, *284*, 131391. [CrossRef] [PubMed]
16. Yang, F.; Chen, Y.; Nan, H.; Pei, L.; Huang, Y.; Cao, X.; Xu, X.; Zhao, L. Metal Chloride-Loaded Biochar for Phosphorus Recovery: Noteworthy Roles of Inherent Minerals in Precursor. *Chemosphere* **2021**, *266*, 128991. [CrossRef] [PubMed]
17. Cheng, R.; Hou, S.; Wang, J.; Zhu, H.; Shutes, B.; Yan, B. Biochar-Amended Constructed Wetlands for Eutrophication Control and Microcystin (MC-LR) Removal. *Chemosphere* **2022**, *295*, 133830. [CrossRef] [PubMed]
18. Kończak, M.; Huber, M. Application of the Engineered Sewage Sludge-Derived Biochar to Minimize Water Eutrophication by Removal of Ammonium and Phosphate Ions from Water. *J. Clean. Prod.* **2022**, *331*, 129994. [CrossRef]
19. Bare, R.; Struhs, E.; Mirkouei, A.; Overturf, K.; Small, B. Engineered Biomaterials for Reducing Phosphorus and Nitrogen Levels from Downstream Water of Aquaculture Facilities. *Processes* **2023**, *11*, 1029. [CrossRef]
20. Bare, W.F.R.; Struhs, E.; Mirkouei, A.; Overturf, K.; Chacón-Patiño, M.L.; McKenna, A.M.; Chen, H.; Raja, K.S. Controlling Eutrophication of Aquaculture Production Water Using Biochar: Correlation of Molecular Composition with Adsorption Characteristics as Revealed by FT-ICR Mass Spectrometry. *Processes* **2023**, *11*, 2883. [CrossRef]
21. Struhs, E.; Mirkouei, A.; You, Y.; Mohajeri, A. Techno-Economic and Environmental Assessments for Nutrient-Rich Biochar Production from Cattle Manure: A Case Study in Idaho, USA. *Appl. Energy* **2020**, *279*, 115782. [CrossRef]
22. Takeshita, S.; Farzaneh, H.; Dashti, M. Life-Cycle Assessment of the Wastewater Treatment Technologies in Indonesia's Fish-Processing Industry. *Energies* **2020**, *13*, 6591. [CrossRef]
23. Colomb, V.; Colsaet, A.; Ait-Amar, S.; Basset-Mens, C.; Mevel, G.; To, V.; Gac, A.; Koch, P.; Mousset, J.; Salou, T. AGRIBALYSE: The French Public LCI Database for Agricultural Products. *OCL* **2015**, *22*, D104. Available online: https://agritrop.cirad.fr/591319/1/ocl140047-s(1).pdf (accessed on 19 March 2025). [CrossRef]

24. Frischknecht, R.; Jungbluth, N.; Althaus, H.-J.; Doka, G.; Dones, R.; Heck, T.; Hellweg, S.; Hischier, R.; Nemecek, T.; Rebitzer, G.; et al. The Ecoinvent Database: Overview and Methodological Framework (7 pp). *Int. J. Life Cycle Assess.* **2005**, *10*, 3–9. [CrossRef]
25. Adams, P.; James, C.; Speas, C. *Rainbow Trout (Oncorhynchus Mykiss) Species and Conservation Assessment*; U.S. Forest Service 2008. p. 25. Available online: https://www.fs.usda.gov/Internet/FSE_DOCUMENTS/stelprdb5202738.pdf (accessed on 19 March 2025).

processes

MDPI

Article

Performance Evaluation of Modified Biochar as a Polycyclic Aromatic Hydrocarbon Adsorbent and Microbial-Immobilized Carrier

Shuying Geng [1], Shushuai Mao [1], Guangming Xu [2], Aizhong Ding [2], Feiyong Chen [1], Junfeng Dou [2,*] and Fuqiang Fan [3,*]

1 Resources and Environment Innovation Institute, Shandong Jianzhu University, Jinan 250101, China
2 Engineering Research Center of Ministry of Education on Groundwater Pollution Control and Remediation, College of Water Sciences, Beijing Normal University, Beijing 100875, China
3 Center for Water Research, Advanced Institute of Natural Sciences, Beijing Normal University, Zhuhai 519087, China
* Correspondence: doujf@bnu.edu.cn (J.D.); ffan02@bnu.edu.cn (F.F.)

Abstract: Herein, biochars derived from corn stalks, rice husks, and bamboo powder were modified by nitric acid oxidation and sodium hydroxide alkali activation to identify efficient and cost-effective polycyclic aromatic hydrocarbon-adsorbent and microbial-immobilized carriers. The surface characterization and adsorption investigation results suggested that acid/alkali modification promoted the phenanthrene removal ability in an aqueous solution of biochars via facilitating π–π/n–π electron donor–acceptor interactions, electrostatic interactions, hydrogen bonds, and hydrophobic interactions. Subsequently, the degrading bacteria *Rhodococcus* sp. DG1 was successfully immobilized on the rice husk-derived biochar with nitric acid oxidation (RBO), which exhibited the maximum phenanthrene adsorption efficiency (3818.99 $\mu g \cdot g^{-1}$), abundant surface functional groups, and a larger specific surface area (182.6 $m^2 \cdot g^{-1}$) and pore volume (0.141 $m^3 \cdot g^{-1}$). Degradation studies revealed that the microorganisms immobilized on RBO by the adsorption method yielded a significant phenanthrene removal rate of 80.15% after 30 days, which was 38.78% higher than that of the control. Conversely, the polymer gel network-based microenvironment in the microorganism-immobilized RBO by the combined adsorption–embedding method restricted the migration and diffusion of nutrients and pollutants in the reaction system. This study thus introduces an innovative modified biochar-based microbial immobilization technology characterized by a simple design, convenient operation, and high adsorption efficiency, offering valuable insights into material selection for PAH contamination bioremediation.

Keywords: modified biochar; microbial immobilization; polycyclic aromatic hydrocarbons; adsorption–degradation mechanism

1. Introduction

Polycyclic aromatic hydrocarbons (PAHs) are representative persistent organic pollutants, which are widely recognized for their ubiquitous presence in the environment, being generally detected in aquatic environments (0.03–8310 $\mu g \cdot L^{-1}$) [1], sediment (11–66 $\mu g \cdot g^{-1}$) [2], and soil (0.05–543.36 $\mu g \cdot g^{-1}$) [3]. PAHs exert serious threats to both natural ecosystems and human health because of their recalcitrant and toxic properties, including carcinogenicity, teratogenicity, and mutagenicity, which have attracted considerable attention [4]. Currently, there are various PAH remediation techniques (e.g., physical repair, chemical oxidation, bioremediation, and integrated approaches) available for removing PAHs. Physicochemical approaches are efficient, but they are typically expensive and can cause secondary pollution, which poses environmental health risks. In contrast, bioremediation is regarded as an eco-friendly and cost-effective alternative [5]. Among these strategies, immobilized microorganism (IM) technology has emerged as an effective and promising approach for remediating PAH contamination by enabling microorganisms to withstand harsh environments while mitigating competition from indigenous species, thereby enhancing the survival rates and metabolic activities of introduced degrading bacteria. In recent years, biochar-based microbial immobilization technology has become a global research hotspot in PAH remediation [6,7].

Biochar (BC) acts as a provider, supporter, and cooperator in biochar-immobilized microorganism (BM) applications [6]. It exhibits an exceptional pollutant adsorption capacity, thereby reducing the spatial distance between microorganisms and contaminants [8,9]. Factors such as the surface area, pore size distribution, surface functional groups of biochar, molecular structure of adsorbate, and specific interactions between the biochar surface and adsorbate can significantly influence the adsorption process of pollutants [10]. Key mechanisms governing this behavior include the hydrophobic distribution, pore filling, electron donor–acceptor interactions, hydrogen bonding, and hydrophobic interactions, etc. [9,11,12]. As both an adsorbent of pollutants and carrier of immobilized microorganisms, the selection of an appropriate biochar becomes paramount. The physiochemical properties of biochar vary with the types of feedstocks and preparation methods (pyrolysis, gasification, and hydrothermal carbonization, etc.). The modification of biochar can optimize its physiochemical properties, such as increasing the active functional group, degree of aromatization, and the specific surface area, as well as enriching the surface and pore structure, etc., ultimately improving the adsorption performance. Chemical modification (e.g., acid/alkali modification, oxidizing agent modification, metal salts or oxidizing agent modification, and carbonaceous material modification) is available to improve the adsorption performance of biochar [13]. Among them, acid oxidation effectively removes impurities such as metals on the surface of biochar and introduces carboxyl, phenol, carbonyl, and other oxygen-containing functions on the surface of biochar [14]. It can also change the microstructure of biochar and increase its surface area, thus improving the adsorption performance of microorganisms and PAHs [11,15]. In addition, alkali-activated biochar has better surface features, which can promote hydrogen bonding, hydrophobic, π–π, and electrostatic interactions between organic pollutants and biochar [16,17]. Acid/alkali modification is most widely used at present, which has the advantages of a simple design, convenient operation, and high adsorption efficiency and has practical application and popularization value [17–19]. Zhou et al. prepared oxygen-rich biochar derived from waste wood by sulfuric acid treatment and found that the biochar had a targeted adsorption effect on PAH removal, especially symmetrical PAHs [15]. Shin et al. suggested that the reinforced aromatic structure of NaOH-activated spent coffee waste biochar facilitated the π-π interaction [16]. Despite the importance of biochar modification, its application on immobilizing microorganisms has rarely been reported. Based on this, the fundamental knowledge of the adsorption behaviors of PAHs on modified biochar and its degradation behaviors after the immobilization of microorganisms remains incomplete.

Biochar has been reported to be a suitable carrier for microbes [8]. Biochar's porous structure coupled with its large specific surface area and abundant functional groups promotes microorganism adhesion and growth while providing protective habitats conducive to bacterial survival. Adsorption and embedding are the most widely used methods for microbial immobilization in current research. The immobilized microorganisms produced via adsorption methods exist in powder form, which offers advantages such as a low cost combined with simplicity of preparation. Due to its small steric hindrance, it can maintain the microorganisms' high activity, making them suitable for application in non-fluid media such as soil environments. However, microbial displacement could occur because the bond between the microbes and the carrier in adsorption-based immobilization is weak. Combining biochar adsorption with embedding can address this limitation since the grid structure of embedding can prevent the displacement of the microbes from the biochar [8]. Sodium alginate (SA) and polyvinyl alcohol (PVA) are frequently utilized crosslinking materials; notably, SA-PVA co-immobilization has garnered significant interest owing to its high mechanical strength and strong stability [20].

To sum up, the remediation of PAH contamination with biochar-immobilized microorganisms has received significant public interest. However, there remains a substantial gap regarding the use of modified biochar for microorganism immobilization. We hypothesize that the acid/alkali modification of biochar can improve its physicochemical properties, thereby enhancing its PAH removal performance and producing an excellent microbial immobilization carrier. Therefore, the primary objective of this study was to evaluate the feasibility of utilizing modified biochar as an adsorbent material and microbial immobilized carrier. The specific aims include the following: (i) to investigate the adsorption behaviors and mechanisms of pristine and acid/alkali-modified biochars derived from various raw materials toward PAHs, and (ii) to identify the degradation performance and mechanisms of modified biochar-immobilized degradation bacteria targeting PAHs.

2. Materials and Methods

2.1. Biochar Preparation and Chemical Reagents

Corn stalk (CS), rice husk (RH), and bamboo powder (BP) were utilized as raw stocks to produce biochar. With reference to the relevant literature reports and the pyrolysis conditions of commercial biochar production, the raw stocks were pyrolyzed at 500 °C for 4 h in a muffle furnace under a N_2 environment to produce biochar [5,21]. The produced biochar was referred to as CB, RB, and BB, respectively. The biochar was hereafter oven-dried at 60 °C for 24 h and passed through a 60-mesh sieve. Subsequently, all biochars were treated with 2 M nitric acid (HNO_3) or hydroxide sodium (NaOH) at a ratio of 20:1 (v/w) and stirred at 75 °C for 8 h. Then, the supernatant was removed by centrifugation. All modified biochars were then rinsed several times with deionized distilled water until neutral pH, oven-dried, and milled to <0.25 mm. The acid-oxidized biochars produced from CS, RH, and BP were named as CBO, RBO, and BBO, respectively. The alkali-activated biochars produced from CS, RH, and BP were named as CBA, RBA, and BBA, respectively. All biochar samples were stored in a desiccator prior to subsequent experiments.

The characterization methods of biochar samples were similar to the methods described in our other work [22]. Briefly, the surface morphology features of the biochars were investigated via a scanning electron microscope (SEM; Gemini 300, Zeiss, Oberkochen, Germany). The presence and abundance of surface functional groups on the biochars were identified by Fourier transform infrared spectroscopy (FTIR; Affinity-1, Shimadzu, Kyoto, Japan). The crystallographic structures of the biochars were determined using a powder X-ray diffractometer (XRD; X′ pert-Pro MPD, Panalytical, Almelo, The Netherlands). The total pore volume and specific surface areas of the biochars were measured from N_2 adsorption–desorption isotherms using an automatic analyzer (ASAP2460, Micromeritics, Atlanta, GA, USA). The elemental composition (i.e., C, H, and N) of the biochars was determined by an element analyzer (Vario EL cube, Hanau, Germany). The C/H ratios were used as surrogate indices to predict the aromaticity and polarity of the biochars.

Organic solvents used for the extraction and analysis of PAHs, including acetone, *n*-hexane, and dichloromethane, were of chromatographic grade and purchased from J.T. Baker Chemical Company, Phillipsburg, NJ, USA. Other chemicals, such as HNO_3, NaOH, and $CaCl_2$, were of analytical grade and were obtained from Beihua Fine Chemicals Co., Ltd., Shenzhen, China.

2.2. Bacterial Strain and Biochar-Immobilized Microorganism Preparation

The bacterial strain *Rhodococcus* sp. DG1 (China General Microbiological Culture Collection Center No. 18981) preserved in our lab was employed. It was isolated from PAH-contaminated soil via a classical shaken liquid medium enrichment method as described elsewhere [23]. The strain exhibited higher PAH tolerance and better PAH removal efficiency. Before use, the DG1 strain was incubated in LB medium for 12 h. The cells were collected by centrifugation at 4 °C and then washed with 0.85% saline three times and finally re-suspended in the same solution with an OD_{600} value of 1.0.

Biochar having maximum adsorption efficiency of PAHs was selected as a carrier for bacteria immobilization [24]. Afterward, the biochar-immobilized microorganisms were synthesized by physical adsorption and combined adsorption–embedding methods. First, the biochar was mixed with the above bacterial suspension at a ratio of 1:10 (w/v) and shaken at 120 rpm under 30 °C for 24 h. After centrifuging the mixture, the precipitate was freeze-dried and hereafter referred to as ARB. Second, the biochar was mixed with the above bacterial suspension at a ratio of 1:10 (w/v) and shaken at 120 rpm under 30 °C for 2 h. Subsequently, the mixed solution of biochar and bacteria was mixed with polyvinyl alcohol (PVA, 8%, w/v) and sodium alginate (SA, 1%, w/v). The colloidal solution containing the bacteria was added to a 5% (w/v) sterile $CaCl_2$ and saturated boric acid solution by dripping slowly and then left for 12 h at 4 °C to complete gelation. The immobilized bacteria in biochar-gel beads were washed three times with normal saline and hereafter referred to as ERB. The fabrication procedure of the biochar-immobilized microorganisms is shown in Figure 1.

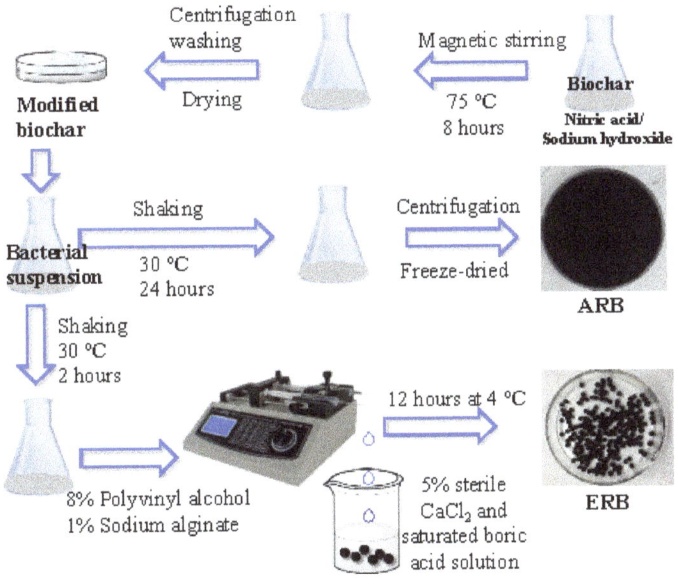

Figure 1. The fabrication procedure of biochar-immobilized microorganisms. The microorganisms immobilized by adsorption method and adsorption–embedding method were labeled as ARB and ERB, respectively.

All operations were performed under strict aseptic conditions. The surface morphology of the BC and BM were investigated using SEM. According to the method proposed by Chen et al. [20], the pelletizing ratio was calculated for the convenience of later calculating the amount of free-living bacteria and immobilized bacterial agents. The count of viable bacteria in the BM was detected using the traditional plate counting method.

2.3. Batch Adsorption Experiments of Biochar in Aqueous Solution

2.3.1. Kinetic Adsorption Experiments

Phenanthrene (PHE) was selected as the model PAH compound in this study. Considering the concentrations of the wastewater and facilitating the study in the laboratory [25,26], a PHE concentration of 2 mg·L^{-1} was chosen. Batch adsorption kinetics experiments of PAH onto various biochars were performed. In each test, 10 mg of each biochar was mixed with 20 mL of PHE solution in a 100 mL brown glass vial. The background solution contained 0.01 M $CaCl_2$ and 200 mg·L^{-1} NaN_3 to maintain a constant ionic strength and to inhibit microbial activity, respectively [27]. After sealing, the samples were kept in the dark, shaken at 200 rpm under 25 °C, and then sampled at predetermined times (0.5, 1, 2, 4, 6, 9, 21.5, 24, 48, 72, 96, 120, 144, and 168 h) to ensure adsorption equilibrium. Subsequently, after centrifugation (4000 rpm, 10 min), the supernatant liquid was filtered through a 0.22 μm nylon millipore membrane, and the filtered solutions were pre-treated and then analyzed via high-performance liquid chromatography (HPLC, DIONEX, UltiMate 3000) for the residual PAH content. Compound separation was accomplished on an Eclipse XDB-C18 column using methanol–ultrapure water (90/10, v/v, 25 min, 1 mL/min) as the mobile phase. Peak detection was at 254 nm, and chromatography was performed at 25 °C. All experiments were conducted in triplicate. Controlled trials were conducted without biochar under otherwise identical conditions.

The amount of adsorbed contaminant was calculated using the following formula:

$$Q_t = \frac{(C_0 - C_t)V}{m}, \quad (1)$$

where Q_t (mg·g^{-1}) is the content of PAH in biochar at time t (h), C_0 (mg·L^{-1}) and C_t (mg·L^{-1}) are the PAH concentration in liquid at initial and time t (h), respectively, V (L) is the volume of the solution, and m (g) is the mass of biochar.

The adsorption kinetics of PAHs were expressed in terms of the pseudo-first-order (PFO), pseudo-second-order (PSO), and intra-particle diffusion (IPD) equations, respectively:

$$Q_t = Q_e\left(1 - e^{-k_1 t}\right), \text{(Pseudo–first–order)} \quad (2)$$

$$Q_t = \frac{k_2 Q_e^2 t}{1 + k_2 Q_e t}, \text{(Pseudo–second–order)} \quad (3)$$

$$Q_t = k_3 t^{0.5} + b, \text{(Intra–particlediffusion)} \quad (4)$$

where Q_e (mg·g^{-1}) is the adsorbed PAH amount per mass unit of biochar at equilibrium; Q_t (mg·g^{-1}) has the same aforementioned meaning; k_1 (1·h^{-1}), k_2 (g·mg^{-1}·h^{-1}), and k_3 (mg·g^{-1}·$min^{-0.5}$) are the adsorption rate constants of the PFO, PSO, and IPD models, respectively.

2.3.2. Isotherm Adsorption Experiments

Isotherm adsorption experiments were conducted using six different concentration levels of PAH (0, 0.2, 0.5, 1.0, 1.5, 2.0 mg·L^{-1}). These experiments were carried out using the same procedures used for the adsorption kinetics experiments as described above. All samples were shaken for 3 days to achieve adsorption equilibration. The adsorption data were fitted by using the Freundlich equation:

$$Q_e = k_4 C_e^n \quad (5)$$

where k_4 and n are the Freundlich constants indicating the relative adsorption capacity and intensity and C_e (mg·L^{-1}) is the solution concentration of PAH at equilibrium.

2.4. Incubation Experiment

The prepared biochar-immobilized microorganisms were used to amend PAHs in contaminated soil, and the degradation performance and kinetics were compared. To prepare the PAH-contaminated soil, an acetone solution of PHE was added and mixed thoroughly into the collected soil. The soil properties after balancing for 3 months were pH 7.1; organic matter 3.54 g·kg^{-1}; total N 1.62 g·kg^{-1}; total P 1.13 g·kg^{-1}; clay 4.59%, silt 35.57%, and sand 59.84%; and PHE: 42 mg kg^{-1}. Subsequently, culture dishes were filled with 1000 g of PHE-contaminated soil that was amended with ARB (3%), ERB, and DG1 inoculant, separately. The additional amount of free strain DG1 and ERB was equivalent to that of ARB according to the calculated immobilization rate [20,24]. The soil without inoculant was used as the control experiment (CK). All treatments were performed in triplicate and incubated in the dark at 25 °C for 30 days. Sterilized deionized water was added to keep the soil moisture at 60% of the maximum water-holding capacity. Soil samples were collected regularly for the determination of residual PHE, and then the degradation removal rate and first-order degradation kinetics of PHE were calculated.

3. Results and Discussion

3.1. Characterization of Biochar

3.1.1. Basic Characteristics

The basic properties of the nine biochars presented varying degrees of differences (Table 1). The pH of the original biochars was weakly alkaline, which is probably due to the separation of alkaline mineral elements contained in the biomass material during the pyrolysis process, leading to an increase in the pH value. After being modified by acid or alkali, the biochar demonstrated correspondingly weak acidic or weak alkaline properties, among which the pH value of BBO was the lowest (4.56). In addition, acid/alkali modification can increase the specific surface area and pore volume of biochar, especially for nitric acid-oxidized biochar. After acid modification, the specific surface area of RB increased from 70.69 $m^2 \cdot g^{-1}$ to 182.6 $m^2 \cdot g^{-1}$, and the specific surface area of BB increased from 43.42 $m^2 \cdot g^{-1}$ to 161.0 $m^2 \cdot g^{-1}$. This was similar to reports in the relevant literature, because activation methods such as nitric acid can remove carbonate and other mineral components from biochar to a large extent, thereby improving the specific surface area and pore volume of biochar [28]. Moreover, the content of C element in biochar materials ranged from 39.51% to 65.86%, and the C/N ratio ranged from 23.84 to 42.44. The contents of C and N elements in biochar are closely related to the growth of microorganisms, and the C/N ratio of 25:1 is considered to be an ideal condition conducive to the growth and metabolism of microorganisms.

Table 1. The basic characteristics of the biochar.

Biochar	pH	Specific Surface Area ($m^2 \cdot g^{-1}$)	Pore Volume ($m^3 \cdot g^{-1}$)	Elements			H/C	C/N
				Carbon (C, %)	Hydrogen (H, %)	Nitrogen (N, %)		
RB [1]	8.11	70.69	0.052	46.23	0.99	1.35	0.02	34.24
RBO [2]	6.34	182.6	0.141	39.51	1.34	1.54	0.03	25.66
RBA [3]	8.52	81.18	0.061	41.36	2.26	1.42	0.05	29.13
BB [4]	8.25	43.42	0.045	57.29	2.08	1.35	0.04	42.44
BBO [5]	4.56	161.0	0.168	52.96	1.97	1.88	0.04	28.17
BBA [6]	9.24	60.58	0.049	54.08	2.11	1.72	0.04	31.44
CB [7]	9.16	22.46	0.037	65.86	1.82	1.97	0.03	33.43
CBO [8]	6.12	69.88	0.053	62.59	2.45	2.03	0.04	30.83
CBA [9]	10.03	50.11	0.050	60.55	1.53	2.54	0.03	23.84

[1] Rice husk-derived biochar. [2] Rice husk-derived biochar modified by acid oxidation. [3] Rice husk-derived biochar modified by alkali activation. [4] Bamboo powder-derived biochar. [5] Bamboo powder-derived biochar modified by acid oxidation. [6] Bamboo powder-derived biochar modified by alkali activation. [7] Corn stalk-derived biochar. [8] Corn stalk-derived biochar modified by acid oxidation. [9] Corn stalk-derived biochar modified by alkali activation.

3.1.2. Structural Characteristics

The FTIR results for the nine biochars are depicted in Figure 2a. The characteristic peaks at approximately 3391 cm^{-1} corresponded to the stretching vibration of hydroxyl (–OH) groups [29]. The presence of the C=C and C=O aromatic ring stretching vibration absorption peaks near 1610 cm^{-1} indicates the existence of carboxyl (vCOOH) or carbonyl (C=O) oxygen-containing functional groups on the surface of biochar, implying that biochars possess a structure in which aromatic carbon compounds play a central role, which can provide π-electrons [26]. It is noteworthy that the absorption peaks of aromatic ring C=C and C=O functional groups also appeared in the acid-modified biochar at 1715 cm^{-1} and 1530 cm^{-1}, especially RBO, which indicates that the nitro-modified biochar contains more characteristic functional groups. For RB, the band centered at 1445 cm^{-1} was assigned to the CO_3^{2-} characteristic peak of the mineral component carbonates in biochar [26,30]. The peak at 1378 cm^{-1} that appeared in BBO was ascribed to the NO_3^- stretching vibration, indicating that there are some residual nitrates in nitro-activated biochar materials, which belonged to the absorbed nitrates during the activation process, and this phenomenon has also appeared in other similar studies [10]. The bands at 1085 cm^{-1} and 1025 cm^{-1} were associated with the O-Si-O stretching vibration, indicating the presence of SiO_2 in biochar, which was similar to the results of other studies [28]. In addition, a stretching vibration peak of C-H appears in the region between 800 and 600 cm^{-1}, indicating the presence of aromatic compounds and heteroaromatic compounds in biochar [31]. The tensile vibration peak of the C-N-C bond existed near 400 cm^{-1} [15]. The FTIR results suggested that the chemical activation of nitric acid or sodium hydroxide changed the surface functional group composition of the original biochar, and the acid-modified biochar was richer in functional groups that provided the specific adsorption sites for PAHs, which may affect the adsorption and removal performance of biochar for pollutants. Moreover, these abundant surface functional groups provided the biochar with good buffering performance [29].

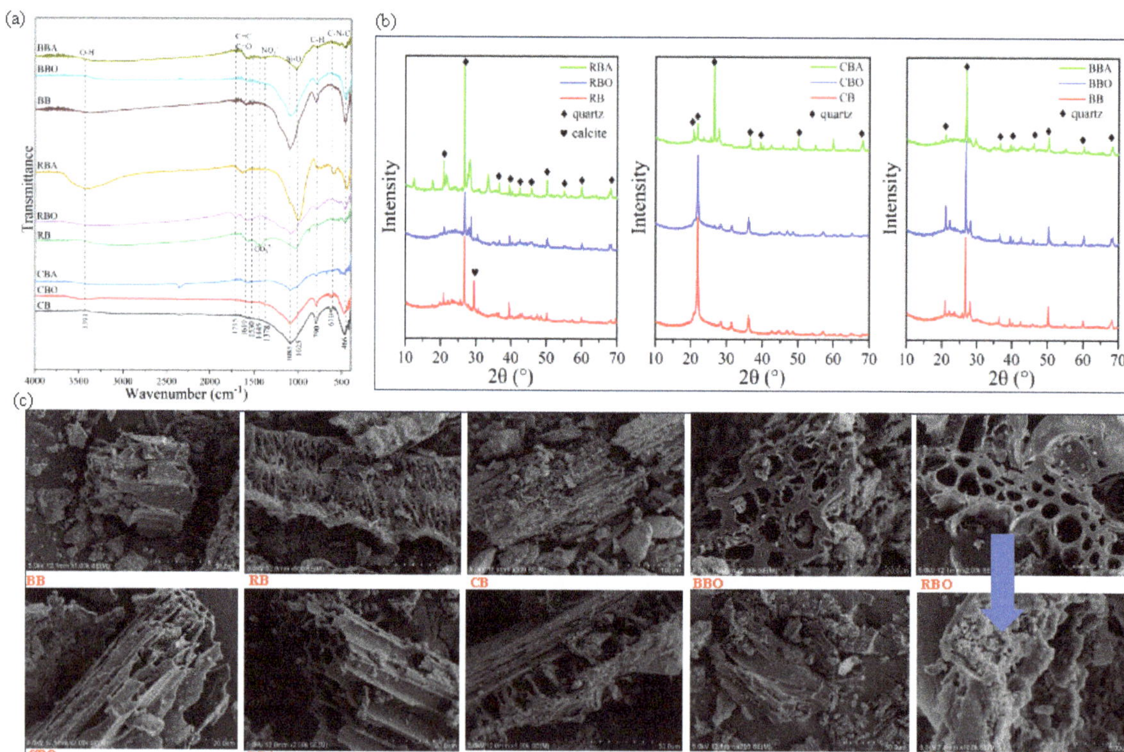

Figure 2. FTIR spectra (**a**), XRD patterns (**b**), and SEM images of biochar (**c**).

As displayed in the XRD patterns of biochar (Figure 2b), a diffraction peak of the nine biochar materials appeared at around 2θ = 23°, which was consistent with the (002) graphene crystal surface [26] and signified that the biochars had a certain crystalline-order degree of carbon materials [30]. Meanwhile, some sharp crystal diffraction characteristic peaks of crystalline substances such as quartz (SiO_2) and calcite ($CaCO_3$) appeared near $2\theta = 20.92°, 21.96°, 26.66°, 29.30°, 36.07°$, and $50.14°$ successively in the XRD patterns of the biochars, which is consistent with previous studies' findings [10,28,30]. However, the presence of alkaline base ions such as $CaCO_3$ is not conducive to adsorption. The $CaCO_3$ content of RB after acid modification was significantly reduced, which was in agreement with the FTIR results. This phenomenon is consistent with previous studies and is mainly due to the fact that protons (H^+) in the acid dissolve the solid-phase carbonate/calcite in the biochar, forming dissolved alkalinity/HCO_3^- and Ca^{2+} [28]. There are a large number of impurities such as calcite ($CaCO_3$) in biochar prepared by direct pyrolysis of raw materials, and acid modification can remove these impurities to a large extent. Therefore, the surface functional groups and the structure of BC were changed significantly by the acid activation process, mainly through the elimination of carbonates [26].

3.1.3. Morphological Characteristics

The SEM results (Figure 2c) indicated that each biochar showed a porous structure, and acid/base modification can significantly alter the morphology and structure of the pristine biochar surface. The surface of CB was seriously damaged, while the surface of other biochars was relatively complete. Especially for RBO, the pore structure was evenly distributed on the surface. These pores are the main structure for biochar to adsorb pollutants, nutrients, and microorganisms. The fine debris on the surface of biochar is the ash produced by the pyrolysis of biomass to produce biochar. This ash blocks the surface pores, and the spatial collapse of biochar with different pore structures after pyrolysis will also cause pore blockage. The ideal carrier for the application of immobilized microorganisms requires that the material should be porous and have a high specific surface area, so as to ensure that the surface pores of the immobilized surface are large enough to maintain the normal physiological metabolism of the bacteria but also to avoid the occurrence of excessive pores resulting in the leakage of degrading bacteria cells. At the same time, the more abundant the pore structure of the biochar materials is, the better its mass transfer permeability, and the metabolites and nutrient substrates in the degradation system can easily enter and exit freely, while the space for the excretion and release of products will not be restricted. The SEM analysis indicated that the pores of RBO were clear and complete, with a more large and dense internal pore structure, which can provide more sufficient growth space for bacteria and can effectively accommodate and immobilize more microbial cells, suggesting that it is a suitable carrier for bacteria immobilization. Based on the above results, it was determined that the activated BC had a better appearance than BC, which may be an important factor in its adsorption capacity.

3.2. *Adsorption Performance of Biochar*

3.2.1. Adsorption Kinetics

The adsorption kinetic curves of biochar are presented in Figure 3. The results indicated that compared with the original biochar, the adsorption properties of the biochar modified by nitric acid or sodium hydroxide were improved to different degrees, and the biochar modified by acid was the most obvious. The adsorption capacity of PHE by different biochars from high to low was RBO, BBO, BBA, RBA, CBO, CBA, BB, RB, and CB. The adsorption capacity of RBO for PHE was the highest (3819 $\mu g \cdot g^{-1}$), which was in accordance with the results of the characteristics discussed above, such as the large specific surface area and abundant oxygen functional groups on the surface. Other studies have also demonstrated that the adsorption performance of biochar after acid treatment can be improved [15]. As shown in Table 2, by comparing the curves and correlation coefficients (R^2) of the PFO and PSO models, it was found that the PSO model can better fit the adsorption process of PHE from different biochars; the R^2 was greater than 90.5%, and the equilibrium absorption capacity obtained by the fitting was relatively close to the experimental value. Therefore, the adsorption of PHE by biochar materials in this study mainly depended on chemical adsorption, indicating that the number of adsorption sites contained on the surface of the biochar was the main factor affecting the adsorption effect. In addition, in order to further determine the rate control steps of PHE adsorption by biochar, the IPD model fitting showed that the correlation coefficient R^2 (0.490–0.814) was relatively low, and all intercept (b) values were non-zero (1037 $\mu g \cdot g^{-1}$–2788 $\mu g \cdot g^{-1}$), which indicated that intra-particle diffusion was not the only factor limiting the adsorption rate during the adsorption process and that other processes such as outer film diffusion also affect the adsorption process.

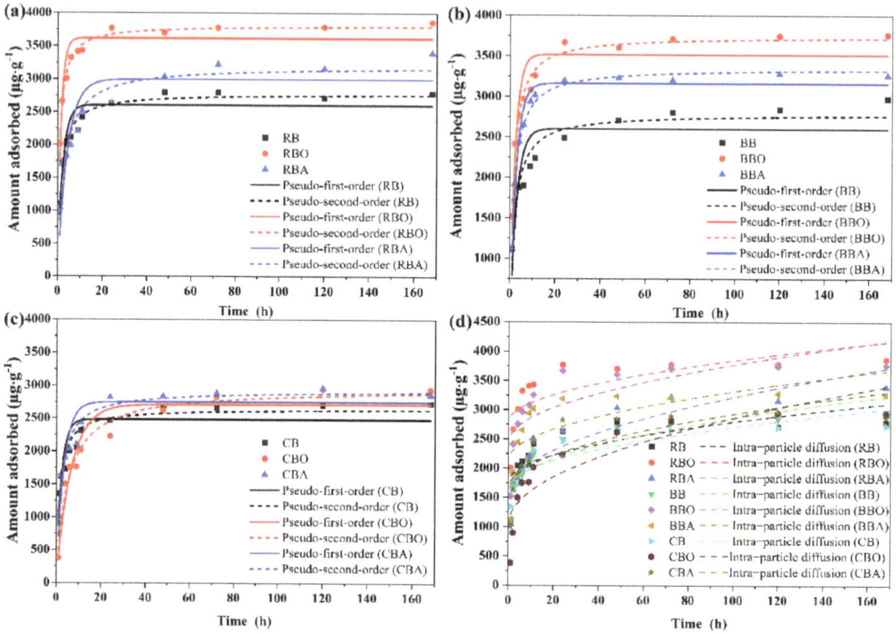

Figure 3. Adsorption kinetics of PHE onto biochars, where (**a–c**) are kinetic model fitting; (**d**) is for IPD model fitting.

Table 2. The fitting parameters for PHE adsorbed by the biochars.

Biochar	PFO			PSO			IPD		
	Q_e ($\mu g \cdot g^{-1}$)	k_1 (h^{-1})	R^2	Q_e ($\mu g \cdot g^{-1}$)	k_2 ($g \cdot \mu g^{-1} \cdot h^{-1}$)	R^2	b ($\mu g \cdot g^{-1}$)	k_3 ($\mu g \cdot g^{-1} \cdot h^{-0.5}$)	R^2
RB	2604	0.438	0.848	2782	2.375×10^{-4}	0.973	1763	103.2	0.607
RBO	3621	0.664	0.862	3819	2.819×10^{-4}	0.988	2788	106.2	0.556
RBA	2998	0.232	0.780	3201	1.137×10^{-4}	0.924	1567	165.6	0.814
BB	2602	0.363	0.723	2796	1.929×10^{-4}	0.919	1597	126.3	0.795
BBO	3521	0.473	0.864	3754	1.915×10^{-4}	0.979	2442	133.4	0.596
BBA	3166	0.397	0.966	3358	1.878×10^{-4}	0.987	2139	118.1	0.490
CB	2481	0.442	0.671	2648	2.582×10^{-4}	0.905	1665	103.9	0.747
CBO	2716	0.150	0.927	2960	6.862×10^{-5}	0.978	1037	182.6	0.776
CBA	2755	0.298	0.837	2932	1.625×10^{-4}	0.959	1654	127.4	0.659

3.2.2. Adsorption Isotherm

The adsorption equilibrium isotherm is helpful in describing how adsorbate molecules are distributed between the liquid and solid phases at equilibrium, which can provide insight into the homogeneity and heterogeneity of the adsorbent surface [27]. Freundlich isotherm models were performed to fit the adsorption data. As shown in Figure 4, with the increase in the initial dose of PHE, the adsorption amount by biochar gradually increased, but the increase range tended to be smaller because with the increase in the PHE equilibrium concentration, the adsorption site on the surface of biochar gradually decreased, and the adsorption amount gradually reached saturation [32]. The PHE adsorption capacity of the nine biochar materials in this study was greatly different, which was consistent with the equilibrium adsorption capacity in the adsorption kinetics experiment. As shown in Table 3, the result indicated a much better fitting of the experimental data by the Freundlich model with R^2 values all greater than 89.4%. The model can accurately describe the adsorption process under experimental conditions, which shows that the adsorption of PHE on biochar mainly occurs on non-uniform surfaces, indicating a complex physicochemical adsorption process [26]. The constant n value obtained by the Freundlich isotherm model was less than 1, indicating that the adsorption process easily occurred in the nine biochar species, and the surface heterogeneity is strong. The n value of modified biochar decreased and the nonlinearity increased, indicating that acid-modified biochar is more prone to PHE adsorption. In the modification process, the carbon in biochar was transformed from an indeterminate form to a transition state, and then from a transition state to an aromatic state, and finally a stable chaotic layer state was formed; thus, the overall structure of the biochar was constantly changing. The adsorption was mainly based on weak interactions such as partition and van der Waals (VDW) forces of the original biochar and was transformed into a common adsorption mechanism mainly based on strong interactions such as electrostatic, chemical, and π–π bonds, supplemented by pore retention. Therefore, when the simple distribution action is no longer the dominant factor of adsorption, the nonlinearity of the adsorption is gradually enhanced.

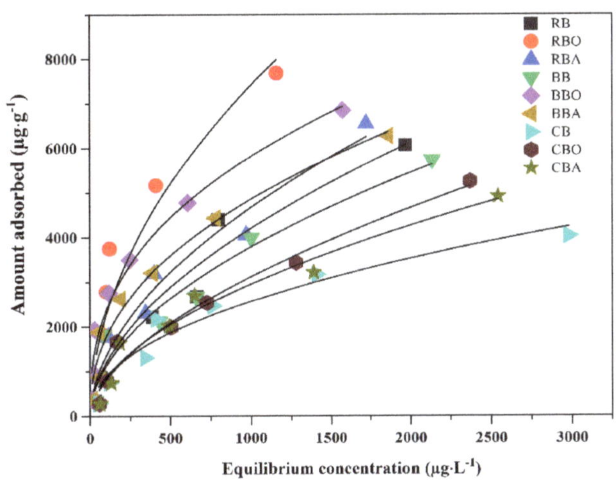

Figure 4. Adsorption isotherms of PHE onto biochar.

Table 3. Isothermal model constants for the adsorption of PHE on biochar.

	RB	RBO	RBA	BB	BBO	BBA	CB	CBO	CBA
k_4	99.79	216.4	115.4	101.3	399.8	241.0	107.4	56.89	76.08
n	0.542	0.512	0.536	0.524	0.388	0.435	0.460	0.580	0.530
R^2	0.937	0.894	0.968	0.951	0.975	0.968	0.946	0.972	0.957

3.2.3. Adsorption Mechanism

The adsorption of biochar included multiple mechanisms [27]. Biochar is highly aromatic and rich in π−π electrons, while PHE contains aromatic π electrons with which the π−π bond can be formed. Thus, PHE could adsorb strongly onto biochar through π−π electron donor–receptor interactions [19]. In this study, the reinforced aromatic structure of the activated biochar facilitated the π−π interactions [16]. In addition, crystal SiO_2

was present in biochar, and according to relevant literature reports, the n−π interaction between the aromatic structure of PHE and the Si-O-Si of biochar could also contribute to the PAH adsorption [10]. Meanwhile, electrostatic adsorption is another relatively strong adsorption mechanism that can be achieved through the variation in the pH value. Nitric acid activation makes the surface of biochar positively charged, which strengthens the electrostatic adsorption with negatively charged microorganisms. As verified by other research, hydrogen bonds are also a pivotal mechanism responsible for the adsorption and immobilization behavior of the PAHs on biochars [29]. The surface O-containing functional groups such as –OH and –COOH of biochar provide crucial sites for the adsorption of PAHs, which could combine into a stable bond [19]. The acid modification method in this study increased the number of functional groups of biochar, promoted the formation of ionic bonds, and enhanced the hydrogen bond and electrostatic adsorption capacity [33]. Additionally, biochar with strong molecular polarity has a high PAH adsorption ability due to the intermolecular dipole–dipole interactions [13,34]. Acid/base modification could increase the surface hydrophobicity of the biochar, exposing more hydrophobic sorption sites and thus achieving an increase in the adsorption capacity for PAHs [17]. Van der Waals (VDW) forces are weak intermolecular forces that can facilitate the adsorption process and play an important role in stabilizing the adsorbed PAH molecules onto the adsorbent [34]. In addition, biochar has an organic phase available for the distribution of PAHs and a developed pore structure, which can inhibit the escape of PAHs after adsorption [19,33]. Previous studies have also found that the adsorption strength of biochar is positively correlated with the surface area. Acid/alkali activation can increase the surface area and porosity of biochar and provide more active adsorption sites for organic pollutants, which is conducive to improving the adsorption capacity of the biochar. Furthermore, it was found that the acid treatment reduced the ash content of the biochar, and related studies reported that the ash could bind to organic matter through specific interactions, which played a negative role in the sorption of biochar [27].

In conclusion, as elucidated in Figure 5, the PHE adsorbed on the surfaces of BC via the π−π/n−π electron donor–acceptor interactions, electrostatic interactions, hydrogen bonds, hydrophobic interactions, and VDW forces, and then it entered into the pores of the biochar through a pore-filling process [32].

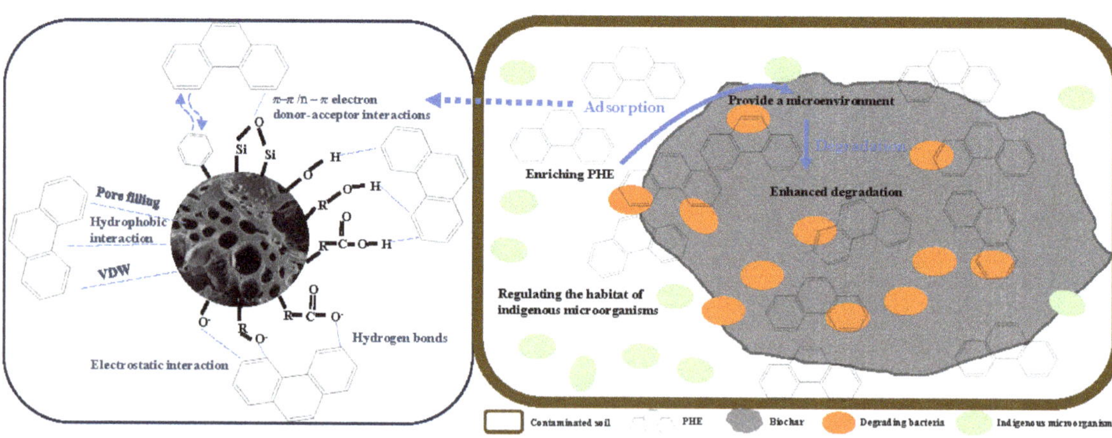

Figure 5. Adsorption and degradation mechanisms of PHE by biochar-immobilized microorganism.

3.3. *Degradation Performance of Biochar-Immobilized Microorganisms*

3.3.1. Degradation Performance

RBO was ultimately selected for loading with bacteria because of its maximum adsorption efficiency. In addition, a relatively larger specific surface area and pore volume were observed in this biochar, which can provide a suitable habitat for microbial colonization and proliferation, indicating a high bacterial adsorption capability [35]. In this study, biochar has a strong ability to fix microorganisms. For ARB, the immobilization rate of biochar to microorganisms was 53%, which was similar to the maximum immobilization rate noted by others [36]. The cell count of strain DG1 attached to biochar was determined by the plate counting method, which was about 0.8×10^8 $CFU \cdot g^{-1}$, and further indicated that the degrading bacteria were effectively fixed on the biochar [37]. The diameter of ERB was about 3.0 mm, and similar studies have reported that the degradation performance of PAHs was better when the particle size was about 3 mm [20]. ERB has a regular shape, full grain, good texture, and strong mechanical strength (97%), indicating that it has good stability and is conducive to maintaining the activity of degrading bacteria for a long time.

The incubation experiment results showed that the degradation rate of PHE by the immobilized bacteria was significantly higher than that by the free bacteria (Figure 6). After 30 days, the degradation rates of PHE by ARB, ERB, and DG1 were 80.15%, 63.92%, and 41.37%, respectively. The degradation process was in good fit ($R^2 > 92.36\%$) with the first-order kinetic model (Table 4). The degradation half-life of free bacteria to PHE was 32.70 days. When ARB or ERB was used, the utilization rate of PHE was improved, and the degradation half-lives were reduced to 10.33 and 16.95 days, respectively. Thus, when the same amount of degrading bacteria DG1 was added, the immobilization technology sped up the start-up rate of the microbial degradation of PHE and enhanced the efficiency of the microbial removal of PHE compared with free bacteria. This may be due to the microenvironment constructed between the immobilized degrading bacteria and the carrier, which can shield the harsh external environment. In addition, biochar with a high adsorbability can adsorb organic molecular PHE and increase the concentration of microorganisms, so that the bacteria can directly use the adsorbed PHE, which causes the reaction to start quickly, thus accelerating the degradation and removal of PHE.

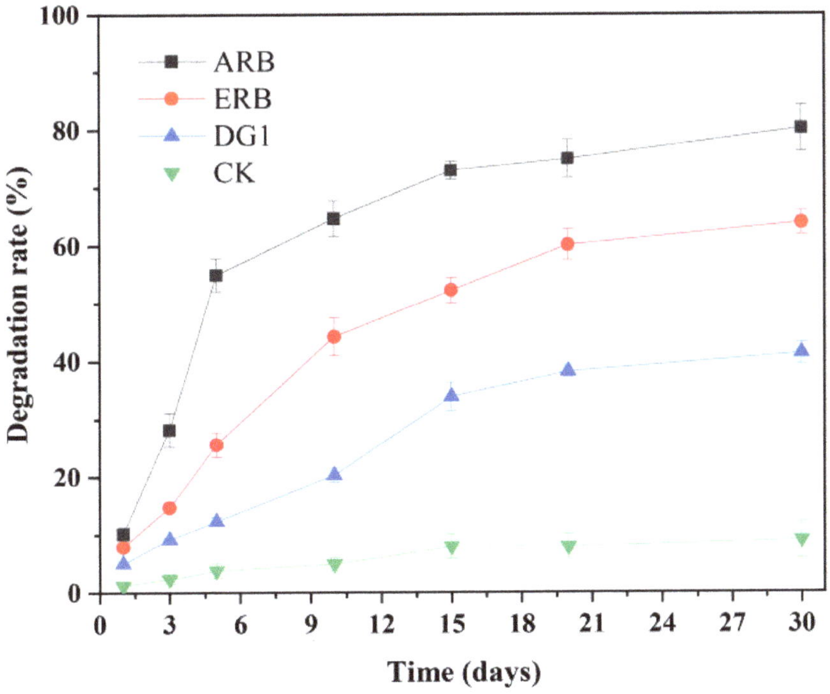

Figure 6. Degradation rates of PHE by different inoculants.

Table 4. First-order kinetic parameters for PHE degradation.

	Regression Equation	Degradation Constant (k, d^{-1})	Half-Life Time ($T_{1/2}$, d)	Regression Coefficient
ARB	$C = 49.21e^{-0.0671t}$	0.0671	10.33	0.9236
ERB	$C = 49.15e^{-0.0409t}$	0.0409	16.95	0.9613
DG1	$C = 49.12e^{-0.0212t}$	0.0212	32.70	0.9660

Moreover, the degradation performance of ARB was higher than that of ERB, mainly because the mass transfer rate of PHE was lower than the biodegradation rate in the gel pellets, and the carbon source of microorganisms could not be replenished in a timely manner, which would inhibit the bio-utilization of PHE. Therefore, the polymer gel network used to provide a microenvironment in the ERB prepared by the embedding method limited the migration and diffusion of nutrients and pollutants in the reaction system, which was more suitable for the remediation of fluid environmental media with a faster mass transfer rate such as wastewater. The ARB in powdered form prepared by the adsorption method can effectively degrade PAHs in both solid degradation systems because of its good mass transfer. For ARB's SEM (Figure 2c), a uniform and dense bacterial layer was found on the biochar surface, which filled the biochar surface and pore entrance, indicating that the degrading strain DG1 was successfully attached and immobilized on the biochar surface [24]. Similar to other studies, the cells of degraded strains clustered and even formed biofilms on the carrier, indicating that bacterial colonization on the biochar followed biofilm formation patterns, such as attachment, adhesion, and proliferation [24,35]. This also confirmed that biochar RBO was a good material for immobilizing strain DG1.

3.3.2. Degradation Mechanism

Biochar, serving as a carrier for immobilized degrading bacteria, plays a crucial role in the process of pollutant remediation. As depicted in Figure 5, biochar provides a protective habitat for the degrading bacteria DG1, shielding them from adverse soil conditions and competition with indigenous microorganisms. This facilitates the adaptability of functional strains to polluted environments, thereby enhancing their capacity to efficiently degrade pollutants in soil. Biochar can also function as a leavening agent, enhancing soil porosity and improving the mass transfer efficiency of water, oxygen, and nutrients within the soil [38]. Moreover, the degradation efficiency of biochar-immobilized microorganisms on pollutants is significantly influenced by the adsorption capacity of the immobilized carrier. Biochar facilitates the adsorption and accumulation of PHE in soil through various physicochemical interactions [34,39]. This process effectively reduces the spatial distance between the degrading bacteria and PHE, mitigates mass transfer limitations, and enhances contact between the degrading microorganisms and pollutants, thereby achieving superior degradation outcomes [40]. Additionally, the adsorption properties of biochar can diminish the bioaccessibility of highly toxic organic contaminants, thus safeguarding the microenvironment of indigenous microorganisms [41]. Furthermore, in the process of degrading pollutants, biochar provides C, N, P, and other nutrient elements for the growth of degrading bacteria DG1 and indigenous microorganisms, further promoting the degradation of PHE by microorganisms [42]. The porous structure of biochar provides a pathway for the diffusion of pollutants and degradation metabolites and provides enough space and oxygen for microorganisms to maintain their normal metabolism. Meanwhile, the enzymes secreted by the fixed degrading bacteria can penetrate into the pores of the biochar and interact with the adsorbed PHE to further degrade it. To sum up, biochar has a complex and diverse interaction with degrading microorganisms, which can affect the adsorption and degradation process of pollutants [43]. The adsorption–degradation process is an important mechanism for the enhanced remediation of pollutants by immobilized bacteria. Giving full play to the synergistic effect of adsorption and degradation with biochar and microorganisms can effectively improve the remediation effect of contaminated soil.

3.4. Implications

Biochars derived from CS, RH, and BP were chemically modified using HNO_3 and NaOH. The modified biochars demonstrated a superior adsorption affinity to the pristine biochars for PAH attributed to the facilitated π–π/n–π interactions, electrostatic interactions, hydrogen bonding, and hydrophobic interactions. Compared with biochar-based material for PHE removal in other studies (Table 5), the modified biochar developed in this study exhibits better adsorption performance, higher cost-effectiveness, and can be popularized and applied. In addition, a modified biochar-immobilized microorganism system with high PHE removal capability was successfully synthesized. This immobilization technique accelerated the start-up rate of PAH biodegradation compared to free bacteria. The polymer gel network-based microenvironment in the combined adsorption–embedding immobilized bacteria restricted the migration and diffusion of nutrients and pollutants in the soil environment, making it particularly suitable for aquatic applications. Meanwhile, the powder form of adsorption-immobilized bacteria, with its superior mass transfer performance, is ideal for soil remediation. Compared with conventional PAH removal methods, the modified biochar-immobilized microorganism method is a more efficient, economical, and environmentally friendly technology. When applied to soil, biochar can have an impact on the soil microenvironment, including the soil physicochemical properties, enzyme activity, etc., as well as the activity of soil microorganisms. However, the introduction of degrading strains may inevitably disturb the indigenous microbial community, and the stability and reusability also need to be further investigated. These factors could potentially limit its sustainable development in practical applications within complex and diverse actual environments. In summary, this study presents an innovative modified biochar-based microbial immobilization technology characterized by its simplicity, ease of operation, and high adsorption efficiency, providing valuable insights for PAH remediation strategies.

Table 5. Removal efficiency of phenanthrene by different biochar-based adsorbents reported in previous studies.

Adsorbent	System	Removal Efficiency	References
Arundo donax-derived biochar-loading copper ions	Constructed wetlands with modified biochar as substrates, the influent phenanthrene was 12.8 ± 1.7 mg·L^{-1}, hydraulic retention time lasted 3 days	94.09%	[44]
Magnetically modified rice husk biochar by a hydrothermal method	Phenanthrene concentration 5–70 mg·L^{-1}, shaken at 150 r·min^{-1}, under 25 °C for 24 h	97.6 mg·g^{-1}/greater than 85%	[32]
Rice straw/wood/bamboo-derived biochar modified by NaOH and washed by HCl	Phenanthrene concentration 0.1–1 mg·L^{-1}, shaken in dark at 130 r·min^{-1} overnight	42.9 mg· g^{-1}	[17]
Willow-derived biochar based on volatile fatty acids	Phenanthrene concentration 0.21 mg·L^{-1}, shaken at 120 r·min^{-1} under ambient temperature in dark	93%	[45]
Phragmites australis-derived biochar produced by ammoniation-hydrothermal method	Phenanthrene concentration 2 mg·L^{-1}, shaken in dark at 25 °C under 125 ± 5 r·min^{-1}	1.97 mg·g^{-1}	[26]

Table 5. *Cont.*

Adsorbent	System	Removal Efficiency	References
Rice husk-derived biochar with nitric acid oxidation	Phenanthrene concentration 2 mg·L^{-1}, shaken in dark at 200 r·min^{-1} under 25 °C	38.2 mg·g^{-1}/95.47%	This study

4. Conclusions

In this study, the HNO_3 oxidation and NaOH activation treatment were used to chemically modify the biochars derived from CS, RH, and BP. The acid- and alkali-modified biochars exhibited enhanced adsorption capacities for PHE ranging from 2865.982 to 3818.99 μg·g^{-1}, compared to the pristine biochars that showed lower capacities of 2729.49 to 2978.058 μg·g^{-1}, attributed to significant improvements in their physicochemical properties such as increased specific surface area, pore volume, and aromaticity. Kinetic parameter calculations indicated that chemisorption was the predominant mechanism governing PHE adsorption onto the biochars. The adsorption mechanisms involved π–π/n–π electron donor–acceptor interactions, electrostatic forces, hydrogen bonding, hydrophobic interactions, VDW, and pore-filling effects. Notably, the RBO, which demonstrated the maximum adsorption efficiency (3818.99 μg·g^{-1}), a relatively larger specific surface area (182.6 m^2·g^{-1}) and pore volume (0.141 m^3·g^{-1}), and abundant surface functional groups, was ultimately selected as a carrier for bacterial immobilization and exhibited effective immobilization performance. The powdered ARB synthesized via adsorption achieved superior degradation efficacy with a PHE removal rate reaching 80.15% within 30 days; conversely, the spherical ERB struggled to attain satisfactory degradation performance in soil media due to restricted mass transfer capabilities. The combined adsorption–degradation process emerged as the primary mechanism facilitating enhanced remediation of pollutants. These findings provide compelling evidence that the nitric acid-oxidized biochar-based microbial immobilization technology represents a promising strategy for the effective removal of PHE contamination. Thus, a green and cost-effective technology for PAH contamination remediation is provided.

Author Contributions: Conceptualization, S.G., G.X. and J.D.; data curation, S.M., G.X. and J.D.; formal analysis, S.G. and F.C.; investigation, S.G., S.M., G.X., A.D., J.D. and F.F.; methodology, S.G. and A.D.; resources, S.G.; supervision, F.F.; validation, A.D. and F.C.; visualization, S.M.; writing—original draft, S.G. and J.D.; writing—review and editing, F.F., funding acquisition, S.G. All authors have read and agreed to the published version of the manuscript.

Funding: This research was funded by the National Natural Science Foundation of China, grant number 52400218; the Shandong Provincial Natural Science Foundation, grant number ZR2023QD063; the Doctoral Research Fund of Shandong Jianzhu University, grant number X24078; the Open Project Program of Engineering Research Center of Groundwater Pollution Control and Remediation, Ministry of Education of China, grant number GW202301; and the Research Fund of Tianjin Key Laboratory of Aquatic Science and Technology, grant number TJKLAST-PT-2022-03.

Data Availability Statement: Data are contained within the article.

Conflicts of Interest: The authors declare no conflicts of interest.

References

1. Mojiri, A.; Zhou, J.L.; Ohashi, A.; Ozaki, N.; Kindaichi, T. Comprehensive review of polycyclic aromatic hydrocarbons in water sources, their effects and treatments. *Sci. Total Environ.* **2019**, *696*, 133971. [CrossRef] [PubMed]
2. Dai, C.; Han, Y.; Duan, Y.; Lai, X.; Fu, R.; Liu, S.; Leong, K.H.; Tu, U.; Zhou, L. Review on the contamination and remediation of polycyclic aromatic hydrocarbons (PAHs) in coastal soil and sediments. *Environ. Res.* **2022**, *205*, 112423. [CrossRef] [PubMed]
3. Idowu, O.; Semple, K.T.; Ramadass, K.; O'Connor, W.; Hansbro, P.; Thavamani, P. Beyond the obvious: Environmental health implications of polar polycyclic aromatic hydrocarbons. *Environ. Int.* **2019**, *123*, 543–557. [CrossRef]
4. Geng, S.; Qin, W.; Cao, W.; Wang, Y.; Ding, A.; Zhu, Y.; Fan, F.; Dou, J. Pilot-scale bioaugmentation of polycyclic aromatic hydrocarbon (PAH)-contaminated soil using an indigenous bacterial consortium in soil-slurry bioreactors. *Chemosphere* **2022**, *287*, 132183. [CrossRef]

5. Wang, H.; Lv, Y.; Bao, J.; Chen, Y.; Zhu, L. Petroleum-contaminated soil bioremediation and microbial community succession induced by application of co-pyrolysis biochar amendment: An investigation of performances and mechanisms. *J. Hazard. Mater.* **2024**, *466*, 133600. [CrossRef] [PubMed]
6. Huang, J.; Tan, X.; Ali, I.; Duan, Z.; Naz, I.; Cao, J.; Ruan, Y.; Wang, Y. More effective application of biochar-based immobilization technology in the environment: Understanding the role of biochar. *Sci. Total Environ.* **2023**, *872*, 162021. [CrossRef] [PubMed]
7. Zhou, X.; Sun, Y.H.; Gao, Y.Z.; Zhou, X.; Sun, Y.; Wang, T.; Tang, L.; Ling, W.; Mosa, A.; Wang, J.; et al. Remediation potential of an immobilized microbial consortium with corn straw as a carrier in polycyclic aromatic hydrocarbons contaminated soil. *J. Hazard. Mater.* **2024**, *469*, 134091. [CrossRef] [PubMed]
8. Dike, C.C.; Hakeem, I.G.; Rani, A.; Surapaneni, A.; Khudur, L.; Shah, K.; Ball, A.S. The co-application of biochar with bioremediation for the removal of petroleum hydrocarbons from contaminated soil. *Sci. Total Environ.* **2022**, *849*, 157753. [CrossRef]
9. Valizadeh, S.; Lee, S.S.; Choi, Y.J.; Baek, K.; Jeon, B.; Andrew Lin, K.; Park, Y. Biochar application strategies for polycyclic aromatic hydrocarbons removal from soils. *Environ. Res.* **2022**, *213*, 113599. [CrossRef] [PubMed]
10. Wang, Y.; Zhang, Y.; Li, S.; Zhong, W.; Wei, W. Enhanced methylene blue adsorption onto activated reed-derived biochar by tannic acid. *J. Mol. Liq.* **2018**, *268*, 658–666. [CrossRef]
11. Peng, P.; Lang, Y.; Wang, X. Adsorption behavior and mechanism of pentachlorophenol on reed biochars: pH effect, pyrolysis temperature, hydrochloric acid treatment and isotherms. *Ecol. Eng.* **2016**, *90*, 225–233. [CrossRef]
12. Nguyen, T.; Sherpa, K.; Bui, X.; Nguyen, V.; Vo, T.; Ho, H.; Chen, C.; Dong, C. Biochar for soil remediation: A comprehensive review of current research on pollutant removal. *Environ. Pollut.* **2023**, *337*, 122571. [CrossRef] [PubMed]
13. Wang, J.; Wang, S. Preparation, modification and environmental application of biochar: A review. *J. Clean. Prod.* **2019**, *227*, 1002–1022. [CrossRef]
14. Wang, H.; Zhao, H.; Zhu, L. Role of Pyrogenic Carbon in Parallel Microbial Reduction of Nitrobenzene in the Liquid and Sorbed Phases. *Environ. Sci. Technol.* **2020**, *54*, 8760–8769. [CrossRef]
15. Zhou, X.; Shi, L.; Moghaddam, T.B.; Chen, M.; Wu, S.; Yuan, X. Adsorption mechanism of polycyclic aromatic hydrocarbons using wood waste-derived biochar. *J. Hazard. Mater.* **2022**, *425*, 128003. [CrossRef] [PubMed]
16. Shin, J.; Kwak, J.; Lee, Y.; Kim, S.; Choi, M.; Bae, S.; Lee, S.; Park, Y.; Chon, K. Competitive adsorption of pharmaceuticals in lake water and wastewater effluent by pristine and NaOH-activated biochars from spent coffee wastes: Contribution of hydrophobic and π-π interactions. *Environ. Pollut.* **2021**, *270*, 116244. [CrossRef] [PubMed]
17. Feng, Z.; Zhu, L. Sorption of phenanthrene to biochar modified by base. *Front. Environ. Sci. Eng.* **2018**, *12*, 1. [CrossRef]
18. Jin, J.; Li, S.; Peng, X.; Liu, W.; Zhang, C.; Yang, Y.; Han, L.; Du, Z.; Sun, K.; Wang, X. HNO_3 modified biochars for uranium (VI) removal from aqueous solution. *Bioresour. Technol.* **2018**, *256*, 247–253. [CrossRef]
19. Qiao, K.; Tian, W.; Bai, J.; Qiao, K.; Tian, W.; Bai, J.; Dong, J.; Zhao, J.; Gong, X.; Liu, S. Preparation of biochar from Enteromorpha prolifera and its use for the removal of polycyclic aromatic hydrocarbons (PAHs) from aqueous solution. *Ecotoxicol. Environ. Saf.* **2018**, *149*, 80–87. [CrossRef] [PubMed]
20. Chen, W.; Zhang, H.; Zhang, M.; Shen, X.; Zhang, X.; Wu, F.; Hu, J.; Wang, B.; Wang, X. Removal of PAHs at high concentrations in a soil washing solution containing TX-100 via simultaneous sorption and biodegradation processes by immobilized degrading bacteria in PVA-SA hydrogel beads. *J. Hazard. Mater.* **2020**, *410*, 124533. [CrossRef] [PubMed]
21. Lv, Y.; Bao, J.; Liu, D.; Gao, X.; Yu, Y.; Zhu, L. Synergistic effects of rice husk biochar and aerobic composting for heavy oil-contaminated soil remediation and microbial community succession evaluation. *J. Hazard. Mater.* **2023**, *448*, 130929. [CrossRef]
22. Xia, M.; Gao, R.; Xu, G.; You, Y.; Li, X.; Dou, J.; Fan, F. Fabrication and investigation of novel monochloroacetic acid fortified, tripolyphosphate-crosslinked chitosan for highly efficient adsorption of uranyl ions from radioactive effluents. *J. Hazard. Mater.* **2022**, *431*, 128461. [CrossRef]
23. Dou, J.; Ding, A.; Liu, X.; Du, Y.; Deng, D.; Wang, J. Anaerobic benzene biodegradation by a pure bacterial culture of Bacillus cereus under nitrate reducing conditions. *J. Environ. Sci.* **2010**, *22*, 709–715. [CrossRef] [PubMed]
24. Sun, T.; Miao, J.; Saleem, M.; Zhang, H.; Yang, Y.; Zhang, Q. Bacterial compatibility and immobilization with biochar improved tebuconazole degradation, soil microbiome composition and functioning. *J. Hazard. Mater.* **2020**, *398*, 122941. [CrossRef] [PubMed]
25. Tulcan, R.; Liu, L.H.; Lu, X.X.; Ge, Z.M.; Rojas, D.Y.F.; Silva, D.M. PAHs contamination in ports: Status, sources and risks. *J. Hazard. Mater.* **2024**, *475*, 134937. [CrossRef] [PubMed]
26. Wang, X.; Guo, Z.; Hu, Z.; Ngo, H.; Liang, S.; Zhang, J. Adsorption of phenanthrene from aqueous solutions by biochar derived from an ammoniation-hydrothermal method. *Sci. Total Environ.* **2020**, *733*, 139267. [CrossRef]
27. Wang, P.; Liu, X.; Yu, B.; Wu, X.; Xu, J.; Dong, F.; Zheng, Y. Characterization of peanut-shell biochar and the mechanisms underlying its sorption for atrazine and nicosulfuron in aqueous solution. *Sci. Total Environ.* **2020**, *702*, 134767. [CrossRef]
28. Mosley, L.M.; Willson, P.; Hamilton, B.; Butler, G.; Seaman, R. The capacity of biochar made from common reeds to neutralise pH and remove dissolved metals in acid drainage. *Environ. Sci. Pollut. Res.* **2015**, *22*, 15113–15122. [CrossRef]
29. Li, N.; Liu, Y.; Du, C.; Wang, Y.; Wang, L.; Li, X. A novel role of various hydrogen bonds in adsorption, desorption and co-adsorption of PPCPs on corn straw-derived biochars. *Sci. Total Environ.* **2023**, *861*, 160623. [CrossRef]
30. Liu, B.; Zhang, Z.; Guan, D.; Wang, B.; Zhou, S.; Chen, T.; Wang, J.; Li, Y.; Gao, B. Qualitative and quantitative analysis for Cd2+ removal mechanisms by biochar composites from co-pyrolysis of corn straw and fly ash. *Chemosphere* **2023**, *330*, 138701. [CrossRef] [PubMed]

31. Del Bubba, M.; Anichini, B.; Bakari, Z.; Bruzzoniti, M.C.; Camisa, R.; Caprini, C.; Checchini, L.; Fibbi, D.; El Ghadraoui, A.; Liguori, F.; et al. Physicochemical properties and sorption capacities of sawdust-based biochars and commercial activated carbons towards ethoxylated alkylphenols and their phenolic metabolites in effluent wastewater from a textile district. *Sci. Total Environ.* **2020**, *708*, 135217. [CrossRef] [PubMed]
32. Guo, W.; Wang, S.; Wang, Y.; Lu, S.; Gao, Y. Sorptive removal of phenanthrene from aqueous solutions using magnetic and non-magnetic rice husk-derived biochars. *R. Soc. Open Sci.* **2018**, *5*, 1723825. [CrossRef]
33. Ahmad, M.; Rajapaksha, A.U.; Lim, J.E.; Zhang, M.; Bolan, N.; Mohan, D.; Vithanage, M.; Lee, S.S.; Ok, Y.S. Biochar as a sorbent for contaminant management in soil and water: A review. *Chemosphere* **2014**, *99*, 19–33. [CrossRef] [PubMed]
34. Li, F.; Chen, J.; Hu, X.; He, F.; Bean, E.; Tsang, D.C.W.; Ok, Y.S.; Gao, B. Applications of carbonaceous adsorbents in the remediation of polycyclic aromatic hydrocarbon-contaminated sediments: A review. *J. Clean. Prod.* **2020**, *255*, 120263. [CrossRef]
35. Xiong, B.; Zhang, Y.; Hou, Y.; Arp, H.P.H.; Reid, B.J.; Cai, C. Enhanced biodegradation of PAHs in historically contaminated soil by M. gilvum inoculated biochar. *Chemosphere* **2017**, *182*, 316–324. [CrossRef]
36. Wahla, A.Q.; Anwar, S.; Mueller, J.A.; Arslan, M.; Iqbal, S. Immobilization of metribuzin degrading bacterial consortium MB3R on biochar enhances bioremediation of potato vegetated soil and restores bacterial community structure. *J. Hazard. Mater.* **2020**, *390*, 121493. [CrossRef] [PubMed]
37. Ma, L.; Hu, T.; Liu, Y.; Liu, J.; Wang, Y.; Wang, P.; Zhou, J.; Chen, M.; Yang, B.; Li, L. Combination of biochar and immobilized bacteria accelerates polyacrylamide biodegradation in soil by both bio-augmentation and bio-stimulation strategies. *J. Hazard. Mater.* **2020**, *405*, 124086. [CrossRef] [PubMed]
38. Li, D.S.; Su, P.J.; Zhang, G.H.; Li, D.; Su, P.; Tang, M.; Zhang, G. Biochar alters the persistence of PAHs in soils by affecting soil physicochemical properties and microbial diversity: A meta-analysis. *Ecotoxicol. Environ. Saf.* **2023**, *266*, 115589. [CrossRef] [PubMed]
39. Ding, Z.; Huang, J.; Chi, J. Contribution of phenanthrene in different binding sites to its biodegradation in biochar-amended soils. *Environ. Pollut.* **2021**, *273*, 116481. [CrossRef]
40. Li, J.; Ou, Y.; Zhang, Y.; Guo, S.; Li, S.; Guo, C.; Dang, Z.; Cao, Z.; Feng, J.; Sun, J. Viability and distribution of bacteria immobilized on Sawdust@silica: The removal mechanism of phenanthrene in soil. *Ecotoxicol. Environ. Saf.* **2020**, *198*, 110649. [CrossRef]
41. Li, X.; Yao, S.; Bian, Y.; Jiang, X.; Song, Y. The combination of biochar and plant roots improves soil bacterial adaptation to PAH stress: Insights from soil enzymes, microbiome, and metabolome. *J. Hazard. Mater.* **2020**, *400*, 123227. [CrossRef] [PubMed]
42. Qi, X.; Gou, J.; Chen, X.; Xiao, S.; Ali, I.; Shang, R.; Wang, D.; Wu, Y.; Han, M.; Luo, X. Application of mixed bacteria-loaded biochar to enhance uranium and cadmium immobilization in a co-contaminated soil. *J. Hazard. Mater.* **2021**, *401*, 123823. [CrossRef]
43. Li, J.; Guo, C.; Liao, C.; Zhang, M.; Liang, X.; Lu, G.; Yang, C.; Dang, Z. A bio-hybrid material for adsorption and degradation of phenanthrene: Bacteria immobilized on sawdust coated with a silica layer. *Rsc. Adv.* **2016**, *6*, 107189–107199. [CrossRef]
44. Shen, X.; Zhang, J.; Xie, H.; Hu, Z.; Liang, S.; Ngo, H.H.; Guo, W.; Chen, X.; Fan, J.; Zhao, C. Intensive removal of PAHs in constructed wetland filled with copper biochar. *Ecotoxicol. Environ. Saf.* **2020**, *205*, 111028. [CrossRef] [PubMed]
45. Bianco, F.; Marcinczyk, M.; Race, M.; Papirio, S.; Esposito, G.; Oleszczuk, P. Low temperature-produced and VFA-coated biochar enhances phenanthrene adsorption and mitigates toxicity in marine sediments. *Sep. Purif. Technol.* **2022**, *296*, 121414. [CrossRef]

processes

MDPI

Article

Spatiotemporal Analysis and Risk Prediction of Water Quality Using Copula Bayesian Networks: A Case in Qilu Lake, China

Xiang Cheng [1,2], Shengrui Wang [1,2,3], Yue Dong [1,2,*], Zhaokui Ni [1,2] and Yan Hong [1,2]

1 Guangdong-Hong Kong Joint Laboratory for Water Security, Beijing Normal University, Zhuhai 519087, China; cx_ecology@163.com (X.C.)
2 Center for Water Research, Advanced Institute of Natural Sciences, Beijing Normal University, Zhuhai 519087, China
3 Yunnan Key Laboratory of Pollution Process and Management of Plateau Lake Watershed, Kunming 650034, China
* Correspondence: dongyue@bnu.edu.cn

Abstract: Lake water pollution under anthropogenic influences exhibits characteristics of high uncertainty, rapid evolution, and complex control challenges, presenting substantial threats to ecological systems and human health. Quantitative risk prediction provides crucial support for water quality deterioration prevention and management. This study employed the Copula Bayesian Network model to conduct a comprehensive risk assessment of water quality in Qilu Lake, China (2010–2020), incorporating inter-indicator correlations and multiple uncertainty sources. Analysis revealed generally "worse" water quality conditions (5.10 ± 1.35) according to established index classifications, with predicted probabilities of reaching "deteriorated" status ranging from 11.80% to 47.90%. Significant spatial and temporal variations in water quality and pollution risk were observed, primarily attributed to intensive agricultural non-point source loading and water resource deficiency. The study established early warning thresholds through key indicator concentration predictions, particularly for the southern region where "deteriorated" risk levels corresponded to specific ranges: TN (3.42–8.43 mg/L), TP (0.07–1.29 mg/L), and COD_{Cr} (27.75–67.19 mg/L). This methodology effectively characterizes lake water quality evolution while enabling risk prediction through key indicator monitoring. The findings provide substantial support for water pollution control strategies, risk management protocols, and regulatory decision-making for lake ecosystem administrators.

Keywords: risk assessment; water pollution; water quality assessment; Bayesian network; Copula function; Qilu Lake

1. Introduction

Lakes constitute essential freshwater ecosystems on Earth, serving critical functions in water purification, climate regulation, and biodiversity maintenance while fundamentally supporting human society and economic development [1]. However, intensifying anthropogenic activities and continuous socioeconomic development have significantly compromised lake ecosystem health [2]. Contemporary lakes face numerous ecological challenges, particularly eutrophication and recurring cyanobacterial blooms, which substantially impact regional ecological security and socioeconomic development [3]. A comprehensive analysis of lake water pollution enables the thorough assessment of ecological environments and associated risks, facilitating the precise identification of water quality issues and the effective implementation of water quality management strategies.

Water quality assessment serves as a fundamental prerequisite for pollution control and water resource management [4]. While single-factor water quality assessment methods offer computational simplicity, they provide limited unilateral results that inadequately represent comprehensive water quality conditions [5]. Consequently, comprehensive evaluation methodologies have gained widespread adoption in surface water quality assessments [6]. Notable applications include the Water Pollution Index (WPI) for Fuxian Lake evaluation [7], Water Quality Index (WQI) for Taihu Lake assessment, Canadian Council of Ministers of the Environment Water Quality Index (CCME-WQI) for Three Gorges Reservoir analysis, and Nemerow Pollution Index (NPI) for urban surface runoff evaluation [8]. This study employs the NPI methodology, which effectively emphasizes severely polluted parameters while accounting for better-quality indicators, thus avoiding arbitrary parameter weighting in computations.

Risk assessment provides an effective framework for environmental management, enabling the identification of critical risk sources [9], outcome prediction [10], and streamlined decision-making processes. Traditional water environmental risk assessment approaches encompass artificial neural networks [11], hierarchical analysis [12], fuzzy integrated evaluation models [13], and gray clustering assessment models [14]. However, these non-probabilistic methods potentially generate ambiguous results, understating the overall risk levels. Probabilistic approaches, particularly Bayesian network (BN) models, offer superior uncertainty quantification through probabilistic causality representation [15]. BN models have demonstrated increasing utility in ecological risk analysis [16], water quality risk assessment [17], and human health risk assessment [18]. These models effectively address high-variability, complex, and uncertain scenarios [19] through directed acyclic graphs (DAGs) and conditional probability tables (CPTs), integrating expert knowledge bases [20] with empirical datasets [21] for comprehensive modeling analyses [22]. To address BN limitations in high-dimensional joint probability distribution construction, Copula functions are introduced, enabling the direct representation of key indicators' joint distributions for multifactor risk analysis.

Qilu Lake, among Yunnan Province's nine plateau lakes, exhibits severe contamination levels. Rapid socioeconomic development and improper water resource utilization within the basin have exacerbated ecosystem degradation, water quality deterioration, and eutrophication [23]. Post-2012, the lake has experienced significant water quality degradation and water volume reduction, with total nitrogen (TN) and chemical oxygen demand (COD_{Cr}) exceeding Class V surface water standards. This study addresses these challenges through the following three primary objectives: (1) the evaluation of water pollution status and spatiotemporal variations using the Nemerow Pollution Index; (2) the implementation of an integrated Copula–Bayesian network model for comprehensive water quality risk assessment, accounting for inter-indicator correlations and multiple uncertainty sources; and (3) the prediction of water quality risk levels and critical indicator concentration ranges through the Bayesian network model to establish effective early warning mechanisms for lake protection and management.

2. Materials and Methods

2.1. Study Area

Qilu Lake is located in Yuxi city, Yunnan Province, on the Yunnan–Guizhou Plateau (24°4′~24°14′ N, 102°33′~102°52′ E; Figure 1). The lake surface elevation is 1796.62 m (highest water level), the surface area is 36.95 km^2, the average depth is 4.0 m, and the residence time is 1.53 years [24]. From 2010 to 2020, the average water level was 1794.65 m, corresponding to an annual storage volume of 110 million m^3 and a water resource volume of 37.24 million m^3. Seasonally, water levels peak between July and September, driven by the summer monsoon's concentrated rainfall, and reach their lowest between March and April due to the dry season's limited precipitation and high evaporation. The region experiences a subtropical southwestern monsoon climate, characterized by a mean annual precipitation of 881 mm and an average temperature of 15.6 °C. As a closed plateau shallow lake, Qilu Lake receives inflow from four tributaries without natural outflow channels. The watershed, supporting a population exceeding 300,000, plays a crucial role in Tonghai County's socioeconomic development through multiple ecosystem services.

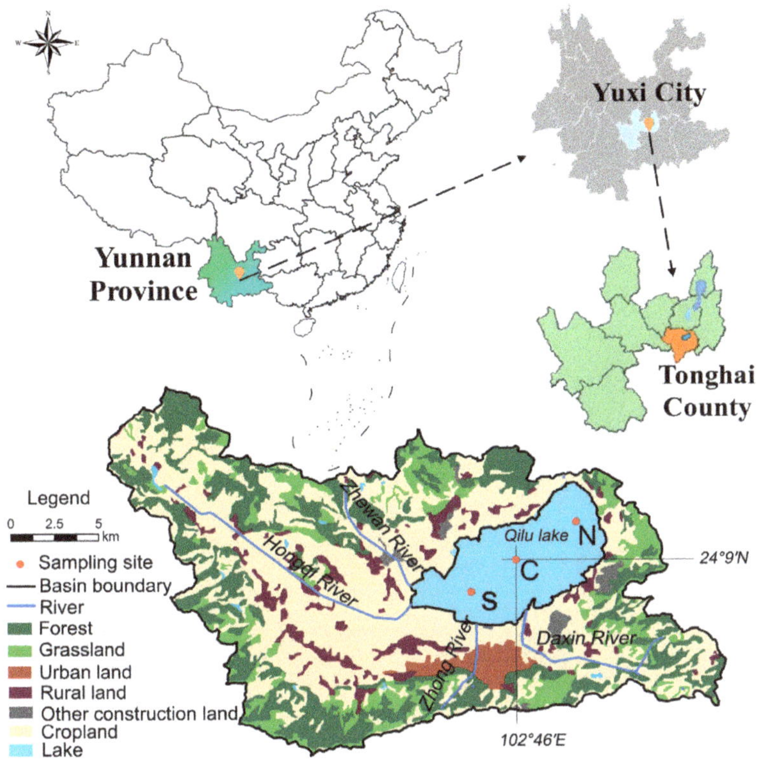

Figure 1. Location of Qilu Lake and distribution of sampling points.

To capture the spatial heterogeneity of water quality in Qilu Lake, three sampling sites were established. Site S is located in the shallow western region (1–3 m depth), adjacent to intensive agricultural land. Site C is situated in the central–eastern mid-depth zone (3–5 m depth), primarily influenced by urban and rural development. Site N lies in the deep eastern region (5–7 m depth), surrounded by forest and grassland buffers. This distribution enables a comprehensive analysis of the interactions between water depth, adjacent land use, and water quality parameters across environmental gradients, contributing to a thorough understanding of lake dynamics.

2.2. Analysis Methods

2.2.1. Water Quality Comprehensive Assessment Method

Water quality data from three routine monitoring stations in Qilu Lake were collected, preserved, and analyzed monthly over an 11-year period (2010–2020). Climatic parameters included precipitation (PP) and air temperature (AT), with observation periods divided seasonally, as follows: spring (March–May), summer (June–August), autumn (September–November), and winter (December–February).

The Nemerow Pollution Index (NPI) was employed for comprehensive water quality evaluation [25]. This method effectively combines average and maximum pollutant values, which are particularly suitable for scenarios where individual contaminant concentrations exceed thresholds while mean values remain compliant. Eight parameters were selected to comprehensively represent surface water quality—physical–chemical indicators (pH and dissolved oxygen), nutrient parameters (total nitrogen, total phosphorus, and ammonia nitrogen), and organic pollution indicators (chemical oxygen demand, permanganate index, and five-day biochemical oxygen demand). These parameters were evaluated against Class

I–V standards (the national surface water quality standards GB3838-2002) [26], establishing six water quality grades (Table 1). The NPI was calculated using the following equations:

$$P_i = \frac{C_i}{S_i}, \tag{1}$$

$$\text{NPI} = \sqrt{\frac{P_{iave}^2 + P_{imax}^2}{2}} \tag{2}$$

where C_i represents the measured concentration of water quality indicators, S_i denotes the standard concentration for each indicator, P_{iave} is the average value of P_i, and P_{imax} is the maximum value of P_i.

Table 1. Grade classification of the water quality condition.

NPI	Feature	Grade
[0–0.8)	The water quality is excellent, and the water environment has not been damaged.	Excellent
[0.8–1.0)	The water quality is good, and the water environment has been slightly damaged.	Good
[1.0–1.8)	The water quality is in general and the water environment has been slightly damaged.	Moderate
[1.8–3.3)	The water quality is poor, and the water environment has been damaged.	Poor
[3.3–6.2)	The water quality is inferior, and the water environment has been dramatically damaged.	Worse
[6.2–+∞)	The water environment has been tremendously deteriorated and has basically lost the use of function.	Deteriorated

2.2.2. Mann–Kendall Test

The Mann–Kendall (M–K) test, a nonparametric statistical method, was implemented to quantify temporal trends in water quality parameters. This distribution-free approach demonstrates a robust performance with outliers. Mutation points were identified at UF and UB curve intersections within ±1.96 confidence intervals. Significant trend changes were indicated by $|UF_i| > 1.96$ [27].

2.2.3. Copula-Based Bayesian Network (CBN)

This study implemented a nonparametric Bayesian network model incorporating Copula functions to conduct a comprehensive, dynamic risk assessment of water quality indicator interactions. The Copula-based Bayesian Network (CBN) model demonstrates superior capabilities in handling nonlinear environmental data compared to conventional regression models, effectively capturing and modeling complex interdependencies among environmental variables. The graphical architecture of CBN facilitates a visual representation of variable relationships, enhancing the interpretability of environmental interactions.

The Bayesian network structure was optimized through an integrated approach combining mutual information and hill-climbing algorithms (MI&HC). This methodology initially establishes the network structure through mutual information analysis, followed by optimization via the hill-climbing algorithm, thereby addressing fundamental limitations of traditional hill-climbing approaches. While mutual information serves as a conventional metric for variable correlation assessment, the computation of joint probability density functions presents significant complexity. The implementation of Copula entropy-based mutual information estimation substantially enhances both computational accuracy and efficiency. This approach leverages the intrinsic relationship between Copula entropy and mutual information, transforming the mutual information estimation challenge into a Copula entropy calculation, thus enabling rapid and precise computation. Detailed operational procedures and mathematical formulations are provided in the Supplementary Materials.

2.3. Data Processing

Data processing and statistical analyses were conducted using multiple specialized software platforms. Primary data management was performed using MS Office 2016, while statistical analyses, including Mann–Kendall trend tests and correlation analyses [28], were executed in R software (version 4.2.0). The Copula-based Bayesian Network (CBN) modeling was implemented through R (version 4.2.0) and UNINET (version 3.1.3), an uncertainty analysis software package developed by Delft University of Technology [29,30]. All statistical analyses maintained a significance threshold of $p < 0.05$.

3. Results

3.1. Comprehensive Characterization of the Water Quality in Qilu Lake

3.1.1. Water Quality Analysis in Qilu Lake

Qilu Lake exhibited severe water quality degradation, with classification inferior to Class V primarily due to elevated COD_{Cr} and TN concentrations. While other parameters remained within Class V standards, TP levels showed an increasing trend in recent years. The annual mean TN concentration was 3.33 mg/L, reaching a maximum of 5.03 mg/L in 2013 (2.5 times the Class V standard). A significant mutation point occurred in 2015, with pre-mutation concentrations increasing from 3.40 mg/L to 4.90 mg/L, followed by a 40.2% post-mutation reduction (Figure 2). Seasonal variations showed maximum concentrations in summer (3.66 mg/L) and minimum in spring (3.07 mg/L), exceeding Class V standards by 3.1 and 2.8 times, respectively. Spatial distribution followed the following pattern: southern (3.83 mg/L) > northern (3.14 mg/L) > central region (3.03 mg/L).

COD_{Cr} exhibited an annual mean concentration of 44.33 mg/L, with peak values (56.08 mg/L) in 2013 exceeding Class V standards by 1.4 times. A mutation point occurred in 2016, followed by a 31.5% decrease, though concentrations increased significantly from 32.6 mg/L to 49.42 mg/L between 2017 and 2020 (Figure 2). Seasonal variations showed elevated levels in spring and summer, with summer maxima exceeding Class V standards by 1.9 times. Regional variations were minimal, though all areas exceeded Class V standards by up to 1.4 times.

NH_3-N concentrations maintained Class II-III levels, averaging 0.45 mg/L with peak concentrations of 0.94 mg/L in 2014 (Figure 2). Seasonal patterns showed the following pattern: autumn (0.53 mg/L) > winter (0.48 mg/L) > summer (0.41 mg/L) > spring (0.40 mg/L). Spatially, concentrations varied, as follows: southern (0.51 mg/L) > northern (0.46 mg/L) > central region (0.40 mg/L); the southern region met Class III standards and others achieved Class II standards.

TP concentrations, while below Class V thresholds, showed an increasing trend, averaging 0.08 mg/L. Concentrations decreased from 0.11 mg/L to 0.05 mg/L (2011–2015) before increasing to 0.12 mg/L by 2020 (Figure 2), reaching Class V classification in 2019–2020. Seasonal variations were minimal (approximately 0.09 mg/L; Class III level), while spatial distribution showed the following pattern: southern (0.10 mg/L) > northern (0.08 mg/L) > central region (0.07 mg/L).

3.1.2. Comprehensive Evaluation of Water Pollution in Qilu Lake

The Nemerow Pollution Index (NPI) of Qilu Lake demonstrated a fluctuating pattern from 2010 to 2020, with maximum (7.79) and minimum (3.33) values in 2013 and 2017, respectively (Figure 3). A significant mutation point occurred in 2015, marking distinct pre- and post-change periods. Pre-mutation water quality deteriorated from "worse" to "deteriorated", primarily due to reduced precipitation, accelerated agricultural development, and increased endogenous pollution. Post-mutation NPI decreased from 7.31 to 4.49, which was attributed to increased precipitation and engineering controls, though water quality began deteriorating again in 2018 due to declining precipitation, inadequate project management, and high agricultural pollution loads.

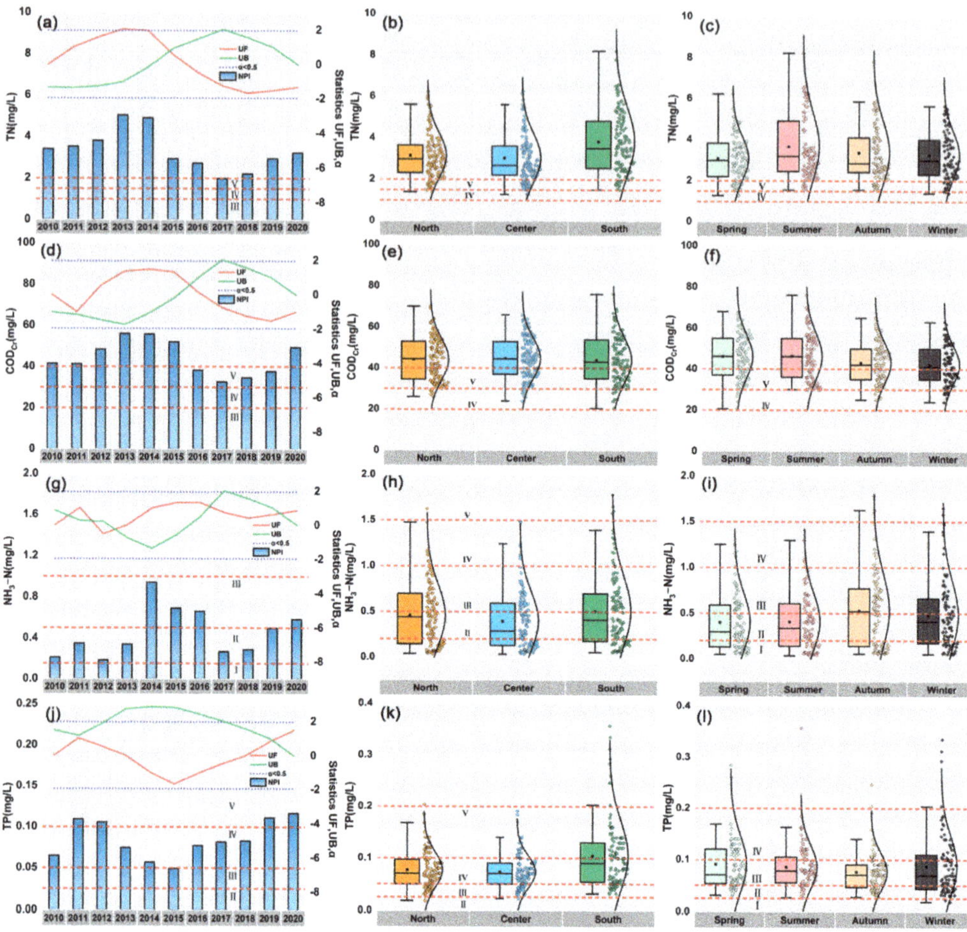

Figure 2. Seasonal and spatial variation characteristics of water quality in Qilu Lake. (**a**,**d**,**g**,**j**) Interannual variations; (**b**,**e**,**h**,**k**) spatial distribution; (**c**,**f**,**i**,**l**) seasonal patterns.

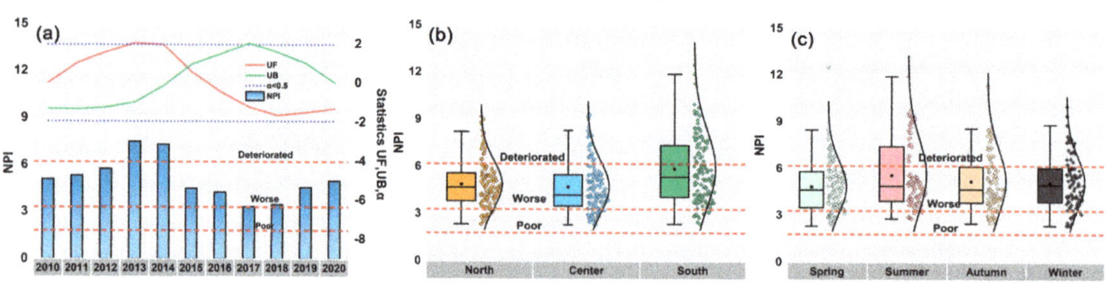

Figure 3. Temporal (**a**,**c**) and spatial (**b**) changes in the NPI value in Qilu Lake.

Seasonal NPI values averaged 4.82, 5.58, 5.16, and 5.09 for spring, summer, autumn, and winter, respectively. Spatial variation showed annual averages of 4.88, 4.71, and 5.89 for northern, central, and southern regions, respectively, with superior water quality in the central region. These variations primarily resulted from uneven seasonal precipitation and regional development patterns, with particularly severe agricultural non-point source pollution in the southern region's primary vegetable cultivation areas.

3.2. Water Quality Risk Prediction in Qilu Lake

The CBN model was employed to conduct prospective risk analysis for predicting water quality risks across various spatial and temporal scales. Results indicated severe water quality issues in Qilu Lake, predominantly characterized by "worse" grade risks (occurrence probability: 45.3–76.5%). The "deteriorated" grade, representing the most critical risk condition, exhibited occurrence probabilities ranging from 11.80% to 47.90% (Figure 4). Temporal analysis revealed significant seasonal variations in water quality degradation risk. Summer demonstrated the highest risk occurrence (mean: 29.3%; range: 17.4–47.9%), while spring exhibited the lowest (mean: 16.7%; range: 11.8–25.7%). The peak summer risk (47.9%) exceeded the spring maximum by 22.2%. Autumn and winter displayed comparable risk profiles. Spatial analysis indicated substantially elevated risks in the southern region (mean: 38.05%; range: 25.7–47.9%) compared to the central region, which exhibited the lowest risk levels (mean: 14.83%; range: 12.6–17.4%). The maximum risk probability in the southern region (47.9%) exceeded the central region's peak by 30.5%, demonstrating significant spatial heterogeneity primarily attributed to seasonal and regional water quality variations.

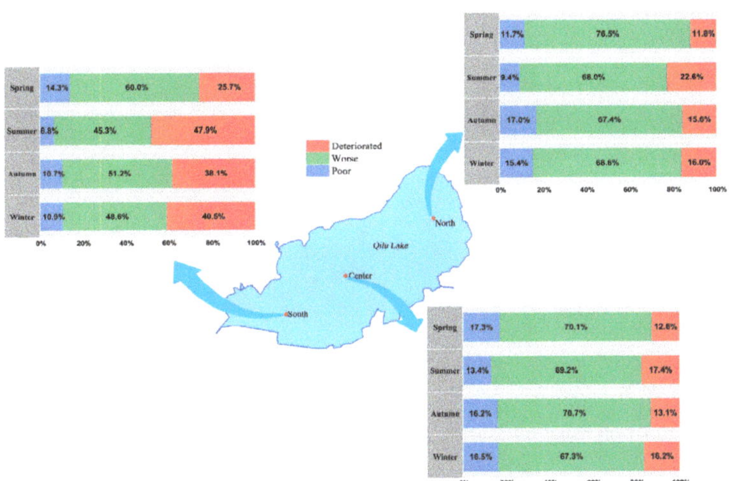

Figure 4. Probability distribution of the water quality risk grade in different seasons for the three locations of Qilu Lake.

Reverse risk analysis utilizing the CBN model was implemented to estimate probability density distributions of specific water quality indicators. Analysis revealed that increasing water quality risk levels corresponded to expanded probability density distribution ranges and elevated mean values. Particular attention was directed toward the "deteriorated" grade risk due to its severe implications. Under "deteriorated" risk conditions in the southern region, water quality parameters exhibited the following ranges: TN (3.42–8.43 mg/L), TP (0.04–0.31 mg/L), NH_3-N (0.07–1.29 mg/L), COD_{Cr} (27.75–67.19 mg/L), COD_{Mn} (7.49–20.16 mg/L), and BOD_5 (4.05–13.04 mg/L) (Figure 5). pH and DO maintained acceptable ranges within their probability densities.

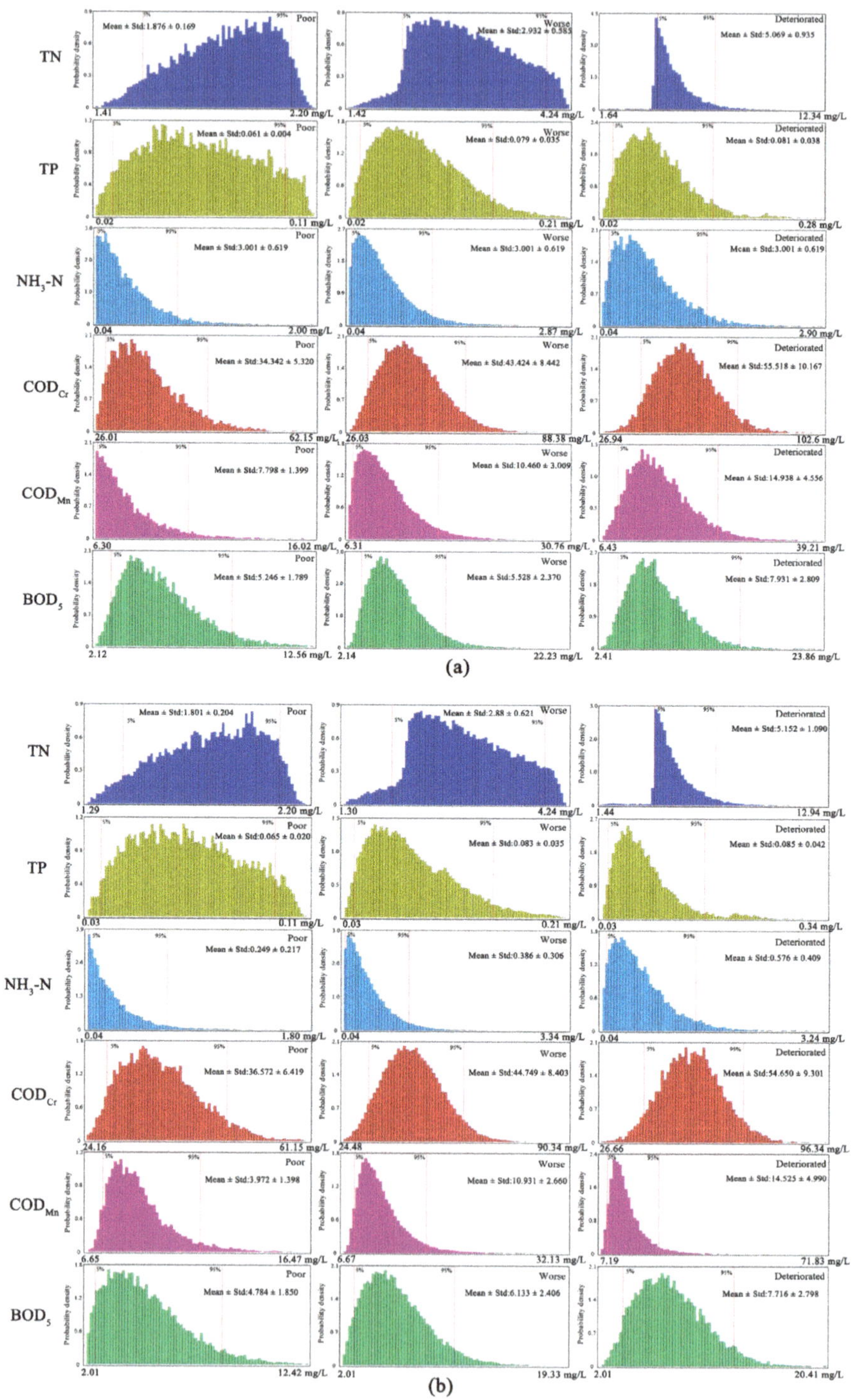

Figure 5. *Cont.*

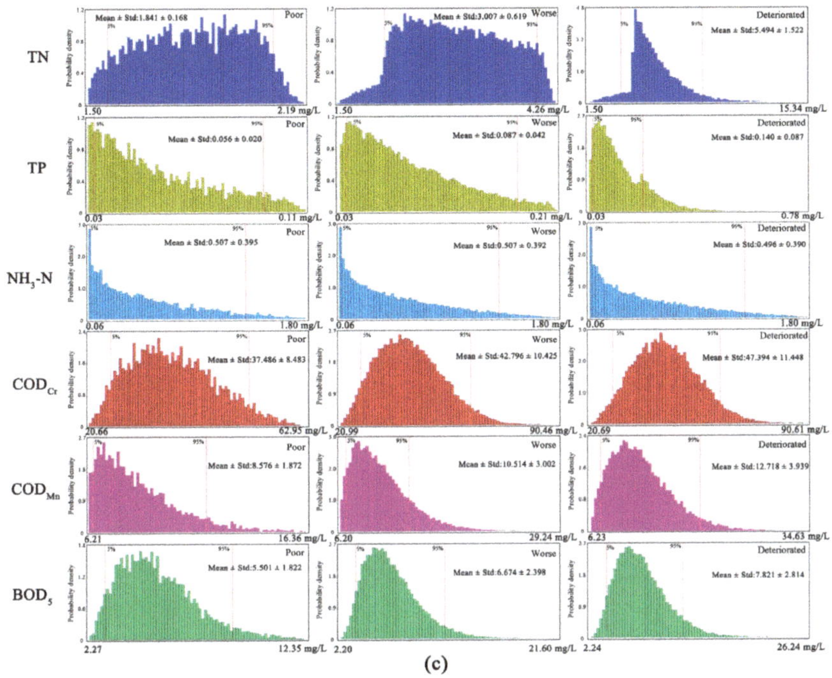

Figure 5. The concentration ranges of TN, TP, NH_3-N, COD_{Cr}, COD_{Mn}, and BOD_5 under different water quality risk grades in the north (**a**), center (**b**), and south (**c**) of Lake Qilu.

The central and northern regions demonstrated more constrained probability density distributions with lower mean values compared to the southern region. Under "deteriorated" risk conditions, the central region exhibited concentrations of TN (4.17–4.26 mg/L), TP (0.04–0.17 mg/L), NH_3-N (0.11–1.38 mg/L), COD_{Cr} (39.79–70.54 mg/L), COD_{Mn} (9.16–23.93 mg/L), and BOD_5 (3.78–12.85 mg/L). The northern region showed similar patterns, as follows: TN (4.19–6.91 mg/L), TP (0.03–0.15 mg/L), NH_3-N (0.11–1.38 mg/L), COD_{Cr} (40.35–73.11 mg/L), COD_{Mn} (8.82–23.46 mg/L), and BOD_5 (4.14–13.07 mg/L). System complexity correlates positively with uncertainty introduction [31], wherein larger standard deviations indicate enhanced risk indicator volatility. The southern region consistently demonstrated higher standard deviations across water quality indicators compared to other regions. These probability density distributions under varying risk scenarios provide essential early warning indicators, enabling lake managers to maintain water quality parameters within acceptable ranges, thereby mitigating potential water quality deterioration risks.

4. Discussion

4.1. Driving Factors of Water Quality Changes in Qilu Lake

The dynamics of water quality indicators effectively reflect variations in lake ecosystem conditions, which are primarily influenced by natural factors and anthropogenic activities [32]. Pearson's correlation analysis revealed significant negative correlations between water volume (WV) and key parameters including NPI, TN, and COD_{Cr} (Figure 6). The Qilu Lake Basin, lacking transit rivers, depends entirely on precipitation for water resources, resulting in constrained water availability and temporal distribution irregularities. Water volume in Qilu Lake showed considerable fluctuations, with a notable minimum in 2013 (approximately 0.5×10^8 m^3) and reaching its peak in 2017 (1.6×10^8 m^3). This variation pattern primarily reflected the precipitation-dependent nature of the lake's water resources, characterized by distinct wet and dry periods throughout the decade. The water quality parameters demonstrated synchronized responses to water volume changes, albeit with varying intensities. COD_{Cr}, TN, and NPI exhibited similar temporal patterns, reaching their maximum values in 2013–2014 (COD_{Cr} = 56 mg/L, TN = 5 mg/L, NPI = 7.5), which was coincident with minimum water volume, followed by a gradual decline during subsequent years. TP showed a different pattern, with an initial peak in 2012 (0.10 mg/L), followed by a significant decrease in 2014 (0.05 mg/L), and a subsequent increasing trend through 2020. NH_3-N maintained relatively stable concentrations with minor fluctuations throughout the study period. Conversely, the improved water quality observed in 2017 coincided with maximum water volume, illustrating the dilution effect during wet periods. This pattern supports the significant negative correlations revealed by Pearson's analysis between water volume and key water quality parameters (NPI, TN, and COD_{Cr}). Furthermore, the basin's water resource development rate exceeds 90%, surpassing 150% during drought periods. Tonghai County, a major vegetable production area, demonstrates continuous increases in cultivation area and yield. Vegetable cultivation demands substantially higher water consumption (16.93 m^3/hm^2) compared to grain crops (approximately seven times greater). During drought periods, reduced precipitation coupled with extensive agricultural irrigation exacerbates water scarcity, accelerating quality deterioration. Conversely, during wet years, reduced human irrigation using lake water led to improved water quality due to greater water availability.

Pollutant discharge significantly influences lake water quality [33,34], with pollution loads identified as primary drivers of quality degradation [35]. Agricultural non-point source pollution emerges as a critical contributor to surface water eutrophication [36]; this is particularly evident in the Qilu Lake watershed, where agricultural activities constitute the predominant pollution source. The watershed is characterized by extensive crop cultivation areas, intensive rotation practices, elevated multiple cropping indices, and substantial chemical fertilizer applications. Chemical fertilizer application intensity in the Qilu Lake Basin reached 516.69 kg/hm^2 in 2017, significantly exceeding the national ecological township standard of 250 kg/hm^2, with persistently high application rates maintained over extended periods. An analysis of agricultural fertilizer runoff revealed significant contributions to total pollutant loads—COD_{Cr} (12.43%), NH_3-N (57.01%), TN (43.05%), and TP (73.98%)—demonstrating the intrinsic relationship between agricultural intensification and lake water quality deterioration. This presents a fundamental conflict between intensive agricultural practices dependent on excessive fertilization and the lake's limited environmental carrying capacity, necessitating urgent modifications to current agricultural methods, particularly regarding fertilization practices and irrigation systems.

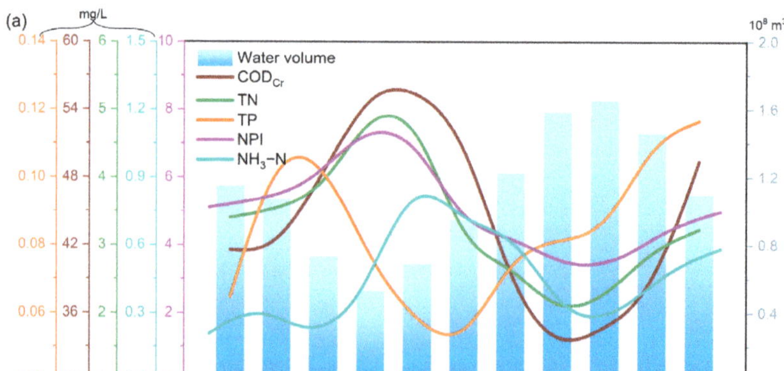

Figure 6. *Cont.*

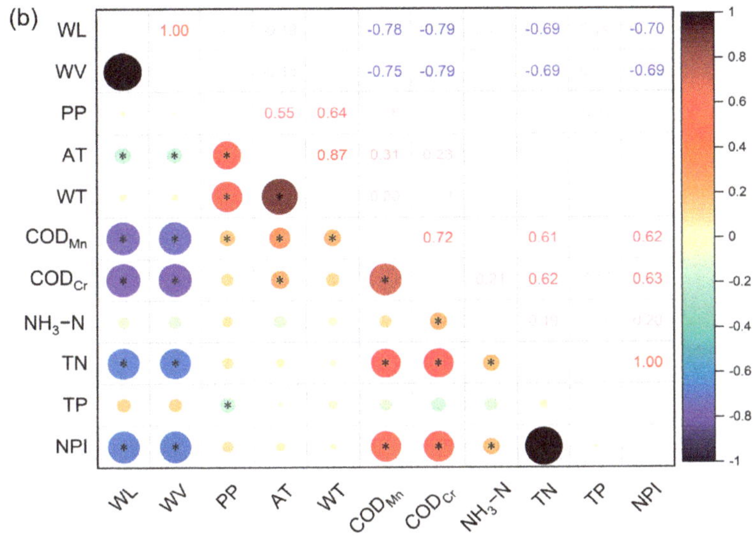

Figure 6. Variation in water volume and water quality index in Qilu Lake from 2010 to 2020 (**a**). Pearson's correlation coefficient between water quality parameters and water volume (**b**).

Lakes, reservoirs, and rivers typically exhibit pronounced seasonal variations in hydrology and water quality [37]. Lake ecosystems demonstrate particular sensitivity to climate change, establishing strong correlations with nutrient loading patterns [38]. Climate change manifests primarily through alterations in precipitation and temperature [39], with rainfall patterns significantly influencing surface and subsurface runoff volumes and subsequent non-point source pollution loads [40]. Our correlation analysis provides robust evidence for the influence of meteorological variables on water quality parameters. The correlation matrix reveals a significant positive correlation between air temperature and water temperature (r = 0.87), demonstrating strong thermal coupling between atmospheric and aquatic environments. Furthermore, air temperature exhibits significant positive correlations with key water quality indicators, including COD_{Mn} (r = 0.31) and COD_{Cr} (r = 0.23), while water temperature shows positive correlations with COD_{Mn} (r = 0.43) and COD_{Cr} (r = 0.19). These findings suggest that water quality deteriorates during warmer seasons. Precipitation patterns show complex relationships with water quality parameters. The correlation analysis indicates that rainfall has positive correlations with COD_{Mn} (r = 0.18) and total nitrogen (r = 0.094)—albeit modest—supporting the hypothesis that increased precipitation may facilitate nutrient loading through surface runoff. The results indicate that water quality reaches its lowest levels and risks peak during summer, characterized by high precipitation and elevated temperatures, while spring demonstrates contrasting conditions, which is consistent with previous research findings [41,42]. The southern region of Qilu Lake, serving as the primary collection area for vegetable cultivation and rural non-point source pollutants, experiences significant water quality challenges. Summer rainfall generates surface runoff, transporting concentrated pollutants into the lake. Direct discharge of agricultural wastewater and domestic sewage through pumping stations, combined with multiple rivers inputs, contributes to sustained poor water quality in the southern region. While external pollution loads can be managed through anthropogenic interventions, persistent eutrophication often results from endogenous pollution in lake sediments [43]. Studies of Dianchi Lake demonstrated high endogenous nitrogen and phosphorus load contributions (72–77%) [44], while Lake Tai research highlighted the role of wind-induced sediment resuspension in enhancing eutrophication through nutrient release [45]. Systematic analysis of sediment pollution in Qilu Lake reveals significant internal loading dynamics, with sediment release contributing 17.64% and 2% of total nitrogen (TN) and phosphorus (TP) loads, respectively. Surface sediment TN concentrations range from 1992 to 7019 mg/kg (mean 3597 mg/kg), with elevated levels (4000–5000 mg/kg) in southern and eastern areas. TP concentrations vary between 981 and 4926 mg/kg (mean 2166 mg/kg), occasionally exceeding TN values. Both nutrients peak at the Hongqi River entrance. Organic matter content (51–143 g/kg; mean 94 g/kg) exceeds national first-class soil fertility standards. Annual sediment release in 2017 reached 339.07 t TN and 3.74 t TP. Effective eutrophication management necessitates addressing internal pollution sources through permanent burial or excavation.

4.2. Implications for Qilu Lake Management Based on Water Quality Risk Prediction

The effective protection of Qilu Lake requires comprehensive water resource management, integrating rational resource allocation, supply–demand optimization, and advanced water conservation and recycling strategies. The hydrochemical dynamics of Qilu Lake, including pollutant dilution and concentration patterns, are intrinsically linked to land use practices and water resource availability within the basin. During wet periods, agricultural runoff and increased surface water transport contribute to elevated concentrations of total nitrogen (TN) and total phosphorus (TP) in the lake, exacerbating eutrophication risks. In contrast, dry periods promote the concentration and sedimentation of pollutants, intensifying chemical oxygen demand (COD_{Cr}) levels and altering nutrient transformation processes. These dynamics underscore the critical need for targeted interventions to mitigate the impacts of basin activities on the lake's water quality.

Agricultural enhancement strategies within the basin play a pivotal role in addressing these challenges, emphasizing the adoption of high-efficiency irrigation technologies and integrated water–fertilizer management systems. Transitioning from conventional flood irrigation to advanced methods such as sprinkler, drip, and micro-irrigation systems directly reduces agricultural water wastage and minimizes nutrient leaching into the lake, thereby lowering TN and TP inputs. Automated operations and precision irrigation protocols further optimize water utilization, ensuring sustainable agricultural productivity while mitigating non-point source pollution. Critical infrastructure improvements, such as the interception of low-pollution surface runoff during rainfall and the prevention of direct sewage discharge, are essential for reducing COD_{Cr} concentrations in the lake. Additionally, water recycling systems for irrigation, combined with subsequent treatment at reclamation facilities, significantly reduce pollutant loads entering Qilu Lake. Enhancing natural purification processes through sedimentation, multi-pond systems, and aquatic vegetation further supports the absorption and degradation of TN and TP, contributing to improved water quality.

The governance framework for Qilu Lake incorporates "source control, emission reduction, pollution interception, and remediation" strategies, explicitly addressing spatial and temporal variations in pollutant distribution and their measurable impacts on lake water quality. In the southern region, where agricultural activities dominate, source control measures focus on reducing chemical fertilizer application and improving nutrient utilization efficiency to mitigate TN and TP inputs, which are strongly correlated with the lake's Nutrient Pollution Index (NPI) [46]. Systematic agricultural transformation, including the replacement of low-efficiency crops and the optimization of fertilizer application protocols, ensures the dual objectives of maintaining crop yields and reducing eutrophication risks [47]. Farmland water management infrastructure, such as comprehensive collection and drainage systems, prevents pollutant leaching during precipitation events, directly reducing TN, TP, and COD_{Cr} levels in the lake. Pollutant interception measures follow a hierarchical approach aligned with contaminant migration pathways, ensuring maximum retention and purification before pollutants reach the lake. Near-source interception systems, such as sedimentation-promoting purification systems within farmland drainage networks, reduce TN and TP loads at their origin. Transportation control through ecological ditch networks further minimizes pollutant transport, while terminal purification via integrated ecological ponds, wetlands, and riparian zones enhances pollutant retention and degradation. This multi-stage system significantly reduces TN and TP concentrations in the lake, mitigating eutrophication risks and improving overall water quality.

The effective management of Qilu Lake necessitates prioritizing the prevention and control of key pollutants, particularly TN, TP, and COD_{Cr}, while implementing early warning systems based on concentration ranges across different water quality risk levels. This approach is especially critical for mitigating water quality risks in vulnerable areas, notably the southern region during summer months, and preventing large-scale pollution events such as algal blooms. Research indicates that recurring cyanobacterial blooms maintain elevated nutrient concentrations [48], enhancing heterotrophic bacterial activity [49] and accelerating nutrient transformation between organic and inorganic states [49,50], thus perpetuating bloom conditions through efficient nutrient cycling. This process establishes a self-reinforcing cycle where increased pollution accelerates nutrient cycling rates, exacerbating eutrophication through continuous nutrient availability. Water quality deterioration typically results from multiple interacting risk factors, with simultaneous exceedance of multiple water quality parameters presenting more significant challenges than single-parameter violations. Spatial analysis reveals strong positive correlations between the NPI and both TN and TP concentrations in the southern region, with diminishing correlations in northern and central regions, reflecting spatial heterogeneity in water quality (Figure 7). Conditional Bayesian Network (CBN) analysis demonstrates weak correlation between NPI and COD_{Cr}, attributed to limited data sensitivity, which is corroborated by minimal variations in COD_{Cr} probability density distributions across risk levels (Figure 6). However, COD_{Cr} exceedance of Class V standards has intensified since 2019, warranting its inclusion alongside TN and TP as critical risk indicators for comprehensive water quality management.

4.3. Uncertainty Analysis

This investigation encompasses two primary sources of uncertainty—structural uncertainty inherent in Bayesian Network (BN) model construction and uncertainty associated with comprehensive water quality metric implementation. The structural uncertainty in the BN framework stems from its construction methodology, which relies on mutual information and mountain climbing algorithms rather than expert-derived field knowledge, potentially overlooking mechanistic relationships between water quality parameters. To validate the model's risk assessment outcomes and ensure result credibility, we conducted 10,000 Monte Carlo simulations. The NPI offers advantages through its computational simplicity, capacity to integrate diverse quantitative parameters, and streamlined state assessment capabilities. However, the absence of mechanistic underpinnings introduces challenges in maintaining assessment result stability. Notable limitations include the exclusion of chlorophyll and transparency measurements from the NPI framework due to the absence of corresponding national standards, though incorporation of these parameters could potentially enhance the study's analytical robustness. These limitations suggest opportunities for methodological refinement in future research.

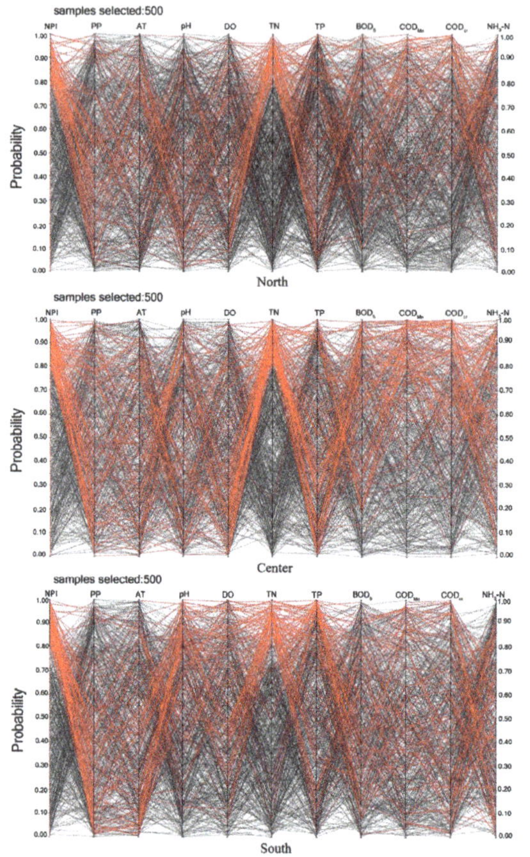

Figure 7. The percentile cobweb plots with high probability values (0.8–1.0, the red lines) of NPI south, center, and north of Qilu Lake. When the NPI has high (or low) probability values, it can be recognized as the key indication if certain other indicators also have high (or low) values.

5. Conclusions

Based on the comprehensive water quality index, M–K test, and the Copula Bayesian Network, this study achieved quantitative water quality pollution identification and risk prediction in Qilu Lake. The results demonstrated that the water quality in Qilu Lake was generally "worse" (NPI, 5.10 ± 1.35), and the continuous deterioration of the water quality led to a surge in risk. The risk probability of reaching the "deteriorated" grade was between 11.80% and 47.90%. The water quality of the lake varied greatly with space, and was observed to be better in the north and center than the south. The southern risk was as high as 38.05% ("deteriorated" grade). Moreover, the water quality presented distinct seasonal variation, with the highest NPI values (5.58) and pollution risk (29.3%, "deteriorated" grade) in summer, followed by autumn (NPI, 5.16), winter (NPI, 5.09), and spring (NPI, 4.82). The water quality and pollution risk were affected closely by high agricultural non-point source load and insufficient water resources. Furthermore, the concentrations of key indicators of different risk levels were predicted as early warning signals. Taking the southern Qilu Lake as an example, when the "deteriorated" level risk occurred, TN (3.42 to 8.43 mg/L), TP (0.07 to 1.29 mg/L), and COD_{cr} (27.75 to 67.19 mg/L) were quantified as early warning intervals for risk control. Based on qualitative cause–effect analysis from anthropogenic influence to water quality risk, the protection of Qilu Lake should reinforce the water resource management and optimize planting mode.

Overall, this study demonstrates a more thorough understanding of the characterization of water quality and pollution risk, which could provide help in making decision from the perspective of quantitative risk early warning indicators and associated anthropogenic influences.

Supplementary Materials: The following supporting information can be downloaded at: https://www.mdpi.com/article/10.3390/pr12122922/s1, Figure S1: Copula-based Bayesian network (CBN) structures for water quality risk assessment in Lake Qilu at the south (a), the center (b), and the north (c) of Qilu Lake. Figure S2: The flowchart of the CBN.; Table S1: Interannual variation in water quality parameters BOD_5, COD_{Mn}, pH, and DO. Table S2: Variation in water quality parameters BOD_5, COD_{Mn}, pH, and DO in different regions. Table S3: Seasonal changes in water quality parameters BOD_5, COD_{Mn}, pH, and DO. Table S4: The best-fitted distribution of the external and internal risk indicators. References cited in Supplementary Materials [51–54].

Author Contributions: Conceptualization, methodology, investigation, and writing—original draft: X.C.; validation, resources, and supervision: S.W.; data curation and writing—review and editing: Y.D.; data curation and supervision: Z.N.; data curation: Y.H. All authors have read and agreed to the published version of the manuscript.

Funding: This research was funded by the Zhuhai Basic and Applied Basic Research Foundation (No. 2320004002582); the Yunnan Key Laboratory of Pollution Process and Management of Plateau Lake-Watershed, grant number 2020-02-2-W2, 2020-124A-W2; and the National High-level Personnel of Special Support Program People Plan, grant number 312232102.

Data Availability Statement: The original contributions presented in this study are included in the article/Supplementary Materials. Further inquiries can be directed to the corresponding author.

Conflicts of Interest: The authors declare no conflicts of interest.

References

1. Cao, T.; Yi, Y.; Liu, H.; Yang, Z. Integrated ecosystem services-based calculation of ecological water demand for a macrophyte-dominated shallow lake. *J. Glob. Ecol. Conserv.* **2020**, *21*, 00858. [CrossRef]
2. Prater, C.; Frost, P.C.; Howell, E.T.; Watson, S.B. Variation in particulate C : N : P stoichiometry across the Lake Erie watershed from tributaries to its outflow. *J. Limnol. Oceanogr.* **2017**, *62*, S194–S206. [CrossRef]
3. Huang, J.; Zhang, Y.; Arhonditsis, G.B.; Gao, J.F. How successful are the restoration efforts of China's lakes and reservoirs? *J. Environ. Int.* **2019**, *123*, 96–103. [CrossRef] [PubMed]
4. Saeed, A.; Hashmi, I. Evaluation of anthropogenic effects on water quality and bacterial diversity in Rawal Lake, Islamabad. *J. Environ. Monit. Assess.* **2014**, *186*, 2785–2793. [CrossRef]
5. Noori, R.; Berndtsson, R.; Hosseinzadeh, M.; Adamowski, J.F.; Abyaneh, M.R. A critical review on the application of the National Sanitation Foundation Water Quality Index. *J. Environ. Pollut.* **2019**, *244*, 575–587. [CrossRef]
6. Wang, G.; Li, J.; Sun, W.; Xue, B.A.Y.; Liu, T. Non-point source pollution risks in a drinking water protection zone based on remote sensing data embedded within a nutrient budget model. *J. Water Res.* **2019**, *157*, 238–246. [CrossRef]
7. Chen, J.; Lyu, Y.; Zhao, Z.; Liu, H.; Zhao, H.; Li, Z. Using the multidimensional synthesis methods with non-parameter test, multiple time scales analysis to assess water quality trend and its characteristics over the past 25 years in the Fuxian Lake, China. *J. Sci. Total Environ.* **2019**, *655*, 242–254. [CrossRef] [PubMed]
8. Chen, R.H.; Li, F.P.; Zhang, H.P.; Jiang, Y.; Mao, L.C.; Wu, L.L.; Chen, L. Comparative analysis of water quality and toxicity assessment methods for urban highway runoff. *J. Sci. Total Environ.* **2016**, *553*, 519–523. [CrossRef] [PubMed]
9. Jiang, H.H.; Cai, L.M.; Wen, H.H.; Hu, G.C.; Chen, L.G.; Luo, J. An integrated approach to quantifying ecological and human health risks from different sources of soil heavy metals. *J. Sci. Total Environ.* **2020**, *701*, 134466. [CrossRef]
10. Péry, A.R.R.; Schüürmann, G.; Ciffroy, P.; Faust, M.; Backhaus, T. Perspectives for integrating human and environmental risk assessment and synergies with socio-economic analysis. *J. Sci. Total Environ.* **2013**, *456–457*, 307–316. [CrossRef]
11. Voza, D.; Vuković, M. The assessment and prediction of temporal variations in surface water quality—A case study. *J. Environ. Monit. Assess.* **2018**, *190*, 434. [CrossRef]
12. Sutadian, A.D.; Muttil, N.; Yilmaz, A.G.; Perera, B.J.C. Using the Analytic Hierarchy Process to identify parameter weights for developing a water quality index. *J. Ecol. Indic.* **2017**, *75*, 220–233. [CrossRef]
13. Che, O.S.F.; Malek, M.A.; Yusoff, M.; Azman, N.H.; Faizal, W.M. Development of river water quality management using fuzzy techniques: A review. *J. Int. J. River Basin Manag.* **2016**, *14*, 243–254.
14. Luo, D.; Ye, L.; Sun, D. Risk evaluation of agricultural drought disaster using a grey cloud clustering model in Henan province, China. *J. Int. J. Disaster Risk Reduct.* **2020**, *49*, 101759. [CrossRef]

15. Marcot, B.G.; Penman, T.D. Advances in Bayesian network modelling: Integration of modelling technologies. *J. Environ. Model. Softw.* **2019**, *111*, 386–393. [CrossRef]
16. Carriger, J.F. A Bayesian Network Approach to Refining Ecological Risk Assessments: Mercury and the Florida Panther (Puma Concolor Coryi). *J. Ecol. Model.* **2020**, *418*, 108911. [CrossRef] [PubMed]
17. Nodoushan, E.J. Monthly Forecasting of Water Quality Parameters within Bayesian Networks: A Case Study of Honolulu, Pacific Ocean. *J. Civ. J.* **2018**, *4*, 188. [CrossRef]
18. Orak, N.H. A Hybrid Bayesian Network Framework for Risk Assessment of Arsenic Exposure and Adverse Reproductive Outcomes. *J. Ecotoxicol. Environ. Saf.* **2020**, *192*, 110270. [CrossRef]
19. Helle, I.; Ahtiainen, H.; Luoma, E.; Hänninen, M.; Kuikka, S. A probabilistic approach for a cost-benefit analysis of oil spill management under uncertainty: A Bayesian network model for the Gulf of Finland. *J. Environ. Manag.* **2015**, *158*, 122–132. [CrossRef]
20. Lehikoinen, A.; Hänninen, M.; Storgård, J.; Luoma, E.; Mäntyniemi, S.; Kuikka, S. A Bayesian Network for Assessing the Collision Induced Risk of an Oil Accident in the Gulf of Finland. *J. Environ. Sci. Technol.* **2015**, *49*, 5301–5309. [CrossRef]
21. Landuyt, D.; Lemmens, P.; D'hondt, R.; Broekx, S.; Liekens, I.; De Bie, T.; Declerck, S.A.T.; De Meester, L.; Goethals, P.L.M. An ecosystem service approach to support integrated pond management: A case study using Bayesian belief networks—Highlighting opportunities and risks. *J. Environ. Manag.* **2014**, *145*, 79–87. [CrossRef]
22. Young, W.A.; Millie, D.F.; Weckman, G.R.; Anderson, J.S.; Klarer, D.M.; Fahnenstiel, G.L. Modeling net ecosystem metabolism with an artificial neural network and Bayesian belief network. *J. Environ. Model. Softw.* **2011**, *26*, 1199–1210. [CrossRef]
23. Lin, S.S.; Shen, S.L.; Zhou, A.; Lyu, H.M. Assessment and management of lake eutrophication: A case study in Lake Erhai, China. *J. Sci. Total Environ.* **2021**, *751*, 141618. [CrossRef] [PubMed]
24. Liu, Y.; Guo, H.; Yang, P. Exploring the influence of lake water chemistry on chlorophyll *a*: A multivariate statistical model analysis. *J. Ecol. Model.* **2010**, *221*, 681–688. [CrossRef]
25. Yu, R.; Zhang, C. Early warning of water quality degradation: A copula-based Bayesian network model for highly efficient water quality risk assessment. *J. Environ. Manag.* **2021**, *292*, 112749. [CrossRef] [PubMed]
26. *GB3838-2002*; Environmental Quality Standards for Surface Water. National Standards of the People's Republic of China: Beijing, China, 2002.
27. Geng, M.; Wang, K.; Yang, N.; Li, F.; Zou, Y.; Chen, S.; Deng, Z.; Xie, Y.H. Evaluation and variation trends analysis of water quality in response to water regime changes in a typical river-connected lake (Dongting Lake), China. *J. Environ. Pollut.* **2021**, *268*, 115761. [CrossRef]
28. Zhang, Q.; Murphy, R.R.; Tian, R.; Forsyth, M.K.; Trentacoste, E.M.; Keisman, J.; Tango, P.J. Chesapeake Bay's water quality condition has been recovering: Insights from a multimetric indicator assessment of thirty years of tidal monitoring data. *J. Sci. Total Environ.* **2018**, *637–638*, 1617–1625. [CrossRef] [PubMed]
29. Pan, Y.; Zhang, L.; Koh, J.; Deng, Y. An adaptive decision making method with copula Bayesian network for location selection. *J. Inf. Sci.* **2021**, *544*, 56–77. [CrossRef]
30. Zha, X. Coupling Bayesian Network and copula theory for water shortage assessment: A case study in source area of the South-to-North Water Division Project (SNWDP). *J. Hydrol.* **2023**, *620*, 129434. [CrossRef]
31. Pan, Y.; Ou, S.; Zhang, L.; Zhang, W.; Wu, X.; Li, H. Modeling risks in dependent systems: A Copula-Bayesian approach. *J. Reliab. Eng. Syst. Saf.* **2019**, *188*, 416–431. [CrossRef]
32. Yang, L.; Song, X.; Zhang, Y.; Han, D.; Zhang, B.; Long, D. Characterizing interactions between surface water and groundwater in the Jialu River basin using major ion chemistry and stable isotopes. *J. Hydrol. Earth Syst. Sci.* **2012**, *16*, 4265–4277. [CrossRef]
33. Dimzon, I.K.D.; Morata, A.S.; Müller, J.; Yanela, R.K.; Lebertz, S.; Weil, S.; Perez, T.R.; Müller, J.; Dayrit, F.M.; Knepper, T.P. Trace organic chemical pollutants from the lake waters of San Pablo City, Philippines by targeted and non-targeted analysis. *J. Sci. Total Environ.* **2018**, *639*, 588–595. [CrossRef] [PubMed]
34. Rios, D.T.; Chandra, S.; Heyvaert, A.C. The importance of small urbanized watersheds to pollutant loading in a large oligotrophic subalpine lake of the western USA. *J. Environ. Monit. Assess.* **2014**, *186*, 7893–7907. [CrossRef] [PubMed]
35. Karakoç, G.; Ünlü Erkoç, F.; Katırcıoğlu, H. Water quality and impacts of pollution sources for Eymir and Mogan Lakes (Turkey). *J. Environ. Int.* **2003**, *29*, 21–27. [CrossRef]
36. Huang, J.; Zhang, Y.; Bing, H.; Peng, J.; Dong, F.F.; Gao, J.F.; Arhonditsis, G.B. Characterizing the river water quality in China: Recent progress and on-going challenges. *J. Water Res.* **2021**, *201*, 117309. [CrossRef]
37. Mwaijengo, G.N.; Vanschoenwinkel, B.; Dube, T.; Njau, K.N.; Brendonck, L. Seasonal variation in benthic macroinvertebrate assemblages and water quality in an Afrotropical river catchment, northeastern Tanzania. *J. Limnol.* **2020**, *82*, 125780. [CrossRef]
38. Shimoda, Y.; Azim, M.E.; Perhar, G.; Ramin, M.; Kenney, M.A.; Sadraddini, S.; Gudimov, A.; Arhonditsis, G.B. Our current understanding of lake ecosystem response to climate change: What have we really learned from the north temperate deep lakes? *J. Great Lakes Res.* **2011**, *37*, 173–193. [CrossRef]
39. Yaghoubi, B.; Hosseini, S.A.; Nazif, S.; Daghighi, A. Development of reservoir's optimum operation rules considering water quality issues and climatic change data analysis. *J. Sustain. Cities Soc.* **2020**, *63*, 102467. [CrossRef]
40. Stockwell, J.D.; Doubek, J.P.; Adrian, R. Storm impacts on phytoplankton community dynamics in lakes. *J. Glob. Chang. Biol.* **2020**, *26*, 2756–2784. [CrossRef] [PubMed]

41. Kong, X.; Ye, S. The impact of water temperature on water quality indexes in north of Liaodong Bay. *J. Mar. Pollut. Bull.* **2014**, *80*, 245–249. [CrossRef]
42. Zhou, Z.Z.; Huang, T.L.; Ma, W.X.; Li, Y.; Zeng, K. Impacts of water quality variation and rainfall runoff on Jinpen Reservoir, in Northwest China. *J. Water Sci. Eng.* **2015**, *8*, 301–308. [CrossRef]
43. Schindler, D.W.; Carpenter, S.R.; Chapra, S.C.; Hecky, R.E.; Orihel, D.M. Reducing Phosphorus to Curb Lake Eutrophication is a Success. *J. Environ. Sci. Technol.* **2016**, *50*, 8923–8929. [CrossRef] [PubMed]
44. Wu, Z.; Liu, Y.; Liang, Z.; Wu, S.; Guo, H. Internal cycling, not external loading, decides the nutrient limitation in eutrophic lake: A dynamic model with temporal Bayesian hierarchical inference. *J. Water Res.* **2017**, *116*, 231–240. [CrossRef] [PubMed]
45. Huang, L.; Fang, H.; He, G.; Jiang, H.; Wang, C. Effects of internal loading on phosphorus distribution in the Taihu Lake driven by wind waves and lake currents. *J. Environ. Pollut.* **2016**, *219*, 760–773. [CrossRef]
46. Liu, X.; Sheng, H.; Jiang, S.; Yuan, Z.; Zhang, C.; Elser, J.J. Intensification of phosphorus cycling in China since the 1600s. *Proc. Natl. Acad. Sci. USA* **2016**, *113*, 2609–2614. [CrossRef]
47. Withers, P.J.A.; Sylvester-Bradley, R.; Jones, D.L.; Healey, J.R.; Talboys, P.J. Feed the Crop Not the Soil: Rethinking Phosphorus Management in the Food Chain. *J. Environ. Sci. Technol.* **2014**, *48*, 6523–6530. [CrossRef]
48. Tang, X.; Gao, G.; Qin, B.; Zhu, L.P.; Chao, J.; Wang, J.; Yang, G. Characterization of Bacterial Communities Associated with Organic Aggregates in a Large, Shallow, Eutrophic Freshwater Lake (Lake Taihu, China). *J. Microb. Ecol.* **2009**, *58*, 307–322. [CrossRef] [PubMed]
49. Gao, G.; Zhu, G.; Qin, B.; Chen, J.; Wang, K. Alkaline phosphatase activity and the phosphorus mineralization rate of Lake Taihu. *J. Sci. China Ser. D* **2006**, *49*, 176–185. [CrossRef]
50. Dai, X.; Yin, H.; Bai, Y. Spatial and temporal variation characteristics of marine wetland ecological environment based on GIS. *J. CCAMLR Sci.* **2018**, *25*, 143–151.
51. Zilko, A.A.; Kurowicka, D.; Goverde, R.M.P. Modeling railway disruption lengths with Copula Bayesian Networks. *J. Transp. Res. Part C Emerg. Technol.* **2016**, *68*, 350–368. [CrossRef]
52. Doguc, O.; Ramirez-Marquez, J.E. A generic method for estimating system reliability using Bayesian networks. *J. Reliab. Eng. Syst. Saf.* **2009**, *94*, 542–550. [CrossRef]
53. Kuhnert, P.M.; Martin, T.G.; Griffiths, S.P. A guide to eliciting and using expert knowledge in Bayesian ecological models. *J. Ecol. Lett.* **2010**, *13*, 900–914. [CrossRef] [PubMed]
54. Sebastian, A.; Dupuits, E.J.C.; Morales-Nápoles, O. Applying a Bayesian network based on Gaussian copulas to model the hydraulic boundary conditions for hurricane flood risk analysis in a coastal watershed. *J. Coast. Eng.* **2017**, *125*, 42–50. [CrossRef]

processes

MDPI

Article

Field Study on Washing of 4-Methoxy-2-Nitroaniline from Contaminated Site by Dye Intermediates

Zhili Wang [1], Kangwen Lao [2], Chen Chen [3], Hong Zhu [3], Yanfei Yang [4], Honghan Chen [1,*] and Hao Pang [3]

1 School of Water Resources and Environment, China University of Geosciences, Beijing 100083, China; wbsw5845@163.com
2 Advanced Institute of Natural Sciences, Beijing Normal University, Zhuhai 519087, China; kangwen_lao@163.com
3 Beijing Zhongdihongke Environmental Technology Co., Ltd., Beijing 100028, China; charlie_chen_85@sina.com (C.C.); zhuzhuf@126.com (H.Z.); zdhkpanghao@163.com (H.P.)
4 Nuclear and Radiological Safety Center of the Ministry of Ecological Environment, Beijing 100082, China; yangyanfei@chinansc.cn
* Correspondence: chenhh56@126.com

Abstract: Dye intermediates are important industrial chemicals; there is a lack of systematic field experiments and relevant validation data regarding the remediation of groundwater contamination by dye intermediates. This study examines the eluting effects of alcohol eluting agents, non-ionic surfactants, and deionized water on 4-methoxy-2-nitroaniline (4M2N) in a contaminated aquifer medium from a historically polluted dye intermediate production site in northwest China. The findings indicate that alcohol eluting agents exhibit superior eluting effects compared to non-ionic surfactants. Under optimized conditions, including 60% n-propanol concentration, a liquid-to-solid ratio of 15:1, two eluting cycles, an elution pH of 3, and a 2 h eluting duration, the eluting concentration of 4-methoxy-2-nitroaniline reached 75.49 mg/kg, exceeding that of the composite eluting agent by two times more and deionized water by three times further. Analysis revealed that the liquid-to-solid ratio and number of eluting cycles are the primary factors influencing eluting efficiency. Field trials conducted using treated groundwater involved injecting 31,560 m^3 of treated groundwater over 152 days, resulting in the extraction of 38,550 m^3 and the removal of about 1887 kg of 4-methoxy-2-nitroaniline. The concentrations of contaminants in both pumping wells and monitoring wells exhibited a certain degree of increase at various times. Field applications of treated groundwater washing facilitated the release of 4-methoxy-2-nitroaniline from the aquifer medium, which significantly enhances remediation efficiency. This provides theoretical support for data analysis and the promotion of similar remediation efforts.

Keywords: 4-methoxy-2-nitroaniline (4M2N); eluting/washing; non-ionic surfactant; alcohol eluting agent; aquifer medium

1. Introduction

China is a leading producer of dye intermediates, essential industrial chemicals extensively utilized in the dye and pharmaceutical sectors [1–3]. The complex production processes generate substantial by-products and waste, which, if inadequately managed, can result in severe soil and groundwater pollution. This investigation targets a historically contaminated dye intermediate production site, where primary raw materials include p-aminobenzyl ether, glacial acetic acid, and nitric acid, yielding 2-amino-4-acetamidobenzyl ether through sequential acetylation, nitration, and reduction reactions. The resultant wastewater and solid waste have severely contaminated soil and groundwater, leading to elevated COD_{Cr} concentrations in groundwater, often reaching tens of thousands of mg/L, and pH levels as low as 1–4. Notably, 4-methoxy-2-nitroaniline constitutes about 81.03% of the organic contaminants in the groundwater, accompanied by other pollutants, including 4-methoxy-3-nitroaniline, p-aminobenzyl ether, 3-chloroaniline, 4-chloroaniline, 2-amino-4-acetamidobenzyl ether, and aniline.

4-methoxy-2-nitroaniline exhibits high toxicity and intense coloration, facilitating its entry into the human body through dermal contact, inhalation, or ingestion, which severely impairs hematopoietic functions in the spleen, liver, and bone marrow. Exposure to this compound can lead to various adverse health effects, including methemoglobinemia, hepatotoxicity, hypoxia, hemorrhagic conditions, and neurotoxicity, posing significant threats to human life. The presence of nitro groups in its molecular structure enhances environmental stability by reducing the reactivity of the benzene ring, resulting in persistent toxicity and potential mutagenic and carcinogenic properties. Consequently, 4-methoxy-2-nitroaniline has been designated a priority pollutant in multiple countries [4].

Research on 4-methoxy-2-nitroaniline has primarily concentrated on its synthesis methodologies [5,6], applications as a pharmaceutical intermediate [7,8], physical properties [9–12], and toxicity assessments [13]. In contrast, remediation strategies for mitigating its contamination in groundwater and the aquifer medium have received relatively little attention, highlighting a significant knowledge gap in addressing its environmental impacts.

Considering the contaminated groundwater characteristics at this site, including diverse pollutants, high concentrations, structural stability of dye intermediates, poor biocompatibility, and pronounced toxicity, in situ remediation has incorporated extraction and treatment techniques. Long-term water extraction can lead to the development of dominant pathways, reducing solid–liquid contact in the absence of injection water disturbances. This reduction gradually decreases extraction efficiency. Furthermore, the absence of sufficient eluting agents hinders the effective removal of strongly adsorptive pollutants. Therefore, the gradual decline in extraction efficiency due to the slow release of pollutants from the aquifer medium necessitates alternative strategies. This study conducts washing experiments targeting 4-methoxy-2-nitroaniline, the primary organic pollutant in the aquifer medium, aiming to enhance extraction and treatment efficiency. The primary objective is to expedite contaminant release from the aquifer medium through targeted eluting agent injection into the aquifer, thereby mitigating the tailing and rebound phenomena commonly associated with extraction and treatment processes.

2. Materials and Methods

2.1. Aquifer Medium Sample Collection and Preservation

The study site is situated in a desert region in northwest China, where a factory operated from 1998 to 2014, producing dye intermediates (2-Amino-4-acetamino anisole). The pollution of soil and groundwater resulted from the discharge of wastewater outside the factory premises. The primary raw materials included p-anisidine, sulfuric acid, acetic acid, and nitric acid. Consequently, groundwater in the area exhibits strong acidity [14] and contains ultra-high concentrations of organic pollutants.

The experimental aquifer medium was obtained from the study site. The unconfined aquifer primarily consisted of Quaternary silty-fine sand, with the water table located at depths of 13.9 to 19.7 m below the surface. The aquifer had an approximate thickness of 25 m, and samples of the aquifer medium were collected from depths ranging between 17 and 35 m below the surface. The collected aquifer medium underwent freeze-drying using lyophilization, in order to ensure aquifer medium properties and minimize the volatilization of organic pollutants. Subsequent processing involved grinding and sieving the dried medium to achieve a uniform particle size of 0.25 mm (60 mesh), thereby ensuring homogeneity for subsequent experiments. The basic physical and chemical properties of the prepared aquifer medium samples are presented in Table 1. The moisture content of the samples was determined to be 0.2%.

Table 1. Basic physical and chemical properties of prepared aquifer medium samples.

Indicators	Conductivity	pH	Organic Matter	Moisture Content
Prepared Aquifer Medium Samples	81.3 mS/m	7.4	7.58 g/kg	0.2%

2.2. Reagent and Instrument

All chemicals such as methanol (Sinopharm Chemical Reagent Co., Ltd., Shanghai, China), ethanol (Sinopharm Chemical Reagent Co., Ltd., Shanghai, China), n-propanol (Sinopharm Chemical Reagent Co., Ltd., Shanghai, China), hydrochloric acid (Sinopharm Chemical Reagent Co., Ltd., Shanghai, China), sodium hydroxide (Sinopharm Chemical Reagent Co., Ltd., Shanghai, China), polyethylene glycol sorbitan monooleate (Tween-80, $C_{24}H_{44}O_6$) (Sinopharm Chemical Reagent Co., Ltd., Shanghai, China), alkyl glycoside (APG, $C_{16}H_{32}O_6$) (Sinopharm Chemical Reagent Co., Ltd., Shanghai, China), and polyethylene glycol tert-octylphenyl ether (Triton X-100, $C_{34}H_{62}O_{11}$) (Sinopharm Chemical Reagent Co., Ltd., Shanghai, China) were of analytical grade, and deionized water was used for experiments. The basic physical and chemical properties of non-ionic surfactants and alcohol-eluting agents are shown in Table 2.

Table 2. Basic physical and chemical properties of reagents used.

Name	CAS	Molecular Formula	Structural Formula	Molecular Weight	Melting Point (°C)	Boiling Point (°C)	Property Description
4-methoxy-2-nitroaniline	97-52-9	$C_7H_8N_2O_3$	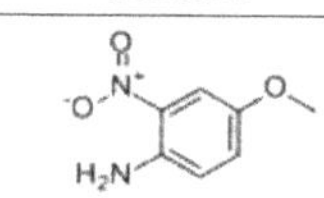	168.15	123–126	338	Orange-red powder. Slightly soluble in water, soluble in ethanol and ether, and slightly soluble in benzene.
methanol	67-56-1	CH_4O	$H_3C—OH$	32.04	−97.8	64.7	Colorless transparent liquid with an irritating odor. Soluble in water and miscible with most organic solvents such as alcohols and ethers.
ethanol	64-17-5	C_2H_5OH	$H_3C—\overset{H_2}{C}—OH$	46.07	−114	78	Colorless liquid with a fruity aroma. Miscible with water and most organic solvents, including ethers, chloroform, glycerol, and methanol.
n-propanol	71-23-8	C_3H_8O	HO⁄⁀CH₃	60.1	−127	97	Colorless transparent liquid with an ethanol-like odor. Miscible with water and most organic solvents such as alcohols and ethers.
Tween-80	9005-65-6	$C_{24}H_{44}O_6$		428.60			Exhibits high hydrophilicity, unaffected by environmental pH. Easily soluble in water, ethanol, vegetable oils, ethyl acetate, methanol, and toluene. Insoluble in mineral oils.
APG	141464-42-8	$C_{16}H_{32}O_6$		320.42			Low surface tension, fully biodegradable in nature, non-toxic, and harmless. Resistant to strong bases, strong acids, hard water, and has strong salt tolerance.
Triton X-100	9002-93-1	$C_{34}H_{62}O_{11}$		646.85			Colorless or nearly colorless transparent viscous liquid. Soluble in water, toluene, xylene, and ethanol. Insoluble in petroleum ether.

In this study, a vacuum freeze-dryer (ZLGJ-18, Zhengzhou Huachen Instrument Co., Ltd., Zhengzhou, China), high-performance liquid chromatography (HPLC, 1220

Infinity II, Agilent, Santa Clara, CA, USA) system, ultrasonic cleaning device (KQ-50B, Kunshan Ultrasonic Instrument Co., Ltd., Kunshan, China), vortex mixer (WIGGENS Vortex3000-Elite, Germany), digital thermostatic shaker (HNY-2102C, Tianjin ONuo Instrument Co., Ltd., Tianjin, China), centrifuge (TGL-20M, Changzhou Jintan Liangyou Instrument Co., Ltd., Changzhou, China), electronic analytical balance (AL104, Mettler Toledo, Greifensee, Switzerland), and pH meter (PB-10, Sartorius, Germany) were used.

2.3. Method of Eluting Experiment

The following non-ionic surfactants and alcohol eluting agents were selected for single eluting agent orthogonal experiments: Triton X-100, Tween-80, and APG, along with methanol, ethanol, and n-propanol. The eluting conditions employed are summarized in Table 3.

Table 3. Eluting conditions for single eluting agents.

Level Factor	1	2	3	4
Non-ionic surfactant concentration (g/L)	0.5	1	2	5
Alcohol eluent concentration (%)	30	60	80	100
Liquid-to-solid ratio (single, non-cumulative)	2.5:1	5:1	10:1	15:1
Eluent time	0.5 h	1 h	2 h	3 h
Eluent pH (hydrochloric acid adjustment)	3	4.5	6	7
Eluting cycles	1	2	3	4

A 3.00 g prepared sample of the contaminated aquifer medium was accurately weighed and distributed among 50 mL centrifuge tubes. Eluting agents were added as specified in Table 3. The sealed tubes were vortexed for 5 s to disperse the prepared sample and then placed horizontally on a temperature-controlled shaker at 20 °C and 200 r/min for the designated duration. Upon completion, the tubes were centrifuged at 3000 r/min for 5 min. The resulting supernatant was collected with a 10 mL syringe, filtered through a 0.22 μm PES membrane, and subjected to HPLC analysis for the 4-methoxy-2-nitroaniline concentration. After decanting the supernatant, a specified volume of eluting agents was added to the sample for subsequent elution cycles, repeating the same steps as before. The concentration of 4-methoxy-2-nitroaniline in each eluent was measured and multiplied by the volume of eluent added per treatment to calculate the mass of 4-methoxy-2-nitroaniline in the eluent. The total mass of 4-methoxy-2-nitroaniline from all elution cycles was then divided by 3 g, which was the mass of the contaminated aquifer medium used in the experiment, to determine the eluted 4-methoxy-2-nitroaniline content.

$$Q = \sum(C_i \times V_i) \div m$$

Q—eluted 4-mehoxy-2-nitroaniline content, relative to solid media, mg/kg;
C—concentration of 4-methoxy-2-nitroaniline in the eluent, mg/L;
V—volume of the eluent per treatment, L;
I—number of eluting cycles;
M—mass of the contaminated aquifer medium used in the experiment, kg.

Based on the orthogonal experiment results, the mixture experiment of non-ionic surfactants was conducted, and the eluting effect was compared with that of the alcohol eluting agent to determine the optimal eluting agents and conditions. The experimental method is detailed in Section 3.3.

In order to promote the release of pollutants on the aquifer medium and verify the applicability of washing technology in the research site, a pilot test was conducted on site, using treated groundwater for injection washing. The test conditions are detailed in Section 3.5.

2.4. Determination Method

The concentration of 4-methoxy-2-nitroaniline in the eluent was analyzed by high-performance liquid chromatography (HPLC, 1220 Infinity II, Agilent). The eluent, following filtration through a 0.22 μm PES membrane filter, was transferred into a brown glass vial for HPLC analysis using an Agilent XDB-C18 column. The mobile phase consisted of water and methanol in a 6:4 ratio, with a flow rate of 1 mL/min and an injection volume of 10 μL. The column temperature was kept at 30 °C, with detection conducted at a wavelength of 232 nm. pH measurements were carried out via the glass electrode method.

3. Results and Discussion

3.1. Alcohol Eluting Agent Eluting Experiments

The outcomes of the alcohol eluting agent eluting experiments are summarized in Tables 4–6. Comparative analysis indicates that the A7 group demonstrated optimal eluting performance in the methanol experiment, characterized by 60% methanol concentration, a 15:1 liquid-to-solid ratio, a 2 h eluting time, pH 3, and two eluting cycles, yielding a 4-methoxy-2-nitroaniline eluting content of 51.29 mg/kg. Conversely, the ethanol experiment showed the B13 group exhibiting superior performance, with 100% ethanol concentration, a 15:1 liquid-to-solid ratio, a 0.5 h eluting time, pH 4.5, and three eluting cycles, resulting in a 4-methoxy-2-nitroaniline content of 69.52 mg/kg. The n-propanol experiment revealed the C7 group as optimal, with 60% n-propanol concentration, a 15:1 liquid-to-solid ratio, a 2 h eluting time, pH 3, and two eluting cycles, yielding 75.49 mg/kg of 4-methoxy-2-nitroaniline.

These findings indicate that n-propanol is the most effective elution agent. The order of influence for n-propanol is liquid-to-solid ratio > number of eluting cycles > eluting solution pH > n-propanol concentration > eluting time. The eluting efficiency of alcohol-based agents is significantly influenced by the alcohol chain length. As the concentration of long-chain alcohols in the system increases, the system's lipophilicity enhances, reducing the curvature of the interfacial film and increasing micelle aggregation. This leads to an improved eluting effect of the eluent. Conversely, short-chain alcohols exhibit stronger hydrophilicity, resulting in slightly lower eluting efficiency compared to that of long-chain alcohols [15–17]. Experimental results indicate that 4-methoxy-2-nitroaniline is more effectively transferred from the aquifer medium to the eluent when n-propyl alcohol is used as the eluting agent.

Table 4. Orthogonal experiment results for methanol eluting.

Experiment No.	Methanol Concentration (%)	Liquid-to-Solid Ratio	Eluting Time (h)	Eluting Solution pH	Eluting Cycles	Elution 4-methoxy-2-nitroaniline Content (mg/kg)
A1	30	2.5:1	0.5	3	1	10.79
A2	30	5:1	1	4.5	2	16.14
A3	30	10:1	2	6	3	22.93
A4	30	15:1	3	7	4	32.46
A5	60	5:1	0.5	6	4	10.91
A6	60	2.5:1	1	7	3	7.92
A7	60	15:1	2	3	2	51.29
A8	60	10:1	3	4.5	1	8.49
A9	80	10:1	0.5	7	2	32.93
A10	80	15:1	1	6	1	18.86
A11	80	2.5:1	2	4.5	4	22.09
A12	80	5:1	3	3	3	22.57
A13	100	15:1	0.5	4.5	3	31.42
A14	100	10:1	1	3	4	28.29
A15	100	5:1	2	7	1	16.88
A16	100	2.5:1	3	6	2	10.44
k1	20.58	12.81	21.51	28.24	13.75	
k2	19.65	16.63	17.80	19.53	27.70	
k3	24.11	23.16	28.30	15.78	21.21	
k4	21.76	33.51	18.49	22.55	23.44	
R	4.46	20.70	10.49	12.45	13.95	

Table 5. Orthogonal experiment results for ethanol eluting.

Experiment No.	Ethanol Concentration (%)	Liquid-to-Solid Ratio	Eluting Time (h)	Eluting Solution pH	Eluting Cycles	Elution 4-methoxy-2-nitroaniline Content (mg/kg)
B1	30	2.5:1	0.5	3	1	11.49
B2	30	5:1	1	4.5	2	15.79
B3	30	10:1	2	6	3	29.67
B4	30	15:1	3	7	4	32.65
B5	60	5:1	0.5	6	4	22.89
B6	60	2.5:1	1	7	3	15.66
B7	60	15:1	2	3	2	26.02
B8	60	10:1	3	4.5	1	14.02
B9	80	10:1	0.5	7	2	32.13
B10	80	15:1	1	6	1	21.67
B11	80	2.5:1	2	4.5	4	27.29
B12	80	5:1	3	3	3	23.15
B13	100	15:1	0.5	4.5	3	69.52
B14	100	10:1	1	3	4	29.77
B15	100	5:1	2	7	1	16.26
B16	100	2.5:1	3	6	2	19.72
k1	22.40	18.54	34.01	22.61	15.86	
k2	19.65	19.52	20.72	31.65	23.42	
k3	26.06	26.40	24.81	23.49	34.50	
k4	33.82	37.47	22.38	24.18	28.15	
R	14.17	18.93	13.28	9.04	18.64	

Table 6. Orthogonal experiment results for n-propanol eluting.

Experiment No.	n-Propanol Concentration (%)	Liquid-to-Solid Ratio	Eluting Time (h)	Eluting Solution pH	Eluting Cycles	Elution 4-methoxy-2-nitroaniline Content (mg/kg)
C1	30	2.5:1	0.5	3	1	2.78
C2	30	5:1	1	4.5	2	18.74
C3	30	10:1	2	6	3	25.19
C4	30	15:1	3	7	4	54.36
C5	60	5:1	0.5	6	4	23.57
C6	60	2.5:1	1	7	3	19.12
C7	60	15:1	2	3	2	75.49
C8	60	10:1	3	4.5	1	9.94
C9	80	10:1	0.5	7	2	23.11
C10	80	15:1	1	6	1	17.91
C11	80	2.5:1	2	4.5	4	4.82
C12	80	5:1	3	3	3	29.06
C13	100	15:1	0.5	4.5	3	40.28
C14	100	10:1	1	3	4	41.41
C15	100	5:1	2	7	1	12.79
C16	100	2.5:1	3	6	2	12.94
k1	25.27	9.91	22.44	37.19	10.85	
k2	32.03	21.04	24.30	18.45	32.57	
k3	18.73	24.91	29.57	19.90	28.41	
k4	26.86	47.01	26.57	27.34	31.04	
R	13.30	37.10	7.14	18.74	21.72	

Alcohol eluting agents reduce the interfacial tension between contaminants and water through wetting, foaming, and emulsification. The optimal concentration of alcohol required must overcome the stable value of the interfacial tension. The polarity order of the tested alcohols, from highest to lowest, is methanol > ethanol > n-propanol [18]. This decrease in polarity effectively reduces interfacial tension, enhancing pollutant mobility and facilitating the transfer of 4-methoxy-2-nitroaniline from the aqueous phase to the eluent [19]. Alcohol-based eluting agents are also effective for polycyclic aromatic hydrocarbons and various nitroaniline compounds, including 4-methoxy-3-nitroaniline and similar substances. Notably, optimal alcohol concentrations decrease with decreasing polarity, attributed to achieving a stable interfacial tension value [20,21]. The liquid-to-solid ratio and number of eluting cycles are the dominant factors influencing alcohol eluting agents' effectiveness. A higher liquid-to-solid ratio yields better outcomes, with optimal eluting cycles ranging between two and three. The optimal pH value for n-propanol eluent determined in this experiment is 3, suitable for an acid-contaminated site. For practical application and safety considerations, the use of a low-concentration n-propyl alcohol solution is recommended for field-scale operations.

3.2. Non-Ionic Surfactant Eluting Experiments

The results of non-ionic surfactant eluting experiments are presented in Tables 7–9. Comparative analysis reveals optimal eluting performance was achieved by the D4 group in the APG experiment, characterized by an APG concentration of 0.5 g/L, a liquid-to-solid ratio of 15:1, an eluting time of 3 h, a pH of 7, and four eluting cycles, resulting in a 4-methoxy-2-nitroaniline eluting content of 34.80 mg/kg. Similarly, the E4 group exhibited optimal results in the Triton X-100 experiment, with conditions including a 0.5 g/L Triton X-100 concentration, a liquid-to-solid ratio of 15:1, an eluting time of 3 h, an eluting pH of 7, and four eluting cycles, yielding 32.40 mg/kg of 4-methoxy-2-nitroaniline. The Tween-80 experiment showed the F4 group achieved optimal performance, corresponding to a Tween-80 concentration of 0.5 g/L, a liquid-to-solid ratio of 15:1, an eluting time of 3 h, a pH of 7, and four eluting cycles, with an eluting content of 31.20 mg/kg of 4-methoxy-2-nitroaniline.

The experimental results demonstrate that APG exhibited superior eluting performance among the three non-ionic surfactants tested. The factors influencing the effectiveness of APG follow the order as follows: solid-to-liquid ratio > number of eluting cycles > APG concentration > elution pH > eluting time. The unique hydrophilic and oleophilic structure of surfactants allows the formation of a layer of non-polar hydrocarbon chains on the water surface, reducing the interfacial tension between phases. When the concentration of surfactant in water surpasses the critical micelle concentration, monomer adsorption at the interface saturates, minimizing interfacial tension. Excess surfactant molecules aggregate in solution, forming micelles with hydrophilic groups oriented outward, facilitating pollutant desorption from soil surfaces and dispersing them into the aqueous phase for enhanced dissolution [22]. Non-ionic surfactants, characterized by lower critical micelle concentrations compared to those of anionic surfactants, exhibit greater solubilization capacity and superior pollutant elution efficiency [23]. In this study, non-ionic surfactants were utilized due to these advantages. While differences in solubilization effects among the three non-ionic surfactants studied were minimal, APG stands out for its superior environmental compatibility, non-toxicity, biodegradability, and enhanced pollutant solubilization. These properties make APG a suitable alternative for on-site applications [24].

The eluting effect of a single surfactant was positively correlated with the liquid–solid ratio and the eluting cycles. With the increase in the liquid–solid ratio and the eluting cycles, the eluting amount of 4-methoxy-2-nitroaniline gradually increased. This finding aligns with previous studies indicating favorable eluting of hydrophobic organic compounds at liquid-to-solid ratios between 4:1 and 20:1 [25]. The optimal liquid-to-solid ratio of 15:1 obtained in this study falls within this recommended range.

The primary mechanisms (Figure 1) governing surfactant interactions with pollutants on the aquifer medium include [26] (1) the incorporation of pollutants into surfactant micelles, reducing aquifer medium surface adsorption; (2) competition between surfactants and pollutants for aquifer medium adsorption sites, decreasing pollutant adsorption; (3) the dispersion of aquifer medium particles by surfactants, increasing available adsorption sites and enhancing pollutant adsorption; and (4) the formation of a secondary adsorption layer through surfactant adsorption on aquifer medium surfaces, re-adsorbing pollutants.

However, increasing surfactant concentration does not enhance eluting effectiveness. At lower concentrations, micelles are not formed or are present in low concentrations, effectively reducing interfacial tension and facilitating eluent entry into pore spaces, thereby promoting 4-methoxy-2-nitroaniline removal. However, as surfactant concentration increases, the number of micelles rises, leading to increased adsorption of the aquifer medium onto micelles, resulting in heightened surfactant loss due to adsorption. Furthermore, micelle adsorption on the medium surface can cause re-adsorption of 4-methoxy-2-nitroaniline, ultimately diminishing eluting effectiveness.

Table 7. Orthogonal experiment results for APG eluting.

Factors	APG Concentration (g/L)	Elution Time (h)	Liquid-to-Solid Ratio	Elution pH	Eluting Cycles	Elution 4-methoxy-2-nitroaniline Content (mg/kg)
D1	0.5	0.5	2.5:1	3	1	7.95
D2	0.5	1	5:1	4.5	2	12.20
D3	0.5	2	10:1	6	3	20.70
D4	0.5	3	15:1	7	4	34.80
D5	1	1	2.5:1	6	4	12.80
D6	1	0.5	5:1	7	3	14.25
D7	1	3	10:1	3	2	15.40
D8	1	2	15:1	4.5	1	16.05
D9	2	2	2.5:1	7	2	7.85
D10	2	3	5:1	6	1	6.65
D11	2	0.5	10:1	4.5	4	24.00
D12	2	1	15:1	3	3	25.65
D13	5	3	2.5:1	4.5	3	9.68
D14	5	2	5:1	3	4	16.00
D15	5	1	10:1	7	1	11.30
D16	5	0.5	15:1	6	2	19.50
k1	18.91	16.43	9.57	16.25	10.49	
k2	14.63	15.49	12.28	15.48	13.74	
k3	16.04	15.15	17.85	14.91	17.57	
k4	14.12	16.63	24.00	17.05	21.90	
R	4.79	1.48	14.43	2.14	11.41	

Table 8. Orthogonal experiment results for Triton X-100 eluting.

Factors	Triton X-100 Concentration (g/L)	Elution Time (h)	Liquid-to-Solid Ratio	Elution pH	Eluting Cycles	Elution 4-methoxy-2-nitroaniline Content (mg/kg)
E1	0.5	0.5	2.5:1	3	1	9.43
E2	0.5	1	5:1	4.5	2	15.60
E3	0.5	2	10:1	6	3	21.00
E4	0.5	3	15:1	7	4	32.40
E5	1	1	2.5:1	6	4	15.50
E6	1	0.5	5:1	7	3	16.65
E7	1	3	10:1	3	2	17.60
E8	1	2	15:1	4.5	1	15.30
E9	2	2	2.5:1	7	2	11.25
E10	2	3	5:1	6	1	11.30
E11	2	0.5	10:1	4.5	4	24.00
E12	2	1	15:1	3	3	26.10
E13	5	3	2.5:1	4.5	3	11.18
E14	5	2	5:1	3	4	17.80
E15	5	1	10:1	7	1	12.80
E16	5	0.5	15:1	6	2	22.20
k1	19.61	18.07	11.84	17.73	12.21	
k2	16.26	17.50	15.34	16.52	16.66	
k3	18.16	16.34	18.85	17.50	18.73	
k4	15.99	18.12	24.00	18.28	22.43	
R	3.61	1.78	12.16	1.76	10.22	

Table 9. Orthogonal experiment results for Tween-80 eluting.

Factors	Tween-80 Concentration (g/L)	Elution Time (h)	Liquid-to-Solid Ratio	Elution pH	Eluting Cycles	Elution 4-methoxy-2-nitroaniline Content (mg/kg)
F1	0.5	0.5	2.5:1	3	1	8.23
F2	0.5	1	5:1	4.5	2	13.30
F3	0.5	2	10:1	6	3	21.60
F4	0.5	3	15:1	7	4	31.20
F5	1	1	2.5:1	6	4	11.90
F6	1	0.5	5:1	7	3	15.30
F7	1	3	10:1	3	2	17.80
F8	1	2	15:1	4.5	1	15.90
F9	2	2	2.5:1	7	2	10.00
F10	2	3	5:1	6	1	7.00
F11	2	0.5	10:1	4.5	4	22.40
F12	2	1	15:1	3	3	26.55
F13	5	3	2.5:1	4.5	3	7.13
F14	5	2	5:1	3	4	16.80
F15	5	1	10:1	7	1	12.00
F16	5	0.5	15:1	6	2	21.00
k1	18.58	16.73	9.31	17.34	10.78	
k2	15.23	15.94	13.10	14.68	15.53	
k3	16.49	16.08	18.45	15.38	17.64	
k4	14.23	15.78	23.66	17.13	20.58	
R	4.35	0.95	14.35	2.66	9.79	

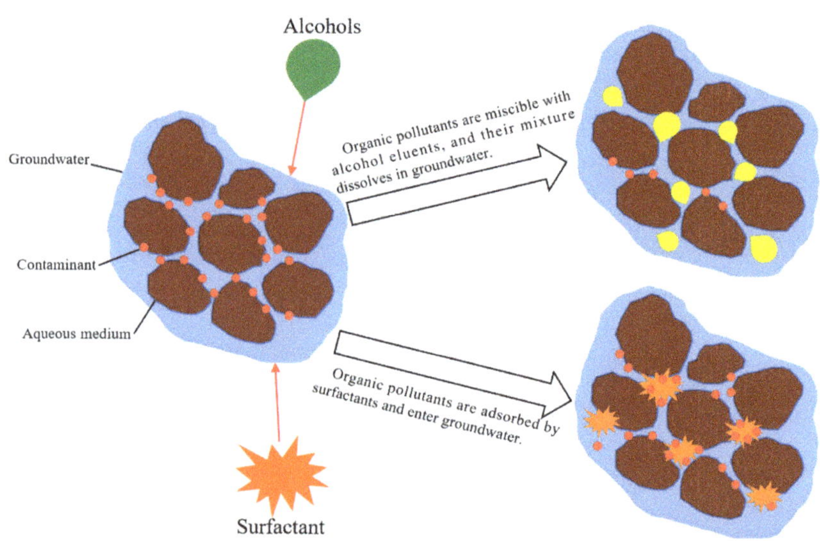

Figure 1. Schematic diagram of alcohol eluting agents and surfactant eluting principles.

3.3. *Determination of Optimal Eluting Agents and Conditions*

APG and Triton X-100 were selected for blending due to their superior eluting performance. The surfactant concentration was standardized at 0.5 g/L, with a fixed liquid-to-solid ratio of 15:1, a pH of 7, and an eluting duration of 3 h, divided into four cycles. Various volume proportions of APG solution were evaluated: 0%, 20%, 40%, 50%, 60%, 80%, and 100%. Three control groups were also established, as detailed in Table 10. The outcomes of these experiments are presented in Figure 2.

Table 10. Control group experimental conditions.

No.	Eluent	Concentration	Liquid-to-Solid Ratio	pH	Elution Time	Eluting Cycles
A	deionized water	/	15:1	7	3 h	4
B	deionized water	/	15:1	3	2 h	2
C	n-propanol	60%	15:1	3	2 h	2

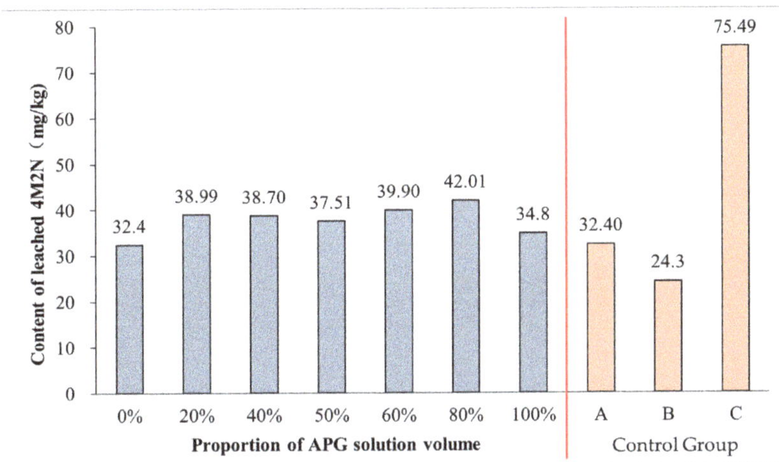

Figure 2. Relationship between 4-methoxy-2-nitroaniline extraction and compound ratio.

The experimental results revealed variations in the effectiveness of combined eluting agents with different APG to Triton X-100 ratios. Optimal eluting occurred when the APG solution comprised 80% of the volume, resulting in a 4-methoxy-2-nitroaniline concentration of 42.01 mg/kg. The APG and Triton X-100 mixture outperformed individual non-ionic surfactants. The likely explanation is that the combination of surfactants leads to the formation of mixed micelles and adsorption layers. The critical micelle concentration of the resulting complex surfactants is lower than that of individual surfactants [27], enhancing the distribution coefficient of pollutants within the micelles and increasing the interaction opportunities between pollutants and micelles. The composite eluting agents exhibit a synergistic enhancement effect among their various components, which demonstrates greater adaptability to different remediation environments and conditions. This synergy can neutralize potential adverse effects caused by individual components, which thus enhances the efficiency of contaminant desorption.

Compared to the control group, significant differences in eluting performance were observed between the combined eluting agent and alcohol-based eluting agents. Under optimal conditions, the eluting efficiency of n-propyl alcohol surpasses that of compound eluting agents, making it the most effective eluent. However, its usage should be carefully evaluated with regard to dosage and potential impacts on human health and the environment. For on-site applications, it is essential to ensure that its use does not pose safety risks or lead to secondary pollution, and it should only be implemented under conditions that guarantee these considerations.

3.4. Deionized Water Washing Experiments

The experimental results indicate that the liquid-to-solid ratio and the number of eluting cycles are the two primary factors influencing eluting effectiveness. The liquid-to-solid ratio directly affects the contact area and contact time between the eluting agent and the contaminants in the aquifer medium, which thus impacts the effectiveness of a single eluting cycle. In contrast, the number of eluting cycles influences the cumulative removal of contaminants. Considering practical feasibility, safety, and economic factors, deionized water was used for washing tests to compare the washing effect of alcohol eluting agents and surfactants. The results are shown in Table 11. Comparative analysis reveals the G4 group achieved optimal washing performance, characterized by a liquid-to-solid ratio of 15:1, a washing time of 3 h, an elution pH of 7, and four washing cycles, yielding 32.40 mg/kg of 4-methoxy-2-nitroaniline. The influencing factors for deionized water washing rank as follows: liquid-to-solid ratio > number of washing cycles > elution pH > washing time.

Table 11. Orthogonal experiment results for deionized water washing.

Factors	Elution Time (h)	Liquid-to-Solid Ratio	Elution pH	Washing Cycles	Elution 4-methoxy-2-nitroaniline Content (mg/kg)
G1	0.5	2.5:1	3	1	7.425
G2	1	5:1	4.5	2	9.5
G3	2	10:1	6	3	20.4
G4	3	15:1	7	4	32.4
G5	1	2.5:1	6	4	10.5
G6	0.5	5:1	7	3	16.8
G7	3	10:1	3	2	16.6
G8	2	15:1	4.5	1	15.3
G9	2	2.5:1	7	2	6.65
G10	3	5:1	6	1	7.3
G11	0.5	10:1	4.5	4	20.4
G12	1	15:1	3	3	24.3
G13	3	2.5:1	4.5	3	7.425
G14	2	5:1	3	4	14.6
G15	1	10:1	7	1	9.9
G16	0.5	15:1	6	2	19.2
k1	15.96	8.00	15.73	9.98	
k2	13.55	12.05	13.16	12.99	
k3	14.24	16.83	14.35	17.23	
k4	15.93	22.80	16.44	19.48	
R	2.41	14.80	3.28	9.49	

The eluting effect of deionized water is slightly lower compared to that of n-propanol. However, its washing efficiency shows no significant difference when compared with the effects of single or compound surfactants. This outcome is likely due to the lithology of the contaminated aquifer medium, which primarily consists of silty-fine sand with poor adsorption properties, facilitating the desorption of pollutants. These findings underscore the effectiveness of deionized water washing under optimized conditions, providing valuable insights for potential field applications.

3.5. On-Site Washing Pilot Test

A pilot-scale test was conducted at a high-concentration pollution area within the contaminated site to validate washing effectiveness. The site configuration is illustrated in Figure 3. The test setup consisted of one injection well for washing, five monitoring wells, and five extraction wells. The distances of the extraction wells and the monitoring wells from the injection well and the slotting locations of the wells are shown in Table 12.

In consideration of the planned implementation of Groundwater Monitored Natural Attenuation (MNA) at the experimental site, the potential impact of n-propanol, despite its low toxicity and environmentally friendly properties [28], on indigenous microorganisms remains uncertain. To ensure construction safety, reduce costs, and minimize adverse effects on native microbial populations, an on-site pilot test was conducted using treated groundwater. After undergoing physicochemical and biochemical treatments (Figure 4), a substantial reduction in pollutant concentrations was observed in the treated groundwater, as shown in Table 13, with the pH value adjusted to approximately 7.0.

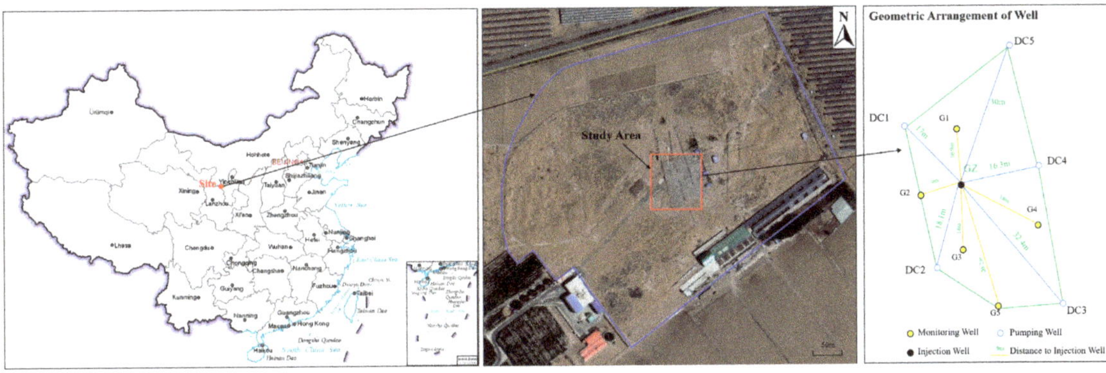

Figure 3. Site location and well layout.

Table 12. Distance to test wells and slotting locations.

Type	Number	Distance from an Injection Well	Screening Location
Washing injection well	GZ	/	8m above the bottom
Monitoring well	G1	10.9 m	Full sieve
	G2	9 m	Full sieve
	G3	14 m	Full sieve
	G4	18 m	Full sieve
	G5	26.2 m	15 m above the bottom
Extraction well	DC1	17 m	3 m above the bottom
	DC2	18.1 m	3 m above the bottom
	DC3	32.4 m	3 m above the bottom
	DC4	16.3 m	3 m above the bottom
	DC5	30 m	3 m above the bottom

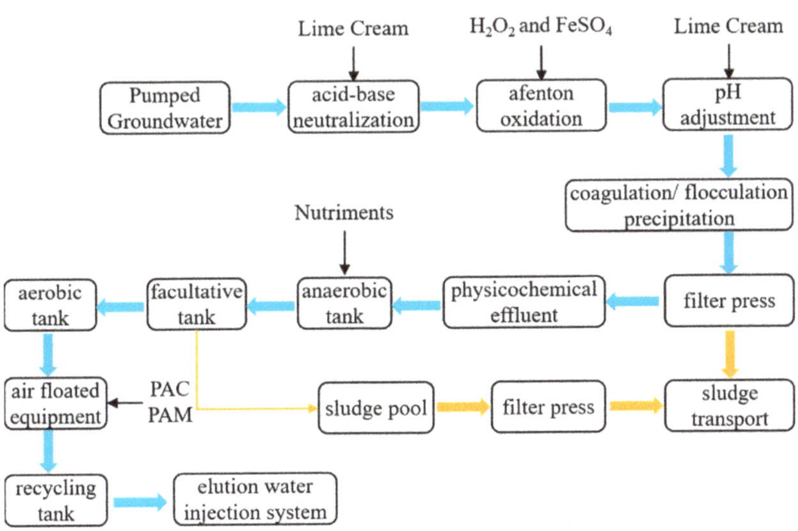

Figure 4. Treatment process for contaminated groundwater.

Table 13. Water quality of injecting water (treated groundwater).

Index	pH	ORP (mV)	CON (μs/cm)	DO (mg/L)	Sulfate (mg/L)	COD_{Cr} (mg/L)	4-chloroaniline (μg/L)	4-methoxy-2-nitroaniline (μg/L)	4-methoxy-3-nitroaniline (μg/L)	3-chloroaniline (μg/L)
Injection Water (Average Value)	6.98	75	6048	6.68	4500	24	10	28.6	292	2.6

This treated groundwater was injected into the aquifer as the eluent. It utilizes physical flushing, dissolution, and the promotion of contaminant migration to transfer contaminants adhered to aquifer medium particle surfaces into the aqueous solution, and they are then removed with pumping, which makes it both economical and environmentally friendly. Continuous steady pressure was applied to the injection well, while the extraction wells operated continuously. Water meters were installed on the extraction wells to measure the flow rate of each well.

Subsequent to water injection initiation, pressure (Figure 5) increased gradually from 0 to 0.88 MPa over the first 15 days. The flow rate (Figure 5) initially peaked at 6.5 m^3/h and then decreased to about 2 m^3/h, before increasing again. By day 31, the flow rate stabilized at around 5 m^3/h, with pressure fluctuating between 0.7 and 0.75 MPa. A rapid pressure decrease to 0.3–0.46 MPa was observed on day 68, accompanied by a flow rate increase to 14–17 m^3/h, which then stabilized. Fluctuations in both flow rate and pressure occurred after 133 days. The water injection period lasted 152 days, with a cumulative injection volume of 31,560 m^3.

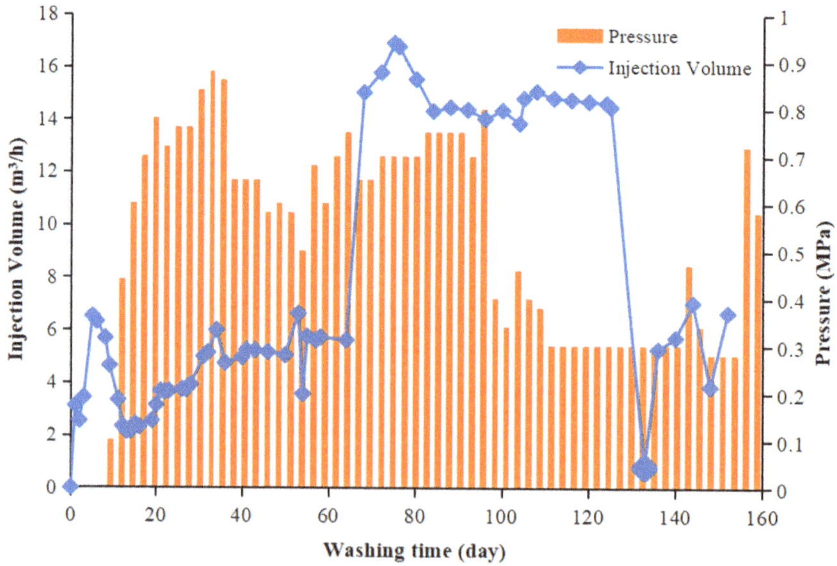

Figure 5. Water injection status.

Daily extraction volumes from individual wells showed fluctuations that stabilized after 80 days (Figure 6). Wells DC1 and DC2, positioned closer to the injection well, yielded higher extraction volumes, spanning 58–881 m^3/d. Conversely, DC3 and DC5, located farther from the injection well, displayed lower extraction volumes, ranging from 1 to 36 m^3/d. DC4 exhibited intermediate daily extraction volumes of 5–155 m^3/d. Over the 140-day period, the cumulative extraction volume from all five wells totaled 38,550 m^3, surpassing the total injection volume.

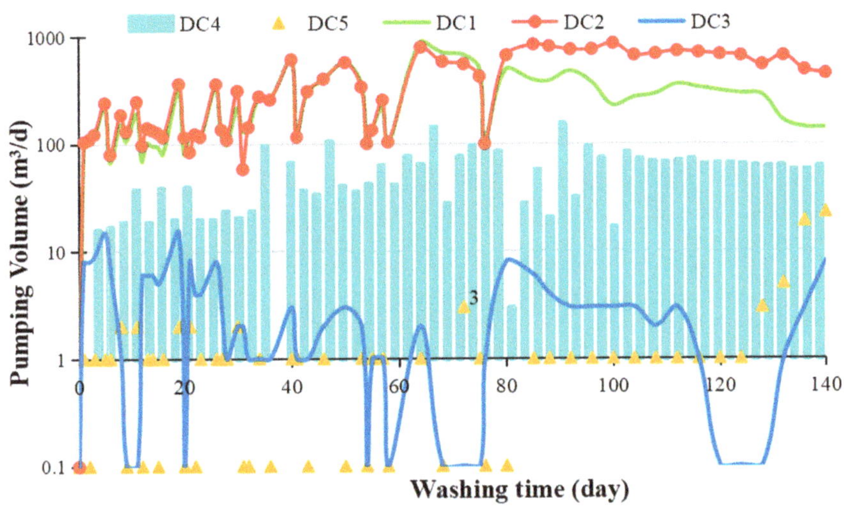

Figure 6. Water extraction status.

Water quality monitoring data from the extraction wells (Figure 7) revealed that DC4, situated closest to the injection well (16.3 m) and downstream of the groundwater flow direction, was most affected by the water injection washing. Pollutant concentrations in DC4 exhibited significant fluctuations, with notable increases observed during days 7–18, 60–74, and 90–122. The maximum concentration reached 116.84 mg/L, exceeding the initial concentration of 45.33 mg/L. After day 90, pollutant concentrations in DC4 remained elevated above initial levels. Calculations based on the average concentration of 4-methoxy-2-nitroaniline extracted from DC4 and the total extraction volume yielded an estimated mass of about 241 kg of 4-methoxy-2-nitroaniline removed during the pilot test period.

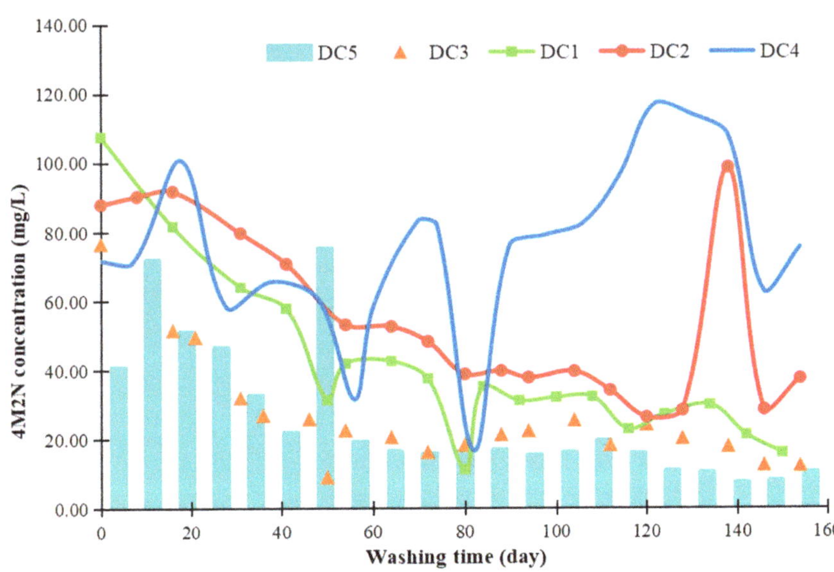

Figure 7. Changes in 4-methoxy-2-nitroaniline concentration in extraction wells.

Extraction wells DC3 and DC5, located at greater distances from the injection well (32.4 m and 30 m, respectively), exhibited a general decline in pollutant concentration, accompanied by fluctuations. Transient increases were observed in DC5 on days 11 and 50, with concentrations reaching 71.87 mg/L and 75.48 mg/L. Between days 54 and 108, concentrations fluctuated minimally, stabilizing between 15.5 and 19.5 mg/L. A gradual decrease occurred from days 116 to 154, with concentrations ranging from 7.4 to 10.85 mg/L. After day 54, DC3 displayed relatively stable fluctuations, averaging about 19.4 mg/L. Over the pilot test period, the cumulative mass of 4-methoxy-2-nitroaniline extracted from DC3 and DC5 was calculated to be 5.26 kg and 3.34 kg, respectively.

Extraction wells DC1 and DC2, situated closer to the injection well (17 m and 18.1 m), were positioned upstream and displayed a declining trend in pollutant concentration, accompanied by fluctuations. DC2 exhibited a gradual increase in pollutant concentration during the initial 16 days, followed by fluctuating levels. A notable spike in pollutant concentration occurred on day 138, reaching 98.36 mg/L. Over the duration of the pilot test, the cumulative mass of 4-methoxy-2-nitroaniline extracted from DC1 and DC2 was calculated to be 518 kg and 1120 kg, respectively.

Water quality monitoring data from the monitoring wells (Figure 8) revealed that upstream monitoring well G2, positioned closest to the injection well (9 m), experienced a substantial increase in pollutant concentration during the initial operational phase (days 1–64), rising from 39.36 mg/L to 158.88 mg/L. This was followed by a notable decrease and stabilization around 18.5 mg/L in the later phase. The side monitoring wells G1, G3, and G5, located progressively further from the injection well, displayed corresponding delays in pollutant concentration increases. Peak concentrations were observed in G1 on day 26 (85.24 mg/L), G3 on day 88 (100.94 mg/L), and G5 on day 112 (512.88 mg/L). Downstream monitoring well G4, situated 18 m from the injection well, exhibited minimal variation in pollutant concentration during the first 104 days, stabilizing around 37.5 mg/L. After day 104, concentrations increased sharply, peaking on day 142 at 390.96 mg/L.

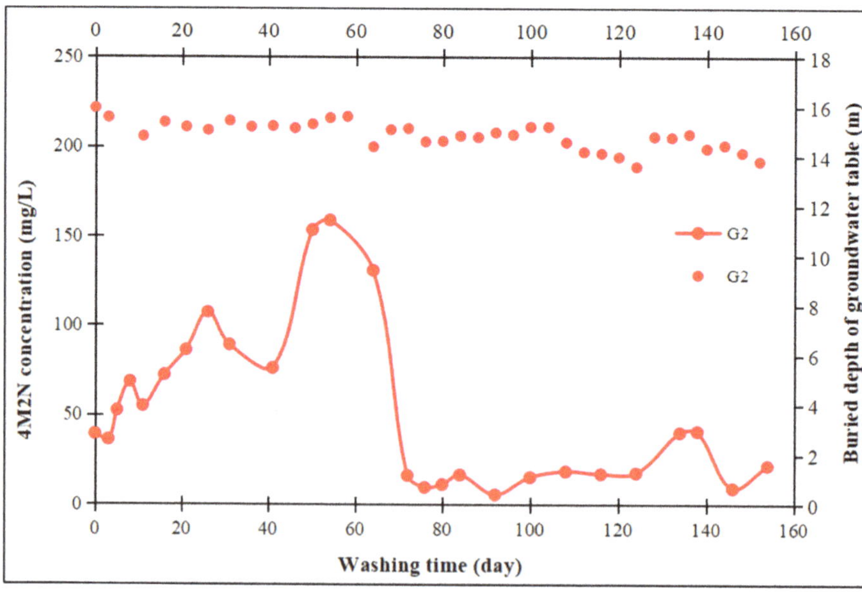

Figure 8. *Cont.*

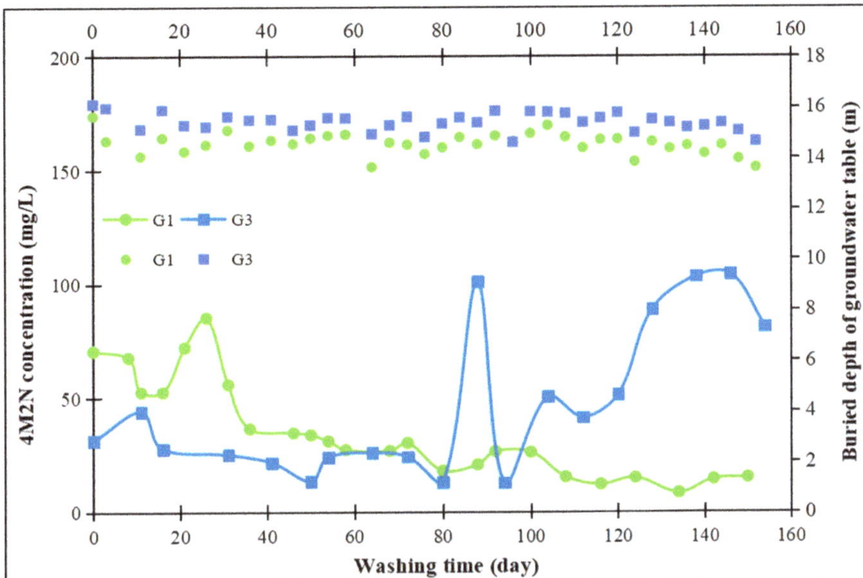

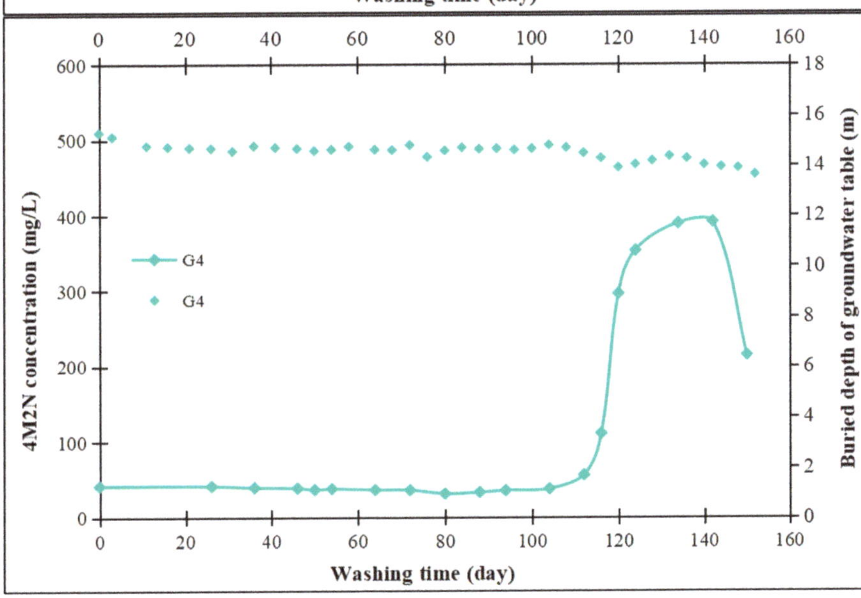

Figure 8. *Cont.*

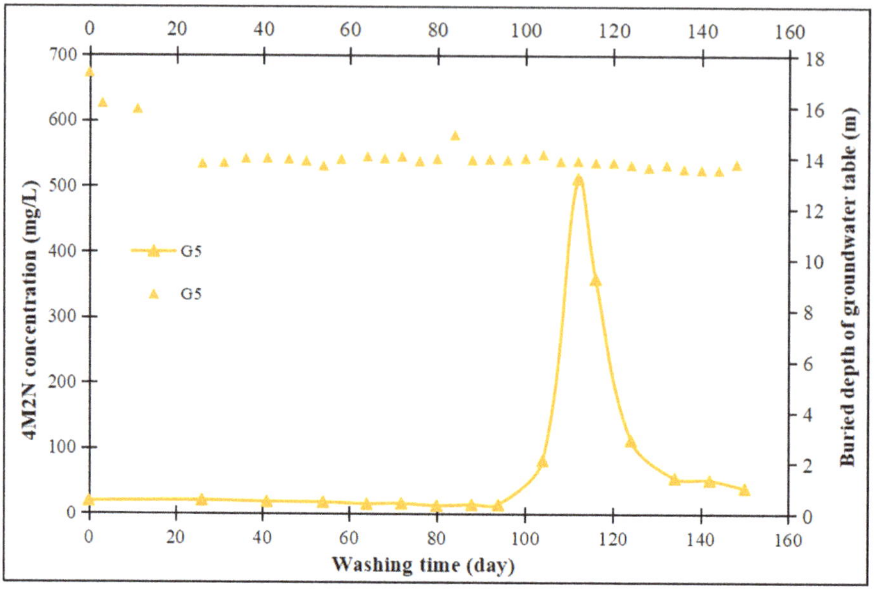

Figure 8. 4-methoxy-2-nitroaniline concentration and water level changes in monitoring wells.

4. Conclusions

(1) The alcohol-based eluting agents applied to leach 4-methoxy-2-nitroaniline (4M2N) from the contaminated aquifer medium showed the following effectiveness: n-propanol > ethanol > methanol. This efficacy is closely related to the polarity of the alcohol eluting agents. The eluting agents reduce the interfacial tension between contaminants and water through wetting, foaming, and emulsification effects, with the optimal concentration of alcohol needed to overcome the stable value of interfacial tension. N-propanol's optimal conditions involve 60% concentration, a 15:1 liquid-to-solid ratio, two eluting cycles, pH 3, and a 2 h eluting duration. Notably, n-propanol exhibited the most effective eluting, achieving a 4-methoxy-2-nitroaniline concentration of 75.49 mg/kg. The liquid-to-solid ratio and number of eluting cycles emerged as the primary factors influencing the efficacy of alcohol-based eluting agents.

(2) The non-ionic surfactants used to leach 4-methoxy-2-nitroaniline from the contaminated aquifer medium exhibited the following effectiveness: APG > Triton X-100 > Tween-80. The optimal concentration of APG was found to be 0.5 g/L, with a liquid-to-solid ratio of 15:1, eluting conducted four times, an elution pH of 7, and an eluting duration of 3 h. The liquid-to-solid ratio and the number of eluting cycles are the two primary factors influencing the efficacy of non-ionic surfactant eluting. The liquid-to-solid ratio directly affects the contact area and contact time between the eluting agent and the contaminants in the aquifer medium, which thus influences the effectiveness of a single eluting event, while the number of eluting cycles affects the cumulative removal of contaminants.

(3) The blending of APG and Triton X-100 as an eluting agent enhances the eluting effect compared to individual non-ionic surfactants. The blended eluting agents exhibited a synergistic enhancement effect among their various components, which demonstrates greater adaptability to different remediation environments and conditions. This synergy can neutralize potential adverse effects associated with individual components, which can enhance the efficiency of contaminant elution. The optimal blending ratio of APG to Triton X-100 was determined to be 4:1.

(4) n-propanol has been identified as the optimal eluting agent for removing 4-methoxy-2-nitroaniline from the aquifer medium. Optimal eluting conditions were determined as a 15:1 liquid-to-solid ratio, two eluting cycles, an elution pH of 3, 60% n-propanol concentration, and a 2 h eluting duration. Under these conditions, n-propanol exhibits significantly enhanced performance, surpassing that of the blended non-ionic surfactant agent by two times more and deionized water eluting by three times further, thereby establishing its efficacy as a superior eluting agent for 4-methoxy-2-nitroaniline removal. This is also suitable for polycyclic aromatic hydrocarbons and other nitroaniline compounds such as 4-methoxy-3-nitroaniline, etc.

(5) Field pilot tests were conducted using treated groundwater washing, which revealed a significant increase in contaminant concentrations in the pumping wells and monitoring wells located downstream of the washing injection well. In contrast, in the upstream and lateral monitoring wells, the time for contaminant concentration increases was correspondingly delayed with increasing distance from the injection well. Water washing facilitates the transfer of contaminants adhering to aquifer medium particle surfaces into the aqueous solution through mechanisms such as physical washing, partial dissolution, and the promotion of contaminant migration. This approach is both cost-effective and environmentally friendly.

Author Contributions: Conceptualization and writing, Z.W.; data curation, K.L. and C.C.; methodology and project administration, Z.W., H.C. and H.Z.; visualization, Z.W. and K.L.; investigation, Y.Y. and H.P. All authors have read and agreed to the published version of the manuscript.

Funding: This work was supported by Ningxia Hui Autonomous Region 2024 Annual Special Project for Central Guiding Local Science and Technology Development: Green and Low-carbon Integrated Technology Research and Demonstration for High-salt Strong Acid Organic Polluted Groundwater Remediation in Zhongwei Industrial Park [grant number 2024FRD05070].

Data Availability Statement: The original contributions presented in this study are included in the article. Further inquiries can be directed to the corresponding author.

Acknowledgments: The authors would like to thank the editor and reviewers for their constructive and valuable comments and suggestions, which significantly improved the quality of this work.

Conflicts of Interest: Authors Chen Chen, Hong Zhu and Hao Pang were employed by Beijing Zhongdihongke Environmental Technology Co., Ltd. The remaining authors declare that the research was conducted in the absence of any commercial or financial relationships that could be construed as a potential conflict of interest.

References

1. Zhu, N.; Gu, L.; Yuan, H.; Lou, Z.; Wang, L.; Zhang, X. Degradation pathway of the naphthalene azo dye intermediate 1-diazo-2-naphthol-4-sulfonic acid using Fenton's reagent. *Water Res.* **2012**, *46*, 3859–3867. [CrossRef]
2. Panke, S.; Wubbolts, M. Advances in biocatalytic synthesis of pharmaceutical intermediates. *Curr. Opin. Chem. Biol.* **2005**, *9*, 188–194. [CrossRef]
3. Swaminathan, K.; Pachhade, K.; Sandhya, S. Decomposition of a dye intermediate, (H-acid) 1 amino-8-naphthol-3, 6 disulfonic acid in aqueous solution by ozonation. *Desalination* **2005**, *186*, 155–164. [CrossRef]
4. Rosli, M.M.; Patil, P.S.; Fun, H.K.; Razak, I.A.; Dharmaprakash, S.M. 4-methoxy-2-nitroaniline. *Acta Crystallogr. Sect. E* **2007**, *63*, 1039–1040. [CrossRef]
5. Hao, Y.; Su, Y. Study on the synthesis of 2-nitro-pmethoxyaniliine by method of solvent. *Coal Chem. Ind.* **2003**, *2*, 25–26.
6. Si, H.; Liu, X.; He, X.; Hou, Z. Synthesis of 2-nitro-4-methoxyaniline. *J. Tianjin Univ. Technol.* **2009**, *25*, 19–21.
7. Liu, D.; Dou, Y.; Gao, Y.; Zhao, S.; Zhuang, C.; Dai, G. The synthesis of the intermediate of omeprazole of N-(2-nitro-4-methoxy) phenylamine. *J. Jiangsu Norm. Univ. (Nat. Sci. Ed.)* **2003**, *21*, 25–27.
8. Khalaji, A.D.; Nikookar, M.; Das, D. Co (III), Ni (II), and Cu (II) complexes of bidentate N, O-donor Schiff base ligand derived from 4-methoxy-2-nitroaniline and salicylaldehyde: Synthesis, characterization, thermal studies and use as new precursors for metal oxides nanoparticles. *J. Therm. Anal. Calorim.* **2014**, *115*, 409–417. [CrossRef]
9. Contreras, R.H.; De Kowalewski, D.G.; Facelli, J.C. The NMR analysis of the methoxy-group conformation in 4-methoxy-2-nitroaniline. *J. Mol. Struct.* **1982**, *81*, 147–149. [CrossRef]
10. Gu, B.; Ji, W.; Huang, X.; Patil, P.S.; Dharmaprakash, S.M. Concentration-dependent two-photon absorption and subsequent excited-state absorption in 4-methoxy-2-nitroaniline. *J. Appl. Phys.* **2009**, *106*, 1245. [CrossRef]

11. Ravikumar, C.; Joe, I.H. Electronic absorption and vibrational spectra and nonlinear optical properties of 4-methoxy-2-nitroaniline. *Phys. Chem. Chem. Phys.* **2010**, *12*, 9452–9460. [CrossRef]
12. El-Mahalawy, A.M.; Wassel, A.R. Structural, optical and photoelectrical characteristics of 4-methoxy-2-nitroaniline for optoelectronic applications. *Mater. Sci. Semicond. Process.* **2020**, *116*, 105124. [CrossRef]
13. Pignatello, J.J.; Oliveros, E.; MacKay, A. Advanced oxidation processes for organic contaminant destruction based on the Fenton reaction and related chemistry. *Crit. Rev. Environ. Sci. Technol.* **2006**, *36*, 1–84. [CrossRef]
14. Yuan, F.; Zhang, J.; Chen, J.; Chen, H.; Barnie, S. In Situ Pumping–Injection Remediation of Strong Acid–High Salt Groundwater: Displacement–Neutralization Mechanism and Influence of Pore Blocking. *J. Water* **2022**, *14*, 2720. [CrossRef]
15. Sripriya, R.; Raja, K.M.; Santhosh, G.; Chandrasekaran, M.; Noel, M. The effect of structure of oil phase, surfactant and co-surfactant on the physicochemical and electrochemical properties of bicontinuous microemulsion. *J. Colloid Interface Sci.* **2007**, *314*, 712–717. [CrossRef]
16. Chai, J.L.; Zhao, J.R.; Gao, Y.H.; Yang, X.D.; Wu, C.J. Studies on the phase behavior of the microemulsions formed by sodium dodecyl sulfonate, sodium dodecyl sulfate and sodium dodecyl benzene sulfonate with a novel fishlike phase diagram. *Colloids Surf. A Physicochem. Eng. Asp.* **2007**, *302*, 31–35. [CrossRef]
17. Zhu, M. Phase Behavior and Solubilization Properties of Middle Phase Microemulsion System Formed by Surfactant Compounding. Master's Thesis, Shandong Normal University, Jinan, China, 2012.
18. Zeng, C.; Lü, J.; Zheng, H.; Chen, X.; Huang, B. Effect of solvent on the solvolysis liquefaction of sawdust with phosphotungstic acid under supercritical condition. *J. Fuel Chem. Technol.* **2016**, *44*, 342–348.
19. Yao, Z.; Lin, K.; Lu, Q.; Xu, K.; Zhang, J. Alcohol eluent for leaching and remediation of soil contaminated by polycyclic aromatic hydrocarbons. *Environ. Pollut. Contr.* **2017**, *39*, 1242–1245.
20. Lu, Y.; Chen, J.; Ling, T. Removal mechanism of polychlorinated biphenyls from soil by cosolvent. *Environ. Sci.* **2010**, *31*, 205–210.
21. Chen, S.; Liu, P.; Zhu, Z.; Liu, Q. Experimental study on surface tension of several alcohol aqueous solutions. *J. Beijing Jiaotong Univ.* **2008**, *32*, 112–115.
22. Yang, X. The Research on the Transport Characteristics of Surfactanthydrophobic Organic Compounds Aggregate in Porous Medium. Doctoral Thesis, Hunan University, Changsha, China, 2022; p. 004709.
23. Cheng, M.; Zeng, G.; Huang, D.; Yang, C.; Lai, C.; Zhang, C.; Liu, Y. Tween 80 surfactant-enhanced bioremediation: Toward a solution to the soil contamination by hydrophobic organic compounds. *Crit. Rev. Biotechnol.* **2018**, *38*, 17–30. [CrossRef]
24. Li, G.; Chen, L.; Ruan, Y.; Guo, Q.; Liao, X.; Zhang, B. Alkyl polyglycoside: A green and efficient surfactant for enhancing heavy oil recovery at high-temperature and high-salinity condition. *J. Pet. Explor. Prod. Technol.* **2019**, *9*, 2671–2680. [CrossRef]
25. Ning, D. Study and Application of Petroleum-Contaminated Soil by Ex-Situ Washing Remediation Technology. Master's Thesis, Hunan Agricultural University, Changsha, China, 2010.
26. Haigh, S.D. A review of the interaction of surfactants with organic contaminants in soil. *Sci. Total Environ.* **1996**, *185*, 161–170. [CrossRef]
27. Bao, L.; Li, J.; Wang, X.; Dong, Q.; Li, X.; Dong, J. Properties of coal-based alkylbenzene sulfonate surfactants compound system. *J/OL. Fine Chem.* **2024**, 1–14. [CrossRef]
28. Ma, S. The Solubilization of Polycyclic Aromatic Hydrocarbons (PAHs) in Phospholipid Microemulsion and Its Application to the Elution and Remediation of PAHs Contaminated Soil. Master's Thesis, Shenzhen University, Shenzhen, China, 2020.

processes

MDPI

Article

Effects of Reclaimed Water Supplementation on the Occurrence and Distribution Characteristics of Antibiotic Resistance Genes in a Recipient River

Xiaohui Zhao [1,2,*], Xiaofei Wang [3,4], Hang Lang [1,2], Panwei Zhang [1,2], Jie Ni [1,2] and Wenqiang Wu [1,2]

1 State Key Laboratory of Simulation and Regulation of Water Cycle in River Basin, China Institute of Water Resources and Hydropower Research, Beijing 100038, China; langhang@iwhr.com (H.L.); zhangpw@iwhr.com (P.Z.); nijie@iwhr.com (J.N.); wwqemail@126.com (W.W.)
2 Department of Water Ecology and Environment, China Institute of Water Resources and Hydropower Research, Beijing 100038, China
3 Faculty of Arts and Sciences, Beijing Normal University, Zhuhai 519087, China; wangxiaofei110107@163.com
4 Guangdong-Hong Kong Joint Laboratory for Water Security, Beijing Normal University, Zhuhai 519087, China
* Correspondence: zhaoxiaohuiforever@163.com

Abstract: Due to the serious scarcity of water resources and the aggravation of water pollution in northern China, replenishing reclaimed water can alleviate the water shortage problem in northern rivers to a certain extent, but has also become an important way for antibiotic resistance genes (ARGs) to spread into rivers. In order to study the characteristics of ARGs in reclaimed water recharge rivers (Wenyu River), the abundance and distribution of ARGs in a typical reclaimed water replenishment river were analyzed by metagenomic sequencing technology combined with river water quality. The results showed that: due to the input of reclaimed water, the water quality characteristics of downstream sampling points of the river were significantly different from those upstream. Water quality factors such as total organic carbon, chemical oxygen demand, chlorophyll, and total nitrogen tended to increase gradually. Quinolones and macrolides were the main types of antibiotics. A total of 1217 ARGs were detected in the reclaimed water river system, including multidrug resistance, macrolide-lincosamide-streptogramin (MLS), tetracycline, glycopeptide, peptide, aminocoumarin, etc. The average abundance of ARGs in reclaimed water was higher than that in rivers. Among them, multidrug ARGs existed most widely, which may gradually become the main trend of ARGs' evolutionary variation. RDA results revealed that the environmental factors EC and DO, as well as tetracycline antibiotics (TCs), may be important environmental factors affecting the distribution of ARGs.

Keywords: reclaimed water; river; antibiotics; antibiotic resistance genes; metagenome

1. Introduction

The progression of societal development has driven industrialization and urbanization, which has been accompanied by a growing demand for water resources. Currently, over 400 cities in China are facing the predicament of inadequate water supply [1]. To alleviate this issue, numerous cities have emphasized the exploration and utilization of new water sources, with wastewater recycling and resource recovery being recognized as crucial avenues to mitigate water scarcity. As an emerging type of supplementary water source, reclaimed water is frequently utilized as the primary source for urban greening and river replenishment due to its abundant and relatively stable water volume [2,3]. However, this practice can have negative consequences. Previous research has reported that reclaimed wastewater can significantly affect the physicochemical properties and biota structure of receiving water bodies, and even pose ecological risks [4], such as the risk of eutrophication caused by excessive nitrogen input and ecological risk caused by environmental hormones [5,6]. Among these, the risk of resistance gene transmission deserves attention.

The increasing emergence and spread of antibiotic resistance genes (ARGs) has become one of the hot issues of global public health concern [7–9]. China, being a major producer and consumer of antibiotics, is facing a serious problem of antibiotic resistance [10]. The massive use of antibiotics induces and accelerates the risk of antibiotic resistance gene generation, transmission, and the formation of drug-resistant bacteria [11]. Antibiotic resistance has emerged in human and animal pathogenic and commensal bacteria [12], and the increase and spread of antibiotic resistance has become a huge problem in the treatment of diseases, resulting in no effective medicines in the face of some deadly infections [12–14]. Unlike traditional chemical pollutants, ARGs exhibit unique environmental behaviors due to their inherent biological properties, such as replicability, dissemination, and environmental persistence [15]. Once these ARGs enter into human or animal pathogenic bacteria, they will pose a potential threat to human public health [9,16]. Studies have shown that resistance genes are widely present in the environment in rivers, soils, and activated sludge from sewage treatment plants [17]. Among them, river ecosystems are the main sites for harboring land-based pollutants, such as the discharge of municipal sewage and farm effluents, leading to the possibility that rivers could be potential reservoirs for resistance genes [18]. Due to factors such as economic and technological development, ARGs carried in reclaimed water are discharged into rivers, which become an important pathway for the diffusion and spread of ARGs into the environment, causing environmental pollution and posing a potential threat to human health, and their environmental behaviors and characteristics have received extensive attention from researchers [19,20].

This study aimed to investigate the distribution characteristics of ARGs, and key environmental factors affecting the distribution of ARGs under reclaimed water recharge stress in the Wenyu River, a core area and "mother river" of Beijing. The Wenyu River is a major sink that receives reclaimed water (2.81×10^8 kiloton/year) from five wastewater treatment plants (WWTPs) that account for the majority of its flow. Therefore, the Wenyu River is an ideal region for investigating the impact of reclaimed water [21]. With the help of modern molecular biology macro-genome sequencing technology and bioinformatic analysis, the diversity and distribution of antibiotic resistance genes in a river replenished by reclaimed water were identified to provide a theoretical basis for the resourceful use of reclaimed water and the control of ecological risks.

2. Materials and Methods

2.1. Study Area and Sampling

The Wenyu River originates in Beijing and has water flowing year-round [22]. It is the upstream reach of the Beijing-Hangzhou Grand Canal, which is located in the core area of Beijing and is regarded as the "Mother River" of Beijing [22]. The Wenyu River is formed by the confluence of three main tributaries, respectively, the Dongsha River, the Beisha River, and the Nansha River, and receives tributaries such as the Qinghe River, the Beixiahe River, and other tributaries along the way, and then merges into the North Canal at the Beiguan Gate in Tongzhou with a total length of 55 km and a watershed area of 2478 km^2 [23,24]. It undertakes the drainage of the Changping District, the Shunyi District, and other parts of the urban area and receives 90% of Beijing's drainage tasks [25]. In August 2022, surface water samples were collected from the main stream of the Wenyu River and the effluents of the five WWTPs, and the information on the sampling sites is shown in Figure 1.

2.2. Sample Collection and Storage

Surface water samples were collected using a stainless steel water collector, which was cleaned three times with ultrapure water before collecting samples, and then rinsed three times with water samples from the sampling area during sampling. Samples for physical and chemical testing of water quality were contained in brown glass bottles soaked in acid and cleaned, and samples for microbial DNA extraction and sequencing were contained in 1.5 L polyethylene sampling bottles cleaned and aseptically processed in advance.

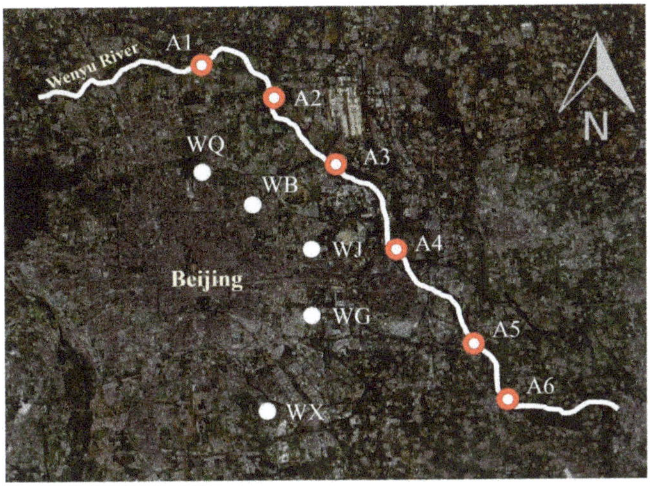

Figure 1. Schematic diagram of study area and sampling point locations (WQ, WB, WJ, WG and WX, represented with the white dot) and river sites (A1–A6, represented with the red circle).

2.3. Chemical Analysis

Water temperature (T), pH, oxidation-reduction potential (ORP), conductivity, and other conventional water quality indicators were measured on-site using a HACH HQ30D (USA). Chemical oxygen demand (COD), total phosphorus (TP), total nitrogen (TN), and other indicators refer to the Monitoring and analysis methods of previously described protocol [21]. Chlorophyll a was determined by spectrophotometry [26], and ammonia nitrogen (NH_3-N), nitrate nitrogen (NO_3-N), and nitrite nitrogen (NO_2-N) were determined by Nessler's reagent photometry [27], UV spectrophotometry [28], and spectrophotometry [29], respectively. The detection of pharmaceuticals and personal care products (PPCPs) in water was carried out using solid phase extraction (SPE) followed by ultraperformance liquid chromatography–triple quadrupole mass spectrometry [30].

2.4. Biological Analysis

Total microbial DNA was extracted from the samples, and the total amount and purity of DNA samples were examined. The hypervariable V3–V4 region of the bacterial 16S rRNA gene in each sample was amplified using a universal primer set of 515F (5′-GTGCCAGCMGCCGCGGTAA-3′) and 909R (5′-CCCCGYCAATTCMTTTRAGT-3′) with a 12 nt unique barcode. PCR amplification was performed according to a previously described method [31]. The PCR products were tested with 1% agarose gel electrophoresis and a NanoDrop Spectrophotometer. The PE150 bipartite sequencing was performed using the Illumina HiSeq Xten platform [32]. Firstly, the quality control of the original sequences was carried out, removing the sequences with a base number greater than 40 bp but a mass fraction lower than 38, removing the sequences with more than 10 divergent nucleotides, and removing the sequences with a 15 bp overlap with the adapter sequence. SOAP-denovo software (version 2.2.4) was used to assemble clean reads [33].

2.5. Data Analysis

Trend analysis was performed using Microsoft Excel (Version 2019). Statistical analysis was carried out with SPSS software (version 20.0). Redundancy analysis (RDA) and non-metric multidimensional scaling analysis (NMDS) was processed and plotted with Canoco 5.0. Principal coordinate analysis (PCoA) and constrained principal coordinate analysis (CP-CoA) were processed and plotted with R (4.0.3) using its amplicon package. SOAPaligner software (SOAP2, version 2.20) was used to compare the retained clean reads with the gene library to calculate the number of matches. Unigene was compared with the NCBI-NR database using DIAMOND software, setting the BLASTP parameter E-value $\leq 1 \times 10^{-5}$, and annotated using the LCA algorithm of MEGEN software (version 1.3.4-1). The sequencing depth should have reached at least 10,000×, and the representative sequences with ≥97% similarity were assigned to the same OTUs. The ARGs were annotated using the comprehensive antibiotic resistance database (CARD) [34].

3. Results and Discussion

3.1. Physicochemical Indices

An in-depth analysis of the physicochemical attributes of water samples collected from various points along the Wenyu River and the sewage treatment facility utilizing principal coordinate analysis (PCoA), revealed notable disparities in water quality between river samples and reclaimed water (Figures 2 and 3). Nevertheless, a noteworthy variation emerged in the water quality at the downstream river sampling stations (A4–A6), as compared to the upstream location (A1), potentially attributed to the substantial influx of reclaimed water. Electrical conductivity (EC), a metric that quantifies a solution's capacity to conduct electricity, serves as a proxy for dissolved mineral content fluctuations [35]. Within this study area, the EC of reclaimed water stood notably higher at 939.40 µs/cm, contrasted with 888.5 µs/cm for river water, reflecting the elevated ionic concentrations inherent in the reclaimed water.

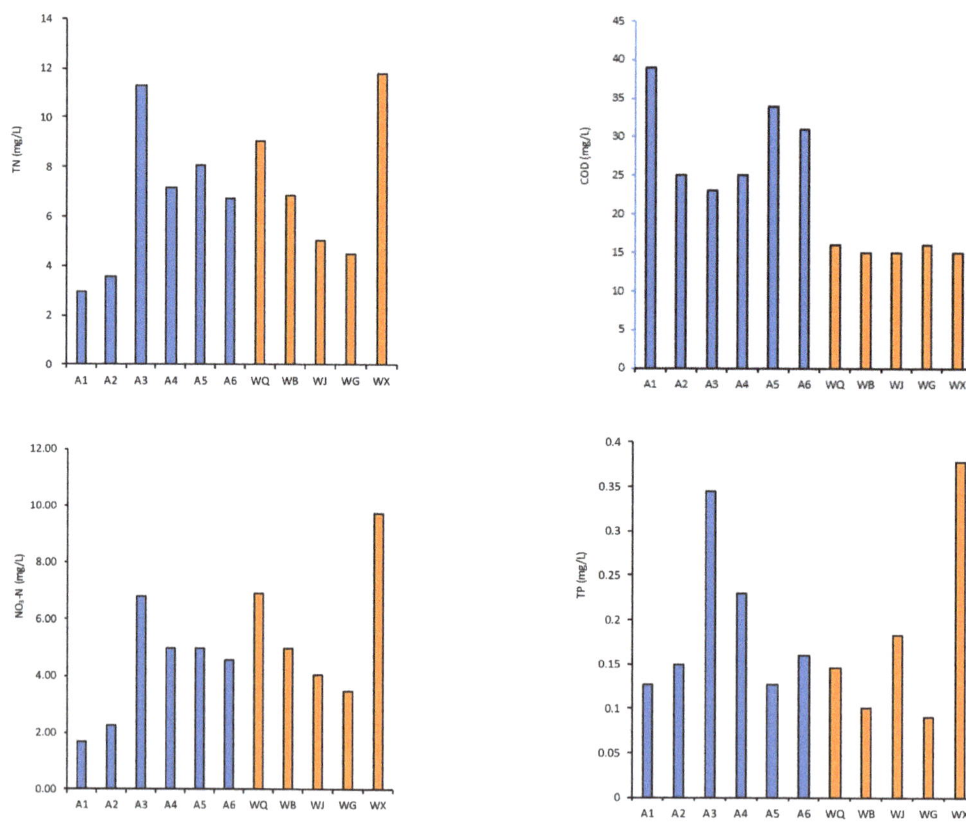

Figure 2. *Cont.*

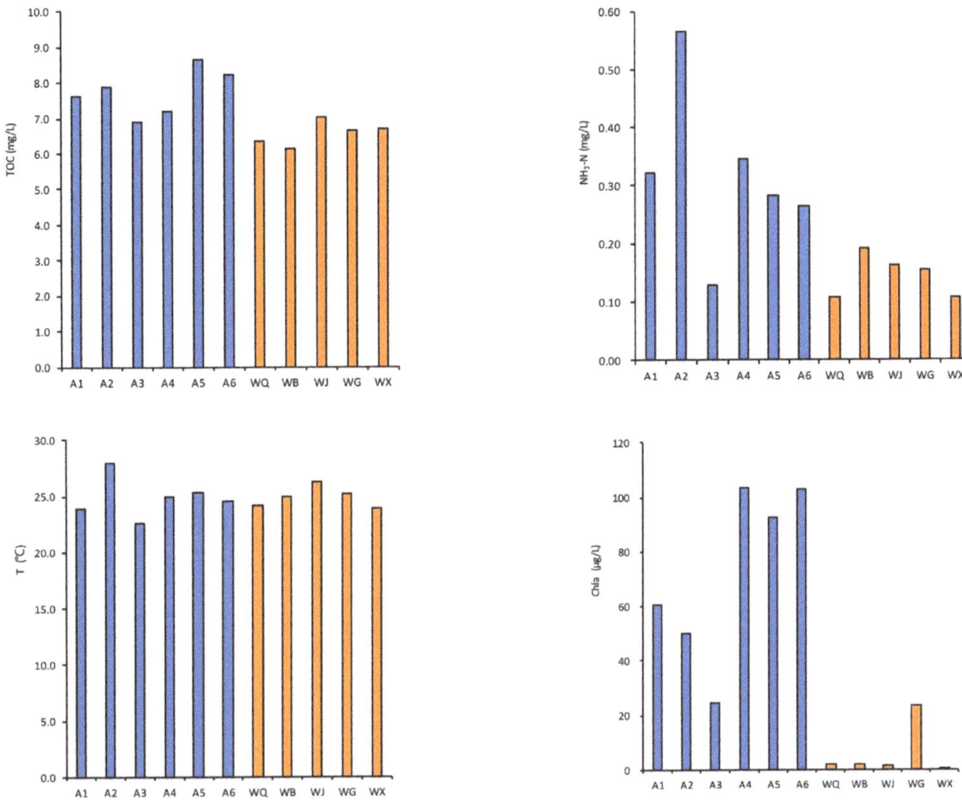

Figure 2. Variation of water quality parameters of the effluents of WWTP sampling sites (WQ, WB, WJ, WG and WX, represented with the color blue) and river sites (A1–A6, represented with the color yellow) in the Wenyu River basin.

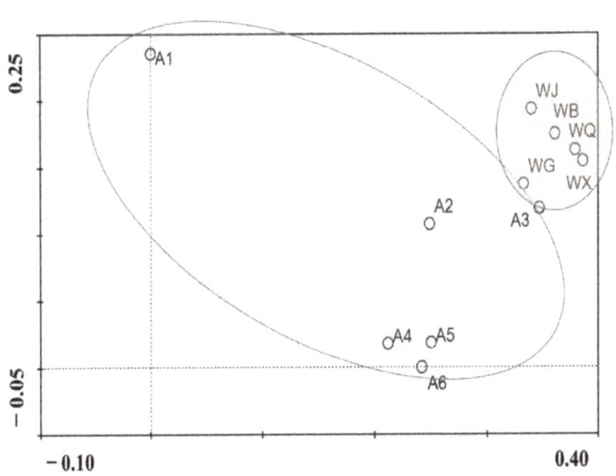

Figure 3. The diversity of the water quality parameters between river (Green ellipse) and WWTP sample (Red ellipse) sites is demonstrated by Principal coordinate analyses (PCoA).

The pH levels of the water samples ranged from 7.85 to 8.84, indicating a weakly alkaline nature with robust buffering capabilities. Conversely, the effluents from the reclaimed water plants exhibited a pH spectrum of 7.63 to 8.91 across the tributaries. Intriguingly, the pH of the Wenyu River water demonstrated a gradual increase from upstream to downstream, with its average value exceeding that of the reclaimed water plants' effluents.

The mean chlorophyll a (Chla) content in the river amounted to 72.31 μg/L, peaking at 103.32 μg/L. Parallel to this, the dissolved oxygen (DO) levels followed a comparable pattern, with the river water system maintaining a generally higher DO concentration than the effluent. The total organic carbon (TOC) and chemical oxygen demand (COD) concentrations exhibited an increasing trend from the upper reaches to the lower reaches along the mainstream. The COD concentration in the effluents from the sewage treatment plants within the study area is lower than that in the river, primarily attributed to the adoption of advanced sewage treatment processes. Through physical, chemical, and biological methods, organic matter in sewage can be effectively removed, leading to a marked reduction in COD values. Furthermore, the sewage treatment plants adhere to stringent discharge standards, ensuring that the quality of their effluents meets or exceeds environmental protection requirements. The DO concentration in rivers is influenced by a myriad of natural factors, including water temperature, atmospheric pressure, flow velocity, and the photosynthetic activity of aquatic plants, among others. Conversely, the DO concentration in the effluents of sewage treatment plants typically hinges on the sewage treatment processes employed. Because of the consumption of DO in the secondary sedimentation tank and aerobic tank in the sewage treatment process, the DO concentration in the river may be higher than that in the effluents from the sewage treatment plants.

In terms of nutrient salt concentration in water bodies, the total nitrogen (TN) concentration range in river water was 2.94–11.30 mg/L. The effluents emanating from the reclaimed water facility exhibited an average concentration of 7.45 mg/L, surpassing the river's average of 6.63 mg/L. Along the course of the Wenyu River, a discernible trend of rising total nitrogen levels is observed, progressing from the upstream reaches towards the downstream end. This augmentation primarily stems from the influx of nitrogen-laden reclaimed water. Nitrate nitrogen (NO_3-N) concentration had a similar trend to that of TN, with river NO_3-N values (mean 4.21 mg/L) lower than reclaimed water effluents (mean 5.81 mg/L). The concentration of ammonia nitrogen (NH_3-N) in the river water (mean 0.32 mg/L) was higher than that in the reclaimed water (mean 0.15 mg/L). Total phosphorus (TP) concentrations in the river ranged from 0.13 to 0.34 mg/L, which was similar to the reclaimed water effluents' concentration levels (0.09 to 0.38 mg/L).

3.2. *Characterization of Antibiotic Distribution in a River Replenished by Reclaimed Water*

The main antibiotic concentrations in the water bodies of the Wenyu River and the effluents from the sewage treatment plants were shown in Figure 4. Eleven antibiotics were detected to varying degrees at all sampling points of the Wenyu River and the effluents from the sewage treatment plants. The average antibiotic concentration at the sampling points of the Wenyu River was 64.5 μg/L, with the highest concentration of 146.9 μg/L at A3. The average antibiotic concentration at the sampling points for the effluents from the sewage treatment plants was 84.1 μg/L, with the highest antibiotic concentration of 124.9 μg/L in WQ. Among different kinds of antibiotics, quinolones accounted for the highest proportion, accounting for 59.6%, followed by macrolide (34.5%), chloramphenicol (3.5%) and tetracycline (2.3%); at the reclaimed water sampling points, quinolones accounted for 84.7%, followed by macrolides (10.2%), tetracyclines (2.8%) and chloramphenicol (2.3%). This result is consistent with the conclusions of relevant research on recycled water in China, such as Li et al., who studied quinolones and found they were the dominant antibiotics detected in 45 wastewater treatment plants in 23 cities in China, followed by macrolides, and sulfonamides. High concentrations of quinolones in sewage sludge indicated that antibiotics are widely used and extensive pollutants in China [36]. In this study, the detected concentration of quinolone antibiotics in the Wenyu River water ranged from 21.4 to 76.5 μg/L, with ofloxacin and norfloxacin exhibiting higher average concentrations and 100% detection rates, reaching 7.9 μg/L and 21.9 μg/L, respectively. The detected concentration of macrolide antibiotics ranged from 10.2 to 60.4 μg/L, with clindamycin having a higher average concentration (9.2 μg/L) and a 100% detection rate. In the effluents of sewage treatment plants, the detected concentration of quinolone antibiotics ranged from 30.6 to 108.7 μg/L, with ofloxacin and norfloxacin exhibiting higher average concentrations and 100% detection rates, reaching 27.5 μg/L and 35.6 μg/L, respectively. The detected concentration of macrolide antibiotics ranged from 0.99 to 14.0 μg/L, with clindamycin having a higher average concentration (5.9 μg/L) and a 100% detection rate.

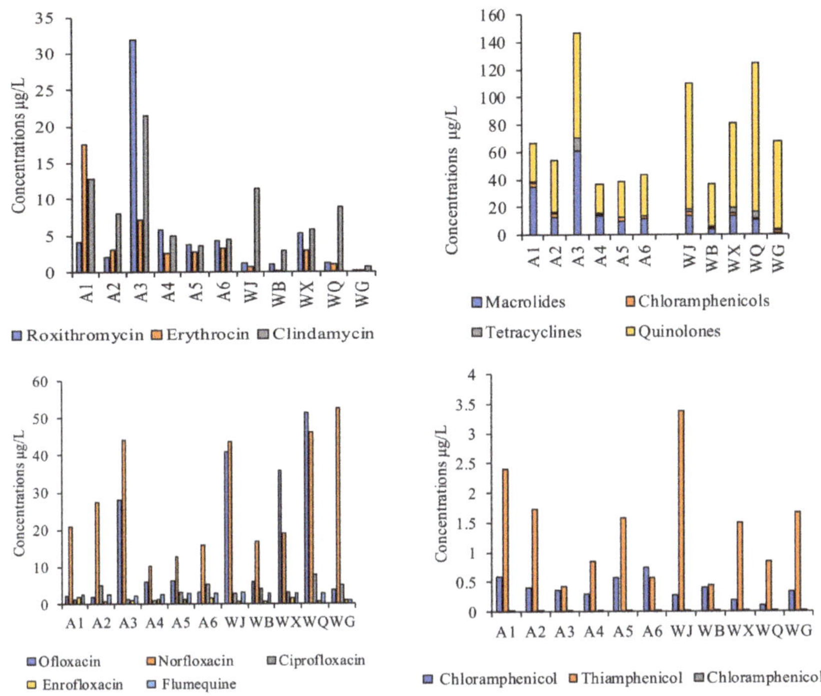

Figure 4. Concentrations of antibiotics in different sampling sites.

3.3. Diversity of Antibiotic Resistance Genes in a River Replenished by Reclaimed Water

A total of 1217 ARGs were detected in the reclaimed water river system, which were classified into 21 categories according to antibiotic resistance categories, including multidrug, macrolide-lincosamide-streptogramin (MLS), tetracycline, glycopeptide, peptide, aminocoumarin, and others (Figure 5). The abundance of different types of resistance genes was as follows: multidrug (643,362 reads/L) > MLS (175,918 reads/L) > tetracycline (148,575 reads/L) > glycopeptide (126,129 reads/L) > peptide (94,859 reads/L) > aminocoumarin (68,281 reads/L).

The abundance of ARGs was also different at different sampling points. The average ARGs' gene abundance in river water (73,950 reads/L) was lower than that in reclaimed water (75,289 reads/L). Meanwhile, numerous previous studies have found that the abundance of ARGs in the effluents of WWTPs was much higher than that in river water. Hang et al., for example, studied the effects of reclaimed water recharge on bacterial communities and ARGs in urban landscape water bodies [37], and found that the amount of specific ARGs in urban landscape water bodies replenished by reclaimed water was relatively increased. This may be explained by the effects of horizontal gene transfer (HGT) and wastewater treatment processes on ARGs. The effluent is prepared from sewage through the advanced treatment process. During the treatment process, a considerable number of

ARGs may be greatly removed. It also had been suggested that most of the microorganisms and ARGs in the reclaimed water came from the pipe network and may exist in dark conditions during transportation and migration, making their response to sunlight irradiation more drastic, leading to gradual attenuation after entering the receiving river. However, it is worth noting that domestic wastewater containing a large amount of residual antibiotics and ARGs may create an ideal environment for the physical transportation, acquisition and spread of ARGs from wastewater treatment systems into rivers.

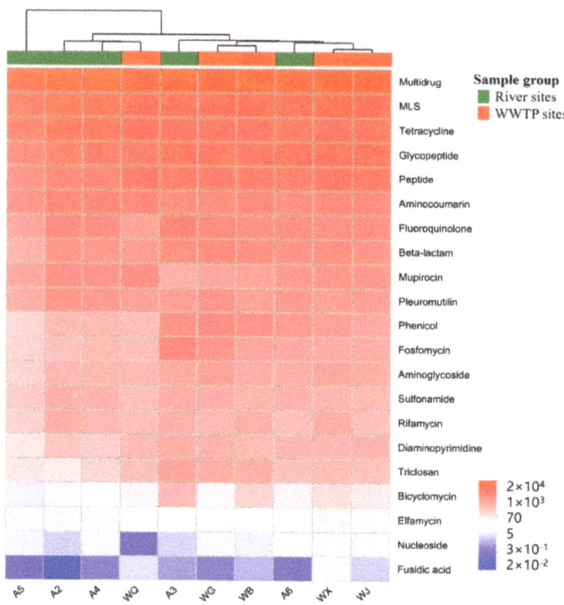

Figure 5. Heatmap of ARGs' abundance at different sampling points.

Among the detected resistance gene classes, multidrug and MLS were the two most abundant classes at all sites, and all had lower average abundance in river water than in reclaimed water. Combined with the main antibiotic classes detected in river surface water, the degree of antibiotic residues can affect the production, migration, and spread of some ARGs to a certain extent. The ARGs detected in the river and reclaimed water bodies mainly contained 442 β-lactamase resistance genes, 188 multidrug resistance genes, 132 aminoglycoside resistance genes, 98 MLS resistance genes, 70 glycopeptide resistance genes, 62 tetracycline resistance genes, etc. Among them, β-lactamase accounted for the highest proportion (36.3%), followed by multidrug resistance genes (15.4%), aminoglycoside antibiotics (10.8%), MLS (8.1%), glycopeptides (5.8%), and tetracyclines (5.1%). The number of β-lactamase species is 36%, but its abundance is only 2.9%, and the number of aminoglycoside antibiotics reached 10.8%, but their abundance was only 0.8%. Resistance genes not only had a high abundance (40.5%) but also a high number of detected species (36.3%). This showed that under the long-term selection of multiple antibiotics, microorganisms in rivers and reclaimed water bodies had screened and evolved ARGs that were resistant to multiple antibiotics and more adaptable to complex and changeable environments, thus promoting the survival and reproduction of multi-resistant ARGs and multi-antibiotic resistant microorganisms.

3.4. Diversity of Multi-Resistant ARGs in a River Replenished by Reclaimed Water

As a category with a relatively high number of species and abundance, multi-antibiotic resistance genes (MARGs) were detected with an average abundance of 643,362 reads/L in the surface water of the Wenyu River and reclaimed water (Figure 6 and Supplementary Table S1). Specifically, the average abundance in the surface water of the Wenyu River was 620,105 reads/L, while that in reclaimed water was 666,618 reads/L. A total of 188 MARGs were identified in both water bodies, with 185 species detected in the surface water of the Wenyu River and an equal number found in reclaimed water. Notably, the surface water of the Wenyu River harbors unique MARGs such as H-NS, LAP-2, and mdtO, whereas AAC(6′)-Ib-cr and *Klebsiella pneumoniae* KpnF are exclusive to reclaimed water. Regarding the abundance ranking of MARGs in the Wenyu River, the most prevalent is msbA (42,972 reads/L), followed by evgS (254,984 reads/L), smeS (33,809 reads/L), mtrA (31,880 reads/L), and rpoB2 (22,505 reads/L). The abundance of multi-antibiotic resistance genes (MARGs) in reclaimed water was msbA (44,956 reads/L) > evgS (30,578 reads/L) > adeL (29,028 reads/L) > smeS (21,804 reads/L) > arlS (21,427 reads/L). Based on the clustering analysis of MARGs, significant differences were observed in the microbial community structures between river water and reclaimed water. Specifically, the compositions of MARGs in samples A2 and A4 from the Wenyu River are similar to those in the reclaimed water WQ, whereas the downstream site A6 in the Wenyu River shares similar MARG compositions with reclaimed water samples WX and WJ. This indicates that the presence of wastewater treatment plants exerts an influence on the microbial functional structure of local river segments.

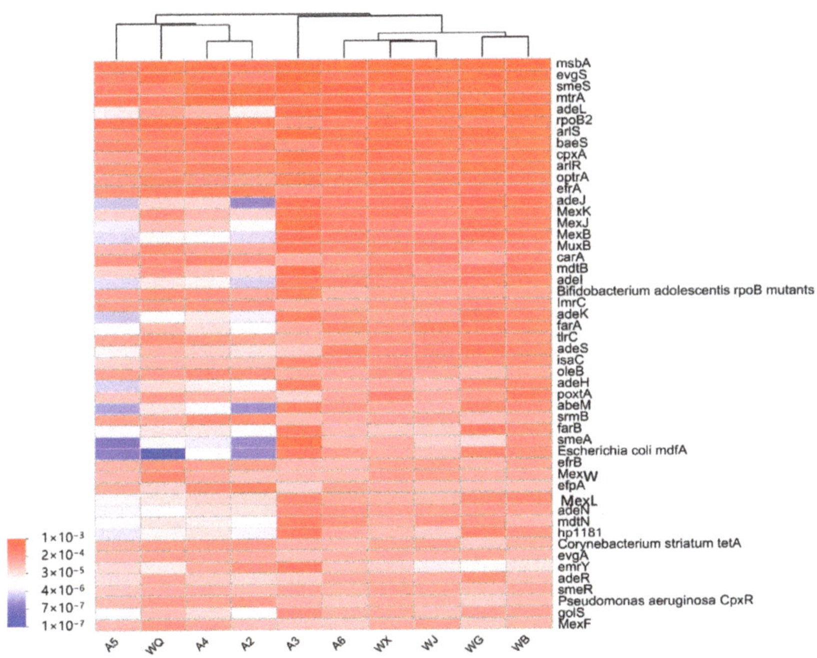

Figure 6. Heatmap of multi-resistant ARGs' abundance at different sampling points.

The resistance mechanisms of multi-antibiotic resistance genes (MARGs) in the surface water samples collected from the Wenyu River and reclaimed water sites are statistically annotated and illustrated in Figure 7. The major resistance mechanisms of MARGs can be categorized into eight types, including antibiotic efflux (140 species, 78%), antibiotic target protection (25 species, 14%), antibiotic target alteration (7 species, 4%), antibiotic target alteration/antibiotic target replacement (2 species, 1%). Across different water bodies, the abundance of these eight resistance mechanisms consistently follows antibiotic efflux > antibiotic target protection > antibiotic target alteration/antibiotic target replacement > antibiotic target alteration. The top five most abundant MARGs in reclaimed water all belong to the antibiotic efflux mechanism, while the most abundant five MARGs in the Wenyu River comprise both antibiotic efflux and antibiotic target alteration/antibiotic target replacement. There is a significant difference in abundance between antibiotic efflux and other resistance mechanisms ($p < 0.01$).

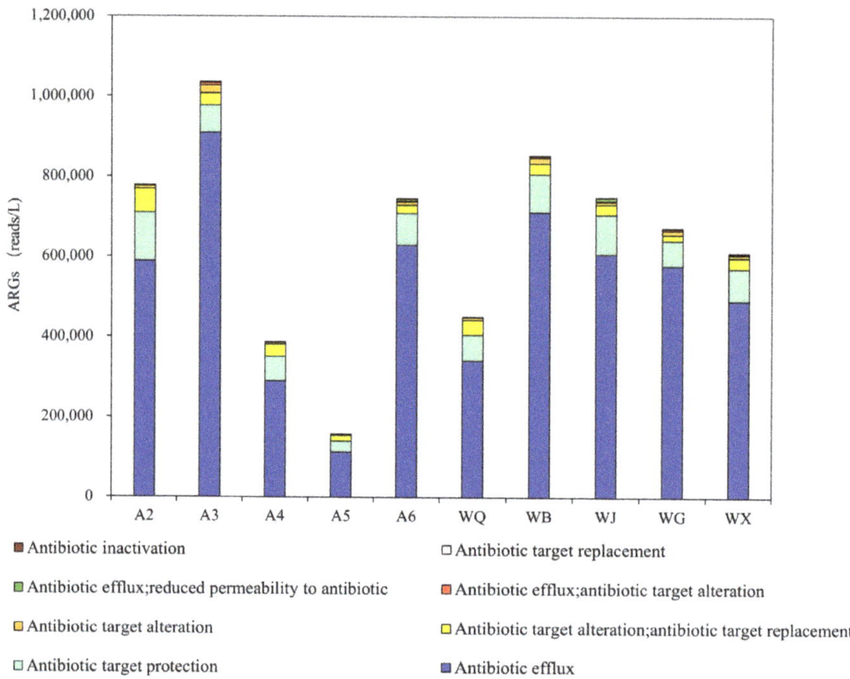

Figure 7. Mechanism of multi-drug resistance genes in the surface water of the Wenyu River and reclaimed water sampling sites.

3.5. Correlation Analysis of Antibiotic Resistance Gene Distribution and Environmental Factors

Correlation analysis between the distribution of ARGs and multiple environmental factors (including temperature, COD, pH, DO, electrical conductivity (EC), and antibiotics) was conducted using the RDA method. The results revealed that ARGs explained 94.65% and 3.05% of the variations along Axes I and II, respectively, suggesting a close relationship between ARGs' abundance and environmental factors to a certain extent. As depicted in Figure 8, the physicochemical factor EC and tetracycline antibiotics (TCs) were intimately associated with the distribution of resistance genes in the surface water of the study area, aligning with previous research findings. Notably, DO emerged as a crucial parameter influencing ARGs' distribution, as it reflects the self-purification capacity of water. This observation concurs with previous reports of a strong negative correlation between DO and various ARGs, such as sul2, blaTEM, and four tetracycline resistance genes. In this study, DO also showed a negative correlation with multiple ARGs, including rifamycin resistance genes. The figure further indicated a positive correlation between TCs and multi-resistance genes ($r = 0.64848$, $p < 0.05$), while the relationship with tetracycline resistance genes was not significant. EC displayed a positive correlation with both multi-resistance and aminocoumarin resistance genes. Overall, the combined effects of antibiotics and physicochemical factors significantly influence the distribution of ARGs.

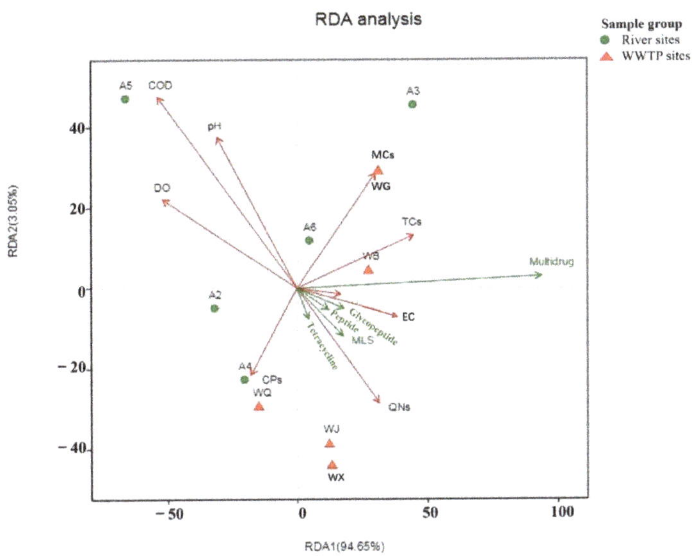

Figure 8. RDA analysis of total ARGs and environmental factors in the surface water of sampling sites in the Wenyu River and reclaimed water.

4. Conclusions

(1) As a result of significant quantities of recycled water being introduced, the water quality downstream of the river noticeably varies from that of the upstream. With the input of reclaimed water, the total organic carbon (TOC), chemical oxygen demand (COD), chlorophyll (Chla), and total nitrogen (TN) in the surface water of the Wenyu River gradually increased.

(2) Eleven types of antibiotics were detected in the river water and sewage treatment plants' effluents. Among them, quinolones and macrolides were the main types of antibiotics. The distribution of ARGs in river water bodies was significantly different from that of reclaimed water, and the abundance of ARGs in river water bodies was lower than that of reclaimed water, with multi-resistance genes the most widespread class, which may gradually become the main trend of evolutionary variation of ARGs.

(3) Through the correlation analysis between the distribution of ARGs and environmental variables, the environmental factors EC and DO, as well as tetracycline antibiotics (TCs) among antibiotics, may be important environmental factors affecting the distribution of ARGs.

Supplementary Materials: The following supporting information can be downloaded at: https://www.mdpi.com/article/10.3390/pr12081717/s1, Table S1: The ARGs conferring resistance to different antibiotic classes, multiple drugs, and MGEs detected in the effluents of WWTPs and river sites.

Author Contributions: Conceptualization and Writing, X.Z.; data curation, H.L.; funding acquisition, J.N.; methodology and project administration, X.W. and W.W.; supervision, P.Z. All authors have read and agreed to the published version of the manuscript.

Funding: This research was funded by the National Key R&D Program of China, grant number [2022YFB2602000].

Data Availability Statement: Data will be made available on request.

Conflicts of Interest: The authors declare no conflict of interest.

References

1. Geng, Y.; Wang, M.; Sarkis, J.; Xue, B.; Zhang, L.; Fujita, T.; Yu, X.; Ren, W.; Zhang, L.; Dong, H. Spatial-temporal patterns and driving factors for industrial wastewater emission in China. *J. Clean. Prod.* **2014**, *76*, 116–124. [CrossRef]
2. Barker-Reid, F.; Fox, E.M.; Faggian, R. Occurrence of antibiotic resistance genes in reclaimed water and river water in the Werribee Basin, Australia. *J. Water Health* **2010**, *8*, 521–531. [CrossRef]
3. Cai, Y.; Bi, Y.; Tian, B.; Cheng, L.; Zhou, S.; Qi, Q. Water quality characteristics and ecological risk evaluation of a landscaped river replenished by three reclaimed water sources in Qingdao, China. *Environ. Sci. Pollut. Res.* **2024**, *31*, 35609–35618. [CrossRef]
4. White, S.A. Pharmaceuticals and Persistent Organic Micropollutants in Reclaimed Irrigation Water. In Proceedings of the 2013 ASHS Annual Conference, Palm Desert, CA, USA, 24 July 2013.
5. Du, C.; Guo, W.; Li, G.; Bai, M.; Zhu, Q.; Tian, Z.; Li, M.; Zhao, C.; Zhang, L. Biomanipulation as a strategy for minimizing ecological risks in river supplied with reclaimed water. *Environ. Res. Sect. A* **2023**, *228*, 115801. [CrossRef] [PubMed]
6. Zhong-Hong, C.; Yong-Min, M.A.; Wei, H.U. Endocrine Disrupting Compounds in Reclaimed Water Produced by a Certain Water Plant: Some Preliminary Results. *J. Environ. Health* **2005**, *22*, 171–173.
7. Zhu, G.; Xue, C.; Fan, H.; Wu, N. Research progress of antibiotic resistance genes in environment. *Prev. Med.* **2020**, *32*, 1121–1125.
8. He, R.; Yuan, K.; Lin, L.; Yang, Y.; Zou, S.; Luan, T.; Chen, B. Functional metagenomics: One of the most robust tools for discovering new antibiotics resistance genes. *Environ. Chem.* **2019**, *7*, 1548–1556.
9. Zhang, Z.; Zhang, Q.; Wang, T.; Xu, N.; Lu, T.; Hong, W.; Penuelas, J.; Gillings, M.; Wang, M.; Gao, W.; et al. Assessment of global health risk of antibiotic resistance genes. *Nat. Commun.* **2022**, *13*, 1553. [CrossRef] [PubMed]
10. Fan, Y.; Hao, F.; Wang, W.; Lu, Y.; He, L.; Wang, G.; Chen, W. Multicenter cross-sectional observational study of antibiotic resistance and the genotypes of *Propionibacterium acnes* isolated from Chinese patients with acne vulgaris. *J. Dermatol.* **2016**, *43*, 406–413. [CrossRef]
11. Mahaney, A.; Franklin, R. Persistence of wastewater-associated antibiotic resistant bacteria in river microcosms. *Sci. Total Environ.* **2022**, *819*, 153099. [CrossRef]
12. Caroline, P.; Orosco, W.G.; Deon, K.J.; Marisa, C. Genotyping and antimicrobial resistance in Escherichia coli from pig carcasses. *Pesqui. Veterinária Bras.* **2017**, *37*, 1253–1260.
13. Lian, C.; Piksa, M.; Yoshida, K.; Persheyev, S.; Pawlik, K.J.; Matczyszyn, K.; Samuel, I.D.W. Flexible organic light-emitting diodes for antimicrobial photodynamic therapy. *Npj Flex. Electron.* **2019**, *3*, 18. [CrossRef]
14. Nan, W.U. Tetracycline Residues and Tetracycline Resistance Gene Pollution in Soil:A Review. *Asian J. Ecotoxicol.* **2010**, *5*, 618–627.
15. Matarese, V. Kinds of Replicability: Different Terms and Different Functions. *Axiomathes* **2022**, *32*, 647–670. [CrossRef]
16. Dafale, N.A.; Srivastava, S.; Purohit, H.J. Zoonosis: An Emerging Link to Antibiotic Resistance Under "One Health Approach". *Indian J. Microbiol.* **2020**, *60*, 139–152. [CrossRef]
17. Xu, J.; Xu, Y.; Wang, H.; Guo, C.; Qiu, H.; He, Y.; Zhang, Y.; Li, X.; Meng, W. Occurrence of antibiotics and antibiotic resistance genes in a sewage treatment plant and its effluent-receiving river. *Chemosphere* **2015**, *119*, 1379–1385. [CrossRef]
18. Szmolka, A.; Wami, H.; Dobrindt, U. Comparative Genomics of Emerging Lineages and Mobile Resistomes of Contemporary Broiler Strains of *Salmonella* Infantis and *E. coli*. *Front. Microbiol.* **2021**, *12*, 642125. [CrossRef]
19. Guan, X.; Guo, Z.; Wang, X.; Xiang, S.; Sun, T.; Zhao, R.; He, J.; Liu, F. Transfer route and driving forces of antibiotic resistance genes from reclaimed water to groundwater. *Environ. Pollut.* **2023**, *330*, 121800. [CrossRef]
20. Wang, C.; Hong, P.Y. Genome-Resolved Metagenomics and Antibiotic Resistance Genes Analysis in Reclaimed Water Distribution Systems. *Water* **2020**, *12*, 3477. [CrossRef]
21. Zhao, X.; Xie, E. Reclaimed water influences bacterioplankton and bacteriobenthos communities differently in river networks. *Water Res.* **2023**, *243*, 120389. [CrossRef]
22. Linjin, L.; Baohui, M.; Rui, P. Water Quality Evaluation of Wenyu River Based on Single Factor Evaluation and Comprehensive Pollution Index Method. *Nat. Environ. Pollut. Technol.* **2021**, *20*, 1041–1046. [CrossRef]
23. Mengyuan, Z.; Li, Y.; Fan, S.; Dong, L.; Hao, P. Study on Greenway Plant Landscape Based on Bird Habitat Conservation—A Case Study of Wenyu River—North Canal Greenway in Beijing. In Proceedings of the Fábos Conference on Landscape and Greenway Planning, Amherst, MA, USA, 28–30 March 2019.
24. Qi, W.; Liu, H.; Pernet-Coudrier, B. Polycyclic aromatic hydrocarbons in wastewater, WWTPs effluents and in the recipient waters of Beijing, China. *Environ. Sci. Pollut. Res.* **2013**, *20*, 4254–4260. [CrossRef] [PubMed]
25. Wang, Z.Y.; Liu, Y.; Wang, J.L.; Zhao, J.Z. Factor Analysis of Heavy Metals in the Sediment in Changping Section of the Wenyu River and Their Pollution Assessment. *Res. Environ. Sci.* **2013**, *26*, 838–843.
26. Ergun, E.; Demirata, B.; Gumus, G.; Apak, R. Simultaneous determination of chlorophyll a and chlorophyll b by derivative spectrophotometry. *Anal Bioanal Chem.* **2004**, *379*, 803–811. [CrossRef]
27. Xian, G.; Fen, C.; Lan, Q. Application Research of Environmental Monitoring on Ammonia Nitrogen Determination. *J. Green Sci. Technol.* **2017**, *18*, 117–119.
28. Ji, X. Comparison on nitrate nitrogen by detections of RP-HPLC and thymol UV-Vis spectrophotometry. *J. Chang. Inst. Technol. (Nat. Sci. Ed.)* **2014**, *15*, 1–4.
29. Gou, Y. Determination of total nitrogen in water samples by spectrophotometry using phenol after alkaline peroxodisulfate digestion. *Bunseki Kagaku* **2001**, *50*, 481–486. [CrossRef]

30. Lara-Martín, P.A.; González-Mazo, E.; Pintado-Herrera, M.G.; Baena-Nogueras, R.M. Determination of Pharmaceuticals in Coastal Systems Using Solid Phase Extraction (SPE) Followed by Ultra Performance Liquid Chromatography—Tandem Mass Spectrometry (UPLC-MS/MS). *Curr. Anal. Chem.* **2016**, *12*, 183–201.
31. Xie, E.; Zhao, X.; Li, K.; Zhang, P.; Zhou, X.; Zhao, X. Microbial community structure in the river sediments from upstream of Guanting Reservoir: Potential impacts of reclaimed water recharge. *Sci. Total Environ.* **2021**, *766*, 142609. [CrossRef]
32. Shahbazi, Z.; Rostami, G.; Hamid, M. New heritableATRXmutation identified by whole exome sequencing and review. *Egyptian J. Med. Hum. Genet.* **2022**, *23*, 19. [CrossRef]
33. Wang, Z.; Chen, J.; Liu, W.; Luo, Z.; Wang, P.; Zhang, Y.; Zheng, R.; Shi, J. Transcriptome Characteristics and Six Alternative Expressed Genes Positively Correlated with the Phase Transition of Annual Cambial Activities in Chinese Fir (*Cunninghamia lanceolata* (Lamb.) Hook). *PLoS ONE* **2013**, *8*, e71562. [CrossRef] [PubMed]
34. Liu, L.; Shen, P.; Zheng, B.; Yu, W.; Xiao, Y. Comparative Genomic Analysis of 19 Clinical Isolates of Tigecycline-Resistant *Acinetobacter baumannii*. *Front. Microbiol.* **2020**, *11*, 1321. [CrossRef] [PubMed]
35. Jain, N.; Yevatikar, R.; Raxamwar, T.S. Comparative study of physico-chemical parameters and water quality index of river. *Mater. Today Proc.* **2022**, *60*, 859–867. [CrossRef]
36. Li, W.; Shi, Y.; Gao, L.; Liu, J.; Cai, Y. Occurrence, Distribution and Potential Affecting Factors of Antibiotics in Sewage Sludge of Wastewater Treatment Plants in China. *Sci. Total Environ.* **2013**, *445–446*, 306–313. [CrossRef]
37. Zhang, C.M.; Liang, J.; Liu, W.Y. Comparative study on the bacterial diversity and antibiotic resistance genes of urban landscape waters replenished by reclaimed water and surface water in Xi'an, China. *Environ. Sci. Pollut. Res.* **2021**, *28*, 41396–41406. [CrossRef] [PubMed]

processes

MDPI

Article

Effects of Irrigation with Slightly Algae-Contaminated Water on Soil Moisture, Nutrient Redistribution, and Microbial Community

Han Zheng [1], Xiaofei Wang [2,3], Chen Huang [4], Zhe Bao [5,*], Xiaohui Zhao [6], Zhanming Tan [7] and En Xie [1,*]

1 College of Water Resources and Civil Engineering, China Agricultural University, Beijing 100083, China; zhenghan171@cau.edu.cn
2 Faculty of Arts and Sciences, Beijing Normal University, Zhuhai 519087, China; wang20021204@yeah.net
3 Guangdong-Hong Kong Joint Laboratory for Water Security, Beijing Normal University, Zhuhai 519087, China
4 Songliao River Water Resources Commission of Ministry of Water Resources, Ministry of Water Resources of the People's Republic of China, Jilin 130000, China; 18301619816@163.com
5 Rural Energy and Environment Agency, Ministry of Agriculture and Rural Affairs, Beijing 100125, China
6 Department of Water Ecology and Environment, China Institute of Water Resources and Hydropower Research, Beijing 100038, China; zhaoxiaohuiforever@163.com
7 College of Horticulture and Forestry, Tarim University, Alar 843300, China; tlmdxtzm@taru.edu.cn
* Correspondence: baozhe1983@126.com (Z.B.); xe@cau.edu.cn (E.X.)

Abstract: The presence of excessive algae in water is always considered as a negative factor in irrigation. However, the optimal balance between algal removal and retention in irrigation water when the algal biomass is controllable remains unknown. Therefore, this study explores the impact of low-level algal presence (*Scytonema javanicum*) on soil and microbial activity through controlled soil column experiments. Soil moisture was measured, and 16S rRNA gene amplicons sequencing was applied to characterize the microbial community. Slight community changes indicated no negative impact on the local microbial community of *S. javanicum*. Enzyme assays and quantitative polymerase chain reaction (qPCR) revealed that algae improved soil moisture retention, and enhanced the nutrient content of the topsoil. The decrease in moisture in the treatment group (from 27.53% to 26.42%) was significantly reduced ($p < 0.05$) compared to the control (from 27.55% to 25.17%), while the contents of ammonium (NH_3-N) and total nitrogen (TN) in the treatment (0.70 mg/kg and 0.54 g/kg) were also higher ($p < 0.05$) than that of the control (0.43 mg/kg and 0.49 g/kg). The results of the abundance of functional gene suggested algae facilitated nitrogen fixation and nitrification. This research offers innovative insights for diversifying the sources of irrigation water.

Keywords: slightly algae-contaminated water; soil moisture content; nitrogen cycle; microbial community; irrigation

1. Introduction

Agricultural irrigation accounts for more than 63% of the total water consumption in China [1]. Although an increase in the irrigation water use coefficient has kept China's annual agricultural water consumption in a stable range, pollution of irrigation water sources [2] and increases in food demand [3] have made irrigation water resources increasingly scarce [4]. Expanding the source of irrigation water to include sources such as reclaimed water [5,6], brackish waters with low levels of pollution, is a promising strategy to address this problem. Among these strategies, irrigation with water containing low levels of algae has high economical and ecological potential. Water with high algal content is conventionally regarded as unsuitable for irrigation because of the potential toxicity of algae and [7] the risk of equipment clogging [8]. For example, microcystins released by algae can reduce the potential nitrification rate of soil and affect enzyme activity [9]. This necessitates additional treatment before such water is used, which increases cost. The pretreatment of slightly contaminated water containing small amounts of algae is even more time-consuming and laborious. However, algae-containing water often contains considerable amounts of nutrients that could be beneficial for improving soil fertility and quality [10]. This highlights the necessity of employing waters containing low levels of algal pollution wisely.

Algae could also improve the soil environment in various ways. The desert algal biocrust technique is a common tool for desertification control that can be used to maintain soil moisture content, improve soil nutrients, and regulate soil temperature [11–14]. Algal crust formation serves as a vital link between surface organisms and soil biological components and has ecological significance in areas with dry and nitrogen-deficient soils [15,16]. While algae actively exchange materials with their external environment through microbial life activities, they also induce changes in the microbial community structure. Microorganisms in turn regulate nutrient cycling and determine soil structure by participating in physiological and biochemical processes in soils, ultimately altering the structural characteristics and physicochemical properties of the algae [17]. The interaction between the two and their joint succession determines the growth of algae [18–20]. In addition, microorganisms play an important role in the nitrogen cycle [21] and are involved in the decomposition of organic matter and in humus formation [22,23]. Current research on genes related to soil N cycling has been focused on the relative abundance of individual genes involved in soil nitrification, denitrification and nitrogen fixation, such as *nifH*, *amoA*, *nirS*, and *nosZ* [24]. The soil N cycle is driven by functional genes carried by microorganisms involved in each key process. In the process, *nifH* genes converted N_2 to NH_3-N through nitrogen fixation, *amoA* and *nxrA* genes converted NH_3-N to NO_3-N through nitrification, and *narG*, *nirK*, *nirS*, and *nosZ* genes finally converted NO_3-N to N_2 through denitrification [25], completing the whole cycle.

However, the positive role of algae in agricultural irrigation has been largely overlooked in the current literature, which mainly focuses on the control of desertification. Previous studies have shown that using algae as biological fertilizers can help increase soil nutrient content, thereby reducing the use of inorganic and organic fertilizers [26]. Hence, it is believed that algae could have beneficial effects on irrigated fields in various ways. Moreover, the direct use of water resources with controllable algae content for irrigation could expand irrigation water sources and reduce irrigation costs.

This study examines the effects of irrigation with water containing low levels of algae on soil moisture content, microbial communities, fertility, enzyme activity, and nitrogen cycle functional potential. The target algae was *Scytonema javanicum*, a desert alga commonly used for biocrust [27]. An indoor simulated treatment method involving soil columns was used in this study, and the objectives were to (1) assess and analyze the impact of algal water irrigation on soil moisture content retention; (2) explore the effect of irrigation with water containing low levels of algae on the nutritional status of different soil layers; and (3) investigate the effect of irrigation with water containing low levels of algae on the soil nitrogen cycle. This article intends to clarify the feasibility of irrigating agricultural fields' waters containing low levels of algae.

2. Materials and Methods

2.1. Soil Sample

Soil samples were collected from the Hetao irrigation area in Inner Mongolia, China (107.276963°E, 40.732998°N), which has a mid-temperate continental monsoon climate, and the agricultural sector in this region relies mainly on diversion irrigation. Figure S1 shows the locations of the sampling points. The soil column was sampled in layers up to a depth of 96 cm using a ring knife and preserved at 4 °C for laboratory pretreatment. A Malvern 3000 laser particle size meter was used to analyze the soil particle composition of the soil samples at different depths, and Table S1 presents the results.

2.2. Algal and Column Experimental Design

Scytonema javanicum (FACHB-887) was used as the target algal species in this experiment. This cyanobacterium was obtained from the Freshwater Algal Species Bank of the Chinese Academy of Sciences and isolated from a desert beetle. The algal suspensions were prepared by culturing the algae in BG11-N medium in an illuminated incubator at 25 °C and 2000 Lux light intensity for 15 days.

Six Plexiglas columns were employed with a height of 100 cm and an inner diameter of 15 cm. Instrument line access ports with a diameter of 1.5 cm were provided on the sides at 10 cm, 50 cm, and 90 cm from the top of the soil column for connecting to the moisture content sensor and collecting soil samples. Soil samples were placed in the columns in sequence according to the stratification depth at the time of sampling, simulating actual soil conditions. The moisture content sensor (HOBO H21-USB, Onset, Cape Cod, MA, USA) was inserted vertically into the line interface. Moisture content data were collected every 15 min during the experimental period. A schematic of the experimental setup is shown in Figure S2. The experiments were conducted in the laboratory hall of the China Agricultural University, Beijing. The ambient temperature was relatively stable during the experiment. Three independent, parallel experiments were conducted at the same time.

After the soil column was constructed, all columns were stabilized for 1 week to redistribute the moisture content between the layers of the soil column to avoid excessive fluctuations in moisture content caused by filling and transportation processes that could affect the subsequent experimental operations. Following the stabilization period, each column was irrigated with 113 mm (2 L) water, either with or without *S. javanicum* (biomass 3 mg/L), and were subsequently assigned to either the HZ group (with *S. javanicum*) or the HB group (without *S. javanicum*). The initial algae concentration in the HZ group was 3 mg/L. Moisture content and other relevant parameters were monitored for a period of 10 days. Thereafter, to investigate the impact of algae on soil water evaporation, specific time points were chosen for various soil columns, which reflected similar soil moisture content (approximately 27.5%) at the onset of the experiment. During the evaporation experimental period, there was no free water on the surface of the soil. The goal was to observe the decline in soil moisture content over the ensuing 7 days.

2.3. Chemical Analysis

Soil samples were collected from the sample port at the beginning and end of each experimental phase. To increase the representativeness of the samples, three replicates were taken from each port and homogenized. The samples were then labeled, air-dried, pulverized, sieved, blended, bottled, and stored in a frozen state until further analysis. The soil parameters measured in this study included NH_3-N, NO_3-N, TN, and total phosphorus (TP), which were measured following the standard methods of the American Public Health Association by an ultraviolet-visible spectrophotometer (UV7, Mettler Toledo Inc., Zurich, Switzerland). Total organic carbon (TOC) was measured by a TOC analyzer (vario TOC cube from Elementar, Germany). The activity of alkaline phosphatase (AKP) and urease (UE) was determined using a multimode plate reader (Synergy 2, Biolog Inc., Hayward, CA, USA). A soil alkaline phosphatase activity assay kit and soil urease activity assay kit were selected for analysis. The activity of AKP was expressed as micrograms of phenol produced per gram of soil reacted in 24 h at 37 °C, and the activity of UE was expressed as micrograms of NH_3-N produced per gram of soil reacted in 24 h at 37 °C.

2.4. Molecular Analysis

The FastDNA® SPIN Kit for Soil (MP Biomedicals, Santa Ana, CA, USA) was used for the extraction of total DNA following the instructions. The V3–V4 highly variable region of the bacterial 16S rRNA gene was selected for PCR amplification with the primers 338F (5′ACTCCTACGGGAGGCAGCAG-3′) and 806R (5′-GGACTACHVGGGTWTCTAAT-3′). PCR products were examined by 1% agarose gel electrophoresis and then purified and recovered by the Agencourt AMPure XP Nucleic Acid Purification Kit (Beckman Coulter

Life Sciences, Beijing, China). After purification, the quality and concentration of PCR products was tested using NanoDrop (ND-1000, Thermo, Waltham, MA, USA). Finally, the products were sequenced on the Illumina MiSeq 2500 platform. The raw sequence data reported in this paper have been deposited in the China National Center for Bioinformation, Beijing, China (GSA: CRA011824) [28]. The operations such as trimming adapters, quality filtering, and removing chimeras on the raw data were performed on EasyAmplicon [29,30]. Subsequently, the clean sequences are grouped into distinct operational classification units (OTUs) at 97% identity. Representative sequence for the OTU is then annotated with the Silva138 database.

Real-time fluorescence quantitative PCR (QuantStudio3, Thermofisher, Waltham, MA, USA) was used to quantify the abundance of eight functional genes (*nifH*, *amoA*, *amoA*, *nxrA*, *narG*, *nirS*, *nirK*, and *nosZ*) involved in the nitrogen cycle. Their roles in the nitrogen cycle are shown in Table S2. The primers and sequences for the functional genes are shown in Table S3. The qPCR amplification system and amplification procedure are shown in Tables S4 and S5.

2.5. Data Analysis

Formula (1) was used for calculating the cumulative infiltration of soil water:

$$h = M \times m/(\rho \times S) \times 1000 \quad (1)$$

where h is the cumulative infiltration, mm. M is the value of soil moisture content change, %. m is the mass of dry soil, kg. S is the cross-sectional area of the column, m^2. ρ is the density of water, kg/m^3.

The physicochemical data were processed with Excel and analyzed for significance with SPSS 22.0 and GraphPad Prism 9 software. Data are presented as the mean ± standard error. Statistical analyses and visualization were performed using the software R (4.1.2) and Origin (v.2021). Principal coordinate analysis (PCoA) and redundancy analysis (RDA) were also conducted and plotted using R software with PCAtools and the vegan package. The Mantel analysis was then performed, and the results were plotted using the linkET package in R software. A species correlation matrix was generated using the psych package in R, and network plotting was performed using Gephi (0.9.7).

3. Results

3.1. Soil Moisture Content Characteristics

The moisture content characteristics of each soil column at different depths are illustrated in Figure 1. The initial moisture content of each layer did not differ significantly between the treatment and control groups. After irrigation, the soil moisture content of both groups increased in each layer (ranging from 16.60–36.02% to 14.30–22.91%). The cumulative depth of infiltration was 61.15 mm, 30.02 mm, and 7.34 mm in the treatment group from top to bottom, respectively, and 54.18 mm, 39.11 mm, and 5.84 mm in the control group. The treatment group did not show significant differences in moisture content in the 85 cm soil layer before and after the experiment ($p > 0.05$), but the treatment group showed significant differences in the 5 cm ($p < 0.01$) and 45 cm soil layers ($p < 0.05$), while the control group did not show significant differences in moisture content in each soil layer.

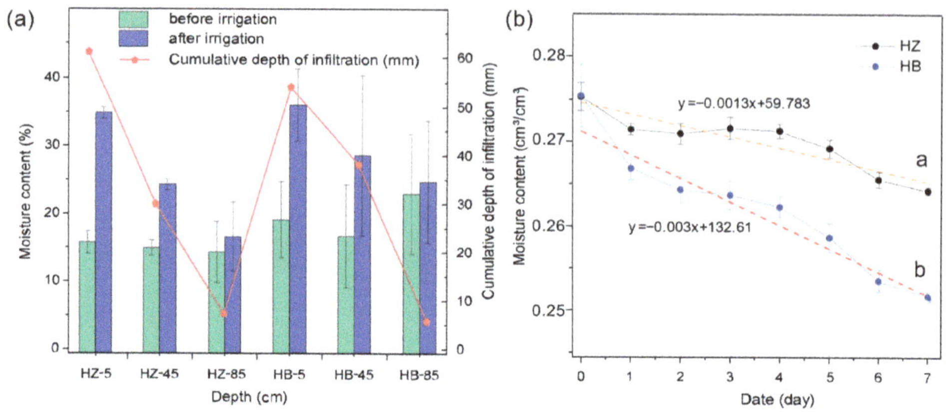

Figure 1. Effect of algal growth on soil water content. (**a**) Variation in soil water content and infiltration in each group during the infiltration stage. (**b**) Variation in soil water content with time for each group during the evaporation stage. The letters represent the groups with different treatments; HZ is the experimental group with algae inoculated on the surface of the soil column and HB is the control group without algae.

Figure 1b illustrates the variation in soil water with time for the soil columns during the evaporation phase. The average moisture contents of the three layers gradually decreased from 27.53 cm^3/cm^3 to 26.42 cm^3/cm^3 over 8 days in the treatment group and from 27.55 cm^3/cm^3 to 25.17 cm^3/cm^3 in the controls. Through calculations, the cumulative evaporative loss of the control group during the experimental period was 22.84, which is 2.14 times that of the treatment group (10.66 mm). With the same initial soil moisture content, at the end of the 8-day experimental period, the treatment group demonstrated exceptional soil moisture retention characteristics ($p < 0.05$), which showed that the application of *S. javanicum* was indeed beneficial for maintaining soil moisture. Moreover, after irrigation, there was no significant difference in the increase of moisture content at the 85 cm depth between the treatment and control groups ($p > 0.05$); this suggests that the better soil moisture retention in the treatment group is primarily due to the reduction in the evaporation of soil moisture. Figure S3 also presents a comparative analysis of the overall soil moisture content between the treatment group and the control group at the evaporation stage. At almost all spots above the 1:1 reference line, indicating that under the experimental conditions, the moisture content of the treatment group is significantly higher than that of the control. Furthermore, a linear function with an intercept on the y-axis was obtained through fitting, which crosses with the 1:1 reference line at x = y = 0.277, indicating that at corresponding soil moisture content below 0.277 cm^3/cm^3, the application of *S. javanicum* is beneficial for maintaining soil moisture.

3.2. Soil Nutrient Characteristics

As illustrated in Figure 2, the content of TOC, TP, and NO_3-N remained relatively consistent across all treatments. However, TN, NH_3-N were significantly higher in the HZ treatment (mean values of 0.54 g/kg and 0.70 mg/kg, respectively) compared to the HB treatment (mean values of 0.49 g/kg and 0.43 mg/kg, respectively) at the end of the experiment ($p < 0.05$). This suggests that the presence of *S. javanicum* enhances nitrogen retention, primarily in the form of ammonia nitrogen, which is less prone to loss through runoff [31]; such an increase is advantageous for soil nutrient fixation and mitigates the risk of non-point source pollution [32]. The activities of AKP and UE, both integral to the nitrogen and phosphorus transformation processes in soil, were also examined. In general, AKP activity did not exhibit significant changes before and after the experimental period. However, at the end of the experiment, a notable increase in UE activity was observed in the HZ-a treatment group, which showed significantly higher activity (226.40 μg/(g·d)) compared to the beginning of the treatment (HZ-b, 149.14 μg/(g·d), $p < 0.05$) and the control group (HB-a, 157.45 μg/(g·d), $p < 0.05$), particularly in the topsoil level, where UE activity peaked at 312.62 μg/(g·d), approximately 1.5 times that of the control group. This finding aligns with the results observed for NH_3-N, as urease has been established to be intimately linked to the production of this nitrogenous compound [33].

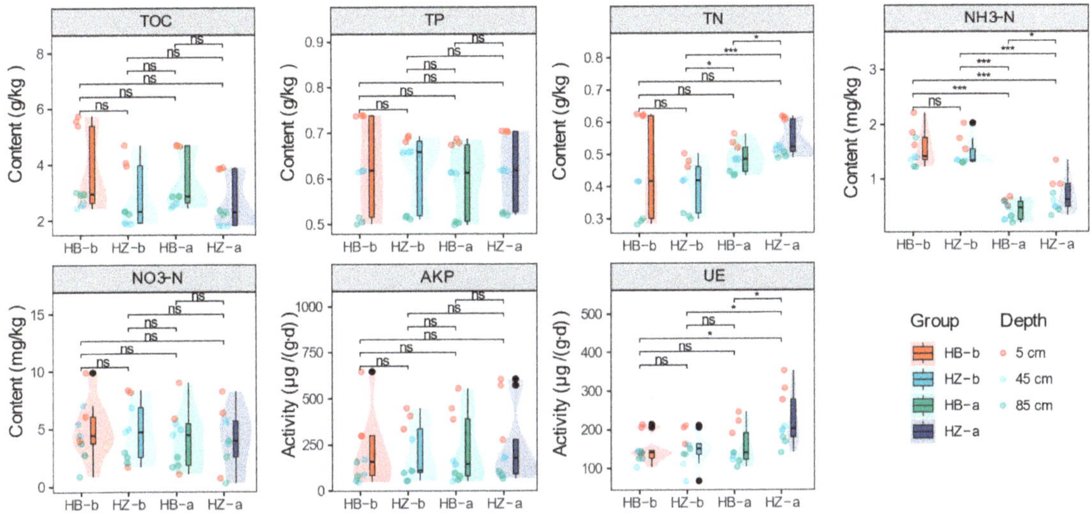

Figure 2. The nutrient characteristics of soil in treatment and control group. Asterisk represents a significant difference, where * represents $p < 0.05$, *** represents $p < 0.001$, the black dots represent outliers.

The spotlight then shifts to the topsoil, where all nutrient characteristics and enzyme activities were found to be most pronounced in the 5 cm layer (Figure S4), indicating its critical role in crop growth [34]. The control group experienced a significant decrease in TOC (from 5.57 g/kg to 4.69 g/kg, $p < 0.01$), TN (from 0.62 g/kg to 0.54 g/kg, $p < 0.01$), NH_3-N (from 1.90 mg/kg to 0.60 mg/kg, $p < 0.01$), and TP (from 0.74 g/kg to 0.68 g/kg, $p < 0.001$). In contrast, the treatment group, despite a significant reduction in NH_3-N (from 1.77 mg/kg to 1.05 mg/kg, $p < 0.05$), which still had levels 1.75 times higher than those of the HB-a group, TN (from 0.48 g/kg to 0.61 g/kg, $p < 0.001$), TP (from 0.68 g/kg to 0.71 g/kg, $p < 0.05$), and UE (from 193.52 μg/(g·d) to 312.62 μg/(g·d), $p < 0.05$), showed an increase, while TOC, NO_3-H, and AKP did not exhibit significant changes in the topsoil of the HZ treatment group. At the experiment's conclusion, the nutritional status of the topsoil between the HZ-a and HB-a groups displayed significant differences, except for AKP and NO_3-N. Especially, HB-a had a higher TOC content (4.69 g/kg) than the HZ-a group (3.87 g/kg, $p < 0.001$). However, the latter group exhibited higher levels of NH_3-N (1.05 mg/kg, $p < 0.05$), TN (0.61 g/kg, $p < 0.01$), TP (0.71 g/kg, $p < 0.01$), and UE (312.62 μg /(g·d), $p < 0.05$) compared to HB-a (0.60 mg/kg, 0.54 g/kg, 0.68 kg, and 220.15 μg/(g·d), $p < 0.05$). These results indicated that the application of *S. javanicum* positively impacts nutrient retention, particularly nitrogen in the topsoil, which is beneficial for crop growth; it is hypothesized that this beneficial effect may be attributed to the elevated activity of UE [35].

3.3. Soil Microbial Communities and Structure

The alpha dilution curves using the microbial diversity of each sample sequencing volume at different sequencing depths (Figure S5) was constructed, and the results showed that the species coverage of the samples was above 90%, indicating that the results of the sequencing experiments represented the actual composition of microorganisms in the samples. Table S6 represents the α-diversity indices of soil microorganisms in each soil layer before and after the experiment. The Chao 1, goods_coverage, and observed_species indices showed significant decreases in each soil layer before and after the experiment ($p < 0.01$), while the changes in other indices were not obvious.

Figure 3 shows the microbial abundance and diversity of soil samples from the treatment and control groups before and after the experiment. In this study, the top 10 phyla and 20 genera in terms of abundance were selected, as shown in Figure 3. The dominant microorganisms in each sample at the phylum level were Proteobacteria, Acidobacteriota, Actinobacteriota, and Chloroflexi; the combined abundance of the four phyla accounts for more than 50% of all the microorganisms.. After irrigation, Proteobacteria (ranging from 17.98–37.48% to 33.42–37.96%) showed an overall significant increase ($p < 0.05$), while the percentage of Actinobacteriota (ranging from 9.97–13.71% to 8.11–9.05%) in the treatment group was significantly lower ($p < 0.05$). In addition, the relative abundance of Actinobacteriota showed significant differences between the treatment (10.01–13.71%) and control groups (8.00–9.15%, $p < 0.05$). Analysis of the abundance of species clades within different soil layers revealed that the abundance of Actinobacteriota (ranging from 12.15–13.71% to 8.65–9.05%) decreased significantly at 85 cm ($p < 0.05$), and the abundance of Chloroflexi increased significantly at 5 cm (ranging from 7.15–8.69% to 8.26–9.71%) and decreased significantly at 45 cm (ranging from 10.98–12.20% to 8.75–9.71%) after the experiment ($p < 0.05$).

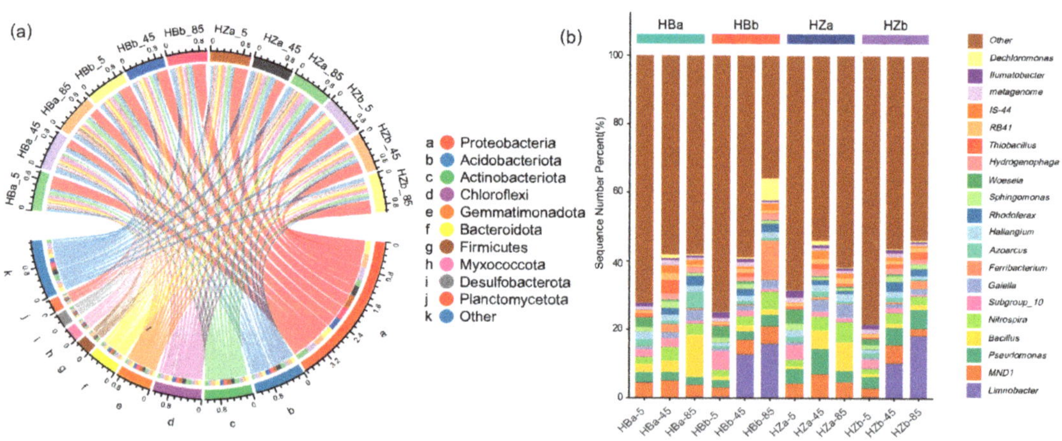

Figure 3. The composition of microbial communities within soils at the phylum and genus level. (**a**) Chord diagram showing correspondence between grouping and species composition at the phylum level, taking the top 10 phyla of the microbial community, and (**b**) stacked histograms showing the community structure of species at the genus level.

As shown in Figure 3b, the microorganisms that dominated at the genus level mainly included *Limnobacter*, MND1, *Pseudomonas*, *Bacillus*, and *Nitrospira*. The average relative abundance of *S. javanicum* only accounts for 0.02%. By comparing the abundance of strains before and after the experiment in different groups, only the treatment group showed a significant decrease ($p < 0.05$) in MND1 abundance (ranging from 4.32–7.02% to 2.06–5.33%), while the other groups showed no significant change overall. By comparing the strain abundance in different soil layers before and after the experiment, the abundance of *Limnobacter* ($p < 0.05$) and *Pseudomonas* ($p < 0.05$) increased significantly and the abundance of MND1 decreased extremely significantly at the 5 cm soil layer ($p < 0.01$). It is worth noting that *Limnobacter* was almost exclusively found in the deep soil layers (45 cm and

85 cm layers) before irrigation, and was not detected in the soil surface after irrigation ($p < 0.05$).

Network analysis of the top 50 genera in terms of abundance yielded a total of 446 edges, of which 280 were positively correlated and 166 were negatively correlated (Figure S6). According to the degree of nodes, the key network nodes were *Pelagibius*, *Gaiella*, TX1A-55, *Halomonas*, SM1A02, etc., and their degrees were greater than 30. Therefore, the significant change in the abundance of these genera could remarkably affect the structure of the soil microbial community. Among them, *Pelagibius*, TX1A-55 and *Halomonas* belonged to Proteobacteria, and *Gaiella* belonged to Actinobacteriota, all of which were the top three phyla in terms of abundance. In contrast, Planctomycetota, to which SM1A02 belongs, was among the top 10 phyla in terms of abundance.

Variations in the composition of bacterial communities across various soil depths were observed under different experimental treatments and at different times relative to irrigation events (Figure 4). According to the sample distance matrix PCoA analysis, the principal components PCo1 and PCo2 contributed 49.5% and 12.8%, respectively, to the total variations. The microbial composition was separated in the second axis. The soil layer at 5 cm, 45 cm, and 85 cm depths each formed distinct clusters. Notably, the 5 cm layer was more distant from the deeper layers, indicating significant differences in the microbial community structure between the topsoil and the subsoil ($p < 0.01$). This statement is supported by Figure S8. This may be due to the higher moisture content and greater susceptibility of the topsoil soil to external environmental influences than the deeper soils, resulting in some changes in microbial community structure. In contrast, no significant differences were observed between the treatment and control groups, indicating that the presence of algae did not significantly affect the overall microbial community of the soil and thus did not disrupt the community composition of the soil. Further analysis of the distance matrix at the 5 cm soil level showed that the distance before and after the experiment was smaller in the treatment group (0.376) than in the control group (0.401), indicating that the microbial community was more altered in the control group. In conclusion, the above results show that the addition of algae did not significantly alter the soil microbial community and contributed to the maintenance of the original microbial community.

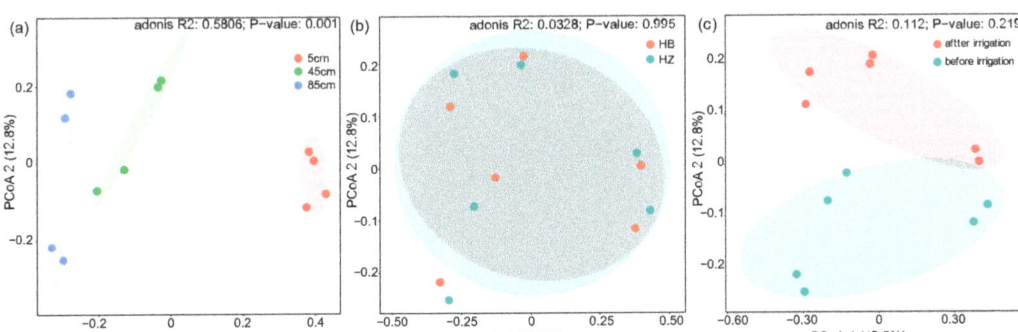

Figure 4. Principal coordinates analysis (PCoA) of samples based on Bray–Cutis distance. (**a**) Analysis of different depths; (**b**) intergroup analysis of experimental and control groups; (**c**) pre- and post-experimental analysis.

The RDA was employed to assess the impact of the individual environmental factor on the structure and composition of soil microbial communities (Figure S7), and the results showed that the explanatory power of the effects of RDA1 and RDA2 on community structure and composition were 62.55% and 20.28%, respectively. The environmental factors were selected as depth (D), moisture content, TOC, TN, TP, NH_3-N and NO_3-N, AKP, and UE, and the species information was selected as the results of the top 10 genus levels in absolute abundance of each sample. *Bacillus* showed a significant negative correlation with NH_3-N. *Pseudomonas*, also a denitrifying bacterium, showed a significant positive correlation with both TN and NO_3-N. Furthermore, according to the arrow length, it is easy to see that *Limnobacter* was the most influential microbial genus, which was most affected by moisture content, while there was no significant difference in the effect of environmental factors.

3.4. Functional Genes Involved in the Soil Nitrogen Cycle

The distribution of nitrogen cycle-related functional genes across various soil layers, contrasting the treatment and control groups, is presented in Figure S9. By the conclusion of the experiment, significant reduction was only observed ($p < 0.05$) in the overall abundance of the *narG* gene (from 2.22×10^7 to 6.48×10^7 copies/g down to 0.30×10^6 to 6.07×10^7 copies/g) and *nirK* genes (from 4.78×10^6 to 8.42×10^6 copies/g down to 1.07×10^6 to 8.23×10^6 copies/g). In the control group, no significant changes were observed in the abundance of any functional genes at the 0.05 significance level when comparing pre-irrigation (HB-b) and post-irrigation (HB-a) samples. In contrast, within the treatment group, a significant decrease was noted for the *nirK* gene ($p < 0.05$). Variability was also noted with depth. The abundance of *nifH* and *nirS* genes at the 5 cm depth notably increased post-experiment ($p < 0.05$), while the abundance of *nirK*, *narG* and *nirS* at the 85 cm depth significantly decreased ($p < 0.05$). In addition, the abundance of genes *amoA*, *narG*, *nifH*, *nxrA* in the topsoil of the treatment group were significantly higher than that of the control group after the experiment ($p < 0.05$).

Figure 5 illustrates a Principal Coordinates Analysis (PCoA) based on Bray–Curtis distances, which accounts for 87% of the total variation in functional gene composition. The analysis indicated no significant impact of experimental treatments on the functional genes, either before and after the irrigation or with and without the presence of *S. javanicum*. However, it is noteworthy that the functional genes in the 5 cm, 45 cm, and 85 cm soil layers distinctly formed three main clusters along the second principal axis ($p < 0.05$), suggesting that soil depth is the predominant factor influencing the overall distribution of functional genes, and the topsoil is the hotspot of nitrogen cycling.

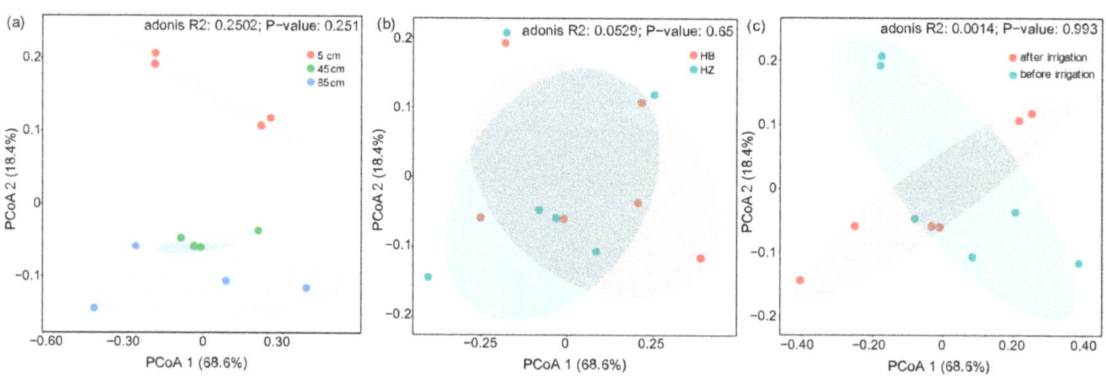

Figure 5. Principal coordinates analysis (PCoA) of functional genes based on Bray–Cutis distance. (**a**) Analysis of soil layers at different depths; (**b**) intergroup analysis of experimental and control groups; (**c**) pre- and post-experimental analysis.

3.5. Factors Influencing Microbial Communities

The correlation between environmental factors and microbiota is illustrated in Figure 6. Most environmental factors are correlated with each other; among them, TN, NO_3-N, and TP all showed significant correlations with depth ($p < 0.05$), with r values of −0.7293, −0.8284, and −0.9303, respectively. TOC and TP both showed significant correlations with TN ($p < 0.05$), with r values of 0.7497 and 0.6327, respectively. TP and NO_3-N also showed a significant correlation ($p < 0.05$), with an r value of 0.8221. Functional genes were significantly correlated with depth only ($p < 0.05$), with an r value of 0.2558. Bacteria were significantly correlated with depth and NH_3-N ($p < 0.01$) and significantly correlated with NO_3-N ($p < 0.05$), with r values of 0.3507, 0.4557, and 0.3620, respectively. Mantel analysis of the treatment group showed that bacteria were significantly correlated with NH_3-N ($p < 0.05$), with an r value of 0.5294. No significant correlation was found between bacteria and environmental factors in the control group. In general, both bacteria and functional genes were affected by depth, and the presence of algae essentially did not have a significant effect on the soil; only NH_3-N was correlated, which is consistent with the previous changes in soil nutrients.

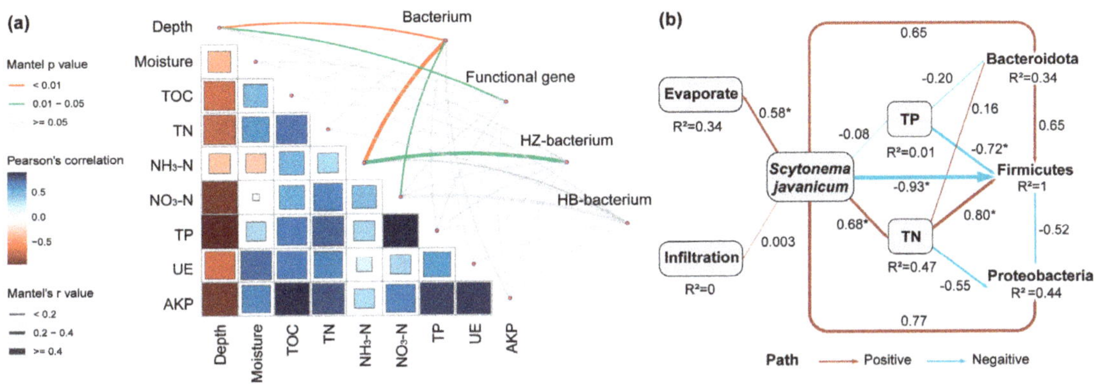

Figure 6. The influence of environmental factors on the bacterial communities. (**a**) Mantel test for correlations between the different environmental factors and the whole and subgroup microbial community, and functional gene. Where the size of the squares and the color shades represent the size of the correlation, the thickness of the line represents the Mantel correlation and the different colors represent significant differences. (**b**) Path analysis based on structural equation model shows direct and indirect effects between environmental variables and the keystone microorganisms. Where * represents $p < 0.05$. Blue for negative effects; red for positive effect; the number next to the path is the path coefficient (normalized regression weight), and the width of the line is directly proportional to the strength of the path coefficient. The model parameters are assumed as follows: Fisher's C = 6.259, P = 0.181, AIC = 73.484.

The RDA analysis at phyla level revealed that Bacteroidota, Firmicutes, and Proteobacteria exert the most significant explanatory influence. Based on the structural equation model (SEM), the path analysis was conducted to determine the direct and indirect effects of environmental factors on these three phyla (Figure 6b). The findings revealed that the presence of algae significantly influenced moisture, TN, and the abundance of Firmicutes. Specifically, algae presence had no significant positive effect on moisture infiltration but greatly enhanced moisture retention during the evaporation phase. Moreover, algae had a direct positive effect on TN ($p < 0.05$). Algae also indirectly impacted the microbial community through TN and TP. Among them, TP negatively affected Bacteroidota and Firmicutes ($p < 0.05$), while TN had a positive effect on both Bacteroidota and Firmicutes ($p < 0.05$). Proteobacteria was basically unaffected by changes in TN and TP. The findings suggest that the prevalence of *Scytonema javanicum* predominantly influences the nutritional profile of soil, which in turn has a significant impact on the composition of the dominant microbial communities.

4. Discussion

4.1. The Effect of Algae on the Physicochemical Properties of Soils

Previous studies have demonstrated that algal exudates [10] or algal products, such as compost [36] and biochar [37], can enhance soil water retention. The results of this work indicate that the presence of algae cells also played a positive role in maintaining soil moisture content; they allow more infiltrating water to remain in the topsoil. The combined action of algal and microorganisms [38,39] formed water-stable soil aggregates' protective layer and increased the soil surface roughness, facilitating a reduction in the evaporation rate and maintaining soil moisture content [40,41]. In addition, the growth of algae in the topsoil can bind bacterial flocs and soil aggregates, and results in the formation of a dense structure [42] that inhibits evaporation [43,44]. Therefore, it is believed that the presence of algae in irrigation water helps to retain soil moisture content and reduce water loss, reducing the frequency and cost of irrigation.

The growth of *S. javanicum* may also affect soil nutrient characteristics through changes in the moisture content and microecological properties of the soil, which are important for the growth of crops. This finding is in line with the report of Alavarez [45]; it is believed that the cyanobacterium *Anabaena cylindrica* UTEX 1611 had a significant positive impact on the soil nutrients, especially the nitrogen content. Based on the changes in soil nutrients, it is presumed that the growth of algae contributes to the maintenance of TOC, TN, TP, and other nutrients in the surface soil [46,47]. Based on the specific conditions of the soil column surface and the characteristics of algae, this result may occur because the growth and reproduction of algae improved the physical structure of the surface soil, increased soil porosity and permeability, altered soil pH, and provided a good environment for microbial metabolic activities [48]. Additionally, *S. javanicum* has the capacity to regulate the redox potential of soil by modulating its moisture levels, thereby influencing the microbial activity, which aligns with the perspective of Khoshru [49]. This alteration can subsequently exert a significant impact on the nitrification and denitrification processes that transpire within the soil environment [50].

However, in the 5 cm soil layer, the TN of the treatment and control groups showed opposite changes briefly, which increased in the treatment group ($p < 0.05$), but decreased in the control, although it is not significant. The previous study [51–53] has proven that that nitrogen fixation does not compensate for N loss during denitrification, so the decrease in TN is expected in the topsoil of the control group. But the significant increasing of TN content within the topsoil of the treatment group can be predominantly attributed to the direct effects of *S. javanicum*, potentially operating through three ways. Firstly, the presence of algae may enhance nitrogen fixation; secondly, it could potentially reduce the process of denitrification; and thirdly, the decomposition of algal cells could release nitrogen-contained organic matter. In this study, the conditions with low levels of algal pollution were simulated, characterized by minimal biomass content. Consequently, the decomposition of deceased algal cells is unlikely to induce significant changes in the nitrogen content. Compared to the control group, the treatment group showed a higher relative abundance of denitrification-related functional genes, *narG* and *nxrA*, indicating that *S. javanicum* did not inhibit the denitrification activity of the broader microbial community. On the other hand, the capability of fixing atmospheric nitrogen of *S. javanicum* has been confirmed by Wu et al. [54]. Consequently, it is postulated that the principal driver behind the observed increase in nitrogen within the treatment group is the elevated nitrogen fixation capacity; this is substantiated by the notable increase of the *nifH* gene in the topsoil of the treatment group.

Analysis of the changes in TP revealed an interesting situation: the difference between the treatment group and control group at the 5 cm soil layer was approximately equal to the sum of the differences between the treatment group and the control group at the 45 cm and 85 cm soil layers, with the concentration of TP in the treatment group being higher at the 5 cm layer and that in the control group being higher in the 45 and 85 cm layers. This could be attributed to the presence of *S. javanicum*, which appears to mitigate the translocation of TP from the irrigation water towards the deeper strata of the soil column. This is consistent with the view of Liu et al. [55], that soil mulches and plant residues can effectively mitigate the loss of nutrients, particularly dissolved P, from soil to groundwater.

As a carbon source, TOC is involved in the process of nitrification and denitrification, providing energy and electron donors for the reactions to proceed [56,57], so all groups showed a decreasing trend in TOC after the experiment. However, the reduction in TOC in the treatment group was less than that in the control group at the 5 cm soil layer; this might be due to the metabolization and degradation of *S. javanicum* which can increase the TOC content in the environment [58]. Moreover, it has been shown that the accumulation rate of soil TOC can be accelerated by the degree of coverage and degradation of algae and plants, which supports the results of this study [59,60]. However, it is worth noting that the extracellular polysaccharides secreted by *S. javanicum* also played the important role that increases soil associativity and thus contributes to TOC accumulation [61,62]. Li et al. [63] also observed the important role of algal extracellular polymers that enhanced carbon accumulation in marine environments.

UE can hydrolyze urea to produce NH_3-N, which provides substrates for nitrification and denitrification processes [64]. The results of Yahya [65] showed that the activity of UE is inhibited by the higher concentrations of NH_3-N, which supports the results of higher UE in the treatment group. In addition, the enzyme activity of the surface soil was significantly greater than that of the deep soil, and the enzyme activity of the treatment group was stronger; it indicates the potential higher nutritional conversion efficiency in the topsoil of the treatment group, which is probably because the algae and microorganisms could secrete enzymes into the surrounding environment, directly or indirectly promoting the biogeochemical cycling of nutrients.

4.2. The Effect of Algae on Soil Microbial Communities and Functional Genes

Nitrifying and denitrifying microorganisms are crucial for the N-cycling; the results of the network analysis also showed that the key genera within the soil microbial community mainly belonged to the phylum associated with nitrification and denitrification. According to previous research [66], most denitrifying bacteria belong to Proteobacteria [66], which also dominated in this work. In both the treatment and control groups, Proteobacteria did not show significant changes before and after the irrigation, and was shown as not significantly affected by TN or TP. Therefore, it is speculated that the presence of B may have more of an impact on the ecological function of the microbial community, rather than the microbial community itself.

In this study, the presence of *S. javanicum* increased the number of copies of functional genes for nitrification, prompting more NH_3-N to be oxidized to NO_3-N and providing more reactants for denitrification. The presence of *S. javanicum* would increase the topsoil moisture content and enhance the anaerobic soil environment, which is conducive to the progress of denitrification [50]. Therefore, there was an overall trend showing promotion of more NO_3-N to undergo the denitrification process; this is consistent with the trend of changes in NO_3-N content in soil. The evidence is an increase in the abundance of *narG* and *nxrA* genes in the 5 cm soil layer of the treatment group compared to the control ($p < 0.01$); both genes were closely related to the denitrification process [67].

The *nirS* and *nirK* genes play equivalent functions [68], but showed completely different trends. Overall, *nirK* was decreased after the irrigation treatment, but *nirS* was increased. However, the decrease in the abundance of functional genes mainly occurs in deeper soil layers, especially 85 cm, but for the topsoil, the total abundance of *nirS* and *nirK* genes increased, especially the *nirS* which was almost doubled to 11.60×10^7 copies/g from 5.38×10^7 copies/g. The changes in the *nirK* gene copy number were significantly lower than those in the *nirS* gene ($p < 0.01$). This may be because *nirS*-containing microorganisms are more dominant in the environment, and thus the *nirS* gene abundance is less influenced by the environment [69,70]. Chen et al. [71] also found that *nirK*-containing microorganisms show higher sensitivity to mineral nitrogen compounds than *nirS*-containing microorganisms [72–74]. In addition, there are many other functional genes and undiscovered biochemical reactions involved in the nitrogen cycle, which may also have an impact on the results of this experiment [75,76].

5. Conclusions

In this work, the impact of irrigating with slightly algae-contaminated water on various soil properties was investigated, including its hydrological and nutrient profiles, enzymatic activities, microbial community composition, and the prevalence of nitrogen cycle-associated functional genes. The findings indicate that the application of *S. javanicum* does not disrupt the equilibrium of the local microbial community, while positively influencing soil moisture retention. Following the experimental treatments, a notable reduction in water content was observed in the control group (25.17%) compared to the treatment group (26.42%), mainly attributed to diminished evaporative losses (10.66 mm and 22.84 mm in the treatment versus control group, respectively). Additionally, the incorporation of *S. javanicum* has led to a significant reallocation of soil nutrients, particularly nitrogen, essential for plant growth, with both ammonia nitrogen and total nitrogen levels in the treatment (0.70 mg/kg and 0.54 g/kg) markedly exceeding those in the control group (0.43 mg/kg and 0.49 g/kg). In sum, employing slightly algae-contaminated water as an alternative to conventional irrigation sources is a practical and beneficial approach. It diversifies irrigation options and alleviates the financial and time-related pressures of irrigation, thereby easing the strain on agricultural water resources. However, it is important to recognize that current research is limited to a single type of exogenous cyanobacteria, which may affect the general applicability of our conclusions. Future studies should expand the range of cyanobacteria species to gain a more comprehensive understanding.

Supplementary Materials: The following supporting information can be downloaded at: https://www.mdpi.com/article/10.3390/pr12081639/s1, Figure S1: Sampling point location diagram. Figure S2: Experimental setup diagram (a) Soil column design drawings; (b) actual arrangement of soil columns. Figure S3: The comparison of soil water content between experimental and control groups. Figure S4: The nutrient characteristics of each layer of soil. Asterisk represents a significant difference, where * represents $p < 0.05$,** represents $p < 0.01$,*** represents $p < 0.001$. Figure S5: The rarefaction curves of various groups. Figure S6: The microbial network for the top 50 genera in terms of abundance. Different colored nodes represent different clades and the size of the nodes is related to the number of links. Red lines represent positive correlations between two nodes; green lines represent negative correlations. Figure S7: Dendrogram of the differences in relative abundance of microbial communities in each soil layer. Figure S8: Redundancy analysis (RDA) between environmental factors, the top 10 genera (a) and phyla (b) of the microbial community and different samples. Blue arrows indicate environmental factors; red arrows indicate microorganisms; green circles indicate samples without experimental treatment; brown circles indicate samples with experimental treatment; green names indicate control groups; brown sample names indicate experimental groups. Figure S9: Effect of algal growth on the copy number of soil functional genes. Including functional genes for (a) *amoA*, (b) *amoB*, (c) *narG*, (d) *nifH*, (e) *nirK*, (f) *nirS*, (g) *nosZ*, and (h) *nxrA*.; Table S1: Particle size composition of soil particles at different depths. Table S2: Major functional genes involved in the soil nitrogen cycle. Table S3: Primers and sequences of functional genes. Table S4: System of qPCR amplification. Procedure of PCR and qPCR amplification. Table S5: Procedure of PCR and qPCR amplification. Table S6: The α-diversity index of each sample.

Author Contributions: Conceptualization, E.X. and Z.B.; formal analysis, H.Z., X.W,. and C.H.; funding acquisition, E.X. and Z.T.; methodology, E.X. and H.Z.; project administration, E.X. and Z.B.; resources, E.X., Z.B. and X.Z.; supervision, E.X. and X.Z.; investigation H.Z. and X.W.; writing—original draft, H.Z., X.W. and C.H.; visualization H.Z. and X.W.; writing—review and editing, E.X. and Z.B. All authors have read and agreed to the published version of the manuscript.

Funding: This research was supported by the Xinjiang Production and Construction Corps Science and Technology Bureau (No. 2023AB071) and the Ministry of Science and Technology of the People's Republic of China (No. 2023YFD1900803 and 2021YFE0192500).

Data Availability Statement: Raw data of microbial community are available in the National Genomics Data Center at the link https://ngdc.cncb.ac.cn/gsa/browse/CRA011824 (accessed on 14 July 2023). All other data will be available from the corresponding author on reasonable request.

Acknowledgments: This research was supported by the Bintuan Science and Technology Program (No. 2023AB071) and National Key R&D Program of China (No. 2023YFD1900803 and 2021YFE0192500). The authors would also like to thank Dr. Naomichi Tanaka and Dr. Akira Harimoto for their input on this paper.

Conflicts of Interest: The authors declare that they have no known competing financial interests or personal relationships that could have appeared to influence the work reported in this paper.

References

1. Liu, K.; Bo, Y.; Li, X.; Wang, S.; Zhou, G. Uncovering Current and Future Variations of Irrigation Water Use Across China Using Machine Learning. *Earth's Future* **2024**, *12*, e2023EF003562. [CrossRef]
2. Bhardwaj, A.; Kumar, M.; Alshehri, M.; Keshta, I.; Abugabah, A.; Sharma, S.K. Smart water management framework for irrigation in agriculture. *Environ. Technol.* **2024**, *45*, 2320–2334. [CrossRef]
3. Siebert, S.; Kummu, M.; Porkka, M.; Doell, P.; Ramankutty, N.; Scanlon, B.R. A global data set of the extent of irrigated land from 1900 to 2005. *Hydrol. Earth Syst. Sci.* **2015**, *19*, 1521–1545. [CrossRef]
4. Bakhoum, G.S.; Sadak, M.S.; Thabet, M. Induction of tolerance in groundnut plants against drought stress and Cercospora leaf spot disease with exogenous application of Arginine and Sodium nitroprusside under field condition. *J. Soil Sci. Plant Nutr.* **2023**, *23*, 6612–6631. [CrossRef]
5. Chen, W.; Lu, S.; Jiao, W.; Wang, M.; Chang, A.C. Reclaimed water: A safe irrigation water source? *Environ. Dev.* **2013**, *8*, 74–83. [CrossRef]
6. Lyu, S.; Chen, W. Soil quality assessment of urban green space under long-term reclaimed water irrigation. *Environ. Sci. Pollut. Res.* **2016**, *23*, 4639–4649. [CrossRef]
7. Moreira, C.; Vasconcelos, V.; Antunes, A. Cyanobacterial Blooms: Current Knowledge and New Perspectives. *Earth* **2022**, *3*, 127–135. [CrossRef]
8. Yamamoto, T.; Fujiyama, H.; Miyamoto, K.; Hatanaka, J.; Wen, G.; Asae, A. Optimizing water cleaning system for microirrigation in the Tohaku irrigation project of Japan. In *National Irrigation Symposium Proceedings of the 4th Decennial Symposium, Phoenix, Arizona, USA, November 14–16*; American Society of Agricultural Engineers: St. Joseph, MO, USA, 2000.
9. Cao, Q.; Steinman, A.D.; Su, X.; Xie, L. Effects of microcystins contamination on soil enzyme activities and microbial community in two typical lakeside soils. *Environ. Pollut.* **2017**, *231*, 134–142. [CrossRef]
10. Bakhoum, G.S.; Tawfik, M.M.; Kabesh, M.O.; Sadak, M.S. Potential role of algae extract as a natural stimulating for wheat production under reduced nitrogen fertilizer rates and water deficit. *Biocatal. Agric. Biotechnol.* **2023**, *51*, 102794. [CrossRef]
11. Xiao, B.; Hu, K.; Ren, T.; Li, B. Moss-dominated biological soil crusts significantly influence soil moisture and temperature regimes in semiarid ecosystems. *Geoderma* **2016**, *263*, 35–46. [CrossRef]
12. Sandoval, P.A.; Lucia, C.S.; Manuel, M.N.; Garcia-Oliva, F.; Alarcon, A.; Adriana, M.S.; Esperon-Rodriguez, M. Biocrusts, inside and outside resource islands of Mimosa luisana (Leguminosae), improve soil carbon and nitrogen dynamics in a tropical semiarid ecosystem. *Eur. J. Soil Biol.* **2016**, *74*, 93–103. [CrossRef]
13. Lan, S.; Zhang, Q.; He, Q.; Yang, H.; Hu, C. Resource utilization of microalgae from biological soil crusts: Biodiesel production associated with desertification control. *Biomass Bioenergy* **2018**, *116*, 189–197. [CrossRef]
14. Canton, Y.; Chamizo, S.; Rodriguez-Caballero, E.; Lazar, R.; Roncero-Ramos, B.; Raul, R.J.; Sole-Benet, A. Water Regulation in Cyanobacterial Biocrusts from Drylands: Negative Impacts of Anthropogenic Disturbance. *Water* **2020**, *12*, 720. [CrossRef]
15. Wang, Z.; Liu, Y.; Zhao, L. Development of fungal community is a potential indicator for evaluating the stability of biological soil crusts in temperate desert revegetation. *Appl. Soil Ecol.* **2020**, *147*, 103404. [CrossRef]
16. Fisher, K.; Jefferson, J.; Vaishampayan, P. Bacterial Communities of Mojave Desert Biological Soil Crusts Are Shaped by Dominant Photoautotrophs and the Presence of Hypolithic Niches. *Front. Ecol. Evol.* **2020**, *7*, 518. [CrossRef]
17. Coban, O.; De, D.G.B.; van der Ploeg, M. Soil microbiota as game-changers in restoration of degraded lands. *Science* **2022**, *375*, abe0725. [CrossRef] [PubMed]
18. Kheirfam, H. Increasing soil potential for carbon sequestration using microbes from biological soil crusts. *J. Arid. Environ.* **2020**, *172*, 104022. [CrossRef]
19. Housman, D.C.; Powers, H.H.; Collins, A.D.; Belnap, J. Carbon and nitrogen fixation differ between successional stages of biological soil crusts in the Colorado Plateau and Chihuahuan Desert. *J. Arid. Environ.* **2006**, *66*, 620–634. [CrossRef]
20. Zhang, Y. The microstructure and formation of biological soil crusts in their early developmental stage. *Chin. Sci. Bull.* **2005**, *50*, 117–121. [CrossRef]
21. Yoon, S.; Cruz-Garcia, C.; Sanford, R.; Ritalahti, K.M.; Loeffler, F.E. Denitrification versus respiratory ammonification: Environmental controls of two competing dissimilatory NO_3^-/NO_2^- reduction pathways in Shewanella loihica strain PV-4. *ISME J.* **2015**, *9*, 1093–1104. [CrossRef]

22. Neumann, D.; Heuer, A.; Hemkemeyer, M.; Martens, R.; Tebbe, C.C. Importance of soil organic matter for the diversity of microorganisms involved in the degradation of organic pollutants. *ISME J.* **2014**, *8*, 1289–1300. [CrossRef] [PubMed]
23. Wang, M.; Qu, L.; Ma, K.; Yuan, X. Soil microbial properties under different vegetation types on Mountain Han. *Sci. China-Life Sci.* **2013**, *56*, 561–570. [CrossRef] [PubMed]
24. Sun, R.; Wang, F.; Hu, C.; Liu, B. Metagenomics reveals taxon-specific responses of the nitrogen-cycling microbial community to long-term nitrogen fertilization. *Soil Biol. Biochem.* **2021**, *156*, 108214. [CrossRef]
25. Grassmann, C.S.; Mariano, E.; Diniz, P.P.; Borges, B.M.F.; Borges, C.D.; Tsai, S.M.; Rosolem, C.A. Functional N-cycle genes in soil and N2O emissions in tropical grass-maize intercropping systems. *Soil Biol. Biochem.* **2022**, *169*, 108655. [CrossRef]
26. Ammar, E.E.; Aioub, A.A.A.; Elesawy, A.E.; Karkour, A.M.; Mouhamed, M.S.; Amer, A.A.; El-Shershaby, N.A. Algae as Bio-fertilizers: Between current situation and future prospective. *Saudi J. Biol. Sci.* **2022**, *29*, 3083–3096. [CrossRef] [PubMed]
27. Zhu, Q.; Wu, L.; Li, X.; Li, G.; Li, J.; Li, C.; Zhao, C.; Wang, F.; Du, C.; Deng, C.; et al. Effects of ambient temperature on the redistribution efficiency of nutrients by desert cyanobacteria- Scytonema javanicum. *Sci. Total Environ.* **2020**, *737*, 139733. [CrossRef] [PubMed]
28. Chen, T.; Chen, X.; Zhang, S.; Zhu, J.; Tang, B.; Wang, A.; Dong, L.; Zhang, Z.; Yu, C.; Sun, Y.; et al. The Genome Sequence Archive Family: Toward Explosive Data Growth and Diverse Data Types. *Genom. Proteom. Bioinform.* **2021**, *19*, 578–583. [CrossRef] [PubMed]
29. Wei, Y.J.; Wu, H.Y.; Zhang, X.D.; Liang, Y.B.; Shi, D.Y.; Wang, L.; Li, H.B. Comparative analysis of chlorine-resistant bacteria after chlorination and chloramination in drinking water treatment plants. *J. Hazard. Mater.* **2024**, *4*, 100138. [CrossRef]
30. Liu, Y.X.; Chen, L.; Ma, T.; Li, X.; Zheng, M.; Zhou, X.; Chen, L.; Qian, X.; Xi, J.; Lu, H.; et al. EasyAmplicon: An easy-to-use, open-source, reproducible, and community-based pipeline for amplicon data analysis in microbiome research. *iMeta* **2023**, *2*, e83. [CrossRef]
31. Li, T.; Zhang, Y.; He, B.; Wu, X.; Du, Y. Nitrate loss by runoff in response to rainfall amount category and different combinations of fertilization and cultivation in sloping croplands. *Agric. Water Manag.* **2022**, *273*, 107916. [CrossRef]
32. Cao, Y.; Bai, M.; Han, B.; Butterly, C.; Hu, H.; He, J.; Griffith, D.W.T.; Chen, D. NH_3 and greenhouse gas emissions during co-composting of lignite and poultry wastes and the following amendment of co-composted products in soil. *Environ. Technol.* **2024**, *1*, 1–14.
33. Lan, T.; He, X.; Wang, Q.; Deng, O.; Zhou, W.; Luo, L.; Chen, G.; Zeng, J.; Yuan, S.; Zeng, M.; et al. Synergistic effects of biological nitrification inhibitor, urease inhibitor, and biochar on NH_3 volatilization, N leaching, and nitrogen use efficiency in a calcareous soil–wheat system. *Appl. Soil Ecol.* **2022**, *174*, 104412. [CrossRef]
34. Chen, T.; Zhang, Y.; Xia, M.; Wang, Q. Soil properties and functional genes in nitrogen removal process of bioretention. *Environ. Technol.* **2024**, *45*, 2268–2283. [CrossRef] [PubMed]
35. Rodriguez, M.J.; Saggar, S.; Berben, P.; Palmada, T.; Lopez-Villalobos, N.; Pal, P. Use of a urease inhibitor to mitigate ammonia emissions from urine patches. *Environ. Technol.* **2021**, *42*, 20–31. [CrossRef] [PubMed]
36. González Fernández, L.A.; Castillo Ramos, V.; Sánchez Polo, M.; Medellín Castillo, N.A. Fundamentals in applications of algae biomass: A review. *J. Environ. Manag.* **2023**, *338*, 117830. [CrossRef] [PubMed]
37. Pak, T.; Gomari, K.E.; Bose, S.; Tonon, T.; Hughes, D.; Gronnow, M.; Macquarrie, D. Biochar from brown algae: Production, activation, and characterisation. *Bioresour. Technol. Rep.* **2023**, *24*, 101688. [CrossRef]
38. Yue, Y.; Cheng, L.; Sun, Y.; Pang, Y.; Wu, B.; Shi, L.; He, J.; Jia, X. Changes of soil water budget in the area covered by biological soil crusts in Mu Us sandy land, China. *J. Appl. Ecol.* **2022**, *33*, 1861–1870.
39. Chamizo, S.; Mugnai, G.; Rossi, F.; Certini, G.; De Philippis, R. Cyanobacteria Inoculation Improves Soil Stability and Fertility on Different Textured Soils: Gaining Insights for Applicability in Soil Restoration. *Front. Environ. Sci.* **2018**, *6*, 49. [CrossRef]
40. Eldridge, D.J.; Freudenberger, D.; Koen, T.B. Diversity and abundance of biological soil crust taxa in relation to fine and coarse-scale disturbances in a grassy eucalypt woodland in eastern Australia. *Plant Soil* **2006**, *281*, 255–268. [CrossRef]
41. David, Q.; Jonah, M.G.; John, M.; Jason, C.Q. Geographical assessment of open pond algal productivity and evaporation losses across the United States. *Algal Res.* **2021**, *60*, 102483.
42. Sun, L.; Bai, Z.; Yang, Q.; Fu, R.; Li, H.; Li, X. In situ assessment of the initial phase of wastewater biofilm formation: Effect of the presence of algae in an aerobic bacterial biofilm system. *Water Res.* **2024**, *253*, 121283. [CrossRef] [PubMed]
43. Zhou, L.; Lamusa, A. Influence mechanism of development of biological soil crust on evaporation process. *J. Arid Land Resour. Environ.* **2011**, *25*, 193–200.
44. Chamizo, S.; Canton, Y.; Rodriguez-Caballero, E.; Domingo, F. Biocrusts positively affect the soil water balance in semiarid ecosystems. *Ecohydrology* **2016**, *9*, 1208–1221. [CrossRef]
45. Alvarez, A.L.; Weyers, S.L.; Gardner, R.D. Cyanobacteria-based soil amendments in the soil-plant system: Effects of inoculations on soil nutrient and microbial dynamics under spring wheat growth. *Algal Res.* **2024**, *77*, 103326. [CrossRef]
46. Elbert, W.; Weber, B.; Burrows, S.; Steinkamp, J.; Budel, B.; Andreae, M.O.; Poschl, U. Contribution of cryptogamic covers to the global cycles of carbon and nitrogen. *Nat. Geosci.* **2012**, *5*, 459–462. [CrossRef]
47. Jassey, V.E.J.; Walcker, R.; Kardol, P.; Geisen, S.; Heger, T.; Lamentowicz, M.; Hamard, S.; Lara, E. Contribution of soil algae to the global carbon cycle. *New Phytol.* **2022**, *234*, 64–76. [CrossRef] [PubMed]
48. Lan, S.; Ouyang, H.; Wu, L.; Zhang, D.; Hu, C. Biological soil crust community types differ in photosynthetic pigment composition, fluorescence and carbon fixation in Shapotou region of China. *Appl. Soil Ecol.* **2017**, *111*, 9–16. [CrossRef]

49. Khoshru, B.; Khoshmanzar, E.; Asgari Lajayer, B.; Ghorbanpour, M. Soil moisture–mediated changes in microorganism biomass and bioavailability of nutrients in paddy soil. In *Plant Stress Mitigators*; Ghorbanpour, M., Adnan Shahid, M., Eds.; Academic Press: Cambridge, MA, USA, 2023; pp. 479–494.
50. Liu, M.; Xia, L.; Liu, R.; Gao, Z.; Han, C.; Feng, J.; Wang, J.; Qu, W.; Xing, T. Degradation of High-Concentration Nitrate Nitrogen in Groundwater: A Laboratory Study. *J. Chem.* **2021**, *2021*, 4797946. [CrossRef]
51. Yao, X.; Zhang, L.; Zhang, Y.; Zhang, B.; Zhao, Z.; Zhang, Y.; Li, M.; Jiang, X. Nitrogen Fixation Occurring in Sediments: Contribution to the Nitrogen Budget of Lake Taihu, China. *J. Geophys. Res.-Biogeosci.* **2018**, *123*, 2661–2674. [CrossRef]
52. Shatwell, T.; Koehler, J. Decreased nitrogen loading controls summer cyanobacterial blooms without promoting nitrogen-fixing taxa: Long-term response of a shallow lake. *Limnol. Oceanogr.* **2019**, *64*, S166–S178. [CrossRef]
53. Xu, H.; McCarthy, M.J.; Paerl, H.W.; Brookes, J.D.; Zhu, G.; Hall, N.S.; Qin, B.; Zhang, Y.; Zhu, M.; Hampel, J.J.; et al. Contributions of external nutrient loading and internal cycling to cyanobacterial bloom dynamics in Lake Taihu, China: Implications for nutrient management. *Limnol. Oceanogr.* **2021**, *66*, 1492–1509. [CrossRef]
54. Wu, L.; Zhang, C.; Vadiveloo, A.; Montes, M.L.; Xia, L.; Song, S.; Fernandez, M.A.; Lan, S. Efficient nutrient recycling from wastewater to deserts: A comparative study on biocrust cyanobacteria performance. *Chem. Eng. J.* **2024**, *491*, 151927. [CrossRef]
55. Liu, J.; Macrae, M.L.; Elliott, J.A.; Baulch, H.M.; Wilson, H.F.; Kleinman, P.J.A. Impacts of Cover Crops and Crop Residues on Phosphorus Losses in Cold Climates: A Review. *J. Environ. Qual.* **2019**, *48*, 850–868. [CrossRef]
56. Sun, D.; Jiang, X.; Wu, Q.; Zhou, N. Intragenomic Heterogeneity of 16S rRNA Genes Causes Overestimation of Prokaryotic Diversity. *Appl. Environ. Microbiol.* **2013**, *79*, 5962–5969. [CrossRef]
57. Bothe, H.; Schmitz, O.; Yates, M.G.; Newton, W.E. Nitrogen Fixation and Hydrogen Metabolism in Cyanobacteria. *Microbiol. Mol. Biol. Rev.* **2010**, *74*, 529–551. [CrossRef]
58. Nguyen, M.L.; Westerhoff, P.; Baker, L.; Hu, Q.; Esparza-Soto, M.; Sommerfeld, M. Characteristics and reactivity of algae-produced dissolved organic carbon. *J. Environ. Eng.* **2005**, *131*, 1574–1582. [CrossRef]
59. Li, X.; Zhang, P.; Su, Y.; Jia, R. Carbon fixation by biological soil crusts following revegetation of sand dunes in arid desert regions of China: A four-year field study. *Catena* **2012**, *97*, 119–126. [CrossRef]
60. Su, Y.; Li, X.; Chen, Y.; Zhang, Z.; Li, Y. Carbon fixation of cyanobacterial–algal crusts after desert fixation and its implication to soil organic carbon accumulation in desert. *Land Degrad. Dev.* **2013**, *24*, 342–349.
61. Song, X.; Bo, Y.; Feng, Y.; Tan, Y.; Zhou, C.; Yan, X.; Ruan, R.; Xu, Q.; Cheng, P. Potential applications for multifunctional microalgae in soil improvement. *Front. Environ. Sci.* **2022**, *10*, 1035332. [CrossRef]
62. Redmile-Gordon, M.; Gregory, A.S.; White, R.P.; Watts, C.W. Soil organic carbon, extracellular polymeric substances (EPS), and soil structural stability as affected by previous and current land-use. *Geoderma* **2020**, *363*, 114143. [CrossRef]
63. Li, J.; Xu, M.; Wang, J.; Lan, C.; Lai, J. Effects of nutrient limitation on cell growth, exopolysaccharide secretion and TEP production of Phaeocystis globosa. *Mar. Environ. Res.* **2023**, *183*, 105801. [CrossRef]
64. Kappaun, K.; Piovesan, A.R.; Carlini, C.R.; Ligabue-Braun, R. Ureases: Historical aspects.; catalytic.; and non-catalytic properties—A review. *J. Adv. Res.* **2018**, *13*, 3–17. [CrossRef]
65. Yahya, M.N.; Gökçekuş, H.; Orhon, D.; Keskinler, B.; Karagunduz, A.; Omwene, P.I. A Study on the Hydrolysis of Urea Contained in Wastewater and Continuous Recovery of Ammonia by an Enzymatic Membrane Reactor. *Processes* **2021**, *9*, 1703. [CrossRef]
66. Yang, Z.; Zhou, Q.; Sun, H.; Jia, L.; Zhao, L.; Wu, W. Metagenomic analyses of microbial structure and metabolic pathway in solid-phase denitrification systems for advanced nitrogen removal of wastewater treatment plant effluent: A pilot-scale study. *Water Res.* **2021**, *196*, 117067. [CrossRef]
67. Yuan, D.; Zheng, L.; Liu, Y.X.; Cheng, H.; Ding, A.; Wang, X.; Tan, Q.; Wang, X.; Xing, Y.; Xie, E.; et al. Nitrifiers Cooperate to Produce Nitrous Oxide in Plateau Wetland Sediments. *Environ. Sci. Technol.* **2023**, *57*, 810–821. [CrossRef]
68. Shrewsbury, L.H.; Smith, J.L.; Huggins, D.R.; Carpenter-Boggs, L.; Reardon, C.L. Denitrifier abundance has a greater influence on denitrification rates at larger landscape scales but is a lesser driver than environmental variables. *Soil Biol. Biochem.* **2016**, *103*, 221–231. [CrossRef]
69. Chen, X.; Jiang, H.; Sun, X.; Zhu, Y.; Yang, L. Nitrification and denitrification by algae-attached and free-living microorganisms during a cyanobacterial bloom in Lake Taihu.; a shallow Eutrophic Lake in China. *Biogeochemistry* **2016**, *131*, 135–146. [CrossRef]
70. Dai, H.; Zhu, R.; Sun, B.; Che, C.; Hou, L. Effects of Sea Animal Activities on Tundra Soil Denitrification and nirS- and nirK-Encoding Denitrifier Community in Maritime Antarctica. *Front. Microbiol.* **2020**, *11*, 573302. [CrossRef]
71. Chen, Z.; Ma, J.; Liu, Y.; Zhao, J.; Ma, J.; Yu, Q.; Zou, P.; Lin, H.; Wang, Q. Differential responses of soil *nirS*- and *nirK*-type denitrifying microbial communities to long-term application of biogas slurry in a paddy soil. *Appl. Soil Ecol.* **2023**, *182*, 104711. [CrossRef]
72. Li, J.; Wang, J.; Hu, H.; Cai, Z.; Lei, Y.; Li, W.; Zhang, M.; Li, Z.; Zhu, Y.; Cui, L. Changes of the denitrifying communities in a multi-stage free water surface constructed wetland. *Sci. Total Environ.* **2019**, *650*, 1419–1425. [CrossRef]
73. Shi, R.; Xu, S.; Qi, Z.; Huang, H.; Liang, Q. Seasonal patterns and environmental drivers of *nirS*- and *nirK*-encoding denitrifiers in sediments of Daya Bay.; China. *Oceanologia* **2019**, *61*, 308–320. [CrossRef]
74. Zheng, Y.; Hou, L.; Liu, M.; Gao, J.; Yin, G.; Li, X.; Deng, F.; Lin, X.; Jiang, X.; Chen, F.; et al. Diversity.; Abundance.; and Distribution of nirS-Harboring Denitrifiers in Intertidal Sediments of the Yangtze Estuary. *Microb. Ecol.* **2015**, *70*, 30–40. [CrossRef] [PubMed]

75. Kuypers, M.M.M.; Marchant, H.K.; Kartal, B. The microbial nitrogen-cycling network. *Nat. Rev. Microbiol.* **2018**, *16*, 263–276. [CrossRef] [PubMed]
76. Graf, D.R.H.; Zhao, M.; Jones, C.M.; Hallin, S. Soil type overrides plant effect on genetic and enzymatic N_2O production potential in arable soils. *Soil Biol. Biochem.* **2016**, *100*, 125–128. [CrossRef]

processes

MDPI

Article

Candida parapsilosis CMGB-YT Biosurfactant for Treatment of Heavy Metal- and Microbial-Contaminated Wastewater

Ortansa Elisabeta Csutak [1,2], Nicoleta-Oana Nicula [3,4], Eduard-Marius Lungulescu [3], Virgil Emanuel Marinescu [3], Ioana Catalina Gifu [5] and Viorica Maria Corbu [1,2,*]

1 Department of Genetics, Faculty of Biology, University of Bucharest, Aleea Portocalelor 1-3, 060101 Bucharest, Romania; ortansa.csutak@bio.unibuc.ro
2 Research Institute of University of Bucharest (ICUB), B.P. Hasdeu Street 7, 050568 Bucharest, Romania
3 National R & D Institute for Electrical Engineering ICPE-CA, Splaiul Unirii 313, 030138 Bucharest, Romania; nicoleta.nicula@icpe-ca.ro (N.-O.N.); marius.lungulescu@icpe-ca.ro (E.-M.L.); virgil.marinescu@icpe-ca.ro (V.E.M.)
4 Faculty of Biology, University of Bucharest, Splaiul Independentei 91-95, 050095 Bucharest, Romania
5 National Institute for Research & Development in Chemistry and Petrochemistry—ICECHIM, 202 Spl. Independentei, 060021 Bucharest, Romania; catalina.gifu@icechim-pd.ro
* Correspondence: viorica-maria.corbu@bio.unibuc.ro

Abstract: During the last few decades, water pollution has become a growing concern at international level. To date, only a few *Candida parapsilosis* strains were successfully used in environmental remediation. In the present article, the strain *C. parapsilosis* CMGB-YT was studied for its ability to assimilate hydrophobic substrates and to produce biosurfactants with antimicrobial activity and positive effects on heavy metal removal from contaminated wastewaters. The strain *C. parapsilosis* CMGB-YT was grown on yeast peptone (YP) media with 1% *n*-decane, *n*-dodecane, *n*-tetradecane, *n*-hexadecane, as well as commercial sunflower and olive oils. The production of the biosurfactant was evaluated using the emulsification index ($E_{24\%}$). The surface properties and emulsifying stability of the biosurfactant were determined. The effect of the biosurfactant on the cell growth of two strains of *Rhodotorula mucilaginosa* and on their removal capacity of lead (0.032 g/L) and cadmium (0.030 g/L) ions from synthetic wastewater were also studied. The antimicrobial potential of 20 mg/mL and 40 mg/mL biosurfactant was established in the presence of pathogenic *Candida krusei* strains. *C. parapsilosis* CMGB-YT assimilated *n*-hexadecane with good rates over 216 h and produced an anionic glycolipidic biosurfactant with stable $E_{24\%}$ towards long-chain carbon compounds at different temperatures, with an alkaline pH and high salinity (10% NaCl). The biosurfactant reduced the surface tension to 53.58 ± 0.42 mN/m, while the critical micellar concentration (CMC) was reached at 4.2% biosurfactant. The crude biosurfactant (5%) enhanced *R. mucilaginosa* growth in heavy metal-contaminated wastewater, increased chemical oxygen demand (COD) removal of up to 80%, and improved Cd^{2+} removal by 10%. Additionally, the concentrated biosurfactant effectively prevented *Candida krusei* biofilm formation. In conclusion, the biosurfactant produced by *C. parapsilosis* CMGB-YT demonstrates promising potential for the efficient treatment of wastewater contaminated with heavy metals and microbial pathogens.

Keywords: *Candida parapsilosis*; hydrocarbons; vegetable oils; biosurfactant; heavy metals; wastewater treatment

1. Introduction

The alarming rate at which water resources are declining due to pollution or inadequate exploitation raises the problem that a better understanding of the factors involved in the maintenance of the ecological equilibrium of the aqueous habitats must be cultivated, representing the basis for the development of new techniques aimed to restore the sustainable water supply [1]. During the last few decades, the use of microbial cultures (bacteria, yeasts, filamentous fungi) or some of their metabolic products or byproducts (enzymes, fatty acids, biosurfactants) in various bioremediation technologies is regarded as an ecological solution to the chemical treatment of polluted environments [2,3].

Some of the wastewaters or effluents originating from agricultural waste streams, the petroleum industry, vegetable oil mills, paper, paint, and the food processing or pharmaceutical industry comprise recalcitrant compounds such as hydrocarbons, fatty acids distillates, oils or heavy metals. Most of these water pollutants are tested for degradation using microorganisms with intrinsic metabolic pathways adapted to assure their efficient assimilation under lab or batch conditions, and their possible conversion into biocompounds with biotechnological applications [4].

The *n*-alkanes are constituents of petroleum (crude-oil) or petroleum-derived products, and their presence is associated with a disturbance of the composition in minerals, nutrients or heavy metals of the soil, the abnormal development of plants, as well as respiratory problems, nervous system diseases, and fatigue or hemocytotoxicity in human and animals, respectively [5,6]. The yeast species from *Candida* genera (e.g., *C. tropicalis*, *C. maltosa*, *C. oleophila*) can degrade with a preference for long chain *n*-alkanes, which are degraded through the mono-terminal pathway to lipids or to acetyl-coenzyme A, respectively, or can assimilate fatty acids or oils (fatty acid esters) through peroxisomal β-oxidation [7].

The ability of the yeasts to grow on fatty acids and vegetable oils (including sunflower or olive oil) as sole carbon substrates is related to the synthesis and secretion of lipases which are responsible for breaking down the ester bonds within the triglycerides. The olive oil contains oleic acid (18:1 *cis*-9) in a proportion of up to 84% and linoleic acid (18:2 *cis*-9,12) up to 20%, in comparison to the sunflower oil, which contains mainly (up to 60%) linoleic acid and, at a smaller content, oleic acid (approx. 30%). While vegetable oils are known inducers for lipase synthesis in many yeast species, such as *Rhodotorula glutinis*, *Yarrowia lipolytica* or *Candida rugosa* [8], few data are available concerning the ability of *C. parapsilosis* strains to degrade vegetable oils as sole carbon substrates. During their study, Briand et al. [9] showed that the lipase synthesized by the strain *C. parapsilosis* CBS 604 has high activity in the presence of esters containing long-chain unsaturated fatty acids with a *cis*-Δ9 double bond. Genome studies [10] revealed that *C. parapsilosis* genome harbors genes for four putative secreted lipases (*LIP1/2*, *OLE1* and *FAS2*), with further studies indicating that *OLE1* and *FAS2* genes are involved in the assimilation of oleic acid and palmitic acid, the latter being one of the saturated fatty acids found both in olive and sunflower oil [11].

C. parapsilosis is mostly studied in relation with non-albicans candidemias in immunocompromised patients, or as a contaminant of wastewaters from medical units and of water sources related to household activities [12–14]. Nevertheless, some studies proved the efficiency of *C. parapsilosis* strains isolated from natural environments, such as dyes or petroleum-contaminated wastewaters, in the remediation of textile industrial wastewater, the degradation of olive oil, fish oil, tributyrin or even plastic polymers (polystyrene or poly(vinyl chloride)) [15–17].

Strains belonging to different *Candida* or which were previously classified as belonging to the *Candida* genus (*Pseudozyma apicola*, *Yarrowia lipolytica*, *Starmerellabomicola*, *Pseudozyma antarctica*, *C. rugosa*, *C. tropicalis*) are reported to assimilate hydrocarbons (including *n*-alkanes) and fatty acids or vegetable oils, and to convert them into biosurfactants with a glycolipid structure, mainly sophorolipids [18]. Due to their amphiphilic structure consisting of a hydrophilic head and a hydrophobic tail, the biosurfactants are able to lower the surface tension at the interface of two non-miscible phases: aqueous and hydrophobic. Their high biodegradability, specificity and stability position them as effective ecological agents in the bioremediation of habitats contaminated with petroleum, diesel, kerosen, oils and heavy metals [19,20]. Although most works deal with soil bioremediation, the efficient removal of heavy metals from water mixtures, sludge or effluents using biosurfactants has begun to gain more interest during the last few years [1,21].

Another application of high interest for yeast biosurfactants is represented by their potential biomedical use. Many *Candida* synthesized sophorolipids are known antimicrobials, interacting with the cell membrane and causing the leakage of intracellular enzymes, damage to the reactive oxygen species (ROS)-mediated endoplasmatic reticulum stress mechanism or alterations of the mitochondrial membrane in human-pathogenic bacteria or fungi, respectively [22,23]. Also, these sophorolipids can determine bacterial biofilm disruption or inhibit hyphal growth and induce septa formation in pathogenic fungi [24,25].

The present work deals with the assimilation of hydrophobic substrates by the yeast strain *C. parapsilosis* CMGB-YT (*n*-alkanes, commercial sunflower oil and olive oil) and the synthesis of a biosurfactant, as well as the complex characterization of the obtained biosurfactant and its use in the removal of heavy metals and emerging human pathogens belonging to *Candida* genus from polluted synthetic wastewater, as a basis for developing further remediation technologies aimed at achieving sustainable water.

2. Materials and Methods

2.1. Yeast Strains

The yeast strain *Candida parapsilosis* CMGB-YT (the Collection of Microorganisms of the Department of Genetics, Faculty of Biology, University of Bucharest, Romania) was previously isolated and purified from oil-polluted soil (Pitesti County, Romania). The strains *Candida parapsilosis* CMGB-YT, *Rhodotorula mucilaginosa* CMGB188, *Rhodotorula mucilaginosa* CMGB-G1, *Candida krusei* CMGB94 and *Candida krusei* CMGB-Y8 were grown on yeast peptone glucose (YPG) medium (g/L: yeast extract 5, peptone 10, glucose 2; pH: 6.8 ± 0.2) and stored at −70 °C using YPG medium supplemented with 20% glycerol. All the strains were grown on YPGA medium (YPG with 1% agar-agar) for 20 h prior to each experiment.

2.2. *C. parapsilosis* CMGB-YT Growth Using Hydrophobic Substrates as Sole Carbon Source

The ability to assimilate hydrophobic substrates was tested by monitoring cell growth. A fresh culture *C. parapsilosis* CMGB-YT, grown for 20 h on YPG medium, was centrifuged 10 min, 6500 rpm, and the sediment was diluted in sterile water to obtain an inoculum of 1×10^8 cells/ mL (cells were counted using a Thoma chamber). A volume of 100 µL of inoculum (final cell density 1×10^6 cell/mL) was cultivated on YP medium (g/L: yeast extract 10, peptone 10; pH: 6.8 ± 0.2) supplemented with 1% carbon source: *n*-alkanes (*n*-decane, *n*-dodecane, *n*-tetradecane, *n*-hexadecane) (Sigma-Aldrich, St. Louis, MO, USA), respectively, commercial vegetable oils (refined sunflower oil "Untdelemn de la Bunica"–Romania, 10.5% saturated fatty acids; extra virgin olive oil "Monini"–origin: Italy, 14% saturated fatty acids). Culture samples were retreived after 72, 144 and 216 h, and the growth was determined by cell counting using a Thoma chamber. The experiment was performed in triplicates, with the cell counting process repeated twice for each replica.

2.3. Biosurfactant Production–Emulsification Activity

Biosurfactant production was evaluated using the emulsification index ($E_{24\%}$). For this purpose, the cell-free broth (crude biosurfactant) from a culture of *C. parapsilosis* CMGB-YT grown on YP medium with 10 g/L of each *n*-alkane for 144 h at 28 °C and 150 rpm, was mixed (3:2 *v:v*) with petroleum (FLUKA) and commercial refined sunflower oil ("Untdelemn de la Bunica")—single fry, respectively, then the mixture was vortexed at 2500 rpm for 2 min, and maintained at 22 °C for 24 h. The emulsification index was determined according to Cooper and Goldenberg [26].

2.4. Biosurfactant Characterization

2.4.1. Spectroscopic Analysis

For the purpose of the analysis, volumes of 20 mL of crude biosurfactant placed on watch glass were incubated in a lab oven at 40 °C until the total evaporation of the liquid. The chemical structure of the crude biosurfactant was assessed by UV–Vis (UV–Vis Jasco V-570, Jasco Inc., Japan) and FTIR (JASCO FTIR 4200, Jasco Inc., Tokyo, Japan). For the UV–Vis technique we used diluted biosurfactant liquid samples and spectroscopic cells with an optical path of 10 mm, while for the FTIR measurements we used dried biosurfactant samples and the ATR (Attenuated Total Reflectance) technique.

2.4.2. Thermal Analysis

The thermal behavior of the biosurfactant was investigated using differential scanning calorimetry (DSC). The analysis was performed on a Setaram DSC 131 EVO (Paris, France) to evaluate its response to temperature changes. The sample, obtained by evaporation/dehydration, was heated in an air atmosphere (50 mL/min) from 30 °C to 340 °C at a heating rate of 10 °C/min. Approximately 2 mg of the biosurfactant was placed in a 30 μL aluminum crucible for the analysis.

2.4.3. Biosurfactant Surface Properties

Oil displacement test

The oil displacement test is a simple and quick method used to assess the activity of biosurfactants. For this test, deionized water (30 mL) was placed in 7 cm diameter Petri dish [27]. Then, 200 μL hydraulic oil was dropped onto the surface of water. Finally, 50 μL of biosurfactant solutions was dropped onto the oil layer. The activity of the biosurfactant was assessed after 60 s.

Surface tension and critical micellar concentration

The surface tension and critical micellar concentration (CMC) of *C. parapsilosis* biosurfactant were determined using a K100 Kruss tensiometer (KRÜSS Optronic GmbH, Hamburg, DE) with the Du Noüy ring method. A series of aqueous surfactant solutions at various concentrations were prepared and equilibrated at room temperature (22 °C) to ensure equilibrium at the air/water interface. Surface tension data were plotted against the surfactant concentration, with the CMC identified at the concentration where the slope of the plot changed, indicating micelle formation.

Ionic character

The agar double diffusion technique was used to determine the ionic character of the biosurfactant. A Petri plate was prepared, with 10 g/L agar in water containing two rows of wells. One row was filled with the biosurfactant solution, and the other with known charged molecules: sodium dodecyl sulfate (SDS, anionic, 20 mM) and barium chloride ($BaCl_2$, cationic, 50 mM). The formation of precipitation lines between the wells, indicative of biosurfactant interaction with oppositely charged molecules, was monitored for 48 h at room temperature.

2.4.4. Stability Studies

The stability of the emulsifying activity of the crude biosurfactant was determined at different temperatures (4 °C, 22 °C, and 70 °C, respectively), pH values (4.0; 6.0; 10.0; 12.0) and sodium chloride (NaCl) final concentrations (2%, 4%, 6%, 8% and 10%), by calculating E_{24}% towards sunflower oil (single fry) similar to the method from 2.3. [28,29].

2.5. Biosurfactant Effect on Removal of Some Pollutants from Synthetic Wastewater

2.5.1. Growth Curves in Synthetic Wastewater

The growth curves of the yeast strains *R. mucilaginosa* CMGB-G1 and *R. mucilaginosa* CMGB 188 were determined in synthetic wastewater contaminated and uncontaminated with Pb^{2+} and Cd^{2+} ions, both in the absence or presence of crude biosurfactant. For this purpose, the 96-microwell plates were used with a final volume of 400 μL/sample inoculated with 1% suspension of *R. mucilaginosa* CMGB-G1, respectively, *R. mucilaginosa* CMGB 188 with an initial concentration of 2 McFarland (for Sw, Pb and Cd with P1, and P2, respectively), adding 5% (*v*/*v*) crude biosurfactant (for Bio C, Bio C + Pb, Bio C + Cd) (Table 1). The growth of the yeast cultures was determined during the 120th hour of the experiment by readings of the optical density (OD) at 570 nm using a Synergy HTX Multimode reader (BioTEK, Winooski, VT, USA).

Table 1. Sample coding and description.

Sample Code	Description
P1	*Rhodotorula mucilaginosa CMGB-G1*
P2	*Rhodotorula mucilaginosa CMGB 188*
Sw	Synthetic wastewater
Pb	Synthetic wastewater + Pb^{2+}
Cd	Synthetic wastewater + Cd^{2+}
Bio C	Synthetic wastewater + Biosurfactant from *C. parapsilosis* CMGB-YT
Bio C + Pb	Synthetic wastewater + Biosurfactant from *C. parapsilosis* CMGB-YT + Pb^{2+}
Bio C + Cd	Synthetic wastewater + Biosurfactant from *C. parapsilosis* CMGB-YT + Cd^{2+}

For synthetic wastewater preparation we adapted the formulations mentioned previously [28,30,31], comprising: 0.2 g/L D-glucose, 0.2 g/L sucrose, 0.07 g/L yeast extract, 0.067 g/L $(NH_4)_2SO_4$, 10.91 mg/L NH_4Cl, 4.43 mg/L KH_2PO_4, 0.021 g/L $MgSO_4 \cdot 7H_2O$, 2.68 mg/L $MnSO_4 \cdot H_2O$, 0.03 mg/L $NaHCO_3$, 0.0197 mg/L $CaCl_2$, 0.14 mg/L $FeCl_3 \cdot 6H_2O$, 0.330 mg/L $NaNO_3$, and 0.390 mg/L $NaNO_2$. The removal efficiency of heavy metal ions by the yeast strains was assessed using synthetic wastewater contaminated with Pb^{2+} (0.032 mg/L, $(CH_3COO)_2Pb$) and Cd^{2+}(0.030 g/L, $Cd(NO_3)_2$) ions. The concentrations of heavy metal ions were determined using an inductively coupled plasma mass spectrometry technique (ICP-MS).

2.5.2. Determination of Physico-Chemical Parameters of Synthetic Municipal Wastewater

The physico-chemical parameters of the synthetic wastewater samples were recorded at the starting point and at the end of the experiment (120 h): the pH (using a pH meter HI10832, Hanna Instruments, Ins., Woonsocket, RI, USA), conductivity (using a Multiparameter HI2020-01 Edge, Hanna Instruments, Ins., Woonsocket, RI, USA) and dissolved oxygen (Multiparameter Oxi340i, Weiheim, Germany) were recorded.

2.5.3. Removal of Pollutants

Volumes of 450 mL of uncontaminated heavy metal (Pb^{2+} and Cd^{2+}) and contaminated synthetic wastewater were inoculated with 1% of *R. mucilaginosa* CMGB-G1, *mucilaginosa* CMGB 188, and *R.* with the initial concentration of 2 McFarland, respectively. The samples Bio C, Bio C + Pb and Bio C + Cd were obtained by adding 5% (*v*/*v*) crude biosurfactant in the inoculated synthetic wastewater. All the samples were incubated at 28 ± 2 °C for 120 h and the efficiency of nutrients and heavy metal removal was determined at the beginning and at the end of the experiment.

The efficiency of the removal of the organic substances was assessed by the determination of the chemical oxygen demand (COD), using a Gerhardt Chemical Oxygen Analyzer (C. Gerhardt GmbH & Co., Bonn, Germany), and according to ISO 6060 standard [ISO6060, 1989].

The efficiency of the removal of Pb^{2+} and Cd^{2+} ions was assessed using an ELAN DRC-e ICP-MS (PerkinElmer, Waltham, MA, USA), while quantitative evaluation was conducted for each element individually using the calibration curve method.

2.6. *Antimicrobial Activity*

2.6.1. Growth Inhibition Assays

The dried biosurfactant, prepared as described previously (Section 2.4.1), was used for obtaining 160 mg/mL and 40 mg/mL solutions of concentrated biosurfactant in sterile distilled water, which was further passed through 0.24 µm syringe filters for sterilization.

Suspensions of *C. krusei* CMGB94 and *C. krusei* CMGB-Y8 were prepared from 18 h-cultures grown on YPGA medium at 37 °C. Then, 10 µL of suspension (0.3×10^7 cells/mL determined using a Thoma chamber) was mixed in microtiter wells with 100 µL of concentrated biosurfactant to reach 20 mg/mL and 40 mg/mL final biosurfactant concentrations, respectively. The mixtures were incubated at 37 °C and the cell growth was determined by reading the O.D. at 570 nm using an automatic plate reader (Multi-mode reader Synergy HTX, BioTek) each hour up to 21 h [28,32]. The positive control was represented by YPG grown cultures, while the sterility control consisted of YPG broth with final concentrations of 20 mg/mL or 40 mg/mL concentrated biosurfactant.

2.6.2. Anti-Adhesion Activity Tests

The effect of the concentrated biosurfactant on the microbial adherence of *C. krusei* CMGB94 and *C. krusei* CMGB-Y8, respectively, to inert substratum, was assayed in 96-well plates (untreated polystyrene) using the crystal violet microtiter method. The two strains were grown in presence of 20 mg/mL and 40 mg/mL biosurfactant, similar to the method described above. After 24 h at 37 °C, the unattached cells were removed using sterile distilled water and the stabilization of the adherent cells to the well was carried out with 110 μL methanol solution (99%). After stabilization, the wells were emptied and dried at room temperature. The attached cells were fixed with 110 μL crystal violet solution (1%) added to each well, and incubated for 15 min. After washing the plates with running tap water (to remove the excess stain), the adherent cells were solubilized by adding 110 μL of 33% acetic acid (*v*/*v*). The O.D. was read at 595 nm [32,33]. For the samples, this was conducted in triplicate. The control wells contained only YPG broth (without inhibitory compound).

The microbial adherence inhibition percentages (PICA%) were determined in the presence of concentrated biosurfactants, and calculated using Equation (1) according to [33].

$$\text{PICA\%} = (\text{As}/\text{Ac}) \times 100 \quad (1)$$

where As is the absorbance of the sample (adherence in the presence of concentrated biosurfactants) and Ac is the absorbance of the control well (adherence in the absence of the inhibitory compound).

2.7. Statistical Analysis

For the synthetic wastewater treatment tests, the analyses were performed in triplicate, and the results were presented as the mean ± standard deviation (SD) of three independent experiments. The statistical significance was analyzed by Student's *t*-test. A value of p less than 0.05 was considered significant. The statistical analysis was performed with Microsoft Excel 2021 software (Microsoft Corporation, Redmond, WA, USA) or with GraphPad Prism (Dotmatics, Bishop's Stortford, UK).

The antimicrobial properties of the biosurfactant were analyzed using GraphPad Prism v9. Data were expressed as mean ± SD for triplicate samples. For multiple comparisons, data were analyzed using ordinary two-way ANOVA with post hoc tests (Tukey correction for the growth inhibition percentage and Dunnett's test for the PICA%), and individual variances were calculated to compare the biological activities of the biosurfactant to the control strain. The significance level was set to $p < 0.05$.

3. Results and Discussion

3.1. Hydrophobic Substrate Assimilation

The ability of the strain *C. parapsilosis* CMGB-YT to assimilate *n*-alkanes and vegetable oils as sole carbon substrates was assessed by monitoring the cell growth over 216 h (9 days). Within the first 72 h, the strain recorded a significant cell growth on all the media (Figure 1), after which differences appeared between the curve profiles obtained on *n*-alkanes and vegetable oils, respectively. Thus, the growth was maintained with a good rate until 144 h for all the *n*-alkanes, and continued in the presence of *n*-tetradecane and *n*-hexadecane until the end of the experiment. On the contrary, after 144 h, a rather important drop was recorded in the media supplemented with *n*-decane and, to a smaller extent, with *n*-dodecane. When vegetable oils were used, the growth followed a descending slope between 72 and 144 h. After this point, the growth was restored, maintaining a higher rate in the presence of sunflower oil compared to olive oil.

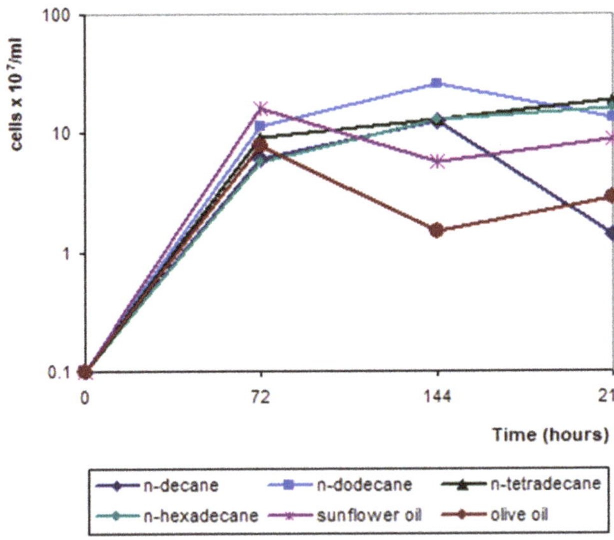

Figure 1. Growth curves of *C. parapsilosis* CMGB-YT cultivated on YP culture media and hydrophobic carbon substrates at 28 °C; 150 rpm.

Until now, rather few experiments were conducted on the *C. parapsilosis* strains able to grow on hydrocarbons or vegetable oils. Thus, an early study [34] showed that the cells of a *C. parapsilosis* strain isolated from oily sludge were able to degrade 50% of the hydrocarbons from the oily sludge within 2 weeks. The process was enhanced by the immobilization of *C. parapsilosis* cells on sterilized granular clay, with the immobilized cells using almost 80% of the hydrocarbons during the same period of time. More recently, ref. [35] reported that the growth medium containing a biomass of *C. parapsilosis* KCC Y1 (from oil sediments from Arabian Gulf costs) lost about 7.7 mg *n*-octadecane and 6.6 mg *n*-docosane within six hours, respectively, with the alkanes being used for lipid biosynthesis. In another study conducted on a *C. parapsilosis* strain isolated from a hydrocarbon-contaminated area, Benmessaoud et al. [36] showed that the strain grew on *n*-hexadecane following an ascending curve similar to the one obtained during our experiment. Also, the strain grew constantly on petroleum within the first four days of incubation and was able to remove approx. 50% of the crude oil from the culture medium within 15 days of incubation. For another *C. parapsilosis* strain isolated from spots surrounding crude oil reservoirs, Al-Otibi et al. [37] recorded a constant growth in the presence of petroleum, with an accumulation of 3.76 g of biomass during a period of 30 days.

The strain *C. parapsilosis* CMGB-YT was isolated from oil-polluted soil, and therefore it was expected to present active metabolic pathways involved in the degradation of paraffins. Our results showed that the strain has a good ability to assimilate *n*-alkanes over six days (144 h), showing a preference for *n*-tetradecane and *n*-hexadecane over an extended period of time (216 h).

Regarding the growth studies using vegetable oils, a strain of *C. parapsilosis* cultivated on a Mineral Salts Medium supplemented with 2% olive oil (containing 76% oleic acid), was able to reduce the fatty acid content of the medium by 6.7% within 20 days [38]. In another study, Popoola and Onilude [38] demonstrated that the addition of 1% of olive oil as the sole carbon source induced the highest growth rates and lipase production in *C. parapsilosis*. Moreover, Ciafardini et al. [39] isolated from commercial extra virgin olive oil two strains of *C. parapsilosis* (DAPES 1890 and 1892), which were inoculated in a liophilized form in olive oil samples supplemented with 45 mg/mL gallic acid, being able to survive up to 18 months in a dark room at 15–17 °C.

In our experiments, *C. parapsilosis* CMGB-YT seemed to better consume the polyunsaturated fatty acids (e.g., linoleic acid), followed by the monounsaturated fatty acids (e.g., oleic acid). The appearance of the growth curves after 144 h of incubation could have two possible explanations. On one hand, some of the assimilated fatty acids are incorporated in the yeast cell into storage or complex membrane lipids that can be used to support cell metabolism and growth under limitative substrate conditions until the end of the experiment [40]. On the other hand, after consuming the preferred substrates, the cells might begin using other compounds (e.g., fatty acids) from oils, thus restoring the cell growth. Further testing involving chemical determination for proving the above statement will be performed as a part of the other studies concerning the metabolic description of *C. parapsilosis* strains.

3.2. Biosurfactant Characterization

3.2.1. Emulsifying Activity

Since the best *C. parapsilosis* CMGB-YT growth rates were obtained during 144 h (six days) of incubation in media with *n*-alkanes, biosurfactant production was evaluated using cultures grown under these conditions. For the evaluation of the emulsification activity of the biosurfactant, we tested two different types of hydrophobic substrates frequently found as industrial or household wastes, respectively: petroleum and sunflower oil—single fry. In all cases, we obtained good $E_{24\%}$ against sunflower oil—single fry, ranging from 32% (for the biosurfactant obtained using *n*-tetradecane) and up to 42% (in the case on *n*-dodecane) (Figure 2). However, only the biosurfactant obtained using *n*-hexadecane-grown cultures was efficient in emulsifying both substrates, with an $E_{24\%}$ of 24% against petroleum and 27% against sunflower oil—single fry, respectively. Therefore, despite the rather reduced $E_{24\%}$, we decided to use this biosurfactant for further experiments. Moreover, the *n*-hexadecane was also one of the two carbon sources for which we determined continuous cell growth over the entire period of our experiment.

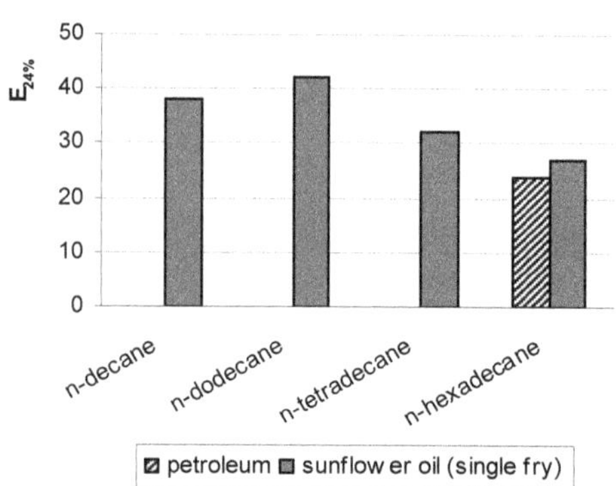

Figure 2. Emulsification index of *C. parapsilosis* CMGB-YT biosurfactant.

The results suggest the important influence of the hydrocarbon substrate used for the production of the biosurfactant on its structure and characteristics, a fact also described by Al-Otibi et al. [37]. During the same study, these authors showed that the highest rate of biosurfactant was obtained using crude oil, and the crude biosurfactant had an $E_{24\%}$ of about 55% against the oil samples used.

The *C. parapsilosis* CMGB-YT biosurfactant obtained using *n*-hexadecane has a good emulsifying activity towards hydrophobic substrates with long-carbon chains, with possible applications in the chemical and food industry, but also in the bioremediation of petroleum- and vegetable oil-polluted environments.

3.2.2. UV–Vis and FTIR Spectroscopy

The understanding of biosurfactants' structural and functional attributes can be achieved through various analytical techniques, including UV–VIS spectroscopy and Fourier-transform infrared spectroscopy (FTIR) [41,42].

The UV–VIS spectroscopy provides information about molecular electronic transitions, offering insights into the chromophores present in biosurfactants. Figure 3 presents the spectra of the *C. parapsilosis* CMGB-YT biosurfactant with an absorption maximum at about 259 nm, due to the presence of their aromatic rings. Aromatic rings, due to their conjugated π electron systems, exhibit an intense absorption at about 200 nm and a less intense absorption at about 260 nm [43].

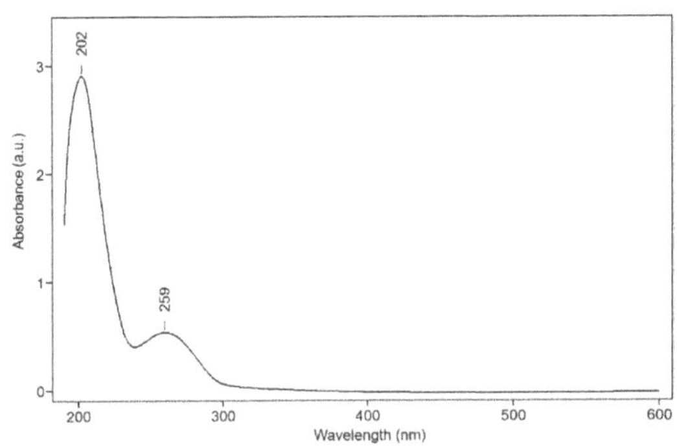

Figure 3. UV–Vis spectra of the *C. parapsilosis* YT-CMGB biosurfactant.

The FTIR spectrum (Figure 4) reveals the biosurfactant's structure, dominated by a broad O-H stretching peak (2200–3600 cm^{-1}) indicative of carboxylic acids (COOH) [30,44]. This is further corroborated by the vibrations observed at 1527–1574 cm^{-1} (COO^{-} stretching) and 974–1176 cm^{-1} (O-C-O stretching) [45]. Lower intensity peaks at 3066 cm^{-1} and 1391 cm^{-1} suggest the possible presence of aromatic rings, as indicated by vinyl =C-H and C=C stretching, respectively [46]. The peak at 1633 cm^{-1} could be associated with the stretching vibrations of N-H bond. Additionally, peaks at 2869, 2914, and 2942 cm^{-1} indicate the presence of aliphatic chains within the biosurfactant molecule [30,44]. The FTIR analysis indicates that the biosurfactant is a glycolipid, consistent with other works on the structure of *C. parapsilosis* biosurfactants [47,48].

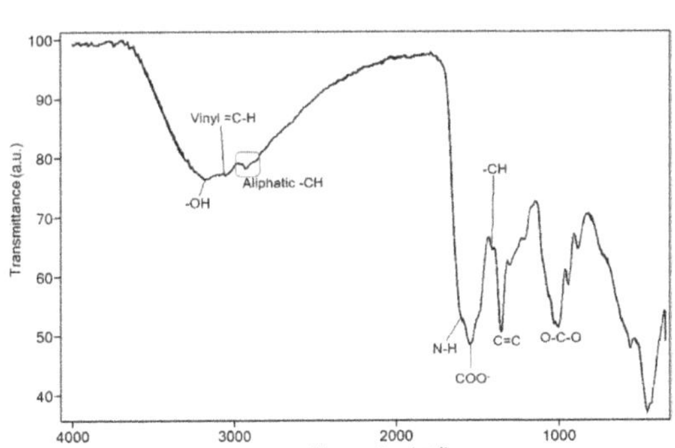

Figure 4. FTIR spectra of the *C. parapsilosis* YT-CMGB biosurfactant.

3.2.3. Thermal Properties

Differential scanning calorimetry (DSC) is a powerful analytical technique used to study the thermal properties of biosurfactants by measuring the heat flow into or out of a sample during heating or cooling. This method provides valuable information about phase transitions, melting points, and thermal stability [49]. The DSC thermogram (Figure 5) recorded on the biosurfactant exhibited three endothermic peaks: the first (80–120 °C) could be associated with the evaporation of water moieties, while the other two peaks (with peaks at 138 °C and 189 °C) most likely represent the melting of different components within the biosurfactant molecule [50]. The exothermic peak starting around 270–280 °C indicates the onset oxidation temperature (OOT), which is an important parameter that reflects the biosurfactant's thermal stability [49,51]. At temperatures lower than the OOT, the biosurfactant is considered stable and less prone to degradation through oxidation. These results are consistent with those obtained for similar biosurfactants [50].

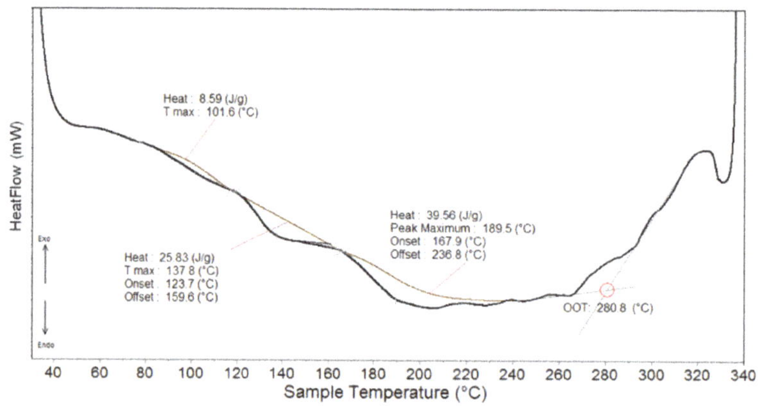

Figure 5. DSC thermogram recorded on the *C. parapsilosis* YT-CMGB biosurfactant.

3.2.4. Surface Properties

The oil displacement test serves as a preliminary screening tool for identifying potential biosurfactant activity by observing the formation of a clear zone around the biosurfactant drop, as illustrated in Figure 6A. This positive result indicates that the biosurfactant effectively lowers the surface tension of water, enhancing its interaction with oil [27]. This finding was corroborated by surface tension analysis, showing that the biosurfactant reduces the surface tension to 53.58 ± 0.42 mN/m at room temperature, significantly lower than the 72.8 mN/m surface tension of distilled water [45].

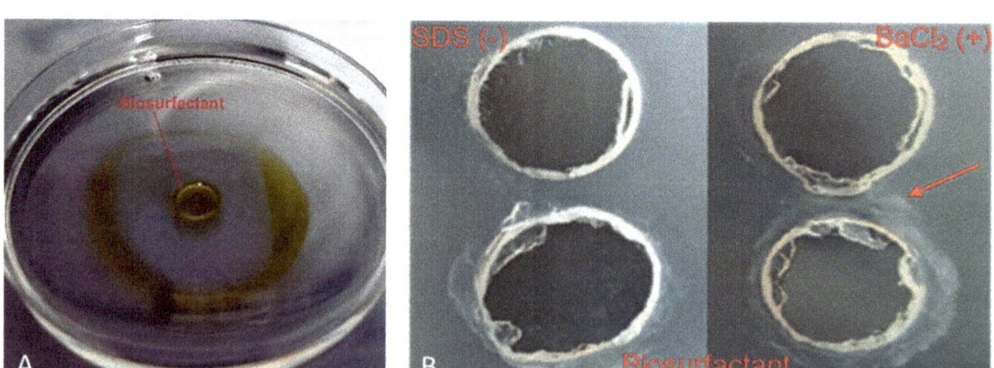

Figure 6. (**A**) Oil displacement test; (**B**) ionic character of the *C. parapsilosis* YT-CMGB biosurfactant.

Agar double diffusion confirmed the anionic character of the *C. parapsilosis* YT-CMGB biosurfactant. A precipitation line (Figure 6B) formed only between the biosurfactant and the cationic barium chloride, but not with the anionic SDS.

Figure 7 depicts the CMC behavior of the *C. parapsilosis* biosurfactant. Initially, as the concentration of the biosurfactant increased, the surface tension decreased significantly, indicating that the biosurfactant molecules were adsorbing at the air–water interface, reducing the surface tension. At a concentration of about 4.2% of biosurfactant, the curve was stabilizing, indicating that the CMC has been reached. Beyond this concentration, the surface tension remained relatively constant despite increasing the surfactant concentration, signifying that any additional surfactant molecules contribute to micelle formation rather than further reducing surface tension, with this behavior being characteristic of surfactants [52,53].

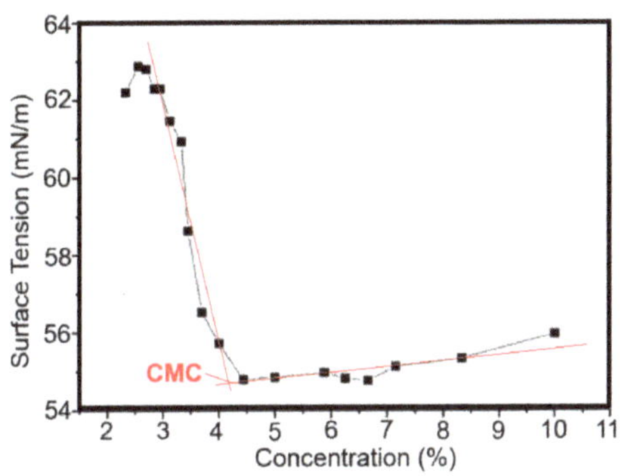

Figure 7. Surface tension isotherms of *C. parapsilosis* YT-CMGB biosurfactant.

3.2.5. Emulsifying Stability

Testing the stability of the emulsifying activity of the biosurfactant is important for further practical applications related to industrial or natural environments, prone to variations of different parameters.

We chose to test the stability of the emulsifying activity of the *C. parapsilosis* YT-CMGB biosurfactant against the sunflower oil—single fry, since for this substrate we obtained the best $E_{24\%}$ during the previous experiment (Section 3.2.1).

The biosurfactant showed high thermostability (Figure 8A), with a slight reduction in the $E_{24\%}$ at high temperature, probably due to its degradation. The emulsifying ability was maintained at alkaline pH values, the significant impact of acidic pH being most probably associated with substantial modifications in the biosurfactant glycopilidic structure [54]. Concerning the salinity effect, similar results were reported in other works on glycolipid biosurfactants from yeasts and bacteria [55,56]. Al-Kashef et al. [48] obtained sophorolipids from a *C. parapsilosis* strain isolated from a chemical polluted environment near Cairo, grown on potato peel waste and frying oil waste. The purified biosurfactant had the highest $E_{24\%}$ (approx. 50%) against soybean oil and approx. 37% against corn oil or motor oil, showing a good stability under thermal and saline stress, while the acidic pH had a similar effect on the emulsifying ability as the one described in our study.

The good thermostability and emulsifying activity at neutral and high pH recommend the *C. parapsilosis* YT-CMGB biosurfactant for developing strategies for the treatment of alkaline wastewaters generated mainly in the meat, food, and beverage processing industry.

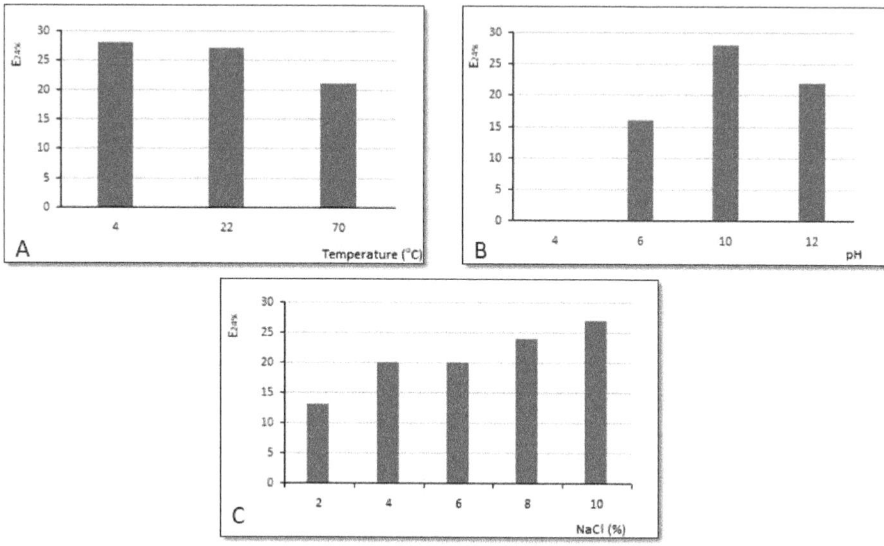

Figure 8. Effect of: (**A**) temperature; (**B**) pH; (**C**) NaCl on the emulsifying activity of *C. parapsilosis* YT-CMGB biosurfactant.

3.3. *Biosurfactant Effect on Synthetic Wastewater Remediation*

3.3.1. Growth Curves of *R. mucilaginosa* Cultures

The yeast species *R. mucilaginosa* (formerly known as *R. rubra*) has been previously described as being resistant to heavy metal exposure and used, with good results, in the treatment of wastewater contaminated with heavy metals, including lead and cadmium [57–62].

The growth curves of the yeasts *R. mucilaginosa* CMGB-G1 (P1) and *R. mucilaginosa* CMGB 188 (P2) were tracked using optical density measurements taken every 24 h over a period of 120 h. This approach enabled an analysis of the yeasts' long-term growth patterns and their responses to specific experimental conditions. The resulting growth curves from this extended observation are depicted in Figure 9.

The growth behavior of the microorganisms varies based on the strain type and experimental conditions, including the presence or absence of Pb^{2+}, Cd^{2+}, and biosurfactants. Generally, both yeast strains exhibit a slight increase in optical density (OD) values within the first 24–48 h, regardless of the conditions. Notably, when biosurfactants are present, OD values increase steadily until the end of the experiment, underscoring their beneficial impact on yeast development. The increase in the number of yeast cells in the presence of the biosurfactant can be explained both by the combined effect of the additional carbon supply (as a food source) and its ability to complex the Pb and Cd ions (the anionic nature of the *C. parapsilosis* YT-CMGB biosurfactant may favor these interactions with these metal ions). By binding these metals, the biosurfactant could reduce their free ion concentration, making them less toxic and potentially even increasing nutrient availability by preventing them from interfering with other yeast cellular processes [63]. For strain P1, the highest OD values after 120 h are observed in the *Bio C + Pb* samples, followed by *Bio C*, and then *Bio C + Cd*. Similarly, for strain P2, the OD values follow a quasi-similar trend, with *Bio C* showing the highest values, followed by *Bio C + Pb*, and then *Bio C + Cd*.

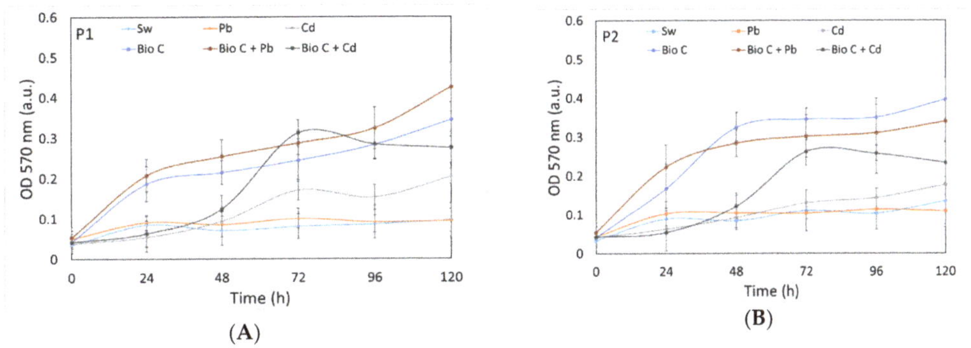

Figure 9. Growth curves in synthetic wastewater (26 ± 1 °C) of (**A**) *R.mucilaginosa* CMGB-G1 (P1) and (**B**) *R. mucilaginosa* CMGB 188 (P2) recorded during the 120 h experiment (Sw: synthetic wastewater; Pb: synthetic wastewater + Pb^{2+}; Cd: synthetic wastewater + Cd^{2+}; Bio C: synthetic wastewater + biosurfactant from *C. parapsilosis* CMGB-YT; Bio C + Pb: synthetic wastewater + biosurfactant from *C. parapsilosis* CMGB-YT + Pb^{2+}; Bio C + Cd: synthetic wastewater + biosurfactant from *C. parapsilosis* CMGB-YT + Cd^{2+}).

3.3.2. Wastewater Physico-Chemical Parameters

Table 2 presents, comparatively, the pH and conductivity of the synthetic wastewater samples, at the start and the end of the experiments, respectively. Temperature was kept quasi-constant at about 26 ± 1 °C. Daily, the samples were aerated using wood air diffuser for about 15 min, in order to increase the amount of dissolved oxygen to about 5 mg/L (the initial amount of dissolved oxygen was 7.5 ± 0.5 mg/L).

Table 2. Variations in the physico-chemical parameters of inoculated synthetic wastewater.

Time (h)		***Rhodotorula Mucilaginosa CMGB-G1* (P1)**					
	Parameters	**Sw**	**Pb**	**Cd**	**Bio C**	**Bio C + Pb**	**Bio C + Cd**
0	Conductivity (µS·cm^{-1})	1078	1040	1075	1249	1242	1248
	pH	7.23	7.12	7.11	7.42	7.39	7.23
120	Conductivity (µS·cm^{-1})	1906	1961	1746	2119	2136	2067
	pH	7.35	6.91	7.49	7.92	7.87	7.35
		***Rhodotorula mucilaginosa CMGB 188* (P2)**					
0	Conductivity (µS·cm−1)	1072	1076	1072	1236	1240	1160
	pH	7.19	7.11	7.11	7.42	7.4	7.19
120	Conductivity (µS·cm−1)	1936	1926	1750	2247	2129	2079
	pH	7.07	6.96	7.39	7.67	7.6	7.07

(Sw: synthetic wastewater; Pb: synthetic wastewater + Pb^{2+}; Cd: synthetic wastewater + Cd^{2+}; Bio C: synthetic wastewater + biosurfactant from *C. parapsilosis* CMGB-YT; Bio C + Pb: synthetic wastewater + biosurfactant from *C. parapsilosis* CMGB-YT + Pb^{2+}; Bio C + Cd: synthetic wastewater + biosurfactant from *C. parapsilosis* CMGB-YT + Cd^{2+}).

Our experiments show a general increase in conductivity for both yeast strains, with samples containing biosurfactants exhibiting the highest values. Interestingly, the pH values decreased during the first 96 h for all samples, regardless of their composition. This was followed by a slight rise until the 120th hour. The lowest pH values were measured in synthetic water containing *Bio C+ Pb* at the 24–48 h timeframe. This occurred for both *R. mucilaginosa* CMGB-G1 (P1, reaching a pH of 6.04, after 24 h) and *R. mucilaginosa* CMGB

188 (P2, reaching a pH of 6.15, after 48 h). These results are consisted with those reported by our group in a previous study [28,30].

3.3.3. Removal Efficiency of Pollutants from Water

Chemical Oxygen Demand (COD)

The chemical oxygen demand (COD) is a key factor in wastewater treatment processes, and provides a measure of the amount of oxygen required to break down organic matter, being an indicator of applied treatment efficiency as well. COD measures all the organic compounds, higher COD meaning more organic pollution [64,65].

The efficiency of COD reduction by the yeast strains *R.mucilaginosa* CMGB-G1 (P1) and *R. mucilaginosa* CMGB 188 (P2) are shown in Figure 10.

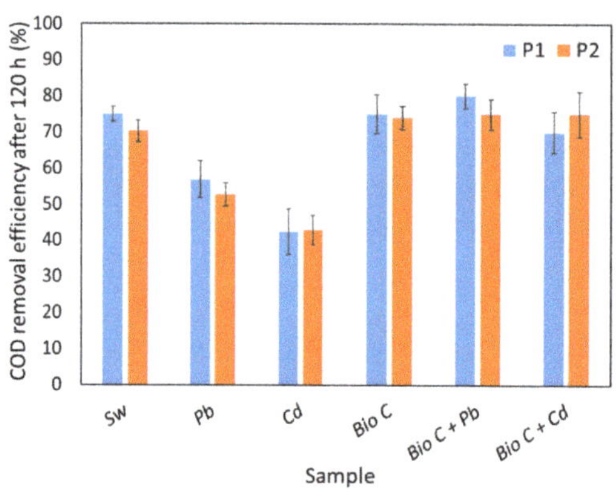

Figure 10. Yeast COD removal efficiency at the end of the experiment (after 120 h, 26 ± 1 °C) (Sw: synthetic wastewater; Pb: synthetic wastewater + Pb^{2+}; Cd: synthetic wastewater + Cd^{2+}; Bio C: synthetic wastewater + biosurfactant from *C. parapsilosis* CMGB-YT; Bio C + Pb: synthetic wastewater + biosurfactant from *C. parapsilosis* CMGB-YT + Pb^{2+}; Bio C + Cd: synthetic wastewater + biosurfactant from *C. parapsilosis* CMGB-YT + Cd^{2+}).

The reduction in COD is significantly influenced by the presence of metal ions in the tested wastewater, with the type of strain studied playing a lesser role. Thus, a COD reduction ranging between 53 and 57% is observed in the presence of Pb ions, and 42–43% in the presence of Cd ions. Remarkably, the addition of biosurfactant to the water greatly enhances COD reduction capacity, achieving an up to 80% reduction, which is comparable to or even exceeds the reduction levels in uncontaminated water (70–75%). Specifically, the *R. mucilaginosa* CMGB-G1 (P1) strain exhibits the highest COD reduction efficiency in the *Bio C + Pb* sample at 80%, followed by 75% in the *Bio C* sample, and 70% in the *Bio C + Cd* sample. Meanwhile, the *R. mucilaginosa* CMGB 188 (P2) strain consistently achieves a COD reduction of approximately 74% across all samples containing a biosurfactant.

These results are comparable to our prior investigations. We previously observed an 80% COD removal efficiency when using similar yeast strains, but with a biosurfactant produced by *Yarrowia lipolytica* [28]. Additionally, studies using five distinct yeast strains achieved a removal rate of around 70% of COD [30].

Heavy Metal Removal

Figure 11 presents a comparison of the efficiency of heavy metal removal, specifically for Pb and Cd, across all tested samples after 120 h, highlighting the influence of sample type on removal efficiency. Although the removal efficiency of Pb ions from the synthetic waters appears to be lower in the presence of the biosurfactant compared to samples without it, the biosurfactant positively impacts the removal capacity of cadmium ions. For instance, the removal efficiency of Cd ions from synthetic wastewater increased from 37% to

42% for *R. mucilaginosa* CMGB 188 (P2) and reached 47% for *R. mucilaginosa* CMGB-G1 (P1). This indicates that while the biosurfactant may not enhance Pb removal, it significantly improves Cd ion removal efficiency in these samples.

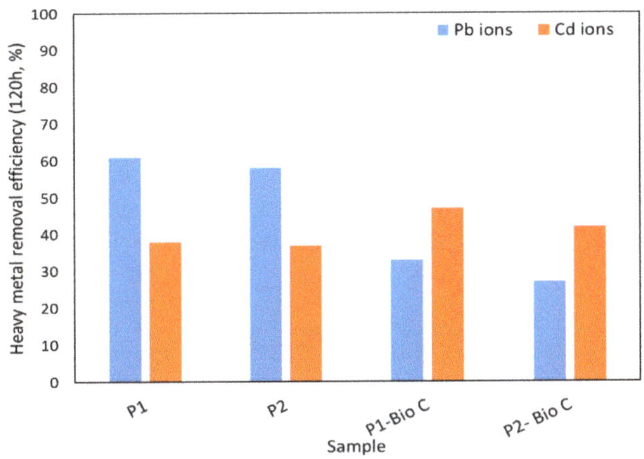

Figure 11. Yeast heavy metals' removal efficiency at the end of the experiment (after 120 h, 26 ± 1 °C) (Sw: synthetic wastewater; Pb: synthetic wastewater + Pb^{2+}; Cd: synthetic wastewater + Cd^{2+}; Bio C: synthetic wastewater + biosurfactant from *C. parapsilosis* CMGB-YT; Bio C + Pb: synthetic wastewater + biosurfactant from *C. parapsilosis* CMGB-YT + Pb^{2+}; Bio C + Cd: synthetic wastewater + biosurfactant from *C. parapsilosis* CMGB-YT + Cd^{2+}).

3.4. Antimicrobial Activity against Candida Strains

The antimicrobial resistance of the intra-hospital pathogens, as well as their propagation in the natural environment, represent a broad international problem for which society needs to find a solution. As a result, a number of cutting-edge substitute tactics for dealing with these (multi-)drug-resistant pathogenic microbes have been discovered [66,67]. The biological activity of biosurfactants includes antimicrobial, antitumor, antiviral, antifungal, hemolytic, and fibrin clotting inhibition. The biosurfactants can break down plasma membranes or increase the permeability of cell membranes, thus being a suitable alternative for testing their potential application in antimicrobial development. The current research analyzes the antimicrobial property of the biosurfactant secreted by *C. parapsilosis* YT CMGB against two strains of *C. krusei*, particularly the lab strain *C. krusei* CMGB94 and the isolated strain *C. krusei* CMGB-Y8 from vulvo-vaginal infections [68]. The two *C. krusei* strains were chosen to be relevant for both intra-hospital infections and the environmental spread of antimicrobial-resistant microorganisms [69,70].

3.4.1. Growth Inhibition Properties of *C. parapsilosis* YT CMGB Biosurfactant

The antimicrobial activity of the biosurfactant was evaluated by monitoring the growth dynamics of the two *C. krusei* strains in the presence of the biosurfactant at final concentrations of 40 mg/mL and 20 mg/mL, followed by determining the percentage of microbial growth at different points along the growth curve.

Figure 12A1,A2 show that the ability of the two *C. krusei* strains to proliferate is only slightly impacted by the concentrated biosurfactant. The existence of any residual carbon source from the biosurfactant precipitation most likely had a positive effect on the growth during the first seven hours of the experiment. However, after a few hours, the growth curve becomes steady and shows no discernible stimulatory or inhibitory effects. After incubating the *C. krusei* CMGB94 strain for 21 h, the exposure to 40 mg/mL biosurfactant reduced the strain growth by nearly 50% (microbial growth percentage of 52.61% compared to the untreated control). This suggests that the biosurfactant may affect *C. krusei* metabolism, but it might require additional time for effective antimicrobial activity. The outcome is not unexpected given that pathogenic *C. krusei* strains are known to have extra stress-coping mechanisms, which most likely enable them to quickly adjust to the stress indicated by the presence of a biosurfactant in the culture medium at low concentrations [71,72]. Until now, there have been a limited number of studies regarding the beneficial role of the *C. parapsilosis* biosurfactants for the biomedical field, mainly because the members of this species emerged as opportunistic and nosocomial pathogens. However, many of the *Candida* species have great potential in biotechnology, including non-pathogenic *C. parapsilosis* strains. There are some studies regarding the potential use of the biosurfactants secreted by *C. parapsilosis* as antimicrobial agents, but their number is limited. According to Garg et al. [46], a *C. parapsilosis* strain isolated from contaminated dairy products showed significant antibacterial activity against both Gram-positive and Gram-negative bacterial strains, namely members of *E. coli* and *S. aureus* species, at 10 and 5 mg/mL concentrations, respectively. Also, Rana et al. [73] investigated the biocompatibility of *C. parapsilosis* biosurfactant-loaded polymeric nanoparticles in a rodent model as a preliminary study for developing suitable platforms for biomedical applications. Furthermore, Mousa et al. [74] examined the potential of *C. parapsilosis*-produced biosurfactants to mitigate bleomycin-induced lung fibrosis, inflammation, and oxidative stress in mice. The results indicated that the biosurfactants could suppress fibrotic responses and lower the production of the extracellular matrix by blocking the TGF-β1/SMAD-3 pathway. Additionally, the results of their investigation showed that the biosurfactant did not damage the kidneys or liver of the test animals or show any acute toxic effects. Thus, the biosurfactant produced by *C. parapsilosis* might be extremely important for the biomedical field.

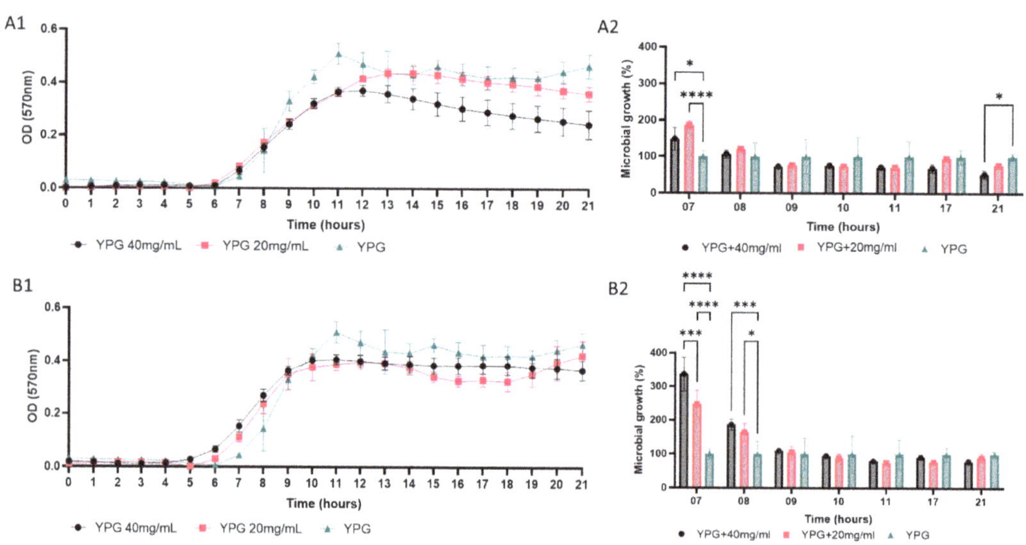

Figure 12. The antimicrobial effect of the biosurfactant produced by *C. parapsilosis* YT-CMGB: (1) the impact on the yeasts' growth behavior; (2) the growth inhibition percentage determined at specific points of time; (**A1**,**A2**)-*C. krusei* CMGB94; (**B1**,**B2**)-*C. krusei* CMGB-Y8 (Tukey multiple comparison test; where * $p < 0.05$, *** $p < 0.001$, and **** $p < 0.0001$).

3.4.2. Anti-Adherence Properties of *C. parapsilosis* YT-CMGB Biosurfactant

The biofilms formed by microbial pathogens poses a significant threat for the biomedical field due to their low permeability to antimicrobial agents. As a consequence, it is imperative to find new treatment approaches able to inhibit the production of biofilms or to improve their penetrability. Due to their amphiphilic properties, the biosurfactants might represent a suitable strategy for preventing microbial adherence to the surfaces of medical devices, thus limiting the first stage of biofilm formation. In this context, we tested the anti-adherence properties of the concentrated biosurfactants for preventing *C. krusei* biofilm formation. According to Figure 13, the biosurfactant produced by *C. parapsilosis* YT-CMGB is highly efficient in preventing both *C. krusei* strains' adherence to inert substratum, and its efficiency is strongly correlated with the concentration tested. Thus, in the case of the lab strain of *C. krusei* CMGB94, the ability to adhere to inert substratum is strongly inhibited in the presence of the biosurfactant, with the PICA% values obtained being 25.09% for the 40 mg/mL concentration and 40.70% for the 20 mg/mL concentration, respectively. A similar effect was observed against the pathogenic *C. krusei* strain, for which the PICA% value obtained for the 40 mg/mL concentration was 35.62%, while for the 20 mg/mL it was 67.24%.

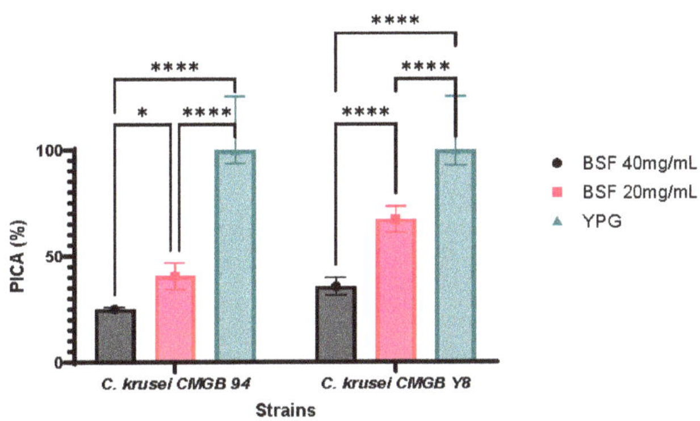

Figure 13. Graphical representation of the inhibitory effect on the adhesion capacity of *C. krusei* CMGB94 and *C. krusei* CMGB-Y8 strains treated with different concentrations of biosurfactant produced by *C. parapsilosis* YT-CMGB (Dunnet's multiple comparison test; where * $p < 0.05$, and **** $p < 0.0001$).

The results obtained for the anti-adherence assessment confirm the assumption that the *C. parapsilosis* YT-CMGB biosurfactant can act both directly on the cells by interfering with their metabolism, or it can act indirectly by changing the hydrophobic characteristics of the inert material.

There are many studies describing the anti-adherence and anti-biofilm properties of microbial biosurfactants, although they do not refer to *C. parapsilosis*-produced biosurfactants. In this respect, the antibacterial activity of glycolipid biosurfactants, such as sophorolipids, has been already proven. In both Gram-positive and Gram-negative bacteria, Diaz de Rienzo et al. [75] examined the antibacterial characteristics and biofilm disruption of sophorolipids derived from *C. bombicola* ATCC 22214. It has been shown that sophorolipids have both fungicidal and biofilm-rupturing properties at low concentrations (50 g L^{-1}). Similarly, *Y. lipolytica* biosurfactants present high efficiency in preventing *S. enterica* biofilm development [76].

4. Conclusions

Water pollution due to accidental spills or intensive industrial development has become a growing concern at an international level. In this respect, many microbial species with intrinsic metabolic biodegradative abilities, including yeasts, are gaining attention as agents for the development of sustainable technologies for water remediation. Although the yeast species *Candida parapsilosis* comprises strains with biotechnological potential, only a few studies have been conducted in this respect. The present work showed that the strain *C. parapsilosis* CMGB-YT assimilates *n*-alkanes and vegetable (sunflower and olive) oils, with a preference for *n*-tetradecane and *n*-hexadecane, over an extended period

(216 h). The *n*-hexadecane, as the sole carbon source, was used for the production of a stable biosurfactant able to emulsify long-carbon chain substrates, with possible applications in the remediation of petroleum- and vegetable oil-polluted environments, as well as in the treatment of alkaline wastewaters generated in the meat, food, and beverage processing industry. The biosurfactant, an anionic glycolipid, showed a good capacity of reducing the surface tension between hydrophobic–hydrophilic phases. The presence of crude biosurfactants in synthetic wastewaters contaminated with Pb^{2+} and Cd^{2+} promoted the growth of heavy metal degrader strains of *R. mucilaginosa* and enhanced their ability to remove cadmium ions. Moreover, the biosurfactant presented high efficiency in stopping *C. krusei* adherence to polypropylene substrate, thus representing a possible strategy to prevent biofilm formation on the surfaces of medical devices. In conclusion, *C. parapsilosis* CMGB-YT presents promising potential for further studies aimed at developing new technologies and strategies for the treatment of heavy metal- and microbial-contaminated wastewater and the restoration of the sustainable water supply.

Author Contributions: Conceptualization, O.E.C. and V.M.C.; methodology, O.E.C., V.M.C., N.-O.N. and E.-M.L.; software, N.-O.N. and V.E.M.; validation, O.E.C. and V.M.C.; formal analysis, O.E.C., N.-O.N., V.E.M. and E.-M.L.; investigation, O.E.C., N.-O.N., V.M.C., E.-M.L. and I.C.G.; resources, O.E.C., N.-O.N. and V.M.C.; data curation, O.E.C. and V.M.C.; writing—original draft, O.E.C., V.M.C., N.-O.N., I.C.G., V.E.M. and E.-M.L.; writing—review and editing, O.E.C. and V.M.C.; visualization, V.E.M., E.-M.L. and I.C.G.; supervision, O.E.C.; project administration, O.E.C. and V.M.C.; funding acquisition, O.E.C. and V.M.C. All authors have read and agreed to the published version of the manuscript.

Funding: This research was financially supported by the ICUB Senior Grant, contract no. 7423/04.07. 2023; by the Research Institute of the University of Bucharest and the ICUB Young Research Grant, contract no. 7412/04.07.2023; and by the Research Institute of the University of Bucharest. N.O.N.; E.M.L also acknowledge the support provided by the Ministry of Research, Innovation and Digitization through contract PN 23140201-42N/2023.

Data Availability Statement: All data are available on request to the corresponding author. A sample of the biosurfactant is available at the Department of Genetics, Faculty of Biology, University of Bucharest.

Conflicts of Interest: The authors declare no conflict of interest.

References

1. Malkapuram, S.T.; Sharma, V.; Gumfekar, S.P.; Sonawane, S.; Sonawane, S.; Boczkaj, G.; Seepana, M.M. A Review on Recent Advances in the Application of Biosurfactants in Wastewater Treatment. *Sustain. Energy Technol. Assess.* **2021**, *48*, 101576. [CrossRef]
2. Azin, E.; Moghimi, H.; Dastgheib, S.M.M.; Darvishi, F. Biovalorization of Wastewater of Fish Canning Process by *Yarrowia lipolytica* for Biodiesel and Animal Feed Supplement Production. *Biomass Conv. Bioref.* **2024**, *14*, 7981–7994. [CrossRef]
3. Hegazy, G.E.; Soliman, N.A.; Farag, S.; El-Helow, E.R.; Yusef, H.Y.; Abdel-Fattah, Y.R. Isolation and Characterization of *Candida tropicalis* B: A Promising Yeast Strain for Biodegradation of Petroleum Oil in Marine Environments. *Microb. Cell Factories* **2024**, *23*, 20. [CrossRef] [PubMed]
4. Sartaj, K.; Prasad, R.; Matsakas, L.; Patel, A. Transforming Recalcitrant Wastes into Biodiesel by Oleaginous Yeast: An Insight into the Metabolic Pathways and Multi-Omics Landscape. *Chem. Eng. J.* **2023**, 145625. [CrossRef]
5. Ince, M.; Ince, O.K. Introductory Chapter: Sources, Health Impact, and Environment Effect of Hydrocarbons. In *Hydrocarbon Pollution and Its Effect on the Environment*; IntechOpen: London, UK, 2019; p. 1.
6. Hu, M. Environmental Behavior of Petroleum in Soil and Its Harmfulness Analysis. In *Proceedings of the IOP Conference Series: Earth and Environmental Science*; IOP Publishing: Bristol, UK, 2020; Volume 450, p. 012100.
7. Fickers, P.; Benetti, P.H.; Waché, Y.; Marty, A.; Mauersberger, S.; Smit, M.S.; Nicaud, J.M. Hydrophobic Substrate Utilisation by the Yeast *Yarrowia lipolytica*, and Its Potential Applications. *FEMS Yeast Res.* **2005**, *5*, 527–543. [CrossRef]
8. Zarevúcka, M. Olive Oil as Inductor of Microbial Lipase. In *Olive Oil-Constituents, Quality, Health Properties and Bioconversions, Intechopen*; IntechOpen: London, UK, 2012; pp. 457–470.
9. Briand, D.; Dubreucq, E.; Grimaud, J.; Galzy, P. Substrate Specificity of the Lipase from *Candida parapsilosis*. *Lipids* **1995**, *30*, 747–754. [CrossRef] [PubMed]

10. Butler, G.; Rasmussen, M.D.; Lin, M.F.; Santos, M.A.; Sakthikumar, S.; Munro, C.A.; Rheinbay, E.; Grabherr, M.; Forche, A.; Reedy, J.L.; et al. Evolution of Pathogenicity and Sexual Reproduction in Eight *Candida* Genomes. *Nature* **2009**, *459*, 657–662. [CrossRef] [PubMed]
11. Nguyen, L.N.; Gacser, A.; Nosanchuk, J.D. Secreted Lipases Supply Fatty Acids for Yeast Growth in the Absence of de novo Fatty Acid Synthesis. *Virulence* **2011**, *2*, 538–541. [CrossRef] [PubMed]
12. Macphail, G.L.P.; Taylor, G.D.; Buchanan-Chell, M.; Ross, C.; Wilson, S.; Kureishi, A. Epidemiology, treatment and outcome of candidemia: A five-year review at three Canadian hospitals. *Mycoses* **2002**, *45*, 141–145. [CrossRef] [PubMed]
13. Pires, R.H.; Santos, J.M.D.; Zaia, J.E.; Martins, C.H.G.; Mendes-Giannini, M.J.S. *Candida parapsilosis* Complex Water Isolates from a Haemodialysis Unit: Biofilm Production and in vitro Evaluation of the Use of Clinical Antifungals. *Memórias do Instituto Oswaldo Cruz* **2011**, *106*, 646–654. [CrossRef] [PubMed]
14. Zupančič, J.; Babič, M.N.; Gunde-Cimerman, N. High Incidence of an Emerging Opportunistic Pathogen *Candida parapsilosis* in Water-Related Domestic Environments. In *Fungal Infection*; IntechOpen: London, UK, 2018.
15. Yalçin, H.T.; Çorbaci, C.; Uçar, F.B. Molecular Characterization and Lipase Profiling of the Yeasts Isolated from Environments Contaminated with Petroleum. *J. Basic Microbiol.* **2014**, *54*, S85–S92. [CrossRef] [PubMed]
16. Mendes, M.; Cassoni, A.C.; Alves, S.; Moreira, P.; Castro, P.M.; Pintado, M.E. Assessment of Drying Conditions for Storage of a Yeast-Based Decolorization Solution for Application in Textile Industrial Wastewater Treatment Plants. *Fungal Biol.* **2023**, *127*, 1111–1117. [CrossRef] [PubMed]
17. Bule Možar, K.; Miloloža, M.; Martinjak, V.; Ujević Bošnjak, M.; Markić, M.; Bolanča, T.; Cvetnić, M.; Kučić Grgić, D.; Ukić, Š. The Potential of AOP Pretreatment in the Biodegradation of PS and PVC Microplastics by *Candida parapsilosis*. *Water* **2024**, *16*, 1389. [CrossRef]
18. Corbu, V.; Csutak, O. Candida–Produced Biosurfactants–Beneficial Agents for Environmental Remediation Biotechnologies. *Rom. Biotechnol. Lett.* **2019**, *24*, 381–387. [CrossRef]
19. Silva, R.d.C.F.S.; Almeida, D.G.; Rufino, R.D.; Luna, J.M.; Santos, V.A.; Sarubbo, L.A. Applications of Biosurfactants in the Petroleum Industry and the Remediation of Oil Spills. *Int. J. Mol. Sci.* **2014**, *15*, 12523–12542. [CrossRef]
20. Eras-Muñoz, E.; Farré, A.; Sánchez, A.; Font, X.; Gea, T. Microbial Biosurfactants: A Review of Recent Environmental Applications. *Bioengineered* **2022**, *13*, 12365–12391. [CrossRef] [PubMed]
21. Silva, R.R.; Silva, Y.A.; Silva, T.A.D.L.; Sarubbo, L.A.; Luna, J.M. Recent Advances in Environmental Biotechnology: Role of Biosurfactants in Remediation of Heavy Metals. *Res. Soc. Dev.* **2022**, *11*, 4411527453. [CrossRef]
22. Kim, K.; Yoo, D.; Kim, Y.; Lee, B.; Shin, D.; Kim, E.K. Characteristics of Sophorolipid as an Antimicrobial Agent. *J. Microbiol. Biotechnol.* **2002**, *12*, 235–241.
23. Haque, F.; Verma, N.K.; Alfatah, M.; Bijlani, S.; Bhattacharyya, M.S. Sophorolipid Exhibits Antifungal Activity by ROS Mediated Endoplasmic Reticulum Stress and Mitochondrial Dysfunction Pathways in *Candida albicans*. *RSC Adv.* **2019**, *9*, 41639–41648. [CrossRef] [PubMed]
24. Banat, I.M.; Satpute, S.K.; Cameotra, S.S.; Patil, R.; Nyayanit, N.V. Cost Effective Technologies and Renewable Substrates for Biosurfactants' Production. *Front. Microbiol.* **2014**, *5*, 697. [CrossRef] [PubMed]
25. Alfian, A.R.; Watchaputi, K.; Sooklim, C.; Soontorngun, N. Production of New Antimicrobial Palm Oil-Derived Sophorolipids by the Yeast *Starmerella riodocensis* Sp. Nov. against *Candida albicans* Hyphal and Biofilm Formation. *Microb. Cell Factories* **2022**, *21*, 163. [CrossRef] [PubMed]
26. Cooper, D.G.; Goldenberg, B.G. Surface-Active Agents from Two *Bacillus* Species. *Appl. Environ. Microbiol.* **1987**, *53*, 224–229. [CrossRef]
27. Degang, L.; Auricchio, G.; Schmidt, V.K.d.O.; Della-Flora, I.K.; de Andrade, C.J. Oil Displacement Properties of Surfactin: A Comparative Study. *Tenside Surfactants Deterg.* **2023**, *60*, 414–418. [CrossRef]
28. Csutak, O.E.; Nicula, N.-O.; Lungulescu, E.-M.; Marinescu, V.; Corbu, V.M. *Yarrowia lipolytica* CMGB32 Biosurfactants Produced Using *n*-Hexadecane: Developing Strategies for Environmental Remediation. *Appl. Sci.* **2024**, *14*, 3048. [CrossRef]
29. Csutak, O.; Stoica, I.; Vassu, T. Evaluation of Production, Stability and Activity of Biosurfactants from Yeasts with Application in Bioremediation of Oil-Polluted Environment. *Rev. Chim.* **2012**, *63*, 973–977.
30. Nicula, N.-O.; Lungulescu, E.-M.; Rîmbu, G.A.; Marinescu, V.; Corbu, V.M.; Csutak, O. Bioremediation of Wastewater Using Yeast Strains: An Assessment of Contaminant Removal Efficiency. *Int. J. Environ. Res. Public Health* **2023**, *20*, 4795. [CrossRef] [PubMed]
31. Corpuz, M.V.A.; Borea, L.; Senatore, V.; Castrogiovanni, F.; Buonerba, A.; Oliva, G.; Ballesteros, F.; Zarra, T.; Belgiorno, V.; Choo, K.-H.; et al. Wastewater Treatment and Fouling Control in an Electro Algae-Activated Sludge Membrane Bioreactor. *Sci. Total Environ.* **2021**, *786*, 147475. [CrossRef] [PubMed]
32. Luna, J.M.; Rufino, R.D.; Sarubbo, L.A.; Rodrigues, L.R.M.; Teixeira, J.A.C.; De Campos-Takaki, G.M. Evaluation Antimicrobial and Antiadhesive Properties of the Biosurfactant Lunasan Produced by *Candida sphaerica* UCP 0995. *Curr. Microbiol.* **2011**, *62*, 1527–1534. [CrossRef]
33. Corbu, V.M.; Gheorghe, I.; Marinaș, I.C.; Geană, E.I.; Moza, M.I.; Csutak, O.; Chifiriuc, M.C. Demonstration of *Allium sativum* Extract Inhibitory Effect on Biodeteriogenic Microbial Strain Growth, Biofilm Development, and Enzymatic and Organic Acid Production. *Molecules* **2021**, *26*, 7195. [CrossRef] [PubMed]
34. Omar, S.H.; Büdecker, U.; Rehm, H.J. Degradation of Oily Sludge from a Flotation Unit by Free and Immobilized Microorganisms. *Appl. Microbiol. Biotechnol.* **1990**, *34*, 259–263. [CrossRef]

35. Radwan, S.S.; Al-Aawadi, H.A.; Khanafer, M. Effects of Lipids on N-Alkane Attenuation in Media Supporting Oil-Utilizing Microorganisms from the Oily Arabian Gulf Coasts. *FEMS Microbiol. Lett.* **2001**, *198*, 99–103. [CrossRef]
36. Benmessaoud, S.; Anissi, J.; Kara, M.; Assouguem, A.; Al-Huqail, A.A.; Germoush, M.O.; Ullah, R.; Ercisli, S.; Bahhou, J. Isolation and Characterization of Three New Crude Oil Degrading Yeast Strains, *Candida parapsilosis* SK1, *Rhodotorula mucilaginosa* SK2 and SK3. *Sustainability* **2022**, *14*, 3465. [CrossRef]
37. Al-Otibi, F.O.; Yassin, M.T.; Al-Askar, A.A.; Maniah, K. Green Biofabrication of Silver Nanoparticles of Potential Synergistic Activity with Antibacterial and Antifungal Agents against Some Nosocomial Pathogens. *Microorganisms* **2023**, *11*, 945. [CrossRef] [PubMed]
38. Popoola, B.M.; Abiodun, O.A. Fatty Acid Methyl Ester Analysis of Olive Oil Degraded by *Candida parapsilosis*. *Appl. Microbiol. Open Access* **2020**, *6*, 168.
39. Ciafardini, G.; Cioccia, G.; Zullo, B.A. Survival of *Candida parapsilosis* Yeast in Olive Oil. *Ann. Microbiol.* **2013**, *63*, 1645–1648. [CrossRef]
40. Klug, L.; Daum, G. Yeast Lipid Metabolism at a Glance. *FEMS Yeast Res.* **2014**, *14*, 369–388. [CrossRef] [PubMed]
41. Ambaye, T.G.; Vaccari, M.; Prasad, S.; Rtimi, S. Preparation, Characterization and Application of Biosurfactant in Various Industries: A Critical Review on Progress, Challenges and Perspectives. *Environ. Technol. Innov.* **2021**, *24*, 102090. [CrossRef]
42. Twigg, M.S.; Baccile, N.; Banat, I.M.; Déziel, E.; Marchant, R.; Roelants, S.; Bogaert, I.N.A. Microbial Biosurfactant Research: Time to Improve the Rigour in the Reporting of Synthesis, Functional Characterization and Process Development. *Microb. Biotechnol.* **2021**, *14*, 147–170. [CrossRef]
43. Antosiewicz, J.M.; Shugar, D. UV–Vis Spectroscopy of Tyrosine Side-Groups in Studies of Protein Structure. Part 2: Selected Applications. *Biophys. Rev.* **2016**, *8*, 163–177. [CrossRef] [PubMed]
44. Saito, K.; Xu, T.; Ishikita, H. Correlation between C=O Stretching Vibrational Frequency and pK(a) Shift of Carboxylic Acids. *J. Phys. Chem. B* **2022**, *126*, 4999–5006. [CrossRef] [PubMed]
45. Safari, P.; Hosseini, M.; Lashkarbolooki, M.; Ghorbani, M.; Najafpour Darzi, G. Evaluation of Surface Activity of Rhamnolipid Biosurfactants Produced from Rice Bran Oil through Dynamic Surface Tension. *J. Pet. Explor. Prod. Technol.* **2023**, *13*, 2139–2153. [CrossRef]
46. Garg, M.; Priyanka; Chatterjee, M. Isolation, Characterization and Antibacterial Effect of Biosurfactant from *Candida parapsilosis*. *Biotechnol. Rep.* **2018**, *18*, 00251. [CrossRef] [PubMed]
47. Khanna, A.; Handa, S.; Rana, S.; Suttee, A.; Puri, S.; Chatterjee, M. Biosurfactant from *Candida*: Sources, Classification, and Emerging Applications. *Arch. Microbiol.* **2023**, *205*, 149. [CrossRef] [PubMed]
48. Al-Kashef, A.S.; Nooman, M.U.; Rashad, M.M.; Hashem, A.H.; Abdelraof, M. Production and Optimization of Novel Sphorolipids from *Candida parapsilosis* Grown on Potato Peel and Frying Oil Wastes and Their Adverse Effect on Mucorales Fungal Strains. *Microb. Cell Factories* **2023**, *22*, 79. [CrossRef] [PubMed]
49. Lungulescu, E.-M.; Setnescu, R.; Ilie, S.; Taborelli, M. On the Use of Oxidation Induction Time as a Kinetic Parameter for Condition Monitoring and Lifetime Evaluation under Ionizing Radiation Environments. *Polymers* **2022**, *14*, 2357. [CrossRef] [PubMed]
50. Singh, P.; Tiwary, B.N. Isolation and Characterization of Glycolipid Biosurfactant Produced by a *Pseudomonas otitidis* Strain Isolated from Chirimiri Coal Mines, India. *Bioresour. Bioprocess.* **2016**, *3*, 42. [CrossRef]
51. Lungulescu, E.M. Contributions to the Study and Characterization of Degradation Processes of the Insulating Polymeric Materials in High-Energy Radiation Fields. Ph.D. Thesis, University of Bucharest, București, Romania, 2014.
52. Esmaeili, H.; Mousavi, S.M.; Hashemi, S.A.; Lai, C.W.; Chiang, W.-H.; Bahrani, S. Chapter 7—Application of Biosurfactants in the Removal of Oil from Emulsion. In *Green Sustainable Process for Chemical and Environmental Engineering and Science*; Inamuddin, T.A., Ed.; Elsevier: Amsterdam, The Netherlands, 2021.
53. Perinelli, D.R.; Cespi, M.; Lorusso, N.; Palmieri, G.F.; Bonacucina, G.; Blasi, P. Surfactant Self-Assembling and Critical Micelle Concentration: One Approach Fits All? *Langmuir* **2020**, *36*, 5745–5753. [CrossRef] [PubMed]
54. Baccile, N.; Cuvier, A.S.; Prévost, S.; Stevens, C.V.; Delbeke, E.; Berton, J.; Soetaert, W.; Bogaert, I.N.; Roelants, S. Self-Assembly Mechanism of pH-Responsive Glycolipids: Micelles, Fibers, Vesicles, and Bilayers. *Langmuir* **2016**, *32*, 10881–10894. [CrossRef] [PubMed]
55. Asfora Sarubbo, L.; Luna, J.; Campos-Takaki, G.M. Production and Stability Studies of the Bioemulsifier Obtained from a New Strain of *Candida glabrata* UCP 1002. *Electron. J. Biotechnol.* **2006**, *9*. [CrossRef]
56. Ali, F.; Das, S.; Hossain, T.J.; Chowdhury, S.I.; Zedny, S.A.; Das, T.; Ahmed Chowdhury, M.N.; Uddin, M.S. Production Optimization, Stability and Oil Emulsifying Potential of Biosurfactants from Selected Bacteria Isolated from Oil-Contaminated Sites. *R. Soc. Open Sci.* **2021**, *8*, 211003. [CrossRef] [PubMed]
57. Salinas, E.; Orellano, M.E.; Rezza, I.; Martinez, L.; Marchesvky, E.; Tosetti, M.S. Removal of Cadmium and Lead from Dilute Aqueous Solutions by *Rhodotorula rubra*. *Bioresour. Technol.* **2000**, *72*, 107–112. [CrossRef]
58. Salvadori, M.R.; Ando, R.A.; Nascimento, C.A.; Corrêa, B. Intracellular Biosynthesis and Removal of Copper Nanoparticles by Dead Biomass of Yeast Isolated from the Wastewater of a Mine in the Brazilian Amazonia. *PLoS ONE* **2014**, *9*, 87968. [CrossRef] [PubMed]
59. Ilyas, S.; Rehman, A.; Coelho, A.V.; Sheehan, D. Proteomic Analysis of an Environmental Isolate of *Rhodotorula mucilaginosa* after Arsenic and Cadmium Challenge: Identification of a Protein Expression Signature for Heavy Metal Exposure. *J. Proteom.* **2016**, *141*, 47–56. [CrossRef] [PubMed]

60. Grujić, S.; Radojevic, I.; Vasić, S.; Čomić, L.; Ostojić, A. Heavy Metal Tolerance and Removal Efficiency of the *Rhodotorula mucilaginosa* and *Saccharomyces boulardii* Planktonic Cells and Biofilm. *Kragujev. J. Sci.* **2018**, *40*, 217–226. [CrossRef]
61. Ignatova, L.; Kistaubayeva, A.; Brazhnikova, Y.; Omirbekova, A.; Mukasheva, T.; Savitskaya, I.; Karpenyuk, T.; Goncharova, A.; Egamberdieva, D.; Sokolov, A. Characterization of cadmium-tolerant endophytic fungi isolated from soybean (*Glycine max*) and barley (*Hordeum vulgare*). *Heliyon* **2021**, *7*, e08240. [CrossRef] [PubMed]
62. Vidal, N.F.; Dávila, J.W. Lead and Cadmium Removal with Native Yeast from Coastal Wetlands. *Open Chem.* **2022**, *20*, 1096–1109. [CrossRef]
63. Mishra, S.; Lin, Z.; Pang, S.; Zhang, Y.; Bhatt, P.; Chen, S. Biosurfactant Is a Powerful Tool for the Bioremediation of Heavy Metals from Contaminated Soils. *J. Hazard. Mater.* **2021**, *418*, 126253. [CrossRef] [PubMed]
64. *ISO6060:1989*; Determination of the Chemical Oxygen Demand. International Organization for Standardization: Geneve, Switzerland, 1989.
65. Dhanjai Sinha, A.; Zhao, H.; Chen, J.; Mugo, S.M. Water Analysis- Determination of Chemical Oxygen Demand. In *Encyclopedia of Analytical Science*; Worsfold, P., Poole, C., Townshend, A., Miró, M., Eds.; Academic Press: Oxford, UK, 2019.
66. Murugaiyan, J.; Kumar, P.A.; Rao, G.S.; Iskandar, K.; Hawser, S.; Hays, J.P.; Mohsen, Y.; Adukkadukkam, S.; Awuah, W.A.; Jose, R.A.M.; et al. Progress in Alternative Strategies to Combat Antimicrobial Resistance: Focus on Antibiotics. *Antibiotics* **2022**, *11*, 200. [CrossRef] [PubMed]
67. MacNair, C.R.; Rutherford, S.T.; Tan, M.-W. Alternative Therapeutic Strategies to Treat Antibiotic-Resistant Pathogens. *Nat. Rev. Microbiol.* **2024**, *22*, 262–275. [CrossRef] [PubMed]
68. Avram, I.; Pelinescu, D.; Stoica, I.; Marutescu, L.; Vassu, T. Phenotypic Profiles of Virulence in Different *Candida* Species Isolated from Vulvovaginal Infections. *Roum. Arch. Microbiol. Immunol.* **2013**, *72*, 225–233.
69. Kacprzak, M.; Neczaj, E.; Okoniewska, E. The Comparative Mycological Analysis of Wastewater and Sewage Sludges from Selected Wastewater Treatment Plants. *Desalination* **2005**, *185*, 363–370. [CrossRef]
70. Hamed, S.Z. Isolation and Identification of Yeasts along Wastewater Treatment Lines at Zagazig Plant. *Ann. Agric. Sci. Moshtohor* **2016**, *54*, 77–84. [CrossRef]
71. Gómez-Gaviria, M.; Mora-Montes, H.M. Current Aspects in the Biology, Pathogeny, and Treatment of *Candida krusei*, a Neglected Fungal Pathogen. *Infect. Drug Resist.* **2020**, *13*, 1673–1689. [CrossRef] [PubMed]
72. de Oliveira Santos, G.C.; Vasconcelos, C.C.; Lopes, A.J.O.; de Sousa Cartágenes, M.d.S.; Filho, A.K.D.B.; do Nascimento, F.R.F.; Ramos, R.M.; Pires, E.R.R.B.; de Andrade, M.S.; Rocha, F.M.G.; et al. *Candida* Infections and Therapeutic Strategies: Mechanisms of Action for Traditional and Alternative Agents. *Front. Microbiol.* **2018**, *9*, 1351. [CrossRef] [PubMed]
73. Rana, S.; Singh, J.; Wadhawan, A.; Khanna, A.; Singh, G.; Chatterjee, M. Evaluation of In Vivo Toxicity of Novel Biosurfactant from *Candida parapsilosis* Loaded in PLA-PEG Polymeric Nanoparticles. *J. Pharm. Sci.* **2021**, *110*, 1727–1738. [CrossRef] [PubMed]
74. Mousa, A.M.; Nooman, M.U.; Abbas, S.S.; Gebril, S.M.; Abdelraof, M.; Al-kashef, A.S. Protective Effects of Microbial Biosurfactants Produced by *Bacillus Halotolerans* and *Candida Parapsilosis* on Bleomycin-Induced Pulmonary Fibrosis in Mice: Impact of Antioxidant, Anti-Inflammatory and Anti-Fibrotic Properties via TGF-B1/Smad-3 Pathway and miRNA-326. *Toxicol. Appl. Pharmacol.* **2024**, *486*, 116939. [CrossRef] [PubMed]
75. Diaz De Rienzo, M.A.; Banat, I.M.; Lajarin Cuesta, M.; Williams, M.; Fracchia, L. Evaluation of the Activity of Sophorolipids against Microbial Biofilms on Medical-Grade Silicone. *New Biotechnol.* **2018**, *44*, S13. [CrossRef]
76. Shatila, F.; Uyar, E.; Yalçın, H.T. Screening of Biosurfactant Production by *Yarrowia lipolytica* Strains and Evaluation of Their Antibiofilm and Anti-Adhesive Activities against *Salmonella enterica* ser. enteritidis Biofilms. *Microbiology* **2021**, *90*, 839–847. [CrossRef]

processes

MDPI

Article

Improved Dujiangyan Irrigation System Optimization (IDISO): A Novel Metaheuristic Algorithm for Hydrochar Characteristics

Jingyuan Shi [1], Dapeng Zhang [1,*], Zifeng Sui [1], Jie Wu [2], Zifeng Zhang [1], Wenjie Hu [1], Zhanpeng Huo [1] and Yongfu Wu [1]

[1] School of Energy and Environment, Inner Mongolia University of Science and Technology, Baotou 014010, China
[2] School of Automotive Engineering, Jiangxi College of Applied Technology, Ganzhou 341000, China
* Correspondence: zhangdapeng@imust.edu.cn

Abstract: Hyperparameter tuning is crucial in the development of machine learning models. This study introduces the nonlinear shrinking factor and the Cauchy mutation mechanism to improve the Dujiangyan Irrigation System Optimization (DISO), proposing the improved Dujiangyan Irrigation System Optimization algorithm (IDISO) for hyperparameter tuning in machine learning. The optimization capabilities and convergence performance of IDISO were validated on 87 CEC2017 benchmark functions of varying dimensions and nine real-world engineering problems, demonstrating that it significantly outperforms DISO in terms of convergence speed and accuracy, and ranks first in overall performance among the seventeen advanced metaheuristic algorithms being compared. To construct a robust and generalizable prediction model for hydrochar element characteristics, this study utilized IDISO and DISO algorithms to fine-tune the parameters of the XGBoost model. The experimental results show that the IDISO-XGBoost model achieved an average prediction performance of 0.95, which represents a 4% improvement over the DISO-XGBoost model. These results indicate that the IDISO algorithm has significant potential and value in practical applications.

Keywords: hydrothermal carbonization; meta-heuristic; XGBoost; single-task prediction; improved Dujiangyan irrigation system optimization (IDISO)

1. Introduction

The global energy shortage and environmental pollution are both becoming more and more serious issues. To achieve China's "carbon peak" and "carbon neutrality" objectives, there is a need for an effective reduction in CO_2 emissions. Exploring renewable and clean energy sources as alternatives to fossil fuels is critical. Biomass is emphasized as the sole viable carbon-based renewable alternative to fossil fuels [1]. Studies indicate that, in certain regions, initial trials exploring its integration into the energy mix are underway. In 2021, the amount of electricity generated from biomass combustion reached 750 TWh, and it is projected that, by 2050, the supply of biomass is expected to increase to 100 EJ [2,3].

Hydrothermal carbonization (HTC) is an emerging thermochemical conversion technology. Biomass is effectively transformed into hydrochar through HTC, with an energy density that is comparable to peat and lignite. This process efficiently recycles agricultural waste and exhibits economic and environmentally friendly characteristics [4]. The complex composition of the biomass material influences the HTC process. It is also affected by the hydrothermal reaction conditions. This process involves nonlinear and strongly coupled multivariate relationships. These factors hinder scholars' deep understanding of the hydrothermal carbonization mechanism. Current studies primarily rely on experimental and simulation methods [5,6]. These methods are used for quantitative analysis to explore the factors affecting the hydrothermal carbonization process. However, the research mainly focuses on one or a few types of biomass materials. The experimental results need more universality, and the findings are challenging to generalize. When modeling using traditional methods, such as computational fluid dynamics, kinetics, and thermodynamics, various assumptions are inevitably made about the complex HTC process. These assumptions are made to simplify the complex models. However, these simplifications may affect the accuracy of the simulation results [7].

Machine learning techniques demonstrate more potential in predicting unknown relationships than traditional methods. Machine learning algorithms are widely applied in predicting industrial problems. They can make compelling predictions without a precise mathematical relationship between the input and output features. They have been proven to be alternatives to traditional modeling techniques for studying and understanding complex processes [8], which demonstrates significant potential in predicting the physicochemical properties of hydrochar.

Rasam et al. [9] pioneered a machine learning model to predict hydrothermal carbonization properties, with the Support Vector Machine (SVM) method showing superior performance over other methods like Decision Tree (DT), Random Forest (RF), and Multi-Layer Perceptron (MLP). Kardani et al. [10] found XGBoost to be the most accurate method in predicting hydrocarbons yield in biomass. Li et al.'s studies [11,12] employed optimized RF and SVM models to assess biomass hydrochar and pyrochar properties, achieving significant accuracy. Furthermore, they demonstrated that DNN models excelled in predicting the physicochemical properties of hydrochar from urban waste, with high accuracy in parameters like HHV and carbon content. Nguyen et al. [13] have also conducted similar work, achieving analogous results.

Machine learning methods are key in predicting hydrochar properties, but model optimization is challenging. Mu et al. [14], in 2022, successfully used Particle Swarm Optimization based on an ANN model to study hydrochar, achieving high accuracy, thereby demonstrating the potential of metaheuristic optimization in such research. Li et al. [15] employed Genetic Algorithms (GA) to optimize Artificial Neural Networks (ANN), achieving superior outcomes compared to those obtained through the response surface methodology (RSM).

Metaheuristic optimization techniques, known for their minimal derivation, flexibility, and their ability to escape local optima [16], are effective for nonlinear problems, finding applications in energy, mechanical, and chemical engineering [17–19]. And, because the optimization process of metaheuristic algorithms does not depend on gradient information [20], they find widespread applications in optimization problems for finding the best parameters. For example, the Dujiangyan irrigation system optimization (DISO) [21] is used to construct a DISO-SVM model [21] to detect the impact of dam displacement on dam operation; Particle Swarm Optimization (PSO) [22] is used to construct PSO-NN [14] and PSO-RF [23] models to predict hydrochar properties; the Grey Wolf Optimizer (GWO) [24] is used to construct a GWO-ELM model [25] for monitoring power quality; the Dandelion Optimizer (DO) [26] is used to improve the efficiency of multilevel inverters [27]; the Jellyfish Search Algorithm (JS) [28] is used to discover unknown parameters in fuel cells [29]; Young's Double-Slit Experiment (YDSE) optimizer [30] is used to construct a YDSE-PWM model for predicting dissolved oxygen levels [31]; the Starling Murmuration Optimizer (SMO) [32] algorithm is used to construct an ADA-SMO model [33] for predicting the strength of the mechanical properties of concrete. The "No Free Lunch" theorem states that specific optimization algorithms are suited only to certain optimization problems [34]. This principle motivates us to create and improve existing optimization algorithms so that they perform better in specific scenarios with specialized difficulties. Researchers have proposed various improved algorithms, such as the Hybrid Whale–Particle Swarm Optimization Algorithm (HWPSO) [35]; adaptive hybrid dandelion optimizer (DETDO) [36]; Ameliorated Young's double-slit experiment optimizer (IYDSE) [37]; enhanced jellyfish search algorithm (EJS) [38]; and efficient hybrid starling murmuration optimizer (DTCSMO) [39].

In this study, the Dujiangyan irrigation system optimization (DISO) [21] is enhanced with nonlinear shrinking factors and Cauchy mutation methods. This improvement aims to overcome its tendencies for converging to local optima and poor performance in high-dimensional spaces. IDISO will be used to optimize the predictive model for the physicochemical properties of hydrochar. The aim is to guide experimental designs. This approach can reduce time and economic costs. Additionally, it aids in revealing the relationships between different parameters in the hydrochar reaction process. It also provides theoretical guidance for the application of hydrochar in fuel chemical engineering.

The main contributions of this paper are as follows:

1. We improved the Dujiangyan irrigation system optimization (DISO) by introducing nonlinear shrinking factors and the Cauchy mutation mechanism, addressing its tendency to become trapped in local optima and its poor performance in high-dimensional spaces.
2. We compared the IDISO algorithm with seventeen state-of-the-art optimization algorithms using twenty-nine CEC2017 benchmark functions across three dimensions (30, 50, and 100) and nine engineering problems. Non-parametric tests indicated that the IDISO algorithm showed significant improvements in terms of convergence speed and accuracy.
3. We developed an IDISO-XGBoost model to predict the physicochemical properties of hydrochar, resulting in a prediction model with high robustness and generalization ability.

Section 2 elaborates on the original Dujiangyan method and its advancements. Sections 3 and 4 involve testing the algorithm's performance. Section 5 involves algorithm performance analysis. Section 6 discusses the application of the refined algorithm in optimizing the predictive model for hydrothermal carbon's physicochemical properties. This manuscript concludes with Section 7, offering conclusions and insights for future research directions.

2. Materials and Methods

2.1. Data Source

The dataset utilized in this study was compiled through the meticulous collection and organization of existing published data, aligning with the rigorous standards of research methodologies. In this dataset, the input features included the operational conditions during hydrothermal carbonization and the elemental and industrial analysis of the biomass itself. The output feature was the basic analysis of hydrochar. A total of 420 datasets were used. The data originated from hydrothermal carbonization research articles on wood biomass, herbaceous biomass, food waste, sewage sludge, and other raw materials. This included common types of biomass materials.

2.2. DISO

The inspiration for the DISO method comes from the Dujiangyan irrigation project. In the DISO algorithm [21], potential solutions of the unknown function are considered as massless, volumeless droplets in the search space. The algorithm primarily consists of four main steps.

The first part is the initialization phase. Then, the global search phase simulates the Fish Mouth Dividing Project, and the formula for updating the velocity of water droplets is as follows:

$$V_{i1} = \begin{cases} \frac{HRO^{\frac{2}{3}} \times HGO^{\frac{1}{2}}}{CF} \times r1 \ r1 < 0.23 \\ \\ \frac{HRI^{\frac{2}{3}} \times HGI^{\frac{1}{2}}}{CF} \times r2 \ r2 \geq 0.23 \end{cases} \tag{1}$$

where r1 and r2 are random numbers between 0 and 1; HRO and HRI are the hydraulic radii of the inner and outer rivers, set at fixed values of 1.2 and 7.2; HGO and HGI represent the hydraulic gradients of the inner and outer rivers, set at fixed values of 1.3 and 0.82; and CF denotes the comprehensive riverbed roughness, the formula for which is as follows:

$$CF_t = CFR \times \text{gampdf}\left(\frac{t}{T}, P, Q\right) \tag{2}$$

where CFR represents the riverbed roughness, set at a fixed value of 9.435; t is the current iteration number; T is the maximum number of iterations; and $\mathrm{gampdf}\left(\frac{t}{T}, P, Q\right)$ is the gamma probability density function, influenced by the values of P and Q, which are set at 0.85 and 2.5, respectively, in the DISO algorithm. The formula for updating the position of the water droplet is as follows:

$$X_{i1+1} = X_{best}(t) + (X_{best} - X_i) \times \mathrm{rand}(0,1) \times V_{i1} + \mathrm{Improve1}_i \quad (3)$$

where $X_{best}(t)$ represents the position of the best-performing candidate solution in the t iteration and $\mathrm{Improve1}_i$ is the recipe for the self-improvement of the water body. The procedure is as follows:

$$\mathrm{Improve1}_i = \left(-1\right)^k \times (X_i - X_{mean}) \times \mathrm{rand}(0,1) \quad (4)$$

In this formula, k is an integer between [0, 1], while the mean represents the average position of the population. After the water body enters the inner river, the water droplet will exhibit a spiral motion. At this point, the position update is influenced by centrifugal and lateral pressures, with the corresponding update formula being as follows.

$$X_{i,j} = \begin{cases} X_{i,j}, RCF_i < LP_i \\ (UB_j - LB_j)\mathrm{rand}(0,1) + LB_j, RCF_i \geq LP_i \end{cases} \quad (5)$$

In this formula, RCF_i and LP_i represent the centrifugal force and lateral pressure experienced by the i_{th} solution, respectively. The formula for this is as follows:

$$RCF_i = WDI \times \cos\left(90 \times \frac{t}{T}\right) \times \sqrt{\sum_{j=1}^{D} \left(X_{best,j} - X_{i,j}\right)^2} \quad (6)$$

$$LP_i = \frac{(WDI \times VDC \times {MLV_i}^2)}{CRC} \quad (7)$$

In this formula, WDI is the density of the water body in the inner river, set at a fixed value of 1.35; VDC is the fluid distribution coefficient, set at 0.46; and MLV_i is the longitudinal average velocity, with the formula being as follows:

$$MLV_i = \mathrm{mean}(Fit) \quad (8)$$

In this procedure, Fit represents the average fitness of the population.

The local development phase simulates the Baopingko Project, and the formula for updating the velocity of the water droplet is as follows:

$$V_{i2} = \begin{cases} \frac{HRI^{\frac{2}{3}} \times HGI^{\frac{1}{2}}}{2CF} \times r3, r3 < p2 \\ \frac{HRI^{\frac{2}{3}} \times HGI^{\frac{1}{2}}}{CF} \times r4, r4 \geq p2 \end{cases} \quad (9)$$

In this formula, r3 and r4 are random numbers between 0 and 1, while p2 is the water dividing ratio at the bottleneck, set at a fixed value of 0.68. The formula for updating the position of the water droplet is as follows:

$$X_{i2+1} = X_{best}(t) + (X_{best} - X_i) \times \mathrm{rand}(0,1) \times V_{i2} + \mathrm{Improve2}_i \quad (10)$$

In this formula, $X_{best}(t)$ represents the position of the best-performing candidate solution in the t iteration, while $Improve2_i$ is the recipe for the self-improvement of the water body. The procedure for this is as follows:

$$Improve2_i = sign(Fit_{idx} - Fit_i) \times (X_{idx} - X_i) \times rand(0,1) \quad (11)$$

In this formula, Fit represents the fitness value of an individual and idx denotes a random individual.

The final phase involves simulating the individual elimination stage of the Feishayan. Fitness values are ranked. Initialization is performed for individuals to be eliminated. This process affects the sand discharge. The elimination ratio per iteration is set at 0.23.

2.3. IDISO

In this section, we introduce the improvements to the original algorithm using the nonlinear shrinking factor and the Cauchy mutation mechanism.

2.3.1. Nonlinear Shrinking Factor

In the DISO algorithm, position updates Equations (3) and (10) primarily depend on the best candidate solution and positions of other solutions in the population. However, this transfer of information could be more practical in coordinating the relationship between the global search and local development. Consequently, a nonlinear shrinking factor is introduced. This is to balance the dynamic equilibrium between global tracking and the local development of individuals. Simultaneously, this approach ensures the convergence of DISO while enhancing its precision. The fractal dimension [40] is introduced as a nonlinear factor in the third term of the position update Formula (5), with the procedure being as follows:

$$C(t) = \log\left(\frac{(C_{max} - C_{min}) \times (T - t)}{T} + C_{min}\right) \quad (12)$$

In this formula, C_{max} and C_{min} are set at fixed values of e^3 and e^2.

A nonlinear factor in the opposite direction is introduced in the second term of the position update Formula (12), with the formula being as follows:

$$Q(t) = e^{-2(1-\frac{t}{T})} \quad (13)$$

The updated position update formula is as follows:

$$X_{i1+1} = X_{best}(t) + (X_{best} - X_i) \times rand(0,1) \times V_{i1} + Improve1_i \times C(t) \quad (14)$$

$$X_{i2+1} = X_{best}(t) + (X_{best} - X_i) \times rand(0,1) \times V_{i2} \times Q(t) + Improve2_i \quad (15)$$

In this formula, $C(t)$ nonlinearly decreases with the increase in the iteration number t, while $Q(t)$ nonlinearly increases correspondingly. During the initial iteration phase, $C(t)$ slowly decays while $Q(t)$ increases gradually. This allows for a more extensive movement range of the water body in the first position update. Consequently, it enables extensive exploration within the search space. During the third position update, the water body can undergo detailed development within a localized area. As the number of iterations increases, $C(t)$ decreases rapidly while $Q(t)$ increases quickly. This results in the initial updates in the later stages of iteration, mainly focusing on local and detailed development. In the third update, the water body can move more quickly towards the global optimum through a more extensive range of motion, allowing for a more effective search for the optimal solution.

2.3.2. Cauchy Mutation Mechanism

The movement mechanism of DISO mainly depends on the information exchange within the population. It tends to fall into local optima and cannot escape from positions of local optimum values. Therefore, the Cauchy mutation mechanism is introduced before the fourth step of eliminating individuals. This enhances the particles' ability to escape from local optima. The formula is as follows:

$$X_{inew} = \frac{\gamma}{\pi(\gamma^2 + (x - a)^2)} \quad (16)$$

$$X_{i3+1} = \begin{cases} X_i, Fit_{inew} < Fit_i \\ X_{inew}, Fit_{inew} \geq Fit_i \end{cases} \quad (17)$$

In this formula, X_{inew} represents the position of the water body after the Cauchy mutation. X_i is the current position of the water body. a and γ are fixed values of 0 and 1. Before individual elimination (in the Feishayan stage), the final position of the water body is determined by comparing the fitness before and after mutation. The flowchart of IDISO is shown in Figure 1.

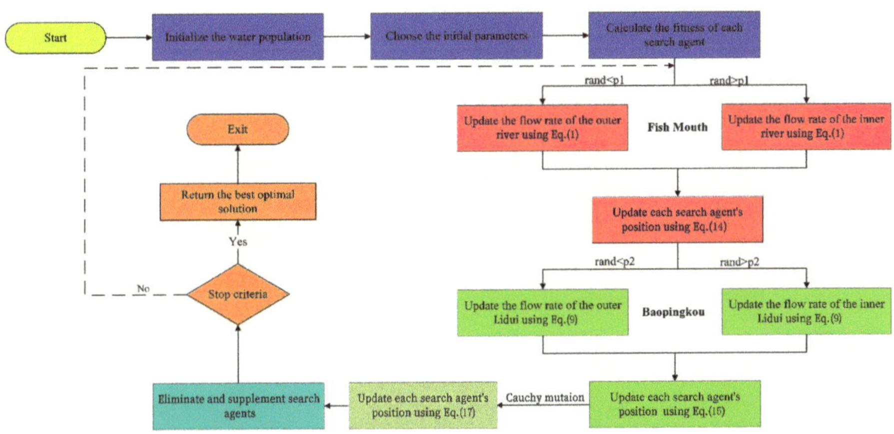

Figure 1. Flowchart of the proposed IDISO algorithm.

2.4. IDISO-XGBoost

Predicting the elemental characteristics of hydrochar is crucial for scholars to understand its reaction process. It is equally essential for guiding experimental and industrial applications.

XGBoost [41] is a machine learning algorithm based on gradient-boosted trees. It introduces regularization terms to control the complexity of the model and employs gradient boosting algorithms for training. Relevant studies indicate that XGBoost exhibits the best performance in predicting structured data problems [42].

Unlike manual tuning methods that are prone to local optima, grid search methods that are time-consuming and labor-intensive, and Bayesian optimization methods that have limitations in hyperparameter optimization, metaheuristic methods have the advantages of not requiring gradient information, strong global search capabilities, and high speed. Therefore, this study proposes the IDISO-XGBoost model, utilizing IDISO to optimize the parameters of XGBoost.

The dataset was divided into a training set, validation set, and test set in an 8:1:1 ratio. The training set was used to train the predictive model. The validation set was used for the hyperparameter tuning of the trained model. The test set was used to test the predictive model. The coefficient of determination (R^2) was used as the evaluation metric for XGBoost and as the fitness value for IDISO. When optimizing the hyperparameters of XGBoost, the IDISO algorithm aimed to maximize the value of the fitness function.

The model selects three hyperparameters that most significantly affect XGBoost's performance as follows: the learning rate, number of trees, and maximum depth of trees. Additionally, it includes five hyperparameters for IDISO's search space as follows: the sample subsampling ratio for model generalization and regularization coefficients to prevent overfitting. Due to the lower number of search dimensions, the number of seeds for the IDISO algorithm is set to 10, with 20 iterations for optimization.

In order to implement the above algorithms, Python version 3.9.13, NumPy version 1.24.2, SciPy version 1.9.1, XGBoost version 1.7.5, scikit-learn (sklearn) version 1.3.0, and pandas version 1.4.4 were employed.

2.5. Performance Evaluation

In this section, we introduce the evaluation of the algorithm using statistical analysis and non-parametric statistics (sign test) methods.

2.5.1. Statistical Analysis

The average ranking method is used to preliminarily evaluate IDISO's performance. This evaluation is specifically applied to CEC2017 and renowned engineering projects.

2.5.2. Non-parametric Statistics (Sign Test)

In average ranking tests, algorithms may have their overall average ranking affected by extreme rankings due to their unsuitability for certain specific problems. Therefore, non-parametric testing methods are used to ascertain IDISO's performance further.

The sign test is a popular and straightforward method for evaluating algorithm performance. It compares the performance of two algorithms in each scenario. The evaluation is based on the number of times one algorithm outperforms the other. The algorithm with a more significant number of overall victories is considered to be the superior algorithm.

3. CEC2017 Results and Discussion

CEC2017 serves as a benchmark test function [43]. It comprises four types of benchmark test functions. Each category focuses on evaluating and testing the different capabilities of the algorithm. For detailed information, refer to Table 1 below. Unimodal function testing assesses the algorithm's in-depth exploitation capability in the search space. Multimodal function testing measures the algorithm's efficiency in exploring complex search spaces. Hybrid function testing focuses on the algorithm's ability to balance between exploration and exploitation tasks. Composite function testing evaluates the overall performance of the algorithm in highly complex search environments.

Table 1. CEC-2017 test functions.

Function Classification	Function Name	Optimal Value
Unimodal Function	Shifted and Rotated Bent Cigar Function (F1)	100
	Shifted and Rotated Zakharov Function (F3)	300
Multimodal Function	Shifted and Rotated Rosenbrock's Function (F4)	400
	Shifted and Rotated Rastrigin's Function (F5)	500
	Shifted and Rotated Expanded Scaffer's F6 Function (F6)	600
	Shifted and Rotated Lunacek Bi_Rastrigin Function (F7)	700
	Shifted and Rotated Non-Continuous Rastrigin's Function (F8)	800
	Shifted and Rotated Levy Function (F9)	900
	Shifted and Rotated Schwefel's Function (F10)	1000
Hybrid Function	Hybrid Function 1 (N = 3) (F11)	1100
	Hybrid Function 2 (N = 3) (F12)	1200
	Hybrid Function 3 (N = 3) (F13)	1300
	Hybrid Function 4 (N = 4) (F14)	1400

Table 1. *Cont.*

Function Classification	Function Name	Optimal Value
Hybrid Function	Hybrid Function 5 (N = 4) (F15)	1500
	Hybrid Function 6 (N = 4) (F16)	1600
	Hybrid Function 6 (N = 5) (F17)	1700
	Hybrid Function 6 (N = 5) (F18)	1800
	Hybrid Function 6 (N = 5) (F19)	1900
	Hybrid Function 6 (N = 6) (F20)	2000
Composite Function	Composition Function 1 (N = 3) (F21)	2100
	Composition Function 2 (N = 3) (F22)	2200
	Composition Function 3 (N = 4) (F23)	2300
	Composition Function 4 (N = 4) (F24)	2400
	Composition Function 5 (N = 5) (F25)	2500
	Composition Function 6 (N = 5) (F26)	2600
	Composition Function 7 (N = 6) (F27)	2700
	Composition Function 8 (N = 6) (F28)	2800
	Composition Function 9 (N = 3) (F29)	2900
	Composition Function 10 (N = 3) (F30)	3000
	Search Range: $[-100, 100]^D$	

This section compares the IDISO algorithm with 17 powerful, state-of-the-art optimization algorithms, including DISO. For detailed information, refer to Table 2 below.

Table 2. Comparative algorithm.

Algorithm Name	Parameters
Arithmetic Optimization Algorithm (AOA) [44]	$MOP_{Max} = 1; MOP_{Min} = 0.2;$ $Alpha = 5; Mu = 0.499$
Black Widow Optimization Algorithm (BWOA) [45]	$-1 < \beta < 1;$ $0.4 < m < 0.9$
Dingo Optimization Algorithm (DOA) [46]	$-2 < beta < 2$
Dynamic Differential Annealed Optimization (DDAO) [47]	$MaxSubIt = 1000; T0 = 2000;$ $ALPHA = 0.995$
Grey Wolf Optimizer (GWO) [24]	None
HarrisHawk Optimization (HHO) [48]	$-1 < E0 < 1$
Jumping Spider Optimization Algorithm (JSOA) [49]	$gravity = 9.80665; v_o = 100$
Reptile Search Algorithm (RSA) [50]	$Alpha = 0.1; Beta = 0.005$
Rat Swarm Optimizer (RSO) [51]	$x = 1; y = 5$
Sine Cosine Algorithm (SCA) [52]	$a = 2$
Sailfish Optimization Algorithm (SFO) [53]	$A = 4; e = 0.001; SFpercent = 0.3$
Skill optimization algorithm (SOA) [54]	None
Supply–Demand-Based Optimization (SDO) [55]	None
Spotted Hyena Optimizer (SHO) [56]	None
Tunicate Swarm Algorithm (TSA) [57]	$P_{min} = 1; P_{max} = 4$
Transient search algorithm (TSO) [58]	$K = 1$
Dujiangyan irrigation system optimization (DISO) [21]	$P = 0.85; Q = 2.5$

The comparison is based on the performance across 29 CEC2017 test functions, covering 30, 50, and 100 dimensions. To ensure the fairness of the experimental results. All algorithms use the same population size (100) and the same number of iterations $(\frac{10000*D}{100})$. The comparison algorithms originate from their original authors and utilize their recommended optimal parameters. To minimize the randomness of the experimental outcomes, each algorithm is executed 30 times. The average fitness value is taken as the evaluation metric.

The comparison results of IDISO with other metaheuristic algorithms are shown in Table 3. Supplementary data provide a detailed display of the comparative algorithms' performance on the three dimensions of the twenty-nine test functions. The results show that among the 17 compared optimization algorithms, DISO's ranking declines as the dimensions increase. It drops from third to the fourth place. In cases of 50 and 100 dimensions, DISO's average performance on CEC2017 test functions is surpassed by the HHO algorithm. This indicates its underperformance in higher dimensions. The improved IDISO algorithm exhibits a superior overall average performance across all three test dimensions when compared to the other 17 benchmarked algorithms. However, the sign test results indicate that IDISO is not superior to other algorithms in every problem. Its overall performance is the strongest. Different optimization algorithms may outperform IDISO in terms of some specific issues, which aligns with the "No Free Lunch theorem".

Table 3. (a) CEC2017 test function statistical analysis result and (b) non-parametric statistics result.

(a)

Algorithm	Average D = 30	Final D = 30	Average D = 50	Final D = 50	Average D = 100	Final D = 100
IDISO	2.41	1	2.17	1	2.06	1
DISO	3.37	3	3.55	4	3.62	4
AOA	9.51	9	10.10	11	11.44	11
BWOA	8.65	8	9.62	9	10.72	10
DOA	5.37	5	5.93	5	7.13	7
DDOA	11	12	11.31	12	11.89	12
GWO	2.44	2	2.24	2	2.34	2
HHO	3.86	4	3.27	3	2.79	3
JSOA	13.68	14	14.06	15	14.17	15
RSA	14.2	15	13.82	14	12.44	13
RSO	10.58	11	9.82	10	9	9
SCA	6.82	7	6.68	6	6.65	5
SFO	15.58	16.5	15.72	16	16.06	16
SOA	6.44	6	7	7	6.93	6
SDO	15.58	16.5	15.86	17	16.24	17
SHO	17.68	18	17.41	18	17.24	18
TSA	10.34	10	9.34	8	7.68	8
TSO	13.37	13	13.03	13	12.51	14

(b)

IDISO	30	50	100	IDISO	30	50	100
AOA	28/1/0	29/0/0	28/1/0	BWOA	29/0/0	29/0/0	29/0/0
DOA	28/1/0	28/1/0	29/0/0	DDAO	29/0/0	29/0/0	29/0/0
GWO	12/17/0	13/16/0	13/16	HHO	23/6/0	23/6/0	23/6/0
JSOA	29/0/0	29/0/0	29/0/0	RSA	29/0/0	29/0/0	29/0/0
RSO	29/0/0	29/0/0	29/0/0	SCA	28/1/0	29/0/0	29/0/0
SFO	29/0/0	29/0/0	29/0/0	SOA	26/3/0	27/2/0	27/2/0
SDO	29/0/0	29/0/0	29/0/0	SHO	29/0/0	29/0/0	29/0/0
TSA	29/0/0	29/0/0	29/0/0	TSO	29/0/0	29/0/0	29/0/0
DISO	17/12/0	20/9/0	23/6/0				

As shown in Figure 2, the accuracy of the F1 function is significantly improved across all three dimensions. This reflects that the improvement strategy has enhanced the algorithm's global search ability in unimodal function search spaces. As a result, it becomes more effective in discovering global optima. There is a significant improvement in the accuracy of the F4 function at 100 dimensions. The improvement strategy enhances the algorithm's exploitation ability in high-dimensional multimodal function search spaces. It also enhances the ability to escape from local optima. There is a significant improvement in the accuracy of the F12 function. This demonstrates the enhanced robustness of the improved algorithm in balancing global exploration and local exploitation. It can effectively find the global optimum within the complex search space of composite functions.

The radar charts of the top four ranked algorithms (IDISO, DISO, GWO, HHO) across three dimensions of the CEC2017 test functions are shown in Figure 3. The radar charts intuitively demonstrate that the IDISO algorithm has the smallest shaded area in all three dimensions, indicating its best performance in the rankings of the test functions. Particularly, in the high-dimensional space of multimodal problems, the IDISO algorithm exhibits better characteristics than in low-dimensional space, suggesting that the improvement methods effectively enhance the performance of DISO in high-dimensional space. Additionally, the radar charts show that the IDISO algorithm has the best stability among all the compared algorithms, indicating that IDISO can be widely applied to various types of problems, especially high-dimensional problems.

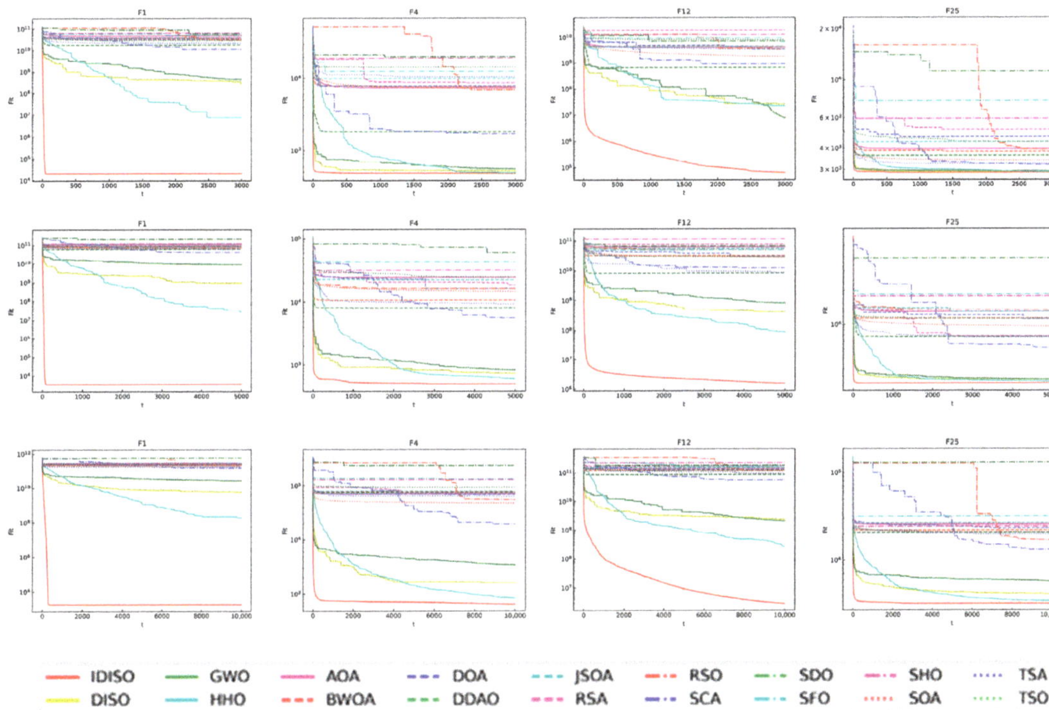

Figure 2. Partial results display of CEC2017 test functions.

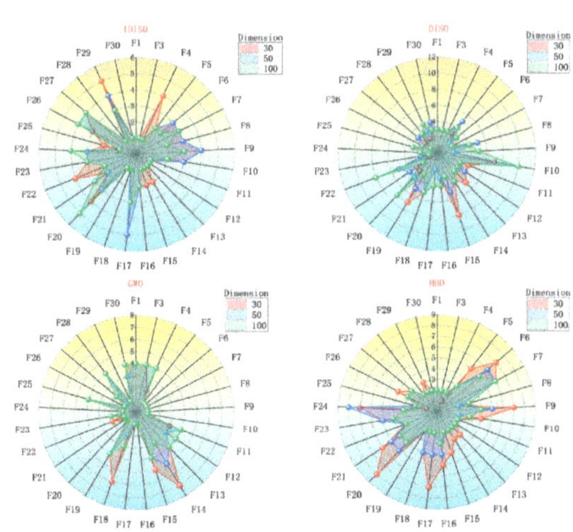

Figure 3. Radar chart of the top four algorithms in the overall ranking of the CEC2017 test.

4. Real-World Engineering Problem Results and Discussion

Nine renowned real-world engineering problems are utilized to test the improved algorithm [59]. The purpose of this is to evaluate its applicability in practical engineering scenarios. A summary of the problems is provided in Table 4. Detailed descriptions can be found at https://github.com/Jingyuango/idiso (accessed on 25 June 2024). The algorithms ranked in the top four across the three dimensions of the CEC2017 test functions are selected for comparison. To ensure the fairness of the experiments, all algorithms use the same population size (100) and the same number of iterations (1000). Each algorithm is executed 30 times. The average fitness value is used as the evaluation metric.

Table 4. Engineering problem description.

Fields	Problem	D	Constraints	Fit_{best}
Mechanical Engineering	Speed reducer (P1)	7	11	2.99442×10^3
	Multiple disk clutch brake design problems (P2)	5	8	3.13660×10^{-1}
	Piston lever (P3)	4	4	8.41270
	Car side impact design (P4)	11	10	2.28430×10^1
Civil Engineering	Cantilever beam (P5)	5	1	1.33996
	Minimize I-beam vertical deflection (P6)	4	2	1.30700×10^{-2}
	Tubular column design (P7)	2	6	2.64864×10^1
	Design of welded beam design (P8)	4	7	1.72485
	Reinforced concrete beam design (P9)	3	2	3.59208×10^2

The test results are presented in Table 5. Detailed test results can be found in the Supplementary Materials. The IDISO algorithm ranks first in overall performance among the nine real-world engineering problems. It is the algorithm with the highest accuracy among all compared algorithms in four problems, which is close to the theoretical optimal value and demonstrates the broad applicability of IDISO in real-world engineering applications.

Table 5. Real-world engineering problem ranking. [Speed Reducer(P1), Multiple disk clutch brake design problems(P2), Piston lever(P3), Car side impact design(P4), Cantilever beam(P5), Minimize I-beam vertical deflection(P6), Tubular column design(P7), Design of welded beam design(P8), Reinforced concrete beam design(P9)].

Algorithm	P1	P2	P3	P4	P5	P6	P7	P8	P9	Average	Final
IDISO	1	2.5	1	2	3	1.5	1	3	1	1.7778	1
DISO	2	2.5	2	1	4	1.5	3	2	2	2.3333	3
GWO	3	1	3	3	1	3	2	1	3	2.2222	2
HHO	4	4	4	4	2	4	4	4	4	3.6667	4

The radar chart of the top four ranked algorithms (IDISO, DISO, GWO, HHO) across the nine engineering problems is shown in Figure 4. The radar chart intuitively demonstrates that the IDISO algorithm has the smallest shaded area among the four compared algorithms, indicating its superior performance. Notably, IDISO achieved the best ranking in five problems and did not rank last in any problem, demonstrating its strong robustness.

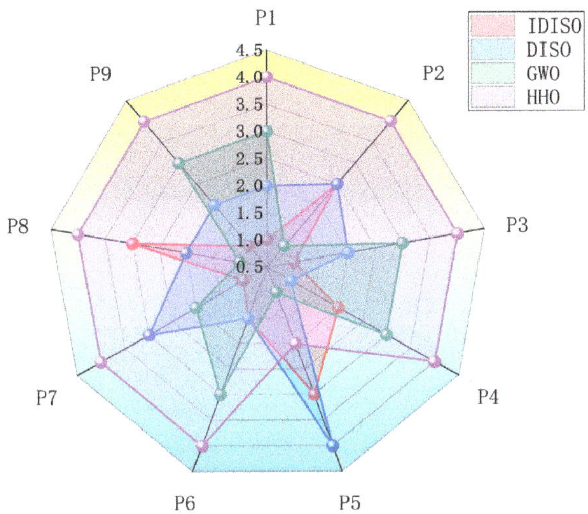

Figure 4. Radar chart of rankings for engineering problems. [Speed reducer (P1), Multiple disk clutch brake design problems (P2), Piston lever (P3), Car side impact design (P4), Cantilever beam (P5), Minimize I-beam vertical deflection (P6), Tubular column design (P7), Design of welded beam design (P8), Reinforced concrete beam design (P9)].

5. IDISO Algorithm Analysis

5.1. Convergence Analysis

The convergence of the IDISO optimization algorithm compared with other algorithms is illustrated in Figure 2. The findings demonstrate that this approach exhibits superior convergence capabilities relative to other optimization methods. In the CEC2017 benchmark suite, IDISO consistently demonstrated rapid convergence to the global optimum across three different dimensions for four types of problems. This performance showcases its capability to quickly and accurately locate global optima in high-dimensional and complex search spaces and the algorithm's broad applicability.

5.2. Exploration and Exploitation Analysis

Merely observing the convergence curves and outcomes of a population does not fully reveal the actual modifications in the algorithm. The ratio of exploration to exploitation in IDISO and DISO populations during iterations is computed. This calculation facilitates a more accurate and comprehensive understanding of the population's behavior. To accurately demonstrate improvements in exploration and exploitation [60], the multimodal function F4 and hybrid function F12 from the CEC2017 test suite are selected as examples in a 30-dimensional context, with the experimental settings being consistent with the aforementioned.

Figure 5 displays the exploration and exploitation curves for IDISO and DISO. The graph illustrates that in multimodal and hybrid function problems, the proportion of exploration by IDISO individuals is higher compared to DISO. The improvements effectively prevent individuals within the population from becoming trapped in local optima, enhancing DISO's ability to escape local optima. Throughout the convergence process, exploration and exploitation oscillate within a certain range, achieving a dynamic balance. This demonstrates IDISO's effective balancing of global exploration and local exploitation.

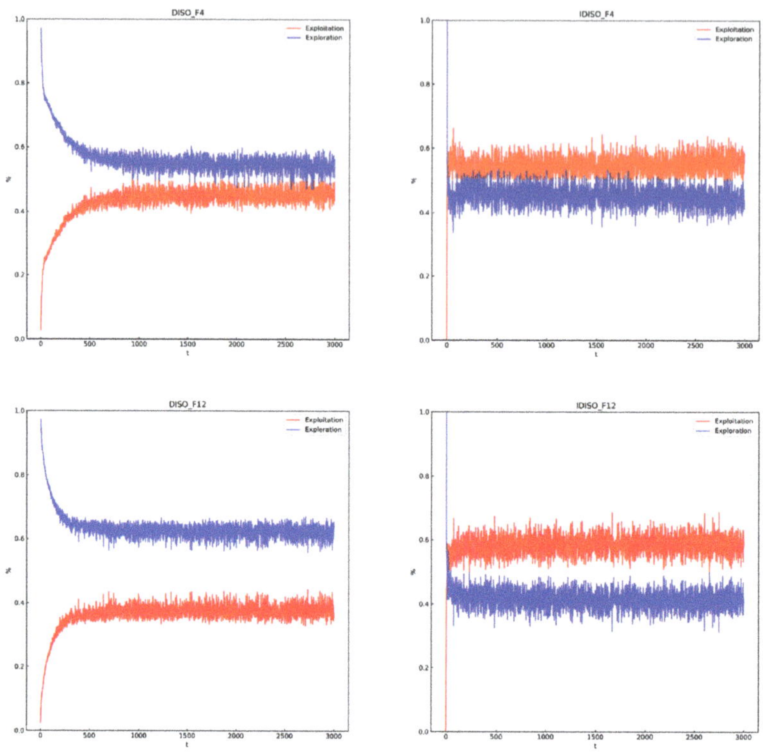

Figure 5. Exploration and exploitation analysis.

5.3. Parameter Sensitivity Analysis

The IDISO algorithm adds the Cauchy variation mechanism and two nonlinear factors to the original algorithm, which has four main parameters as follows: C_{max}, C_{min}, a, and γ. The subsequent section discusses the sensitivity of the IDISO algorithm to these parameters.

C_{max} and C_{min} are significant determinants of the nonlinear contraction factor C(t). Since the primary convergence process of the IDISO optimization algorithm takes place in the initial stage of the iterative process, the analysis is conducted in the first part of the entire convergence process, specifically when $\frac{t}{T} = \frac{1}{100}$. Since the nonlinear shrinkage factor primarily balances the search relationship, we use the F1 function at 30 dimensions for exemplification. The selected range for C_{max} is $\left[10, 15, e^3, e^4, 100, e^5, 200, 300\right]$, and the values for C_{min} are $\left[2, e, 3, 4, 5, 6, e^2, 7\right]$. When C_{max} is a variable, $C_{min} = e^2$. When C_{min} is a variable, $C_{max} = e^3$. The findings are presented in Figure 6a, displaying a decrease in the $\frac{\partial C(\sqcup)}{\partial C_{max}}$ value during the initial stages of the iteration as the C_{max} values increase. This indicates a slower rate of change for $C(t)$. Increasing the step size during the initial position update will result in a more significant number of iterations, but it may also cause the algorithm to cross the local optimum, resulting in a decline in the final results. In the initial iteration phase, when the value of C_{min} decreases, $C(t)$ follows a similar pattern to C_{max} and is highly responsive to any changes in these two variables. As a result, the convergence value range can be extensive. This study reveals that the IDISO algorithm performs best with respect to the influence factor of $C(t)$, when $C_{max} = e^3$ and $C_{min} = e^2$.

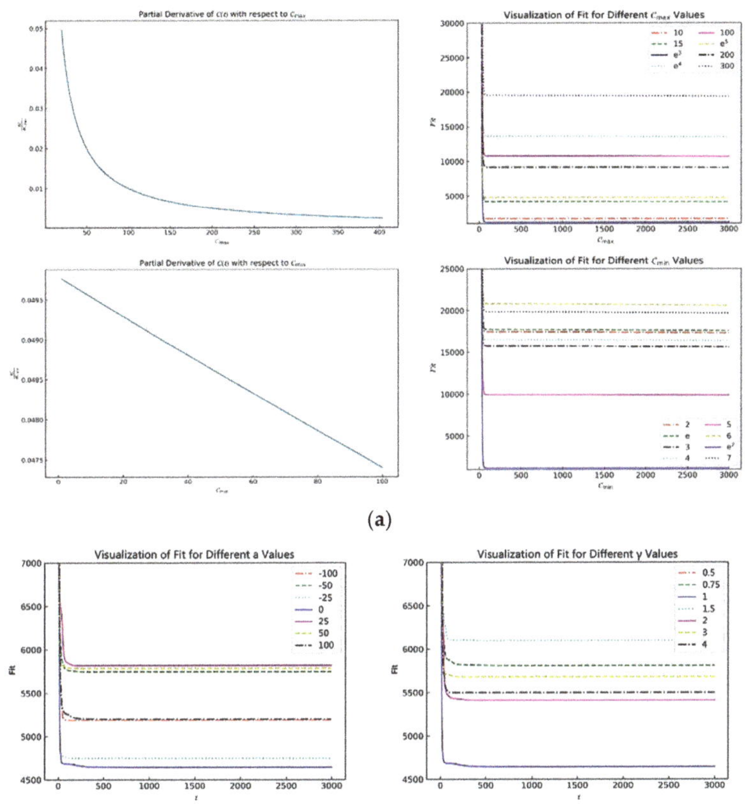

(a)

(b)

Figure 6. (**a**). C_{max} and C_{min} sensitivity analysis. (**b**). a and γ sensitivity analysis.

The Cauchy variation mechanism is influenced by the parameters a and γ, which operate during the final stage of the process. This mechanism enables the water to escape local optima and prevents the population from being trapped as well. a represents the midpoint of the Cauchy distribution and γ represents the distribution's width. Using the example of the 30 dimensions of the F9 function, the Cauchy variation mechanism is added to counter DISO's inclination to converge towards a local optimum. The selected range for a is $[-100, -50, -25, 0, 25, 50, 100]$, and the values for γ are $[0.5, 0.75, 1, 1.5, 2, 3, 4]$. When a is a variable, $\gamma = 1$. When γ is a variable, $a = 0$. The findings presented in Figure 6b demonstrate that γ exhibits greater sensitivity to alterations in the central region of distribution a. Optimal performance is achieved when a and γ are set to 0 and 1, respectively.

5.4. IDISO Complexity Analysis

As shown in Figure 1, the improved algorithm mainly includes new factors in the two positions to update the formulas and the Cauchy variation mechanism before eliminating the water body. This process uses the same time complexity as DISO, both of which are $O(N \times T)$. Therefore, IDISO significantly enhances the accuracy performance of the DISO algorithm in high-dimensional complex functions without increasing the algorithm's running time complexity.

5.5. IDISO Ablation Study

In Figure 7, IDISO1 represents the DISO algorithm with the Nonlinear Shrinking Factor added, and IDISO2 represents the DISO algorithm with the Cauchy mutation mechanism added. From left to right, the figure displays the CEC test functions F1, F4, F12, and F25, which represent unimodal, multimodal, hybrid, and composite problems, respectively. From top to bottom, the figure shows results for dimensions 30, 50, and 100. It can be observed that the performance of the DISO algorithm is inferior to the improved algorithms across all four problem types. In unimodal and hybrid problems, the differences among the three methods are significant, and these differences become more pronounced as the dimensionality increases. In multimodal and composite problems, the accuracy differences among the three methods are not substantial, with distinctions existing primarily in convergence speed. This indicates that the combination of the nonlinear shrinking factor and Cauchy mutation mechanism improves the performance of the DISO algorithm in high-dimensional unimodal and hybrid problems and accelerates the convergence of the DISO algorithm across all problem types.

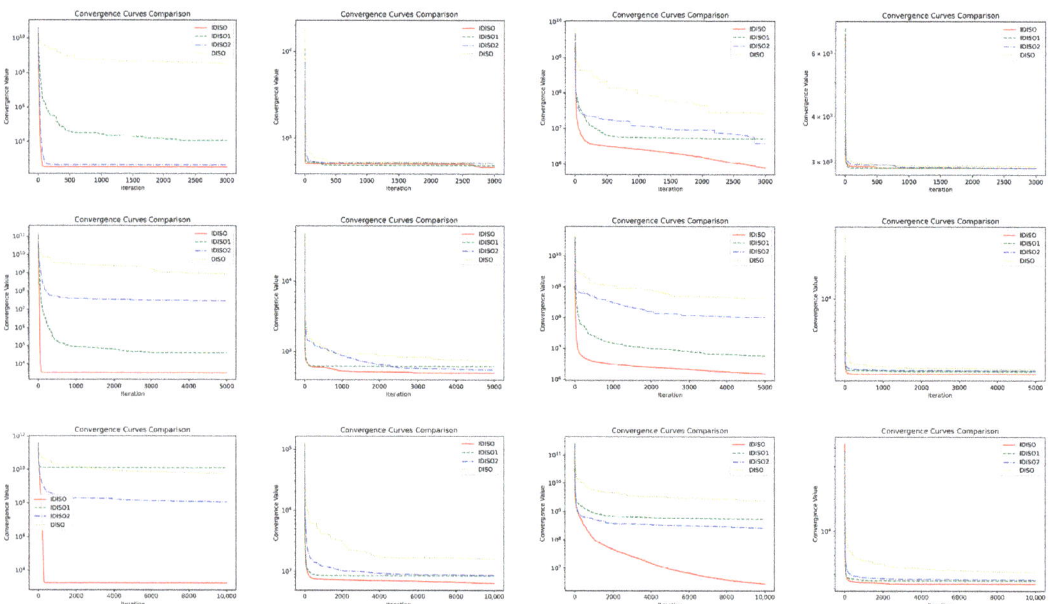

Figure 7. IDISO ablation Study.

6. IDISO-XGBoost Model

The hyperparameters are selected as shown in Table 6, and the comparison of the tuning results of IDISO-XGBoost with DISO-XGBoost is shown in Figure 8. The results indicate that the IDISO-XGBoost model achieved an average prediction performance of 0.95, representing a 4% improvement over the average prediction performance of the DISO-XGBoost model. Compared to the study by Chen et al., which used grid search to optimize the RF model [23], this research achieved an improvement in the R^2 for predicting elemental analysis on the test set from 0.92 to 0.96, representing a 4.4% increase. This advancement proposes a novel method for predicting the elemental analysis of hydrochar and subsequently guiding the design of experimental protocols.

Table 6. Target feature hyperparameter values.

	Learning Rate	Number of Trees	Depth of the Tree	The Proportion of Subsamples Sampled	Regular Term Coefficient	Val_R^2	Test_R^2
upper and lower bounds	[0.01–1]	[100–1000]	[3–10]	[0.01–0.3]	[0.01–0.05]		
IDISO_C	2.9846157×10^{-1}	655	10	3.8879849×10^{-1}	3.3397663×10^{-2}	0.97	0.95
DISO_C	$2.25181533 \times 10^{-1}$	549	9	$1.92181424 \times 10^{-1}$	$4.15709607 \times 10^{-2}$	0.96	0.92
IDISO_H	$2.95662070 \times 10^{-1}$	407	3	$2.53441246 \times 10^{-1}$	$4.78747122 \times 10^{-2}$	0.97	0.96
DISO_H	$4.68936589 \times 10^{-1}$	233	6	$2.15128044 \times 10^{-1}$	$2.02117095 \times 10^{-2}$	0.95	0.93
IDISO_O	7.053795×10^{-2}	659	6	5.638555×10^{-1}	3.864307×10^{-2}	0.97	0.96
DISO_O	$4.65478169 \times 10^{-1}$	569	8	$2.33724901 \times 10^{-1}$	$1.24724460 \times 10^{-2}$	0.93	0.91
IDISO_N	8.1734951×10^{-2}	533	8	5.5392384×10^{-1}	3.5553879×10^{-2}	0.95	0.94
DISO_N	$7.38914069 \times 10^{-1}$	997	8	$2.26746118 \times 10^{-1}$	$4.19970431 \times 10^{-2}$	0.90	0.89

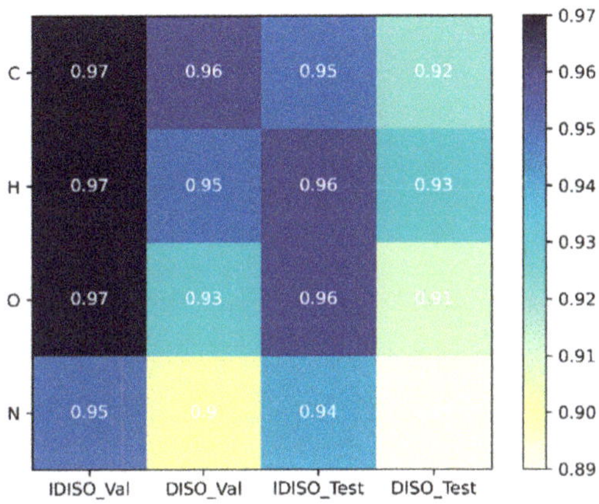

Figure 8. IDISO-XGBoost VS. DISO-XGBoost.

Figure 9 displays the correlation between the actual and predicted values in both the validation and test sets. The green auxiliary line shows that the predicted values match the true values. The figure reveals that only a few points lie far away from the auxiliary line in the C, N, and O prediction models. Conversely, all issues remain near the auxiliary line in the H prediction model. These results demonstrate that the IDISO-XGBoost model can accurately predict the elemental properties of hydrothermal charcoal with good generalization and robustness. This study presents a new approach to predicting such properties and offers a valuable contribution to the field.

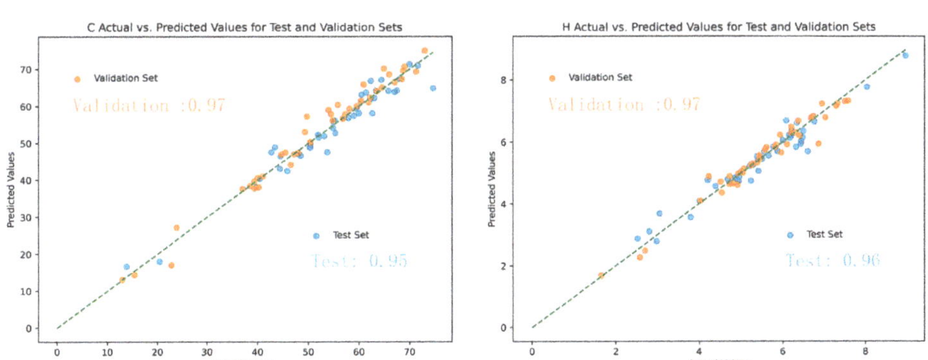

Figure 9. *Cont.*

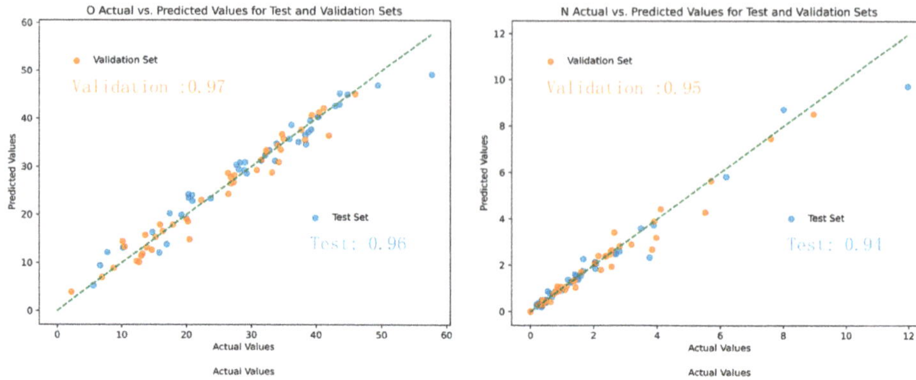

Figure 9. Relationship between true and predicted values for validation and test sets.

7. Conclusions

In this paper, we propose the Improved Dujiangyan Irrigation System Optimization (IDISO) algorithm by introducing the nonlinear shrinking factor and the Cauchy mutation mechanism to enhance the original DISO algorithm. To validate the optimization capabilities and convergence of the IDISO algorithm, we employed 87 CEC2017 benchmark functions of varying dimensions and compared its performance with 17 state-of-the-art metaheuristic algorithms. Additionally, we assessed the search performance of the IDISO algorithm in unknown spaces across nine real-world engineering problems. The results demonstrate that the IDISO algorithm achieved the fastest convergence speed and the highest convergence accuracy compared to the 17 powerful metaheuristic algorithms, significantly outperforming the DISO algorithm. In both the CEC2017 benchmark functions and the nine real-world engineering problems, the IDISO algorithm ranked first in overall performance.

Finally, we utilized the IDISO and DISO algorithms to fine-tune the parameters of the XGBoost model, aiming to construct a robust and generalizable prediction model for the characteristics of hydrochar elements. The results indicate that the IDISO-XGBoost model achieved an average prediction performance of 0.95, representing a 4% improvement over the average prediction performance of the DISO-XGBoost model. This novel approach to predicting hydrochar element characteristics highlights the significant potential and value of the IDISO algorithm in practical applications.

Supplementary Materials: The following supporting information can be downloaded at: https://github.com/Jingyuango/idiso, (accessed on 25 June 2024).

Author Contributions: Conceptualization, D.Z.; Methodology, J.S. and Z.S.; Software, J.S.; Validation, J.W.; Investigation, Z.S., Z.Z. and W.H.; Data curation, J.W. and Z.H.; Writing—original draft, J.S.; Writing—review & editing, D.Z. and Y.W. All authors have read and agreed to the published version of the manuscript.

Funding: This project is supported by the Inner Mongolia Autonomous Region Science and Technology Plan Project (2023YFDZ0031), Fundamental Research Fund for Inner Mongolia University of Science & Technology (2023QNJS129,2023QNJS130,2024RCTD008), Inner Mongolia Natural Science Foundation (2024QN05058), and the Open Research Project of State Key Laboratory of Baiyunobo Rare Earth Resource Researches and Comprehensive Utilization (2021H2275).

Data Availability Statement: The data presented in this study are available on request from the corresponding author.

Acknowledgments: Our profound gratitude is extended to ChatGPT for its assistance. This manuscript has benefited from the facilitation provided by ChatGPT during the refinement process, enhancing the fluidity of the English expression. However, it is imperative to underscore that the final content and mode of expression within this document are a reflection of my personal choice, embodying my perspectives and stylistic approach.

Conflicts of Interest: The authors declare no conflict of interest.

References

1. Xiu, S.; Shahbazi, A. Bio-oil production and upgrading research: A review. *Renew. Sustain. Energy Rev.* **2012**, *16*, 4406–4414. [CrossRef]
2. Lee, D.; Nam, H.; Seo, M.W.; Lee, S.H.; Tokmurzin, D.; Wang, S.; Park, Y.-K. Recent progress in the catalytic thermochemical conversion process of biomass for biofuels. *Chem. Eng. J.* **2022**, *447*, 137501. [CrossRef]
3. Sikarwar, V.S.; Zhao, M.; Fennell, P.S.; Shah, N.; Anthony, E.J. Progress in biofuel production from gasification. *Prog. Energy Combust. Sci.* **2017**, *61*, 189–248. [CrossRef]
4. Mäkelä, M.; Benavente, V.; Fullana, A. Hydrothermal carbonization of industrial mixed sludge from a pulp and paper mill. *Bioresour. Technol.* **2016**, *200*, 444–450. [CrossRef] [PubMed]
5. Hansen, L.J.; Fendt, S.; Spliethoff, H. Impact of hydrothermal carbonization on combustion properties of residual biomass. *Biomass Convers. Biorefinery* **2022**, *12*, 2541–2552. [CrossRef]
6. Samaksaman, U.; Pattaraprakorn, W.; Neramittagapong, A.; Kanchanatip, E. Solid fuel production from macadamia nut shell: Effect of hydrothermal carbonization conditions on fuel characteristics. *Biomass Convers. Biorefinery* **2021**, *13*, 2225–2232. [CrossRef]
7. Wong, K.I.; Wong, P.K.; Cheung, C.S.; Vong, C.M. Modelling of diesel engine performance using advanced machine learning methods under scarce and exponential data set. *Appl. Soft Comput.* **2013**, *13*, 4428–4441. [CrossRef]
8. Jeon, P.R.; Moon, J.-H.; Olanrewaju, O.N.; Lee, S.H.; Ling, J.L.J.; You, S.; Park, Y.-K. Recent advances and future prospects of thermochemical biofuel conversion processes with machine learning. *Chem. Eng. J.* **2023**, *471*, 144503. [CrossRef]
9. Rasam, S.; Talebkeikhah, F.; Talebkeikhah, M.; Salimi, A.; Moraveji, M.K. Physico-chemical properties prediction of hydrochar in macroalgae Sargassum horneri hydrothermal carbonisation. *Int. J. Environ. Anal. Chem.* **2021**, *101*, 2297–2318. [CrossRef]
10. Kardani, N.; Hedayati Marzbali, M.; Shah, K.; Zhou, A. Machine learning prediction of the conversion of lignocellulosic biomass during hydrothermal carbonization. *Biofuels* **2022**, *13*, 703–715. [CrossRef]
11. Li, J.; Pan, L.; Suvarna, M.; Tong, Y.W.; Wang, X. Fuel properties of hydrochar and pyrochar: Prediction and exploration with machine learning. *Appl. Energy* **2020**, *269*, 115166. [CrossRef]
12. Li, J.; Zhu, X.; Li, Y.; Tong, Y.W.; Ok, Y.S.; Wang, X. Multi-task prediction and optimization of hydrochar properties from high-moisture municipal solid waste: Application of machine learning on waste-to-resource. *J. Clean. Prod.* **2021**, *278*, 123928. [CrossRef]
13. Nguyen, V.G.; Sharma, P.; Ağbulut, Ü.; Le, H.S.; Tran, V.D.; Cao, D.N. Precise prognostics of biochar yield from various biomass sources by Bayesian approach with supervised machine learning and ensemble methods. *Int. J. Green Energy* **2023**, *21*, 2180–2204. [CrossRef]
14. Mu, L.; Wang, Z.; Wu, D.; Zhao, L.; Yin, H. Prediction and evaluation of fuel properties of hydrochar from waste solid biomass: Machine learning algorithm based on proposed PSO–NN model. *Fuel* **2022**, *318*, 123644. [CrossRef]
15. Li, P.; Du, Z.; Chang, C.; Zhao, S.; Xu, G.; Xu, C.C. Efficient catalytic conversion of waste peanut shells into liquid biofuel: An artificial intelligence approach. *Energy Fuels* **2020**, *34*, 1791–1801. [CrossRef]
16. Osaba, E.; Villar-Rodriguez, E.; Del Ser, J.; Nebro, A.J.; Molina, D.; LaTorre, A.; Suganthan, P.N.; Coello, C.A.C.; Herrera, F. A tutorial on the design, experimentation and application of metaheuristic algorithms to real-world optimization problems. *Swarm Evol. Comput.* **2021**, *64*, 100888. [CrossRef]
17. Gavrilas, M. Heuristic and metaheuristic optimization techniques with application to power systems. In Proceedings of the 12th WSEAS International Conference on Mathematical Methods and Computational Techniques in Electrical Engineering, Kantaoui, Sousse, Tunisia, 3–6 May 2010; pp. 95–103.
18. Abualigah, L.; Elaziz, M.A.; Khasawneh, A.M.; Alshinwan, M.; Ibrahim, R.A.; Al-Qaness, M.A.; Mirjalili, S.; Sumari, P.; Gandomi, A.H. Meta-heuristic optimization algorithms for solving real-world mechanical engineering design problems: A comprehensive survey, applications, comparative analysis, and results. *Neural Comput. Appl.* **2022**, *34*, 4081–4110. [CrossRef]
19. Houssein, E.H.; Hosney, M.E.; Oliva, D.; Mohamed, W.M.; Hassaballah, M. A novel hybrid Harris hawks optimization and support vector machines for drug design and discovery. *Comput. Chem. Eng.* **2020**, *133*, 106656. [CrossRef]
20. Turgut, O.E.; Turgut, M.S.; Kırtepe, E. A systematic review of the emerging metaheuristic algorithms on solving complex optimization problems. *Neural Comput. Appl.* **2023**, *35*, 14275–14378. [CrossRef]
21. Niu, J.; Ren, C.; Guan, Z.; Cao, Z. Dujiangyan irrigation system optimization (DISO): A novel metaheuristic algorithm for dam safety monitoring. *Structures* **2023**, *54*, 399–419. [CrossRef]
22. Blackwell, T.; Kennedy, J.; Poli, R. Particle swarm optimization. *Swarm Intell.* **2007**, *1*, 33–57.
23. Chen, C.; Liang, R.; Wang, J.; Ge, Y.; Tao, J.; Yan, B.; Chen, G. Simulation and optimization of co-pyrolysis biochar using data enhanced interpretable machine learning and particle swarm algorithm. *Biomass Bioenergy* **2024**, *182*, 107111. [CrossRef]

24. Mirjalili, S.; Mirjalili, S.; Lewis, A. Grey wolf optimizer. *Adv. Eng. Softw.* **2014**, *69*, 46–61. [CrossRef]
25. Subudhi, U.; Dash, S. Detection and classification of power quality disturbances using GWO ELM. *J. Ind. Inf. Integr.* **2021**, *22*, 100204. [CrossRef]
26. Zhao, S.; Zhang, T.; Ma, S.; Chen, M. Dandelion Optimizer: A nature-inspired metaheuristic algorithm for engineering applications. *Eng. Appl. Artif. Intell.* **2022**, *114*, 105075. [CrossRef]
27. Saglam, M.; Bektas, Y.; Karaman, O.A. Dandelion Optimizer and Gold Rush Optimizer Algorithm-Based Optimization of Multilevel Inverters. *Arab. J. Sci. Eng.* **2024**, *49*, 7029–7052. [CrossRef]
28. Chou, J.-S.; Molla, A. Recent advances in use of bio-inspired jellyfish search algorithm for solving optimization problems. *Sci. Rep.* **2022**, *12*, 19157. [CrossRef]
29. Gouda, E.A.; Kotb, M.F.; El-Fergany, A.A. Jellyfish search algorithm for extracting unknown parameters of PEM fuel cell models: Steady-state performance and analysis. *Energy* **2021**, *221*, 119836. [CrossRef]
30. Abdel-Basset, M.; El-Shahat, D.; Jameel, M.; Abouhawwash, M. Young's double-slit experiment optimizer: A novel metaheuristic optimization algorithm for global and constraint optimization problems. *Comput. Methods Appl. Mech. Eng.* **2023**, *403*, 115652. [CrossRef]
31. Dong, Y.; Sun, Y.; Liu, Z.; Du, Z.; Wang, J. Predicting dissolved oxygen level using Young's double-slit experiment optimizer-based weighting model. *J. Environ. Manag.* **2024**, *351*, 119807. [CrossRef]
32. Zamani, H.; Nadimi-Shahraki, M.H.; Gandomi, A.H. Starling murmuration optimizer: A novel bio-inspired algorithm for global and engineering optimization. *Comput. Methods Appl. Mech. Eng.* **2022**, *392*, 114616. [CrossRef]
33. Chao, X. Optimal boosting method of HPC concrete compressive and tensile strength prediction. *Struct. Concr.* **2024**, *25*, 283–302. [CrossRef]
34. Wolpert, D.H.; Macready, W.G. No free lunch theorems for optimization. *IEEE Trans. Evol. Comput.* **1997**, *1*, 67–82. [CrossRef]
35. Laskar, N.M.; Guha, K.; Chatterjee, I.; Chanda, S.; Baishnab, K.L.; Paul, P.K. HWPSO: A new hybrid whale-particle swarm optimization algorithm and its application in electronic design optimization problems. *Appl. Intell.* **2019**, *49*, 265–291. [CrossRef]
36. Hu, G.; Zheng, Y.; Abualigah, L.; Hussien, A.G. DETDO: An adaptive hybrid dandelion optimizer for engineering optimization. *Adv. Eng. Inform.* **2023**, *57*, 102004. [CrossRef]
37. Hu, G.; Guo, Y.; Zhong, J.; Wei, G. IYDSE: Ameliorated Young's double-slit experiment optimizer for applied mechanics and engineering. *Comput. Methods Appl. Mech. Eng.* **2023**, *412*, 116062. [CrossRef]
38. Hu, G.; Wang, J.; Li, M.; Hussien, A.G.; Abbas, M. EJS: Multi-strategy enhanced jellyfish search algorithm for engineering applications. *Mathematics* **2023**, *11*, 851. [CrossRef]
39. Hu, G.; Zhong, J.; Wei, G.; Chang, C.-T. DTCSMO: An efficient hybrid starling murmuration optimizer for engineering applications. *Comput. Methods Appl. Mech. Eng.* **2023**, *405*, 115878. [CrossRef]
40. Guan, Z.; Ren, C.; Niu, J.; Wang, P.; Shang, Y. Great Wall Construction Algorithm: A novel meta-heuristic algorithm for engineer problems. *Expert Syst. Appl.* **2023**, *233*, 120905. [CrossRef]
41. Chen, T.; He, T.; Benesty, M.; Khotilovich, V.; Tang, Y.; Cho, H.; Chen, K.; Mitchell, R.; Cano, I.; Zhou, T. Xgboost: Extreme gradient boosting. *R Package Version 0.4-2* **2015**, *1*, 1–4.
42. Grinsztajn, L.; Oyallon, E.; Varoquaux, G. Why do tree-based models still outperform deep learning on typical tabular data? *Adv. Neural Inf. Process. Syst.* **2022**, *35*, 507–520.
43. Mohamed, A.W.; Hadi, A.A.; Fattouh, A.M.; Jambi, K.M. LSHADE with semi-parameter adaptation hybrid with CMA-ES for solving CEC 2017 benchmark problems. In Proceedings of the 2017 IEEE Congress on Evolutionary Computation (CEC), Donostia, Spain, 5–8 June 2017; pp. 145–152.
44. Hashim, F.A.; Hussain, K.; Houssein, E.H.; Mabrouk, M.S.; Al-Atabany, W. Archimedes optimization algorithm: A new metaheuristic algorithm for solving optimization problems. *Appl. Intell.* **2021**, *51*, 1531–1551. [CrossRef]
45. Hayyolalam, V.; Kazem, A.A.P. Black widow optimization algorithm: A novel meta-heuristic approach for solving engineering optimization problems. *Eng. Appl. Artif. Intell.* **2020**, *87*, 103249. [CrossRef]
46. Bairwa, A.K.; Joshi, S.; Singh, D. Dingo optimizer: A nature-inspired metaheuristic approach for engineering problems. *Math. Probl. Eng.* **2021**, *2021*, 2571863. [CrossRef]
47. Ghafil, H.N.; Jármai, K. Dynamic differential annealed optimization: New metaheuristic optimization algorithm for engineering applications. *Appl. Soft Comput.* **2020**, *93*, 106392. [CrossRef]
48. Heidari, A.A.; Mirjalili, S.; Faris, H.; Aljarah, I.; Mafarja, M.; Chen, H. Harris hawks optimization: Algorithm and applications. *Future Gener. Comput. Syst.* **2019**, *97*, 849–872. [CrossRef]
49. Peraza-Vázquez, H.; Peña-Delgado, A.; Ranjan, P.; Barde, C.; Choubey, A.; Morales-Cepeda, A.B. A bio-inspired method for mathematical optimization inspired by arachnida salticidade. *Mathematics* **2021**, *10*, 102. [CrossRef]
50. Abualigah, L.; Abd Elaziz, M.; Sumari, P.; Geem, Z.W.; Gandomi, A.H. Reptile Search Algorithm (RSA): A nature-inspired meta-heuristic optimizer. *Expert Syst. Appl.* **2022**, *191*, 116158. [CrossRef]
51. Dhiman, G.; Garg, M.; Nagar, A.; Kumar, V.; Dehghani, M. A novel algorithm for global optimization: Rat swarm optimizer. *J. Ambient Intell. Humaniz. Comput.* **2021**, *12*, 8457–8482. [CrossRef]
52. Mirjalili, S. SCA: A sine cosine algorithm for solving optimization problems. *Knowl. -Based Syst.* **2016**, *96*, 120–133. [CrossRef]
53. Shadravan, S.; Naji, H.R.; Bardsiri, V.K. The Sailfish Optimizer: A novel nature-inspired metaheuristic algorithm for solving constrained engineering optimization problems. *Eng. Appl. Artif. Intell.* **2019**, *80*, 20–34. [CrossRef]

54. Givi, H.; Hubalovska, M. Skill Optimization Algorithm: A New Human-Based Metaheuristic Technique. *Comput. Mater. Contin.* **2023**, *74*, 179–202. [CrossRef]
55. Zhao, W.; Wang, L.; Zhang, Z. Supply-demand-based optimization: A novel economics-inspired algorithm for global optimization. *IEEE Access* **2019**, *7*, 73182–73206. [CrossRef]
56. Dhiman, G.; Kumar, V. Spotted hyena optimizer: A novel bio-inspired based metaheuristic technique for engineering applications. *Adv. Eng. Softw.* **2017**, *114*, 48–70. [CrossRef]
57. Kaur, S.; Awasthi, L.K.; Sangal, A.; Dhiman, G. Tunicate Swarm Algorithm: A new bio-inspired based metaheuristic paradigm for global optimization. *Eng. Appl. Artif. Intell.* **2020**, *90*, 103541. [CrossRef]
58. Qais, M.H.; Hasanien, H.M.; Alghuwainem, S. Transient search optimization: A new meta-heuristic optimization algorithm. *Appl. Intell.* **2020**, *50*, 3926–3941. [CrossRef]
59. Talatahari, S.; Bayzidi, H.; Saraee, M. Social network search for global optimization. *IEEE Access* **2021**, *9*, 92815–92863. [CrossRef]
60. Hussain, K.; Salleh, M.N.M.; Cheng, S.; Shi, Y. On the exploration and exploitation in popular swarm-based metaheuristic algorithms. *Neural Comput. Appl.* **2019**, *31*, 7665–7683. [CrossRef]

processes

MDPI

Article

Preparation of Bioaerogel from Iron-Rich Microalgae for the Removal of Water Pollutants

Xinqi Niu [1,2,3], Junhui Si [1,2,3], Binyi Chen [1,2,3], Qianting Wang [1,2,3,*], Sen Zeng [1,2,3,*] and Zhixiang Cui [1,2,3,*]

[1] College of Materials Science and Engineering, Fujian University of Technology, Fuzhou 350118, China; m15225469596@163.com (X.N.)
[2] Fu Jian Provincial Key Laboratory of Advanced Materials Processing and Application, Fuzhou 350118, China
[3] Key Laboratory of Polymer Materials and Products of Universities in Fujian, Fujian University of Technology, Fuzhou 350118, China
[*] Correspondence: cocolark@163.com (Q.W.); zengs@fjut.edu.cn (S.Z.); cuizhixiang2006@126.com (Z.C.)

Abstract: Microalgae-based materials have gained significant attention considering their rich resources, cost-effectiveness, and environmental friendliness. Herein, iron-rich microalgae (Chlorella pyrenoidosa, CP) were treated by hydrothermal reaction under alkaline conditions to remove the protoplast and obtain a hollow shell with an Fe_xO_y core inside. Then, the iron-rich microalgae-based aerogel (Fe-CP aerogel) was fabricated through a freeze-drying process. The as-prepared Fe-CP aerogel exhibited superior adsorption performance, and the maximum adsorption quantity for Cu^{2+} could reach 208.3 mg/g due to the synergistic adsorption of the hollow shell of CP cells and Fe_xO_y core. The Fe-CP aerogel also possessed super-hydrophilicity and displayed high separation efficiency (over 99%) when used for separating different oil/water emulsions. Moreover, the existence of Fe_xO_y endowed the Fe-CP aerogel with photo-Fenton activity, thus exhibiting excellent antifouling performance. The prepared Fe-CP aerogel could be considered an ideal adsorbent, with green, efficient, low-cost, and regenerative properties, and has the potential for the practical treatment of wastewater.

Keywords: microalgae; aerogel; adsorption; wastewater treatment; photo-Fenton

1. Introduction

Water pollution including heavy metal ions, organic dyes, and oily wastewater possessing negative impacts on living organisms has been one of the major concerns and challenges to the environment and economy [1,2]. The accumulation of these complex contaminants in the natural environment will be mutagenic, teratogenic, and carcinogenic to humans; however, eliminating these microcontaminants is difficult [3–5]. Hence, the development of efficient and green approaches to remove the contaminants in water is urgent and significant.

Several methods, such as catalytic degradation, adsorption, chemical precipitation, bio-degradation, and membrane separation, have been proposed [6–8]. Adsorption is considered as a promising method to remove metal ions, organic dyes, and oily wastewater considering its simple operation, high efficiency, and sustainable properties [9]. Until now, many porous materials (e.g., activated carbons, carbon nanotubes, metal oxide, graphene, and metalorganic frameworks) have been extensively explored [10]. Although many of them possessed excellent adsorption capability, most of them exhibited the ubiquitous shortcomings of high cost, non-degradability, complex preparation, and possible secondary pollution [11]. Hence, it is essential to develop adsorbents that possess large adsorption capacity and overcome these drawbacks simultaneously. Moreover, the selectivity against other ions, long-term stability, cyclic regeneration, and antifouling properties of the adsorbent are also very important [12]. Hence, it is essential to develop novel adsorbents that can meet these requirements.

Natural biological adsorbent materials with several merits, such as low cost, abundant resources, degradability, and green qualities, have been regarded as ideal candidates for contaminant removal [12,13]. He et al. reported a carboxylated wood aerogel prepared by delignification and the in situ oxidation treatment of natural balsa wood that exhibited an adsorption quantity of 115 mg/g for Cu^{2+} [14]. Cellulose nanofiber (CNF)/gelatin aerogel was prepared to remove dyes by adsorption, and the maximum adsorption quantities for Congo red and methylene blue were 590.6 and 182.6 mg/g, respectively [15].

Recently, microalgae have notably received considerable attention [16,17]. The abundant microalgae resources provide a large amount of raw materials for the large-scale preparation of microalgae-based adsorbents [18–20]. In particular, harmful microalgae can grow very fast, forming a large amount of phytoplankton and water blooms and leading to the eutrophication of water quality [21]. Therefore, the utilization of microalgae to prepare biological adsorbents that change waste into valuables is of great significance. Microalgae-based materials possess numerous active groups, such as amino, hydroxyl, and carboxyl, which can advance the efficiency of adsorption [22,23]. However, on the one hand, a critical issue of using microalgae is that they are difficult to collect after the adsorption process in solutions [24]. On the other hand, adsorbents are commonly limited by their low regeneration efficiency and thus hard to recycle, especially for organic pollutants [25]. To address these issues, microalgae were prepared in bulk materials or loaded on porous material. For example, Singhal et al. [26] prepared an aerogel constructed by microalgae and chitosan, which exhibited an adsorption quantity of 300 mg/g and 250 mg/g in freshwater and seawater. However, these aerogel adsorbents will be seriously affected by the fouling caused by organic pollutants and microorganisms [27]. Consequently, the development of novel microalgae-based adsorbents combined with high adsorption capacity and antifouling ability is crucial to realizing practical applications.

Aerogels have attracted extensive attention and are explored as novel adsorbent materials due to the unique advantages of high porosity, ultra-low density, and excellent adsorption ability [28]. Herein, an all-natural microalgae-based aerogel was developed as an adsorbent and used for the removal of contaminants from water. The morphology, composition, and structure of the aerogel were characterized. The adsorption and separation performances of the aerogel for different pollutants were evaluated. The designed aerogel is purely based on algae with excellent adsorption ability and antifouling performance and showed great potential application in wastewater treatment.

2. Experimental Section

2.1. Materials

The *Chlorella pyrenoidosa* (CP) was supplied by the Wuhan Hydrobiology Institute of the Chinese Academy of Sciences. BG-11 culture medium was obtained from Shanghai Guangyu Biotechnology Co., Ltd. (Shanghai, China) Ligroin, n-hexane, kerosene, hydrochloric acid (HCl), rhodamine b (RB), and methylene blue (MB) were supplied by Xilong Chemical Reagent Co., Ltd. (Guangzhou, China) Copper sulfate ($CuSO_4$) and sodium dodecyl sulfate (SDS) were provided by Tianjin Zhiyuan Chemical Reagent Co., Ltd. (Tianjin, China).

2.2. Preparation of the Fe-CP Aerogels

CP was cultured in BG-11 medium under illumination in an incubator with a temperature of 25 °C. After a period of cultivation, CP was collected using the coagulation method, and the $FeCl_3$ (100 mg/L) was added and kept for 15 min. Then, the iron-rich CP (Fe-CP) biomass was harvested with centrifugation and dried in an oven at 65 °C. The obtained biomass (10 g) was mixed with alkaline solution (80 mL, pH = 12), and then the SDS solution (2% w/v, 80 mL) was added. The mixture was kept in an oven at 90 °C for 3 h. The treated Fe-CP biomass was recovered and washed with DI water three times. Subsequently, 0.5 g of the obtained biomass was added to the deionized water (50 mL) and heated and kept at 160 °C for 2 h. The obtained biomass was recovered by centrifugation and washed three times, and then added into a customized mold and freeze-dried by freeze-dryer for 12 h, as illustrated in Figure 1.

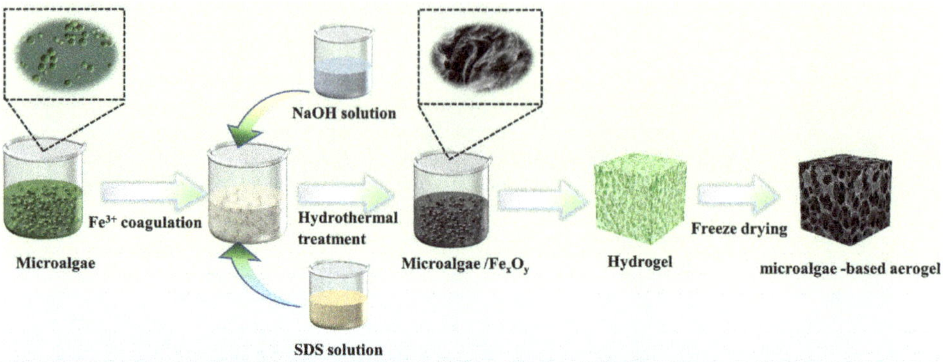

Figure 1. Schematic diagram of the preparation process of the Fe-CP aerogel.

2.3. Characterization

The morphology and composition of the aerogels were analyzed by scanning electron microscopy (SEM, S-3400 N, Tokyo, Japan) and energy dispersive X-ray spectroscopic (EDS, Oxford IE 250, London, UK) mappings, respectively. The chemical structure of the aerogel was characterized by Fourier transform infrared spectroscopy (FT-IR, Nicolet iS50, Waltham, MA, USA). The concentration of dyes was measured using a UV–vis spectrophotometer (UV-2600, Kyoto, Japan) at room temperature. The hydrophilicity of aerogel was determined by a contact angle measuring instrument (DSA25, KRUSS, Hamburg, Germany). The concentrations of Cu^{2+} were determined by ICP-OES (Optima 8300, PerkinElmer, Waltham, MA, USA).

2.4. Adsorption Experiments

Cu^{2+} was selected as a representative ion for investigating the adsorption performance of Fe-CP aerogel. In detail, the Fe-CP aerogel (50 mg) was added to Cu^{2+} solutions (50 mL, pH = 6) with different concentrations of 20, 40, 60, 100, 150, and 200 mg/L to analyze the adsorption isotherm. After shaking for 24 h at 25 °C, the concentration of Cu^{2+} in the solution was measured when it reached the adsorption equilibrium. The isotherm model and maximum adsorption quantity of Cu^{2+} adsorbed into Fe-CP aerogel were studied. Langmuir (Equation (1)) and Freundlich (Equation (2)) models were used to fit the data and further the adsorption processes.

$$c_e/q_e = 1/K_L q_{\max} + c_e/q_{\max} \quad (1)$$

$$q_e = k_F {c_e}^{1/n} \quad (2)$$

where C_e (mg/L) and q_e (mg/g) are the concentration and adsorption capacity at equilibrium. k_L (L·mg^{-1}) is the Langmuir constant, and n and k_F (mg$^{1-1/n}$·L$^{1/n}$·g^{-1}) are constants related to the intensity and capacity of the absorption process.

To investigate the adsorption kinetics, the Fe-CP aerogel (50 mg) was immersed into Cu^{2+} solution (50 mL, 100 mg/L) under stirring conditions. A certain amount of solution was taken out at different time intervals to measure the concentration of Cu^{2+}. The adsorption kinetics of Cu^{2+} onto Fe-CP aerogel were simulated with pseudo-first-order (Equation (3)), pseudo-second-order (Equation (4)), and intra-particle diffusion (Equation (5)) kinetics models, respectively.

$$q_t = q_e\left(1 - e^{-k_1 t}\right) \quad (3)$$

$$t/q_t = 1/k_2{q_e}^2 + t/q_e \quad (4)$$

$$q_t = k_3 t^{0.5} + C \quad (5)$$

where q_e (mg/g) represents the adsorption quantity in equilibrium, k_1 (min^{-1}) and k_2 (g/(mg/min)) are the related rate constants. q_t (mg/g) is the adsorption quantity at time t. k_3 (mg/G·$h^{1/2}$) is the rate constant, and C is the intercept that is calculated from the plotting q_e vs. $t^{1/2}$.

2.5. Separation Performance of Fe-CP Aerogel for Oil/Water and Dye Solution

The oil/water and dye wastewater separation performance of the obtained Fe-CP aerogel was investigated by using a filtering apparatus. The Fe-CP aerogel was horizontally fixed in a glass funnel, and then simulated oily and dye wastewaters were poured into the glass funnel and separated by gravitational force. Ligroin, n-hexane, and kerosene (1 mL) were added into the water (99 mL) containing 10 mg Tween80 and stirred for 12 h to obtain stable oil/water emulsion. MB and RB solution with a concentration of 10 mg/L was prepared as the dye wastewater. The Fe-CP aerogel was horizontally fixed in a glass funnel, and then simulated oily and dye wastewaters were poured into the glass funnel and separated by gravitational force. During the gravity filtration test, the feed solution was added repeatedly to maintain a constant quantity and thus kept a constant flow rate. After filtration for 30 min, the flow rate was calculated. The flux (J, L/(m^2·h) was obtained using Equation (6).

$$Flux = V/S \times T \quad (6)$$

where S represents the actual area during separation, V is the volume of the permeate solution, and T is the time of separation process.

2.6. Photo-Fenton Catalytic Performance of the Fe-CP Aerogel

MB was selected as a model organic pollutant to assess the photo-Fenton catalytic capability of the Fe-CP aerogel. Specifically, the as-fabricated Fe-CP aerogel (50 mg) was added to the MB solution (50 mL, 20 mg/L) and placed in the dark condition for 90 min to achieve the adsorption equilibrium. Subsequently, after the addition of H_2O_2 (50 µL), the photo-Fenton reaction was carried out under simulated sunlight irradiation. In this process, every 10 min, the MB solution (2 mL) was collected and then the concentration measured using a UV–vis spectrophotometer.

Microorganisms widely exist in natural water bodies. Thus, the anti-biofouling property of the Fe-CP aerogel was further investigated by choosing *E. coli* as target microorganisms. *E. coli* was incubated at 37 °C in Luria-Bertani medium overnight and then diluted by fresh medium. The Fe-CP aerogel samples (2 × 2 cm^2) were immersed in the biological solution for 24 h. Then, the Fe-CP aerogel was taken out and treated with photo-Fenton, and the *E. coli* on the Fe-CP aerogel were observed by SEM.

3. Results and Discussion

3.1. Characterization of the Microalgae Aerogel

Figure 2a shows the morphology of CP cells after a hydrothermal treatment in an alkaline condition at a high temperature, which exhibit a shrink morphology with an average size of 3 µm. During this process, the protoplast inside the CP cells was removed, and hollow cell walls whose main component was cellulose were obtained. The amorphous regions of cellulose will undergo alkaline degradation through a peeling-off reaction under high pH conditions [29,30]. As shown in Figure 2b,c, many particles could be observed, which was due to the formation of Fe_xO_y inside the hollow shell. Hence, the porous shell mainly consisting of cellulose and nano-cores consisting of Fe_xO_y were formed simultaneously during the hydrothermal treatment. Then, the Fe-CP aerogel was obtained by freeze-drying. Figure 2d–f exhibit the structure of the Fe-CP aerogel as observed by SEM at different magnifications. The Fe-CP aerogel showed a porous structure with a pore size ranging from about 40 to 60 µm (Figure 2d). Moreover, the prepared Fe-CP aerogel can be easily placed on top of the flower without destroying it, suggesting the ultra-light mass of the Fe-CP aerogel (photograph, inset of Figure 2d).

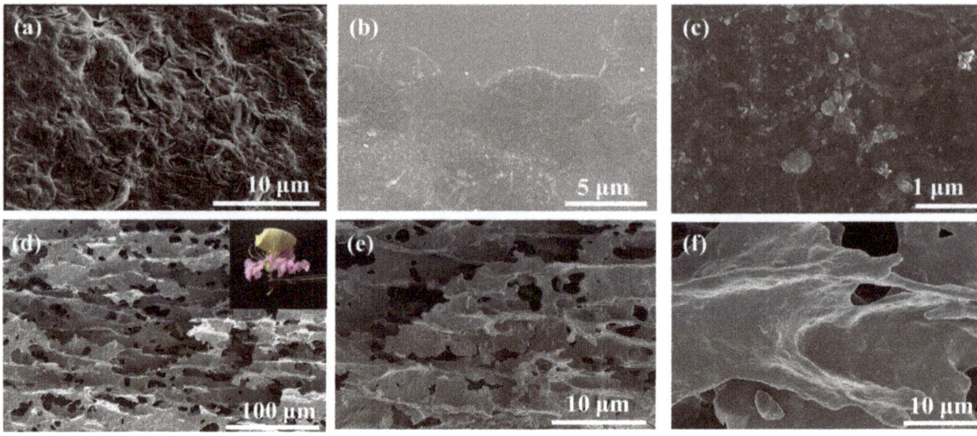

Figure 2. SEM images of the (**a**–**c**) CP cells after hydrothermal treatment under an alkaline medium and (**d**–**f**) the prepared Fe-CP aerogel.

The wettability of the Fe-CP aerogel was illustrated in Figure 3a. The water droplet was rapidly and completely adsorbed by the Fe-CP aerogel, which only took 20 microseconds, and the final water contact angle was 0°, implying that the Fe-CP aerogel possessed excellent hydrophilic properties. Figure 3b exhibits the XRD patterns of the Fe-CP aerogel. The peaks at 34.3°, 37.9°, and 40.3° corresponded to the Fe_2O_3 crystals (PDF# 76-1821), and the two peaks at 31.9° and 32.9° were due to the existence of Fe_3O_4 (PDF# 76-0958) [31].

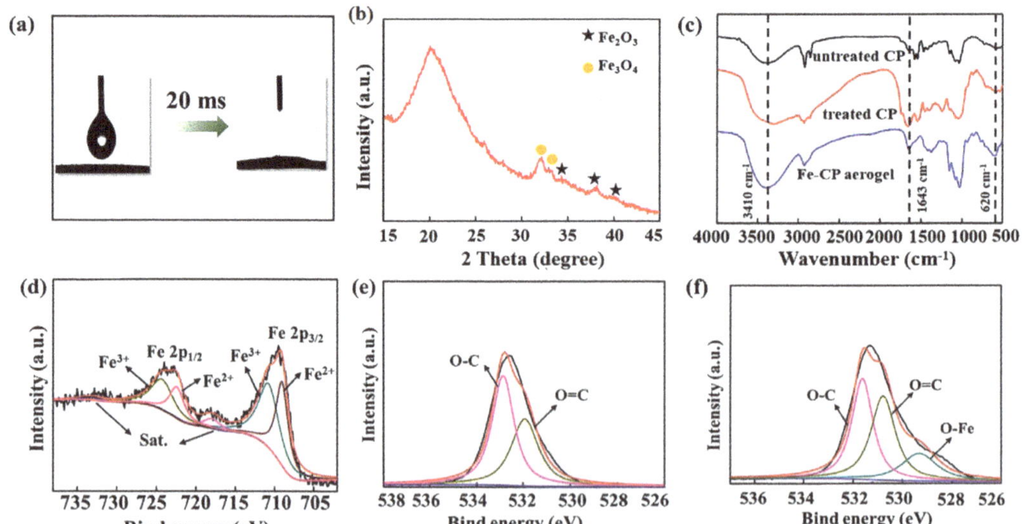

Figure 3. Characterization of the Fe-CP aerogel. (**a**) Wettability, (**b**) XRD, and (**c**) FTIR spectra of the Fe-CP aerogel. XPS patterns of (**d**) Fe 2p and (**e**) O 1s of the Fe-CP aerogel, and (**f**) O 1s of untreated CP.

The FTIR spectra of untreated CP, treated CP, and Fe-CP aerogel are exhibited in Figure 3c. The absorption peak at 3410 cm^{-1} was ascribed to the -OH groups, which became broad for the treated CP and Fe-CP aerogel. The peaks at 2924 cm^{-1} and 1384 cm^{-1} were related to the symmetric bending of the -CH or -CH_2 groups [32]. Compared with untreated CP, the peak at 1643 cm^{-1} became stronger, indicating the increased carboxylic acids after NaOH treatment. Correspondingly, the antisymmetric stretching vibration of the -CH_2 groups at 2924 cm^{-1} was weakened. Moreover, the peak at 620 cm^{-1} was attributed to the Fe-O stretching vibrations, which confirm the form of the Fe_xO_y in the treated CP and Fe-CP aerogel [33].

The chemical environment of the Fe-CP aerogel was further investigated by XPS. Figure 3d exhibits the high-resolution XPS spectra of the Fe element. The fitting characteristic peaks at 710.56 eV and 724.12 eV corresponded to the Fe^{2+}; meanwhile, the peaks at 713.45 eV and 726.25 eV were attributed to the Fe^{3+} [34,35]. In the O 1s spectrum (Figure 3e,f), the peaks at 531.9 eV and 532.8 eV were attributed to the O=C and O-C bonds, respectively. The new peak at 531.52 eV for the O-Fe bond confirmed the formation of Fe_xO_y in the Fe-CP aerogel [36].

3.2. Adsorption Performance of Microalgae Aerogel for Heavy Metal Ion

As exhibited in Figure 4a, the adsorption capacity of the Fe-CP aerogel for Cu^{2+} increased with the increased concentrations and then reached a plateau. The Freundlich and Langmuir models were used to fit the adsorption isotherm data (Figure 4c), with the higher fitting constant of the Langmuir model (0.975) than the Freundlich model (0.897) suggesting that the adsorption process belonged to monolayer adsorption (Table 1) [37]. Accordingly, the maximum adsorption capacity was calculated to be 208.3 mg/g. The adsorption kinetics were investigated through the time-dependent adsorption experiments. Figure 4b shows that the adsorption process presented a fast rate for the Fe-CP aerogel, particularly in the first 60 min, and then achieved adsorption equilibrium within 240 min. Moreover, the adsorption kinetics of Cu^{2+} on the Fe-CP aerogel could fit the pseudo-second-order model better, with an adsorption rate constant (k_2) of 0.47×10^{-3} g $mg^{-1}min^{-1}$ (Figure 4d), suggesting that the Cu^{2+} adsorption into the Fe-CP aerogel was a chemisorption process.

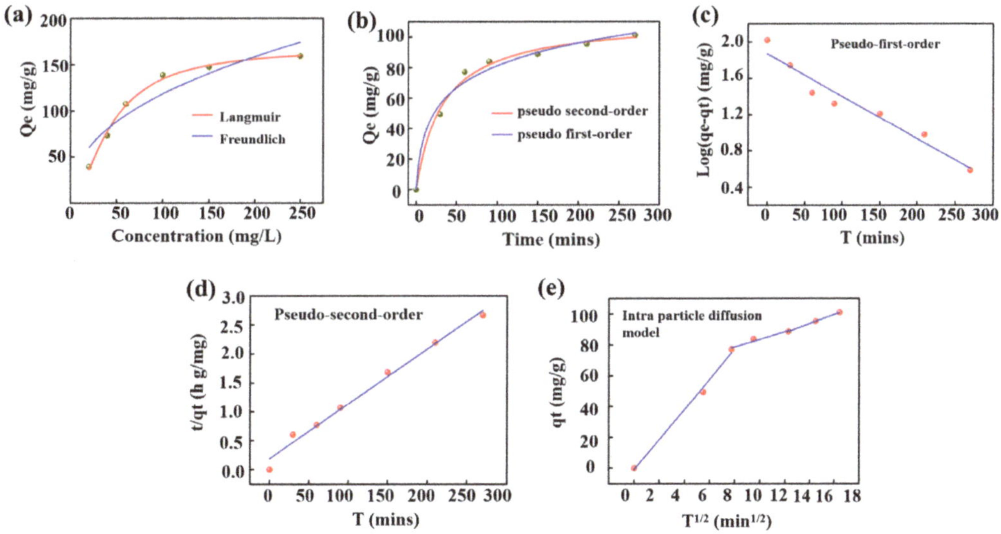

Figure 4. The influence of (**a**) concentration and (**b**) contact time on the absorption capacity of Fe-CP aerogel for Cu^{2+}. (**c**) The pseudo-first-order kinetics model, (**d**) the pseudo-second-order kinetics model, and (**e**) intra-particle diffusion plots for Cu^{2+} adsorption onto Fe-CP aerogel.

Table 1. Isotherm model parameters for adsorption of Cu^{2+} by Fe-CP aerogels.

Kinetics Models	Parameters	
Langmuir	Q_m (mg/g)	208.3
	K_L (L/mg)	0.015
	R^2	0.975
Freundlich	1/n	0.554
	K_f (L/mg)	9.124
	R^2	0.897

The diffusion mechanisms of the Cu^{2+} adsorption process were further studied through the Weber–Morris intraparticle diffusion kinetics model (Figure 4e). Three parts were observed during the adsorption process; thus, the adsorption behavior of Cu^{2+} onto the Fe-CP aerogel was affected by more than one step. The relative association constants and criterion values (R^2) of the intraparticle diffusion kinetics model were summarized in Table 2. The first part exhibited a higher slope, indicating that the Cu^{2+} diffusion to the surface of the Fe-CP aerogel was governed by the film diffusion under the driving force of the concentration difference. Moreover, the first part line avoided the origin ($c \neq 0$), implying that this process was affected by the resistance of the boundary layer [38]. The second part could be associated with the transportation of Cu^{2+} from the external surface into the inner mesopores of the Fe-CP aerogel, which was mainly controlled by the chemical reactions between Cu^{2+} and the adsorption site. In the third part, the line exhibited a small slope, which indicated that the intra-particle diffusion slowed down and reached the equilibrium stage because of the reduced Cu^{2+} concentration in the solution. In summary, the adsorption rate of Cu^{2+} onto the Fe-CP aerogel was mainly controlled by film diffusion and intra-particle diffusion.

Table 2. Parameters of kinetics models for adsorption of Cu^{2+} by Fe-CP aerogels.

Kinetics Models	Parameters		
Pseudo-first-order	Q_e (mg/g)		105.55
	K_1 (min^{-1})		0.011
	R^2		0.948
Pseudo-second-order	Q_e (mg/g)		105.26
	K_2 ($g.mg^{-1}.min^{-1}$)		0.00047
	R^2		0.987
Intra-particle diffusion	Stage 1	K_3	9.799
		C	−0.949
		R^2	0.989
	Stage 2	K_3	2.486
		C	58.902
		R^2	0.899
	Stage 3	K_3	2.951
		C	52.657
		R^2	0.999

The adsorption capacity of the Fe-CP aerogel for Cu^{2+} was compared with the aerogel-based adsorbents reported by the other literature, as summarized in Table 3. The Fe-CP aerogel exhibited superior adsorption capacity for Cu^{2+} compared to the other aerogel-based adsorbents apart from Alg-PEI. In this study, the oxygen-containing groups in the CP and Fe_xO_y can act as electron donors and transfer electrons to heavy metal ions, thus forming ionic and covalent bonds. Although some aerogel-based adsorbents exhibit higher adsorption quantity than that of the microalgae-based aerogel in our study, many of them were prepared through complex preparation processes and expensive raw materials.

Table 3. Comparison of the maximum adsorption capacity of aerogels for copper ions in related studies.

Aerogel Samples	Initial Concentration	pH	Adsorbent Dose	Q_{max} (mg.g^{-1})	Ref.
PEI@wood aerogel	150 mg/L	5	1000 mg/L	59.80	[39]
ZIF-67/BC/CH	N/A	6	1000 mg/L	200.06	[40]
CS-MMT2	10 mg/L	6	1000 mg/L	86.95	[41]
Alg-PEI	0.1 mM	1–7	5000 mg/L	214.40	[42]
CGP3	100 mg/L	3–6.2	2000 mg/L	163.40	[43]
Chitosan aerogels	100 mg/L	N/A	1500 mg/L	116.70	[44]
PEI modified CCS aerogel	50 mg/L	6	400 mg/L	175.56	[45]
NFC/PEI	20 mg/L	5	300 mg/L	175.44	[46]
Fe-CP aerogel	50 mg/L	6	1000 mg/L	208.30	This work

3.3. Oil/Water Emulsion and Dye Wastewater Separation Capacity of Fe-CP Aerogel

The oil/water emulsion separation ability of the Fe-CP aerogel was investigated by separating different oil (ligroin, n-hexane, and kerosene)/water mixtures. As exhibited in Figure 5b, the Fe-CP aerogel displayed a separation efficiency higher than 99.9% for three different oil/water mixtures. Once the emulsion is in contact with the Fe-CP aerogel, water droplets can be quickly absorbed by the Fe-CP aerogel due to the preferential affinity. Thus, the water droplets are captured by the Fe-CP aerogel from the emulsions and then permeate through the Fe-CP aerogel. In this process, a stable thin water film layer that can reject oil droplets will be formed on the surface of the Fe-CP aerogel [47]. Hence, the capillary forces and gravity resulted in the water in the oil/water emulsion continuously passing through the Fe-CP aerogel. As the oil content gradually increases, oil droplets will aggregate with increased size on the surface of the Fe-CP aerogel.

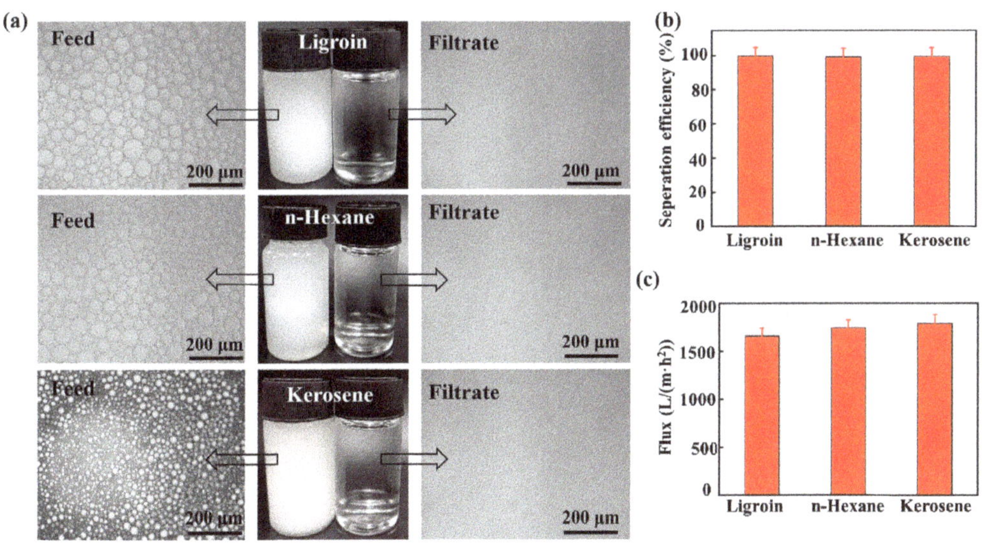

Figure 5. (**a**) The optical microscope images and the digital images of the emulsions before and after separation. (**b**) Separation efficiency and (**c**) flux of the Fe-CP aerogel for three types of oil-in-water emulsions.

There were no oil droplets observed in the transparent filtrate under the optical microscopy (Figure 5a). Moreover, the Fe-CP aerogel also possessed a high flux with a permeation flux of 1661 L/(m^2·h), 1745 L/(m^2·h), and 1794 L/(m^2·h) for water-in-ligroin, water-in-n-hexane, and water-in-kerosene. Therefore, the excellent separation efficiency as well as the flux of the Fe-CP aerogel for oil/water emulsion demonstrated the great potential for oil–water separation application.

Cationic dye (MB) and anionic dye (RB) solutions were prepared and used to study the removal capacity of the Fe-CP aerogel towards dyes. As shown in Figure 6a, the two different dye solutions can be almost completely decolored, with removal efficiency values of 99.1% and 95.9% for MB and RB. This result also accorded with the results of the UV–vis spectra (Figure 6c,d), which revealed remarkable selective adsorption towards the dyes. Therefore, the Fe-CP aerogel has the potential to be used for the adsorption or separation removal of dyes. The Fe-CP aerogel can adsorb dyes through hydrogen bonds and electrostatic interaction. The abundance of amino groups, carboxylate groups, and hydroxyl groups in the Fe-CP aerogel could interact with the N, O, and S atoms of the dyes, thus forming hydrogen bonding and the electrostatic interaction. The oxygen and nitrogen affinity on the Fe-CP aerogel could attract dyes through electrostatic forces. Moreover, the π system of the cellulose chain binds with the π electrons in the aromatic ring of the dye, resulting in π–π stacking interactions and electrostatic attractions.

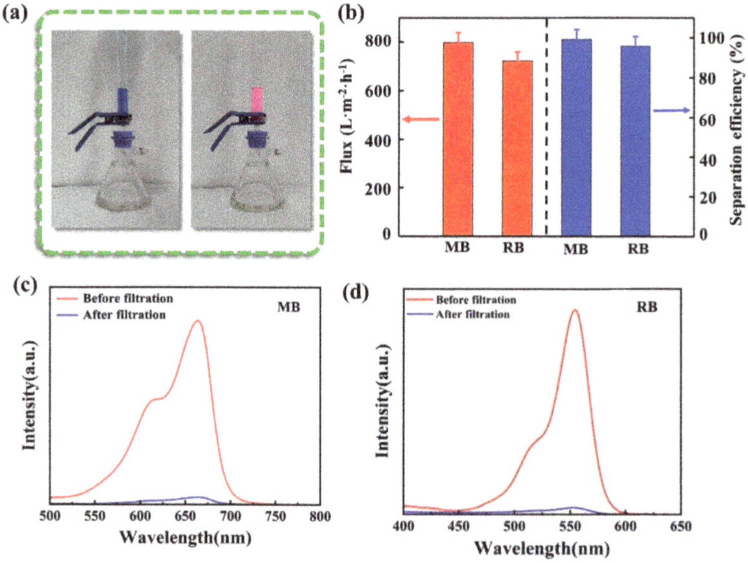

Figure 6. (**a**) The digital images of the dye waste before and after separation by Fe-CP aerogel. (**b**) Separation efficiency and flux of the Fe-CP aerogel for different dye solutions. The change in characteristic absorption peaks of (**c**) MB and (**d**) RB solutions.

3.4. Photocatalytic Capacity of Fe-CP Aerogel

The photo-Fenton catalytic ability of the Fe-CP aerogel was studied through the degradation experiment of MB. Before light irradiation, the mixture of the Fe-CP aerogel and MB solution was placed in a dark environment for 90 min with stirring to achieve the adsorption equilibrium. Figure 7a shows that the MB was degraded quickly under light irradiation. The characteristic absorption peak intensity of MB (664.8 nm) gradually decreased, and the MB almost completely degraded after irradiation for 140 min, indicating a complete MB degradation. These results indicate that the Fe-CP aerogel possessed excellent photo-Fenton degradation ability for organic dyes. Thus, the Fe-CP aerogel could be recycled by photocatalytic regeneration with high efficiency under photo-Fenton treatment.

Moreover, the antifouling ability of the Fe-CP aerogel was further evaluated by studying its antimicrobial performance. Figure 7c,d shows the microstructure of the microorganisms before and after photo-Fenton self-cleaning. Before treatment, the *E. coli* showed a rod-shaped structure. In contrast, after the photo-Fenton treatment, most of the *E. coli* cells became flattened and decomposed/distorted, indicating the antimicrobial effects of the Fe-CP aerogel through photo-Fenton catalysis. These results indicate that the Fe-CP aerogel can effectively kill microorganisms through photo-Fenton self-cleaning. Thus, the Fe-CP aerogel possessed excellent antifouling properties and would contribute to maintaining the adsorption ability in the complex water environment.

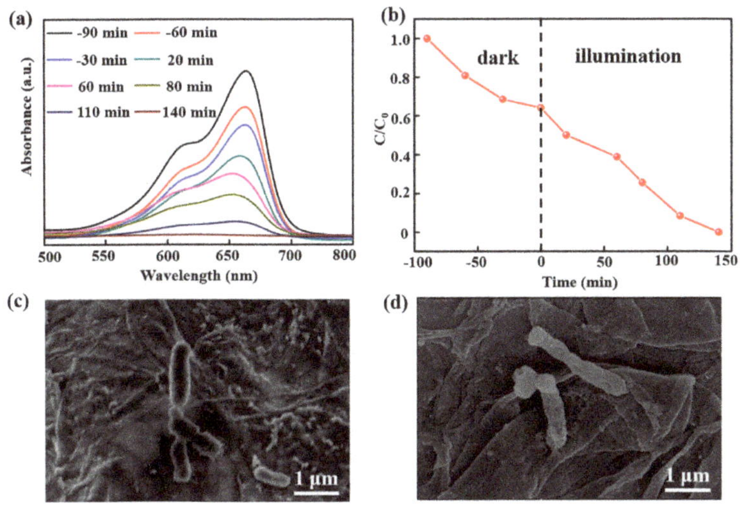

Figure 7. The photo-Fenton performance of the Fe-CP aerogel. (**a**) The changes in UV–vis spectra of MB with time during the photo-Fenton process. (**b**) The concentration changes in MB in the photo-Fenton process. The morphological changes in *E. coli* on the Fe-CP aerogel (**c**) before and (**d**) after photo-Fenton treatment.

The possible mechanisms of the photo-Fenton self-cleaning of the Fe-CP aerogel are exhibited in Figure 8. Under the visible light irradiation, the Fe_xO_y in the Fe-CP aerogel can produce electrons (^-e) and holes (h^+) [48,49]. Subsequently, the photogenerated h^+ could react with H_2O and form hydroxyl radicals (·OH). The electrons (^-e) will also react with oxygen (O_2) and produce superoxide ions (O^{2-}). The newly formed radicals with strong oxidation can degrade the contaminant and kill microorganisms and thus achieve self-cleaning.

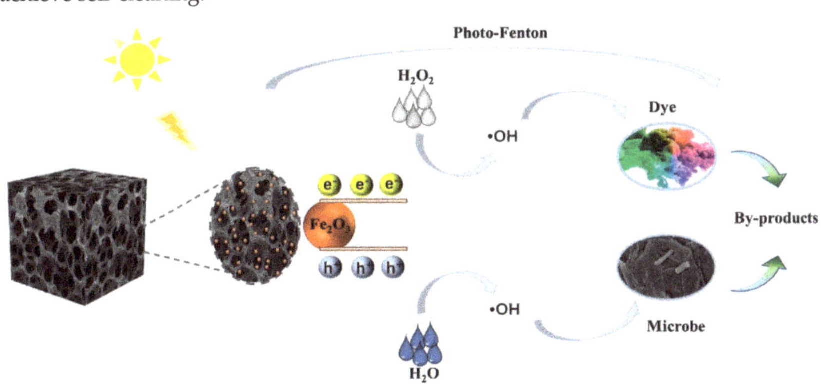

Figure 8. The mechanisms of photo-Fenton antifouling for the Fe-CP aerogel.

4. Conclusions

In summary, we prepared a novel Fe-CP aerogel as an adsorbent for the removal of pollutants from wastewater. Microspheres prepared from the iron-rich microalgae were used to prepare the Fe-CP aerogel. The Fe-CP aerogel possessed a maximum adsorption capacity of 208.3 mg/g for copper ions, and the adsorption process followed the Langmuir model. The Fe-CP aerogel also exhibited a removal efficiency of 99.1% and 95.9% for MB and RB, respectively, indicating excellent adsorption ability for dye. Due to the super-hydrophilicity, the Fe-CP aerogel can also be used for oil/water emulsions and possessed excellent separation performance. Thus, the prepared Fe-CP aerogel with rich resources, excellent adsorption capacity, and photo-Fenton self-cleaning properties demonstrated overall that it may have great potential for the practical treatment of heavy metal ions and dye wastewater.

Author Contributions: Conceptualization, X.N.; methodology, S.Z.; software, B.C.; validation, J.S. and S.Z.; formal analysis, Z.C.; investigation, B.C.; resources, J.S., B.C. and Q.W.; data curation, S.Z.; writing—original draft preparation, X.N. and J.S.; writing—review and editing, X.N. and S.Z.; visualization, J.S.; supervision, Q.W.; project administration, X.N. and Z.C.; funding acquisition, Z.C., S.Z. and Q.W. All authors have read and agreed to the published version of the manuscript.

Funding: This work was supported by the Natural Science Foundation of Fujian Province (Grant No. 2022J01942, 2023J01948); the National Innovation Demonstration Zones and Collaborative Innovation Platform Project for Fuzhou–Xiamen–Quanzhou (Grand No. 3502ZCQXT2023005); the Foreign Cooperation Projects (Industrialization) of Fujian Province (Grant No. 2023I1006, 2021I0023); University–Industry–University Cooperation Project of Fujian Province (Grant No. 2023H6036); The Young and Middle-aged Teacher Education Research Project of Fujian Province (Grant No. JAT220228); and Scientific Research Foundation of Fujian University of Technology (No. GY-Z20039, GY-Z220189).

Data Availability Statement: The data presented in this study are available on request from the corresponding author.

Conflicts of Interest: The authors declare that they have no known competing financial interests or personal relationships that may appear to influence the work reported in this paper.

References

1. Li, X.; Liu, Y.; Zhang, C.L.; Wen, T.; Zhuang, L.; Wang, X.X.; Song, G.; Chen, D.Y.; Ai, Y.J.; Hayat, T.; et al. Porous Fe_2O_3 microcubes derived from metal organic frameworks for efficient elimination of organic pollutants and heavy metal ions. *Chem. Eng. J.* **2018**, *336*, 241–252. [CrossRef]
2. Le, V.G.; Nguyen, M.N.; Nguyen, H.L.; Thai, V.A.; Le, V.R.; Manh, V.Q.; Asaithambi, P.; Woong, C.S.; Nguyen, D.D. Ecotoxicological response of algae to contaminants in aquatic environments: A review. *Environ. Chem. Lett.* **2024**, *22*, 919–939. [CrossRef]
3. Liu, X.G.; Shan, Y.Y.; Zhang, S.T.; Kong, Q.Q.; Pang, H. Application of metal organic framework in wastewater treatment. *Green Energy Environ.* **2023**, *8*, 698–721. [CrossRef]
4. Wei, Z.M.; Su, Q.; Lin, Q.Y.; Wang, X.J.; Long, S.R.; Zhang, G.; Yang, J. Multifunctional oxidized poly (arylene sulfide sulfone)/UiO-66 nanofibrous membrane with efficient adsorption/separation ability in harsh environment. *Chem. Eng. J.* **2022**, *430*, 133021. [CrossRef]
5. Fajal, S.; Dutta, S.; Ghosh, S.K. Porous organic polymers (POPs) for environmental remediation. *Mater. Horiz.* **2023**, *10*, 4083–4138. [CrossRef] [PubMed]
6. Wang, J.L.; Zhuang, S.T. Removal of cesium ions from aqueous solutions using various separation technologies. *Rev. Environ. Sci. Bio/Technol.* **2019**, *18*, 231–269. [CrossRef]
7. Vinothkumar, K.; Jyothi, M.S.; Lavanya, C.; Sakar, M.; Valiyaveettil, S.; Balakrishna, R.G. Strongly co-ordinated MOF-PSF matrix for selective adsorption, separation and photodegradation of dyes. *Chem. Eng. J.* **2022**, *428*, 132561. [CrossRef]
8. Zhang, L.P.; Liu, Z.; Zhou, X.L.; Zhang, C.; Cai, Q.W.; Xie, R.; Ju, X.J.; Wang, W.; Faraj, Y.; Chu, L.Y. Novel composite membranes for simultaneous catalytic degradation of organic contaminants and adsorption of heavy metal ions. *Sep. Purif. Technol.* **2019**, *237*, 116364. [CrossRef]
9. Chen, X.; Cui, J.; Xu, X.R.; Sun, B.J.; Zhang, L.; Dong, W.; Chen, C.T.; Sun, D.P. Bacterial cellulose/attapulgite magnetic composites as an efficient adsorbent for heavy metal ions and dye treatment. *Carbohydr. Polym.* **2020**, *229*, 115512. [CrossRef] [PubMed]
10. Elattar, R.H.; El-Deen, A.K. Porous material-based quechers: Exploring new horizons in sample preparation. *Trends Anal. Chem.* **2024**, *172*, 117571. [CrossRef]

11. Liu, Y.L.; Jin, C.; Yang, Z.Z.; Wu, G.M.; Liu, G.F.; Kong, Z.W. Recent advances in lignin-based porous materials for pollutants removal from wastewater. *Int. J. Biol. Macromol.* **2021**, *187*, 880–891. [CrossRef]
12. Lim, Y.J.; Goh, K.; Goto, A.; Zhao, Y.L.; Wang, R. Uranium and lithium extraction from seawater: Challenges and opportunities for a sustainable energy future. *J. Mater. Chem. A* **2023**, *11*, 22551–22589. [CrossRef]
13. Liu, H.W.; Li, X.Y.; Zhang, X.X.; Coulon, F.; Wang, C.Q. Harnessing the power of natural minerals: A comprehensive review of their application as heterogeneous catalysts in advanced oxidation processes for organic pollutant degradation. *Chemosphere* **2023**, *337*, 139404. [CrossRef]
14. He, W.; Cao, J.Z.; Guo, F.Y.; Guo, Z.H.; Zhou, P.G.; Wang, R.; Liang, S.; Pang, Q.Y.; Wei, B.R.; Jiao, Y.; et al. Nanostructured carboxylated-wood aerogel membrane for high-efficiency removal of Cu (II) ions from wastewater. *Chem. Eng. J.* **2023**, *468*, 143747. [CrossRef]
15. Hu, X.D.; Zhang, T.Y.; Yang, B.; Hao, M.; Chen, Z.J.; Wei, Y.; Liu, Y.B.; Wang, X.X.; Yao, J.B. Water-resistant nanocellulose/gelatin biomass aerogel for anionic/cationic dye adsorption. *Sep. Purif. Technol.* **2024**, *330*, 125367. [CrossRef]
16. Chew, K.W.; Khoo, K.S.; Foo, H.T.; Chia, S.R.; Walvekar, R.; Lim, S.S. Algae utilization and its role in the development of green cities. *Chemosphere* **2021**, *268*, 129332. [CrossRef]
17. Xu, Y.A.; Hellier, P.; Purton, S.; Baganz, F.; Ladommatos, N. Algal biomass and diesel emulsions: An alternative approach for utilizing the energy content of microalgal biomass in diesel engines. *Appl. Energy* **2016**, *172*, 80–95. [CrossRef]
18. Daneshvar, E.; Zarrinmehr, M.J.; Kousha, M.; Hashtjin, A.M.; Saratale, G.D.; Maiti, A.; Vithanage, M.; Bhatnagar, A. Hexavalent chromium removal from water by microalgal-based materials: Adsorption, desorption and recovery studies. *Bioresour. Technol.* **2019**, *293*, 122064. [CrossRef]
19. Francavilla, M.; Pineda, A.; Lin, C.S.; Franchi, M.; Trotta, P.; Romero, A.A.; Luque, R. Natural porous agar materials from macroalgae. *Carbohydr. Polym.* **2013**, *92*, 1555–1560. [CrossRef] [PubMed]
20. Shaikh, W.A.; Chakraborty, S.; Islam, R.U.; Ghfa, A.A.; Naushad, M.; Bundschuh, J.; Maity, J.P.; Mondal, N.K. Fabrication of biochar-based hybrid Ag nanocomposite from algal biomass waste for toxic dye-laden wastewater treatment. *Chemosphere* **2022**, *289*, 133243. [CrossRef]
21. Hu, X.; Liu, W.J.; Ma, L.L.; Yu, H.Q. Sustainable Conversion of Harmful Algae Biomass into a CO_2 Reduction Electrocatalyst for Two-Fold Carbon Utilization. *Environ. Sci. Technol.* **2023**, *57*, 1157–1166. [CrossRef] [PubMed]
22. Wang, S.; Lu, W.M.; Esakkimuthu, S.; Chen, H.; Yang, J.W.; Mu, M.; Gong, X. Life cycle assessment of carbon-based adsorbent preparation from algal biomass. *J. Clean. Prod.* **2023**, *427*, 139269. [CrossRef]
23. Ji, X.Y.; Luo, X.; Zhang, J.B.; Huang, D.Y. Effects of exogenous vitamin B12 on nutrient removal and protein expression of algal-bacterial consortium. *Environ. Sci. Pollut. Res.* **2021**, *28*, 15954–15965. [CrossRef] [PubMed]
24. Baskar, A.V.; Bolan, N.; Hoang, S.A.; Sooriyakumar, P.; Kumar, M.; Singh, L.; Jasemizad, T.; Padhye, L.P.; Singh, G.; Vinu, A.; et al. Recovery, regeneration and sustainable management of spent adsorbents from wastewater treatment streams: A review. *Sci. Total Environ.* **2022**, *822*, 153555. [CrossRef]
25. Zhu, S.Y.; Xu, J.; Wang, B.; Xie, J.X.; Ying, G.D.; Li, J.P.; Cheng, Z.; Li, J.; Chen, K.F. Highly efficient and rapid purification of organic dye wastewater using lignin-derived hierarchical porous carbon. *J. Colloid Interface Sci.* **2022**, *625*, 158–168. [CrossRef] [PubMed]
26. Jiang, X.Y.; Wang, H.Q.; Hu, E.; Lei, Z.W.; Fan, B.L.; Wang, Q.L. Efficient adsorption of uranium from aqueous solutions by microalgae based aerogel. *Microporous Mesoporous Mater.* **2020**, *305*, 110383. [CrossRef]
27. Tee, G.T.; Gok, X.Y.; Yong, W.F. Adsorption of pollutants in wastewater via biosorbents, nanoparticles and magnetic biosorbents: A review. *Environ. Res.* **2022**, *212*, 113248. [CrossRef] [PubMed]
28. Prasad, C.; Jeong, S.G.; Won, J.S.; Ramanjaneyulu, S.; Sangaraju, S.; Kerru, N.; Choi, H.Y. Review on recent advances in cellulose nanofibril based hybrid aerogels: Synthesis, properties and their applications. *Int. J. Biol. Macromol.* **2024**, *261*, 129460. [CrossRef]
29. Bi, L.; Luan, X.; Geng, F.L.; Xu, X.L.; Chen, Y.P.; Zhang, F. Microwave-assisted synthesis of hollow microspheres with multicomponent nanocores for heavy-metal removal and magnetic sensing. *ACS Appl. Mater. Interfaces* **2020**, *12*, 46779–46787. [CrossRef]
30. Liu, X.; Wu, Z.C.; Han, Y.H.; Han, L.J. Characteristic modification of alkalized corn stalk and contribution to the bonding mechanism of fuel briquette. *Energy* **2017**, *133*, 299–305. [CrossRef]
31. Tu, F.; Yu, Y.X.; Wang, Y.; Huang, L.Y.; Ye, D.H.; Fu, Z.Y. Preparation of SiO_2/Fe_2O_3 composite aerogels for thermal insulation enhancement. *Ceram. Int.* **2024**, *50*, 2976–2986. [CrossRef]
32. Han, X.H.; Ding, S.Q.; Zhu, L.J.; Wang, S.R. Preparation and characterization of flame-retardant and thermal insulating bio-based composite aerogels. *Energy Build.* **2023**, *278*, 112656. [CrossRef]
33. Abdullah, J.A.; Jiménez-Rosado, M.; Perez-Puyana, V.; Guerrero, A.; Romero, A. Green synthesis of Fe_xO_y nanoparticles with potential antioxidant properties. *Nanomaterials* **2022**, *12*, 2449. [CrossRef]
34. Grosvenor, A.P.; Kobe, B.A.; Biesinger, M.C.; McIntyre, N.S. Investigation of multiplet splitting of Fe 2p XPS spectra and bonding in iron compounds. *Surf. Interface Anal.* **2004**, *36*, 1564–1574. [CrossRef]
35. Ateia, E.E.; Saeid, Y.A.; Abdelmaksoud, M.K. Synthesis and characterization of $CaFe1.925Gd0.025Sm0.05O_4$/PEG core–shell nanoparticles for diverse applications. *J. Supercond. Nov. Magn.* **2023**, *36*, 1011–1024. [CrossRef]

36. Wang, T.N.; Jiao, Y.H.; He, M.C.; Peng, X.J.; Liu, P.; Ouyang, W.; Lin, C.Y.; Liu, X.T.; Xie, H.J. Efficient removal of antimony by a facile liquid-controlled strategy reinforced hematite-spinel (Fe_2O_3-$MnFe_2O_4$) composite: Construction, simulation and practical evaluation. *Chem. Eng. J.* **2023**, *451*, 138974. [CrossRef]
37. Huo, J.B.; Yu, G.C. Hexacyanoferrate-modified polyvinyl alcohol/graphene oxide aerogel as an efficient and retrievable adsorbent for cesium. *J. Mater. Sci.* **2022**, *57*, 351–365. [CrossRef]
38. Li, Q.M.; Fu, L.C.; Wang, Z.; Li, A.; Shuang, C.D.; Gao, C.Z. Synthesis and characterization of a novel magnetic cation exchange resin and its application for efficient removal of Cu^{2+} and Ni^{2+} from aqueous solutions. *J. Clean. Prod.* **2017**, *165*, 801–810. [CrossRef]
39. Meng, J.W.; Guan, H.; Dai, X.J.; Wang, X.Q. Amino-functionalized wood aerogel for efficient removal of copper ions from water. *Int. J. Polym. Sci.* **2021**, *8*, 4913226.
40. Li, D.W.; Tian, X.J.; Wang, Z.Q.; Guan, Z.; Li, X.Q.; Qiao, H.; Ke, H.Z.; Wei, Q. Multifunctional adsorbent based on metal-organic framework modified bacterial cellulose/chitosan composite aerogel for high efficient removal of heavy metal ion and organic pollutant. *Chem. Eng. J.* **2020**, *383*, 123127. [CrossRef]
41. Ye, X.; Shang, S.S.; Zhao, Y.F.; Cui, S.; Zhong, Y.; Huang, L.J. Ultra-efficient adsorption of copper ions in chitosan–montmorillonite composite aerogel at wastewater treatment. *Cellulose* **2021**, *28*, 7201–7212. [CrossRef]
42. Wang, M.; Yang, Q.; Zhao, X.Q.; Wang, Z.Q. Highly efficient removal of copper ions from water by using a novel alginate-polyethyleneimine hybrid aerogel. *Int. J. Biol. Macromol.* **2019**, *138*, 1079–1086. [CrossRef]
43. Tang, C.X.; Brodie, P.; Li, Y.Z.; Grishkewich, N.J.; Brunsting, M.; Tam, K.C. Shape recoverable and mechanically robust cellulose aerogel beads for efficient removal of copper ions. *Chem. Eng. J.* **2020**, *392*, 124821. [CrossRef]
44. Wang, Q.Z.; Li, Y. Facile and green fabrication of porous chitosan aerogels for highly efficient oil/water separation and metal ions removal from water. *J. Environ. Chem. Eng.* **2023**, *11*, 109689. [CrossRef]
45. Wang, Q.Y.; Li, L.P.; Kong, L.C.; Cai, G.Y.; Wang, P.; Zhang, J.; Zuo, W.; Tian, Y. Compressible amino-modified carboxymethyl chitosan aerogel for efficient Cu(II) adsorption from wastewater. *Sep. Purif. Technol.* **2022**, *293*, 121146. [CrossRef]
46. Li, J.; Zuo, K.; Wu, W.B.; Xu, Z.Y.; Yi, Y.G.; Jing, Y.; Dai, H.Q.; Fang, G.G. Shape memory aerogels from nanocellulose and polyethyleneimine as a novel adsorbent for removal of Cu(II) and Pb(II). *Carbohydr. Polym.* **2018**, *196*, 376–384. [CrossRef]
47. Xu, H.; Zhang, Z.; Jiang, W.; Fang, X.R.; Xia, Z.G.; Niu, H.T.; Zhou, H. Multifunctional amphibious superhydrophilic-oleophobic cellulose nanofiber aerogels for oil and water purification. *Carbohydr. Polym.* **2024**, *330*, 121774. [CrossRef]
48. He, X.H.; Wang, L.L.; Sun, S.M.; Yang, X.H.; Tian, H.Y.; Xia, Z.J.; Li, X.L.; Yan, X.L.; Pu, X.J.; Jiao, Z. Self-assembled synthesis of recyclable g-C_3N_4/NH2-MIL-53(Fe) aerogel for enhanced photocatalytic degradation of organic pollutants. *J. Alloys Compd.* **2023**, *964*, 169391. [CrossRef]
49. Mei, J.; Yuan, G.J.; Ma, Y.S.; Chen, X.; Ren, L.L. Fe-Doped alumina aerogels as a green heterogeneous catalyst for efficient degradation of organic pollutants. *Catal. Lett.* **2019**, *149*, 1874–1887. [CrossRef]

processes

MDPI

Article

Effects of Vegetation Cover Varying along the Hydrological Gradient on Microbial Community and N-Cycling Gene Abundance in a Plateau Lake Littoral Zone

Jing Yuan [1,2,†], Jing Cao [1,2,†], Wanxue Liao [1,2], Feng Zhu [1,2,3], Zeying Hou [1,2] and Zhaosheng Chu [1,2,*]

1 National Engineering Laboratory for Lake Pollution Control and Ecological Restoration, Chinese Research Academy of Environmental Sciences, Beijing 100012, China; yuan.jing@craes.org.cn (J.Y.)
2 State Environmental Protection Key Laboratory for Lake Pollution Control, Chinese Research Academy of Environmental Sciences, Beijing 100012, China
3 College of Water Sciences, Beijing Normal University, Beijing 100875, China
* Correspondence: chuzs@craes.org.cn
† These authors contributed equally to this work.

Abstract: The lake littoral zone is periodically exposed to water due to water level fluctuations, driving the succession and distribution of littoral vegetation covers, which complexly affect nutrient biogeochemical transformation. However, the combined effects of water level fluctuations and other environmental factors on microbial characteristics and functions at the regional scale remain unclear. In this study, typical vegetation cover types along various water levels were chosen to investigate the effects of water level and vegetation cover on the microbial community and functional genes in the Lake Erhai littoral zone. The results showed that water level fluctuations influenced oxygen and nitrogen compound contents due to oxic–anoxic alternations and intensive material exchange. Meanwhile, vegetation cover affected the organic matter and total nitrogen content through plant residues and root exudation supplying exogenous carbon and nitrogen. Along the hydrological gradient, the high microbial diversity and abundant microbes related to nitrogen cycling were observed in interface sediments. It was attributed to the alternating aerobic–anaerobic environments, which filtered adaptable dominant phyla and genera. The abundances of *amoA* AOA, *nirS*, and *amx* were higher than those of the other genes and were strongly related to flooding days and water content. In conclusion, water level fluctuations and vegetation type jointly affect microbial community structure and nitrogen-related functional genes.

Keywords: lake littoral zone; water level fluctuation; vegetation type; microbial community; functional gene

1. Introduction

In recent decades, the rapid growth of agriculture and increased travel around lakesides have significantly increased nutrient runoff into lakes, particularly nitrogen and phosphorus, causing freshwater eutrophication [1,2]. The lake littoral zone, a transitional junction between terrestrial and aquatic ecosystems, acts as a crucial barrier filtering nutrient-rich runoff before it enters the lake [3]. Due to the increased nutrient input and material exchange, this transitional zone has been reported as a hotspot for nitrogen cycling, playing a vital role in controlling lake eutrophication [4]. Microorganisms are the key mediators and drivers of biogeochemical processes in lake littoral zones, responsible for nitrogen decomposition and mineralization [5,6]. Understanding their composition and function in response to the microenvironment in this ecotone is essential for advancing research in this area.

The lake littoral zone is periodically exposed to water due to water level fluctuations, which are critical in shaping the hydrological characteristics of the ecotone [7]. The dynamic hydrologic connectivity resulting from these fluctuations causes the periodic drying and rewetting of ecotone soil and sediment, creating highly heterogeneous conditions with oxic–anoxic alternations [8]. This environmental variation, coupled with changes in the microbial community, affects the degree of the biogeochemical transformation of various nitrogen compounds in the lake littoral zone [9]. A comparative study of the nitrogen-cycling microbial community in the littoral and limnetic sediments of Lake Chaohu in China concluded that microbial abundance and diversity were higher in the littoral zone, identifying it as a nitrogen-cycling hotspot [1]. Previous studies have investigated the coupling of nitrification and denitrification in the riparian zone by examining the specific distribution of microbial communities and functional genes, attributing microbial composition characteristics to water level fluctuations [5,6,10]. However, there is still no consensus on the specific microbial features that change with water level fluctuations and their exact impact. Therefore, more research is needed to understand the complex influence of water level fluctuations and other environmental factors on microbial characteristics and functions at a regional scale.

Periodic inundations in the lake littoral zone drive the succession and distribution of littoral vegetation, creating a unique vegetation landscape [11]. Different vegetation types, as direct participants in the soil environment, impact soil physicochemical properties and further affect nitrogen-cycling microorganisms in the soil and sediment. The rhizosphere of vegetation, crucial for nitrogen cycling, interacts with microorganisms through litter inputs and root exudates, affecting soil aggregate stability, water-holding capacity, and carbon sequestration [7]. Previous studies on the riparian zones of reservoirs have shown differing results, one study based on the Three Gorges Reservoir [6] and another based on the Miyun Reservoir [12]. The unique periodic fluctuations with water level variation from 145 m to 175 m in riparian zones of the Three Gorges Reservoir resulted in water level being the main factor in microbial composition and nitrogen-related gene abundance, while plant types were critical in the riparian zones of the Miyun Reservoir with 6 cm water level variation in two sampling seasons. These differences may be attributed to significant altitude variations (over 30 m) between the riparian zones of the two reservoirs, indicating that the impact of rapid and robust fluctuations in the ecotone is not yet fully understood. Therefore, a more comprehensive study of the complex environment in the lake littoral zone, including upstream and downstream areas with a more refined division, is needed.

Lake Erhai is the second largest plateau freshwater lake in China. Rapid urbanization in the lakeside area over the past few decades has increased the pollution load into the lake [13]. Due to the narrow width of the catchment area and the short surface-inflow path, the purification effectiveness of the lake littoral zone is crucial for controlling eutrophication. Since the construction of hydroelectric power plants in the 1980s, the water level of Lake Erhai has been artificially regulated. The water level is low in the summer (May to June) and high in the autumn (September to November). Over the past two decades, interannual and seasonal water level variations have been 1.0 and 1.2 m, respectively [14]. The long-term artificial regulation of water levels has significantly affected the littoral hydrological regime and environmental conditions. However, there is limited knowledge about the influence of nitrogen removal in the lake littoral zone, particularly concerning the microbial community and N-cycling functional genes.

This study aimed to investigate the impact of water level and vegetation cover on the microbial community and functional genes in the littoral zone of Lake Erhai. The research focused on nine different types of vegetation covers across three vertical spatial zones: the landward zone, the interface zone, and the waterward zone. We examined soil and sediment microenvironment properties, microbial communities, and N-cycling functional genes to (1) characterize the physicochemical properties of soil and sediment under various vegetation covers along the hydrological gradient; (2) explore the response of microbial community structure and N-cycling functional genes to water level fluctuations and vegetation types; and (3) reveal the relationships between water level fluctuations, vegetation types, soil/sediment properties, and microbial compositions.

2. Materials and Methods

2.1. Study Area and Sampling Process

Lake Erhai is a large and deep fault lake in the Yunnan Province, located on the Yunnan–Guizhou Plateau of southwest China (Figure 1a). It has an area of 252 km^2 and a maximum depth of 21 m. [14]. The lake experiences a low-latitude subtropical monsoon climate, with a latitude ranging from 25°36′ N to 25°38′ N and a water level ranging from 1966.0 m to 1964.3 m. The annual mean temperature of the basin is 15 °C, and the temperature of the lake water varies from 10 °C to 20 °C. [13]. In 2022, the water level varied by 0.55 m, ranging from 1965.5 m to 1964.9 m.

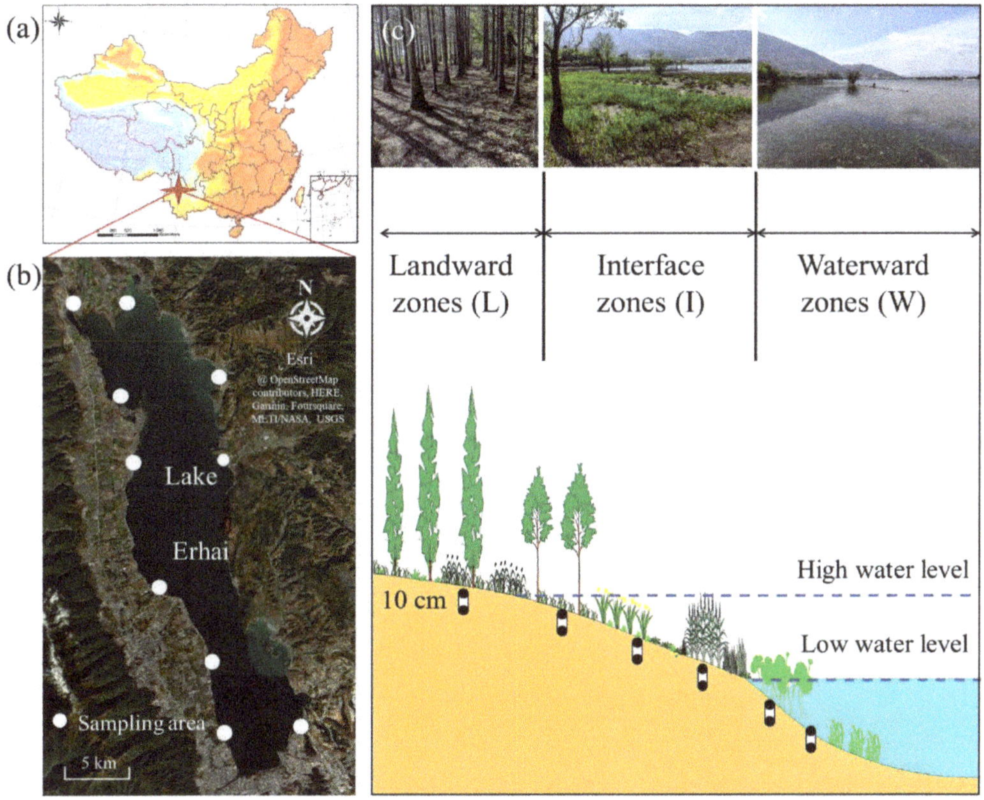

Figure 1. Schematic illustration of the sampling site's location and distribution. (**a**) The location of Lake Erhai in China. (**b**) The location of the sampling area around Lake Erhai. (**c**) The vertical distribution of the sampling sites on the lakeshore. The photographs show the typical sampling sites in various zones with certain plant types.

Ten typical sampling areas were selected surrounding Lake Erhai based on the hydrographic and topographic features of the lake, as shown in Figure 1b. Each sampling area was divided into three vertical zones based on the average annual flooding days to study the hydrological impacts of water level changes (Figure 1c). These zones were as follows: (1) landward zones (rarely flooded, i.e., no flooding in a year); (2) interface zones (intermittently flooded, i.e., 133–270 flooding days in a year); (3) waterward zones (always flooded, i.e., about 365 flooding days in a year). In each zone, the main vegetation cover was recorded and classified into nine types across the three spatial zones. In the landward zones, the three vegetation types were forest (LF), grassland (LG), and inter-forest grassland (LFG). In the interface zones, the three vegetation types were wet grassland (IG), inter-forest grassland (IFG), and emergent plants (IE). In the waterward zones, the three vegetation types were emergent plants (WE), submerged plants (WS), and no plants (WN). At each sampling site, triplicated samples were randomly collected from 1 m × 1 m quadrats. Additional details about each sampling site can be found in Table S1, and the typical photographs are shown in Figure S1.

The samples were collected in August 2022, at a time when nitrogen removal activity in the lakeshore was high [15]. Surface soil and sediment samples (0–10 cm) were collected in the landward and interface zones using a stainless steel corer, and in the waterward zones using a mud bucket. After removing stones and plant residues, each soil/sediment sample was thoroughly mixed and divided into two sub-samples. One sub-sample was stored at −20 °C for the physicochemical property test, while the other was stored at −60 °C for Illumina-MiSeq sequencing and functional gene testing.

2.2. Analysis of Soil and Sediment Physicochemical Properties

The contents of the total nitrogen (TN), total carbon (TC), ammonia (NH_4^+-N), nitrate (NO_3^--N), nitrite (NO_2^--N), soi/sediment organic matter (SOM), pH, and water content (WC) of the soil and sediment were tested as the following methods. Total nitrogen (TN) and total carbon (TC) were measured by an element analyzer (Vario EL, Elementar Analysensysteme, Hanau, Germany). The contents of NH_4^+-N, NO_3^--N, and NO_2^--N were determined by extraction with 2 M KCl followed by the indophenol blue colorimetry, copperized cadmium reduction, and naphthalene ethylenediamine methods, respectively [16]. The SOM content was determined by the oxidation of potassium dichromate ($K_2Cr_2O_7$), followed by titration with iron (II) sulfate ($FeSO_4$) [17]. The pH was measured by a pH meter (solid/water = 2.5:1). The WCs of the soil and sediment were tested by drying the fresh samples at 105 °C for 24 h [18].

2.3. DNA Extraction, Quantitative Real-Time PCR (qPCR), and Sequencing

Microbial DNA was extracted by a Soil Microbe Microprep Kit (ZYMO RESEARCHE R2040, Orange, CA, USA) from 0.5 g fresh soil or sediment samples. Quantitative real-time PCR (qPCR) was conducted to target the nitrogen-related functional genes, i.e., *amoA* AOA and *amoA* AOB for the ammonia oxidizers, *narG* and *napA* for the nitrate reducers, *nirS*, *nirK*, and *nosZ* for the denitrifiers, and 16S rRNA for ANAMMOX bacteria (*amx*), and the primers are listed in Table S2. All the qPCRs were conducted using an instrument (BIO-RAD CFX96, Hercules, CA, USA) with AceQ qPCR SYBR Green Master Mix (Jizhenbio JZ121-02, Shanghai, China). The qPCR reactions were conducted in triplicate with mixtures containing 7.5 μL of SYBR Green Mix (2×), 0.7 μL of primers (each 10 μM), 1 μL of gNDA, and 15 μL ddH_2O. The thermal cycling was conducted by following steps: 95 °C for 5 min, 40 cycles of 95 °C for 10 s, and 60 °C for 30 s. In this study, the qPCR amplification efficiencies were from 93.96% to 104.77%, with the R^2 of calibration curves ≥99%. The Illumina-MiSeq sequencing was conducted by steps described in a previous study by Yuan et al. [3].

2.4. Statistical Analyses

Origin 2021 was employed to calculate Spearman rank correlation coefficients, conduct principal component analysis (PCA) with Bray–Curtis distances, and characterize bar graphs and heatmaps. SPSS 23.0 was used to perform the ANOVA and *t*-test to assess the difference [19]. Canoco 5.0 was used to perform the constrained ordination of redundancy analysis (RDA). Cytoscape 3.9.0 was employed to visualize the network analysis [20]. Mantel test analysis was conducted using the vegan 2.4.3 packages in R 3.6.1 [21].

3. Results

3.1. Physicochemical Characteristics

Various environmental factors in the soil and sediment were measured to investigate their physicochemical characteristics under various water levels and vegetation types, including the contents of total nitrogen (TN), ammonia (NH_4^+-N), nitrate (NO_3^--N), nitrite (NO_2^--N), soi/sediment organic matter (SOM), the ratio of TC to TN (C/N), pH, and water content (WC), as summarized in Table 1. For the distribution of nitrogen, the NH_4^+-N and NO_3^--N contents in the interface sediments were the lowest, while the NO_2^--N and TN were similar to those in the landward soils and waterward sediments, respectively. The SOM contents followed the order of waterward < interface < landward, and the C/N ratios in landward soils and interface sediments were similar, but lower than those in the waterward sediments. Regarding the effects of vegetation type, the TN and SOM contents in the landward soils and interface sediments followed a similar order: grassland (LG and IG) > inter-forest grassland (LFG and IFG) > forest and emergent plants (LF and IE). In the landward soils, the TN and SOM contents were significantly different between grassland (LG) and other plant types (LF and LFG) ($p < 0.05$). In the waterward sediments, the NO_3^--N contents varied significantly between the sites with or without plants ($p < 0.05$), but there was little difference between various vegetation types ($p > 0.05$). Additionally, the NH_4^+-N and NO_3^--N contents in the waterward sediments without plants (WN) were the highest, and the NO_2^--N contents in the inter-forest grassland soils in landward (LFG) were the highest. The water contents in the landward soils (0.32 on average) were lower than in the interface and waterward sediments (both 0.45 on average), and varied significantly between up and down the ordinary water level ($p < 0.05$), but showed little variation along the water depth ($p > 0.05$). The water content under various vegetation covers in the interface sediments showed little variation ($p > 0.05$), while it varied significantly in the landward soils and waterward sediments ($p < 0.05$). The pH of the soils and sediments ranged from 6.16 to 7.42, indicating a slightly acidic state, with little spatial difference across the sampling sites ($p > 0.05$).

Table 1. Physicochemical characteristics of soil and sediment in different sampling sites.

Samples	NH_4^+-N (mg kg^{-1})	NO_3^--N (mg kg^{-1})	NO_2^--N (mg kg^{-1})	TN (mg g^{-1})	SOM (mg g^{-1})	C/N	pH	WC
LF	80.54 ± 43.0^{ab}	12.41 ± 3.8^{b}	25.83 ± 10.4^{a}	1.32 ± 0.5^{bc}	21.06 ± 6.7^{bc}	16.33 ± 7.8^{a}	6.62 ± 0.3^{a}	0.25 ± 0.1^{b}
LFG	42.70 ± 25.5^{b}	7.14 ± 3.8^{b}	34.57 ± 34.61^{a}	2.72 ± 0.8^{abc}	50.24 ± 11.7^{bc}	12.91 ± 1.4^{a}	6.89 ± 0.5^{a}	0.34 ± 0.1^{b}
LG	80.46 ± 45.3^{ab}	28.69 ± 26.4^{ab}	21.87 ± 7.5^{a}	7.01 ± 3.8^{a}	128.50 ± 71.1^{a}	13.72 ± 2.4^{a}	7.03 ± 0.2^{a}	0.35 ± 0.2^{b}
IE	57.81 ± 19.2^{ab}	8.38 ± 7.2^{b}	16.95 ± 5.3^{a}	1.37 ± 1.0^{bc}	26.22 ± 18.0^{bc}	14.63 ± 3.7^{a}	6.95 ± 0.4^{a}	0.43 ± 0.0^{ab}
IFG	76.39 ± 23.5^{ab}	13.77 ± 10.5^{b}	16.80 ± 3.4^{a}	2.26 ± 0.9^{abc}	44.79 ± 20.5^{bc}	13.52 ± 1.3^{a}	6.89 ± 0.4^{a}	0.40 ± 0.1^{b}
IG	39.90 ± 10.4^{b}	14.05 ± 18.4^{b}	15.46 ± 2.8^{a}	5.12 ± 2.1^{ab}	88.66 ± 39.5^{ab}	14.02 ± 2.7^{a}	6.83 ± 0.2^{a}	0.51 ± 0.2^{ab}
WE	53.69 ± 30.6^{b}	12.22 ± 8.8^{b}	16.73 ± 5.0^{a}	1.85 ± 1.1^{bc}	32.39 ± 18.60^{bc}	17.65 ± 6.8^{a}	6.85 ± 0.3^{a}	0.43 ± 0.1^{b}
WS	69.18 ± 47.0^{ab}	16.88 ± 21.1^{b}	16.05 ± 2.3^{a}	1.10 ± 0.7^{c}	18.35 ± 11.2^{c}	18.96 ± 7.6^{a}	6.65 ± 0.3^{a}	0.36 ± 0.1^{b}
WN	116.97 ± 61.6^{a}	73.50 ± 80.9^{a}	22.74 ± 16.9^{a}	4.64 ± 4.6^{ab}	71.04 ± 69.5^{ab}	17.59 ± 6.8^{a}	6.70 ± 0.3^{a}	0.62 ± 0.2^{a}
Landward soil (L)	65.96 ± 39.5^{a}	15.39 ± 16.5^{a}	27.97 ± 21.7^{a}	3.61 ± 3.1^{a}	65.34 ± 58.3^{a}	14.21 ± 4.4^{a}	6.85 ± 0.4^{a}	0.32 ± 0.1^{b}
Interface sediment (I)	58.03 ± 22.9^{a}	12.07 ± 12.0^{a}	16.41 ± 3.7^{b}	2.92 ± 2.1^{a}	53.22 ± 37.1^{a}	14.06 ± 2.5^{a}	6.89 ± 0.3^{a}	0.45 ± 0.1^{a}
Waterward sediment (W)	74.07 ± 50.4^{a}	27.48 ± 45.7^{a}	16.05 ± 9.0^{b}	2.13 ± 2.6^{a}	34.68 ± 39.0^{a}	18.20 ± 6.90^{a}	6.73 ± 0.3^{a}	0.44 ± 0.2^{a}

Note: The first letter in the sample classification is short for landward (L), interface (I), and waterward (W), meaning the spatial classification. The second and third letters refer to the vegetation types. F, forest; FG, inter-forest grassland; G, grassland; E, emergent plants zones; S, submerged plants zones; N, no plant zones. Different letters in the columns indicate significant differences between different sites ($p < 0.05$).

The principal component analysis (PCA) (Figure 2a) showed a clear separation between the landward and waterward sites, with the samples in the interface sites falling in the transition area from landward to waterward. The sample correlation distance heatmap results (Figure 2b) further indicated that the inter-forest grassland (IFG) and emergent plant (IE) in the interface area have more similar environmental characteristics to the vegetation belt in the waterward area (WS and WE), while the terrestrial inter-forest grassland (LFG) and forest land (LF) have more similar environmental characteristics. Furthermore, the areas without plants (WN) differed significantly from those with vegetation cover.

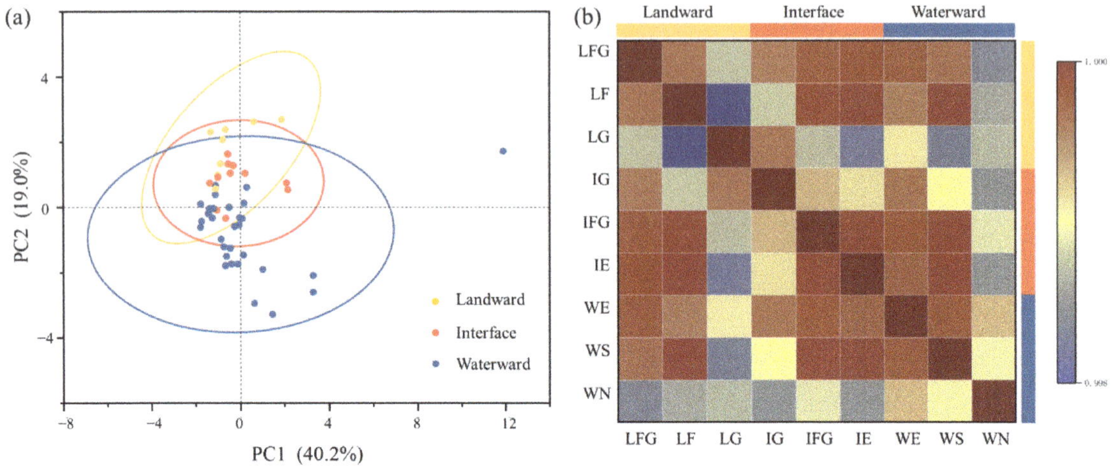

Figure 2. Physicochemical characteristics of the sampling sites showing the vegetation type and space variation. (**a**) The PCA of the soil/sediment physicochemical parameters between the sampling sites, with the 95% confidence intervals represented by colored ellipses. (**b**) The heatmap of the soil/sediment physicochemical parameters.

3.2. Microbial Community Composition and Diversity

3.2.1. Richness and Diversity of Microbial Communities

The sequences and microbial α-diversity of microbial communities are compared in Table S3 and Table S4, respectively. The filtered 16S rRNA sequences ranged from 29,074 to 170,008 per sample (Table S3). To ensure sequencing uniformity, all the reads were denoised into 5456 OTUs, ranging from 1182 to 2306 per sample at 97% identity. In summary, the highest number of OTUs were found in the interface zones, while the lowest were in the land. The microorganisms were clustered according to different plant types from land to waterward zones, as indicated in the results of PLS-DA (Figure S2).

The α-diversity indices (Sobs, ACE, Chao 1, Shannon, and Simpson) were used to describe the richness and diversity of the microbial communities in the soil and sediment samples. The Sobs index indicates the observed richness, while ACE and Chao 1 indicate the statistical richness. The Shannon and Simpson indices represent the statistical diversity [3]. The results of α-diversity indices are presented in Table S4. The highest richness and diversity of microbial communities were observed in the interface, while the lowest was in the land. Specifically, the diversity of IE was the highest, while the richness of IFG was the highest. The diversity and richness of no plants (WN) were both the lowest among all samples.

3.2.2. Microbial Community Structure in Various Soil and Sediment Samples

The sequences obtained from the soil and sediment samples were categorized into 75 phyla, 225 classes, 528 orders, 909 families, and 1888 genera. The microbial composition at the phylum level in different sites is shown in Figure 3a. The findings suggested that the microbial community structures in the soil/sediments were similar, but there were variations in the dominant phyla in the areas with different types of plant cover in terms of their relative abundance. Proteobacteria (with a relative abundance of 18.17–25.45%), Actinobacteriota (9.53–32.88%), Chloroflexi (13.39–21.54%), Acidobacteriota (8.03–14.16%), Firmicutes (5.00–12.29%), and Desulfobacterota (0.85–7.48%) together accounted for over

85% of the relative abundance. Proteobacteria was the most abundant phylum in the interface and waterward sediments. The relative abundance of Acidobacteriota decreased from the land to waterward zones, while the relative abundance of Bacteroidota and Desulfobacterota increased.

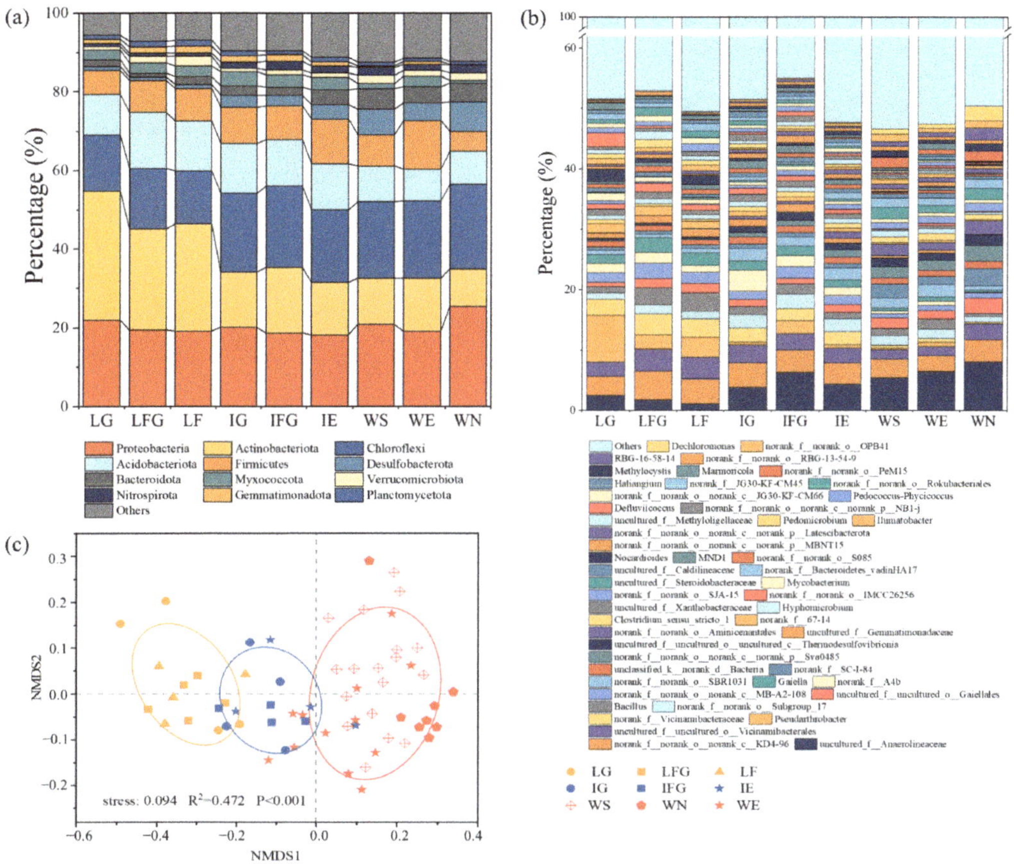

Figure 3. Microbial compositions of the various soil/sediment samples. The relative abundances of the microorganisms (**a**) at the phylum level and (**b**) at the genus level. (**c**) The NMDS analysis of the microbial community at the genus level from the various sample sites, with the 95% confidence intervals represented by colored ellipses.

The microbial structures at the genus level in the various soil and sediment samples are illustrated in Figure 3b. Across all the sites, the dominant genera belonged to *Chloroflexi*, *Acidobacteriota*, *Actinobacteriota*, and *Firmicutes*, which are the dominant phyla and are associated with carbon and nitrogen cycling. The relative abundances of nitrogen-related functional genera were calculated to identify the typical genera in each site, as shown in Figure S3. *Nitrospira* slightly decreased (from 2.86% to 2.19%) while *Thiobacillus* increased (from 3.84% to 4.84%) from the land to water. *Nitrospira* is related to nitrification, and *Thiobacillus* is related to facultative autotrophic denitrification [3]. The relative abundances of *Dechloromonas* in the land soil and water sediment were about 8–10 times higher than those in the interface sediment. The relative abundances of *Bacillus* were the highest (average 1.59%) in the interface sediment, especially in the IFG (average 1.83%) and IG (average 1.97%) zones.

The NMDS analysis based on Bray–Curtis distances was performed to display the variations in the microbial composition at the genus level across the different soil and sediment samples, as depicted in Figure 3c. The ADONIS method was employed to compare differences between the groups, and the results indicated that the variation between the groups was greater than within the groups (R^2 = 0.472, p < 0.001). The sampling sites were clearly grouped based on spatial factors, with the interface samples showing similarity to both the landward and waterward samples. The samples from the areas without plant growth (WN) formed a distinct cluster. Additionally, the heat map of the top 50 genera (Figure S4) further supported the spatial distribution of the genera across the land, interface, and waterward sites.

3.3. Relationship of Microbial Community to the Various Water Levels and Vegetation Types

The mantel correlation analysis was used to identify differences in the microbial communities (at the OUT level) and environmental variables (NH_4^+-N, NO_3^--N, NO_2^--N, TN, SOM, C/N, pH, WC, and flooding days) under various water levels and vegetation covers as shown in Figure 4. As shown in Figure 4a, all the microbial communities under different water levels were found to be related to SOM and TN (r > 0.22, p < 0.05). Additionally, the microbial communities in the land soil were also correlated to WC (r = 0.28, p < 0.05), while those in the interface sediment were related to C/N and flooding days (r > 0.25, p < 0.05), and those in the waterward sediment were associated with WC and pH (r > 0.16, p < 0.05).

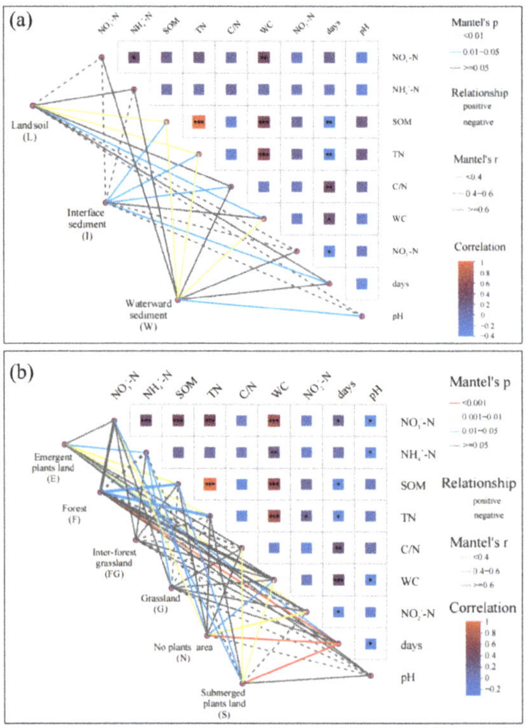

Figure 4. Relationship between the environmental variables and microbial communities. Mantel test for correlation between the environmental variables and microbial communities in (**a**) various water levels and (**b**) vegetation covers. A single, double, and triple asterisk indicate 0.05 < $p \leq$ 0.01, 0.01 < $p \leq$ 0.001, and p < 0.001, respectively.

As shown in Figure 4b, the dominant environmental factors in various plant types affecting the microbial communities varied. The microbial communities in the forest soil (LF and IF) were highly related to SOM and TN ($r > 0.94$, $p < 0.05$), while those in the grassland soil (LG and IG) were correlated with NH_4^+-N ($r = 0.39$, $p < 0.05$). The microbial communities in the emergent plant zone sediment (IE and WE) were significantly influenced by NO_2^--N, TN, and SOM ($0.25 < r < 0.48$, $p < 0.01$), and those in the submerged plant zone sediment (WS) were significantly influenced by C/N and WC ($0.22 < r < 0.24$, $p < 0.01$). The microbial communities in the no plant zone sediment were significantly related to NO_2^--N and NO_3^--N ($0.34 < r < 0.47$, $p < 0.01$).

3.4. Abundances of Nitrogen-Related Functional Genes

To further analyze the microbial communities involved in the nitrogen cycle, the abundances of nitrogen-related functional genes were tested and compared, as shown in Figure 5. Nitrogen-related functional genes were observed in all the soil/sediment samples studied, although their abundances varied significantly among the different sites.

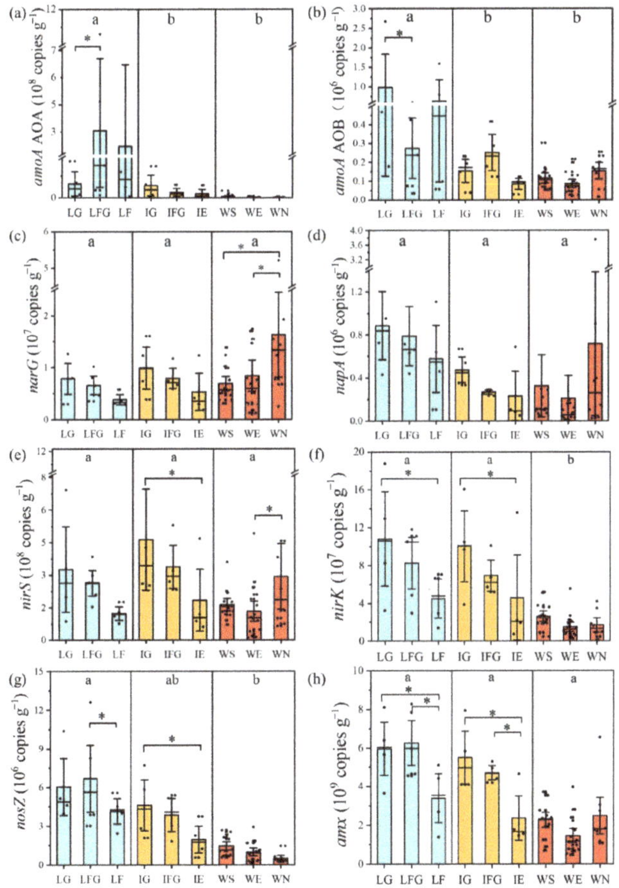

Figure 5. Variations in the abundance of functional genes in the soil/sediment samples. (**a**) *amoA* AOA; (**b**) *amoA* AOB; (**c**) *narG*; (**d**) *napA*; (**e**) *nirS*; (**f**) *nirK*; (**g**) *nosZ*; (**h**) *amx*. The dots represent the data values for each sample. Different letters indicate significant differences between the samples in different groups ($p < 0.05$) and the single asterisk (*) indicates that between the different groups ($p < 0.05$).

The genes *amoA* AOA and *amoA* AOB are related to the ammonia oxidation process, and the abundance and activity of *amoA* AOA and *amoA* AOB are reflected in the gene transcript abundances [6]. The abundance of *amoA* AOA was 2–3 orders of magnitude higher than that of *amoA* AOB, as indicated in Figure 5a,b. In the landward soil, the gene abundances of *amoA* AOA and *amoA* AOB were remarkably higher compared to the interface and waterward sediments. The abundances of the two genes showed opposite trends in the sample sites. The abundance of *amoA* AOA in LFG (3.50×10^8 copies g^{-1}) was significantly higher than in LG (0.60×10^8 copies g^{-1}) ($p < 0.05$), but there was no significant difference among the other sites.

Additionally, genes such as *narG*, *napA*, *nirS*, *nirK*, and *nosZ* encode enzymes involved in the reduction reaction of the denitrification process, and the abundance results are shown in Figure 5c–g, respectively. The abundances of *nirK* and *nosZ* were significantly lower in the land soil and interface sediment ($p < 0.05$), while there was no significant spatial difference among the sites from land to water in the abundances of *narG*, *napA*, and *nirS*. In the land soil, the order of *narG*, *napA*, *nirS*, and *nirK* gene abundances was LG > LFG > LF, with the gene abundances in LF significantly lower than those in LG and LFG. A similar trend was observed in the interface sediment samples as well. In the water sediment, the abundances of narG and nirS in WN were significantly higher than those in WS and WE ($p < 0.05$).

The abundances of the ANAMMOX bacteria (*amx*) were also investigated to illustrate the anaerobic ammonia oxidation process in the soil/sediment. The amx abundances were at least two orders of magnitude higher than those of the other genes (Figure 5h). The *amx* abundances varied little from land to water, but significantly in the different plant types in the landward and interface zones ($p < 0.05$). Specifically, the *amx* abundances in the grassland (LG and IG) and the inter-forest grassland (LFG and IFG) were significantly higher than those in the other plant types (LF and IE) ($p < 0.05$).

3.5. Relationships between Environmental Variables, Microbial Communities, and Functional Genes

To elucidate the response of the microbial community and functional genes to environmental factors under various water levels and plant covers, RDA was performed to identify correlations between the environmental variables (NH_4^+-N, NO_3^--N, NO_2^--N, TN, TC, SOM, C/N, pH, and WC) and the microbial community and functional genes (Figure 6). The results indicated that SOM (18.5%, F = 15.4, $p < 0.01$) and WC (16.4%, F = 10.8, $p < 0.01$) were the primary environmental factors influencing the microbial community at the phylum level (Figure 6a), followed by NO_2^--N (4.1%, F = 3.8, $p < 0.01$) and NO_3^--N (4.1%, F = 3.6, $p < 0.05$). For the functional genes as shown in Figure 6b, SOM (36.5%, F = 31.6, $p < 0.01$) and TC (13.8%, F = 15.0, $p < 0.01$) were the two dominant environmental factors. Spearman's correlation coefficients (Figure S5) further revealed that flooding duration differentiated the land soil, interface sediment, and waterward sediment, with most phyla and functional genes showing significant correlations with flooding days. Specific correlations were identified between the environmental factors and microbial communities and functional genes. Actinobacteriota was negatively related to WC and C/N but positively related to NO_2^--N ($p < 0.05$), while Chloroflexi displayed the opposite trend. Additionally, phyla such as Desulfobacterota, Bacteroidota, Verrucomicrobiota, and Cyanobacteria were negatively correlated with SOM and TN ($p < 0.05$), whereas Desulfobacterota, Nitrospirota, and Nitrospinota were positively influenced by WC and C/N ($p < 0.05$). Moreover, nitrogen cycling-related functional genes were positively correlated with SOM, TC, and TN but negatively related to C/N ($p < 0.05$). The *amoA* AOA and *nosZ* genes were negatively correlated with WC ($p < 0.05$), while *napA* was positively correlated. The gene *narG* showed positive correlations with NH_4^+-N and NO_3^--N ($p < 0.05$), whereas *napA* was positively influenced by NO_2^--N ($p < 0.05$). Additionally, the *amoA* AOB and *nirK* were more correlated with pH.

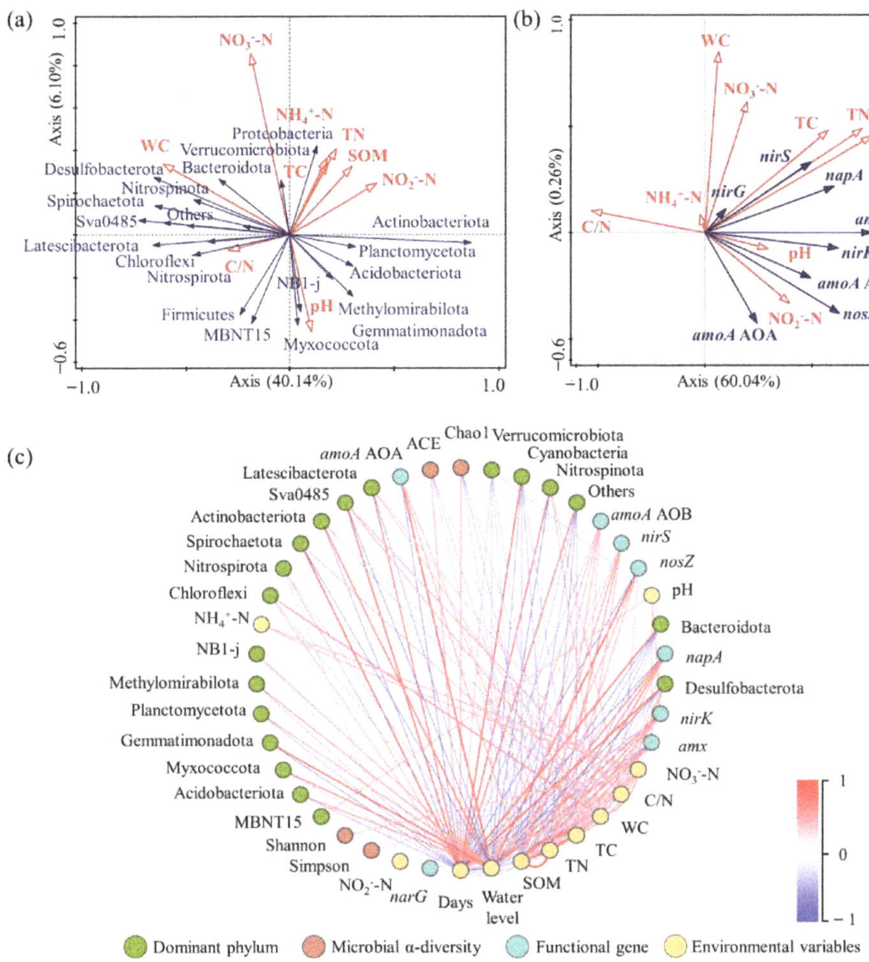

Figure 6. Interaction effects of the various factors. The results of RDA between the environmental variables and (**a**) microbial communities and (**b**) functional genes. (**c**) The network analysis of the various factors. The nodes are colored based on 4 factor types, which are the dominant phylum, the microbial α-diversity, the functional gene, and the environmental variables. The edges represent the Spearman rank correlation coefficients ($p < 0.01$). The red links represent positive correlations, and the blue links represent negative correlations.

The network analysis demonstrated the interaction effects among the environmental variables, dominant phylum, microbial α-diversity, and functional genes (Figure 6c). The edge numbers represent the factors significantly positively or negatively correlated to each node. Flooding days and water level were the primary influencing factors, followed by the environmental variables such as SOM, TN, TC, WC, and NO_3^--N. These findings were consistent with the Mantel test results between the environmental variables and microbial communities (Figure S6). The functional genes, particularly *amx* and *nirK*, were highly influenced by the environmental factors. The dominant phyla *Desulfobacterota* and *Bacteroidota* were the most sensitive to the environmental variation.

4. Discussion

4.1. Relationships between Environmental Variables, Water Levels, and Plant Covers

The fluctuating water level is a primary hydrologic characteristic of the lake littoral zone, directly affecting the physicochemical properties of the soil and sediments [4]. This study demonstrated that water level fluctuations significantly influenced WC and the oxidized and reduced nitrogen compounds ($p < 0.05$), aligning with previous research [6]. Furthermore, this study further examined the effects of vegetation type in various water level zones, given that the rhizosphere microenvironment is a crucial site for N-cycling in the lake littoral zone. Vegetation cover displays a vertical landscape pattern that varies along the hydrological gradient.

The interface zones exhibited transitional characteristics between land and water, combining the physicochemical properties of both zones (Table 1, Figure 2a). Intensive material exchange in these zones, driven by water level fluctuations, affects nitrogen compounds, resulting in lower NH_4^+-N and NO_3^--N contents compared to the landward and waterward zones [4]. In the landward zones, where flooding is rare, vegetation residues not carried away by surface water act as exogenous organic substances, significantly influencing differences among various plant types. In the landward grass zones (LG), high biomass and debris that leach C and N resulted in higher TN and SOM contents compared to the other plant types. In contrast, the evergreen coniferous forest (dominated by *Taxodium distichum* and *Metasequoia glyptostroboides*) in the landward forest zones (LF) has lower TN and SOM due to reduced plant biomass input [12]. Similarly, biomass differences among various plant types in interface grass zones (IG) lead to higher TN and SOM contents than in other zones. In the waterward zones, vegetation types were not the primary factors influencing the sediment properties. Previous studies have reported that nitrogen and carbon deposition occur under consistent inundation without nutrient limitation in most flooding zones [22]. The NH_4^+-N and NO_3^--N contents were higher in the non-vegetated waterward zones (WN) compared to the vegetated zones, as anaerobic conditions in the non-vegetated sediment (WN) promote NH_4^+-N accumulation, and NO_3^--N accumulated without plant consumption.

Flooding results in similar environmental properties between the interface emergent plant zones and inter-forest grassland (IE/IFG) and waterward emergent and submerged plant zones (WE/WS) (Figure 2b). Conversely, similar vegetation configurations in the landscape resulted in more similar environmental characteristics in the landward forest (LF) and landward inter-forest grassland (LFG) (Figure 2b). The varied physicochemical properties across different zones indicated that water level fluctuations and vegetation cover jointly affect the soil and sediment microenvironment.

4.2. Effects of Water Level Fluctuations and Vegetation Cover on the Microbial Diversity and Composition

Water level fluctuations significantly affected the α-diversity of the microbial communities across various zones, with the highest microbial diversity found in the interface sediments and the lowest in land soil (Table S4). This trend differs from a previous study where bacterial community α-diversity increased from water sediments (altitude 145 m) to landward soil (altitude 175 m) in the riparian zone of the Three Gorges Reservoir [6]. The large water level variation (up to 30 m) in the reservoir caused cyclic inundation and drying in the soil and sediments, leading to spatial variation in soil moisture and resulting in different microbial diversity patterns compared to this study [5]. Research on nitrogen cycling in the terrestrial–aquatic continuum suggested that interactions between the oxic and suboxic environments occur on small spatial (centimeters or less) and temporal (<1 d) scales but are substantially separated over larger spatial (tens of meters to kilometers) and temporal (weeks to years) scales [23]. In this study, the oxic–anoxic interface sediments, arising from a 0.51 m water level variation in the interface zones, led to high microbial diversity. Vegetation types had little effect on the microbial diversity, but significant differences were observed between the zones with and without plants. The rhizosphere of vegetation, especially the aerenchyma and shallow rooting of amphibious plants, promotes an oxidative state by providing more air channels and rhizosphere respiration, contributing to a complex microenvironment underwater for the microorganisms in the interface zones [24,25].

The microbial composition at the phylum and genus levels further supported the observed diversity trends (Figure 3a,b). Besides Proteobacteria, the most abundant phylum in the interface sediments, dominant phyla in the landward and waterward zones, such as Actinobacteriota, Bacteroidota, and Desulfobacterota, were also prevalent in the interface zones. These phyla are closely associated with the degradation of complex organic matter and denitrification processes in ecosystems [3,20,26]. Additionally, rare phyla, such as Myxococcota and Gemmatimonadota, were also abundant in the interface sediments, related to dissimilatory nitrate reduction to ammonium (DNRA) and nitrate reduction to nitrite [27,28]. The microbial composition at the genus level confirmed the environmental filtering effect on the microorganisms noted in previous studies, indicating that the species resistant to unfavorable conditions were selected by the frequent environmental variations in the interface zones [5,6]. The dominant genus *Bacillus,* which is strictly aerobic or facultatively anaerobic, is well adapted to the alternating aerobic–anaerobic conditions in the interface sediments. The NMDS results supported the direct response of the microbial spatial distribution to water level fluctuations (Figure 3c).

In addition, the Mantel test results showed that water level fluctuations indirectly affected the spatial distribution of the microbial communities by influencing the nitrogen and carbon content in soil and sediments (Figure 4a). The RDA results and Spearman's correlation coefficients between the dominant phyla and environmental factors further illustrate the specific response of the microbial composition to the environmental variables (Figure 6a,c). SOM and TN were the key environmental factors in the waterward zones due to the high abundance of Bacteroidota and Desulfobacterota, which were common in activated sludge in wastewater treatment systems. Bacteroidota is typically dominant in denitrification systems, while Desulfobacterota is often involved in DNRA under anaerobic conditions [29,30]. WC in the landward zones was lower than in the other zones and was a vital factor influencing the microbial community structures. Additionally, the negative relationship between the dominant phylum Actinobacteriota and WC explained its high abundance in landward soils. The direct influence of vegetation types on the microbial spatial distribution was minimal, but the dominant environmental factors affecting the microbial communities varied among plant types. According to the Mantel test results, vegetation type mainly indirectly affects the microbial community structure by influencing soil physicochemical properties (Figure 4b). The high TN and SOM in the grassland (LG/IG) provide a relative supplement for N-cycling in soil and sediments, and NH_4^+-N was the limiting factor for the microbial communities. Myxococcota (dominant genera such as *Haliangium*) was abundant in the grassland, which is associated with the degradation of organic carbon and nitrogen [31].

4.3. Effects of Water Level Fluctuations and Vegetation Cover on the Nitrogen-Related Functional Genes

The transcript abundances of nitrogen-related functional genes indicated the activity of nitrification and denitrification processes. The ammonia oxidation process in soil and sediments is driven by ammonia-oxidizing archaea (*amoA* AOA) and ammonia-oxidizing bacteria (*amoA* AOB), which occupy distinct niches and are influenced by N-substrate levels [4] and environmental factors such as WC and pH [10]. In this study, *amoA* AOA dominated the ammonia oxidation process at all the sites, with abundances 2–3 orders of magnitude higher than those of *amoA* AOB (Figure 5). This result is consistent with several studies on anaerobic ammonium oxidation in terrestrial–aquatic interfaces, indicating that *amoA* AOA is more tolerant of high moisture and hypoxic environments [12,18]. Additionally, the NH_4^+-N affinity of *amoA* AOA resulted in its high abundance in the land soil, where the NH_4^+-N contents were higher than in the interface sediments [4,6]. The *narG, nirS*, and *nirK* gene copies were higher than the other denitrification genes at all the sites, with *nirS* gene abundances surpassing those of *nirK*, aligning with studies on denitrifying microbial communities in coastal wetlands [10]. Core *nirS*-type aerobic denitrifying genera, such as *Pseudomonas* and *Thiobacillus* [32], were also the dominant functional genera in soil and sediments, with abundance trends consistent with *nirS* gene copies (Figure S3). Furthermore, *amx* gene copies were abundant at most sampling sites. Although the high abundance of *amx* may overestimate the actual abundance of the ANAMMOX bacteria, as the abundances were measured for all the anaerobic ammonia oxidation process-related bacteria instead of specific functional genes, the results still reflect the relative abundance trends across the various sites [10].

The RDA results (Figure 6b) and Spearman rank correlation coefficients (Figure S5) further elucidated the impacts of water level fluctuations and plant types on nitrogen-related microbes. Nearly all the functional genes were associated with flooding days, which represent the spatial position of the sampling sites along elevation. These flooding days were closely linked to WC (Figure 6c and Figure S5), directly influencing oxygen diffusion to the microbes and thereby regulating the nitrification and denitrification processes [6]. As crucial carbon and nitrogen substrates for the nitrogen-cycling process, SOM, TN, and TC were strongly and positively correlated with most functional genes. Consequently, a stronger correlation was observed in the landward grassland (LG) due to the abundant biomass input from residues. Water level fluctuations in the interface zones led to frequent material exchange and minimal variations in soil and sediment pH (average values varying from 6.83 to 6.95). Although pH is a vital environmental factor for pH-sensitive microbes, no significant relationship between the pH and microbial communities was observed in this study. It may be attributed to the minor changes in the pH values of soils and sediments, which ranged from 6.16 to 7.42, with little spatial difference across the sampling sites ($p > 0.05$), not to mention the effect of vegetation covers (Table 1). The similar pH environment was insufficient to be an environmental factor in grouping the nitrogen-related microbes.

4.4. Mechanism of Water Level Fluctuation and Vegetation Type Effecting on the Microbial Communities

Based on the above-mentioned results, Figure 7 illustrates the potential mechanisms driving the effects of water level fluctuations and vegetation types on microbial communities. Water level fluctuations frequently result in a typical drought–rewetting soil and sediment environment, directly filtering adaptive microorganisms. Furthermore, hydrological fluctuations in the lake littoral zone regulate the physicochemical properties of soil and sediment, indirectly influencing microbial composition mainly through WC, nitrogen substrate levels, and pH values. It is noteworthy that WC is closely related to oxygen content in soil and sediments, making water level fluctuations a key determinant of environmental oxygen content. This multifaceted impact of water level fluctuations contributes to the composition of dominant microbial genera and the abundance of functional genes in the lake littoral zone. Vegetation, a vital participant in the nitrogen cycle of the lake littoral zone, directly affects microbial composition and activity through root exudation [27]. The physicochemical factors vary in vegetation-covered soil and sediments, and litterfall from lake littoral vegetation provides exogenous organic substances, further enriching microbial species and functional genera abundances. The vegetation types in this study represent the lakeshore landscape with plant species combinations, exhibiting a vertical landscape patch and pattern along the hydrological gradient [7]. In conclusion, water level fluctuations and vegetation types jointly affect microbial community structure and nitrogen-related functional genes. This study proposes a simplified topological relationship network to highlight key influencing factors and core pathways affecting microbial community structure and functional gene abundances in the lake littoral zone, providing a possible scientific explanation for the impact of water level fluctuations on nitrogen removal capacity in littoral zones.

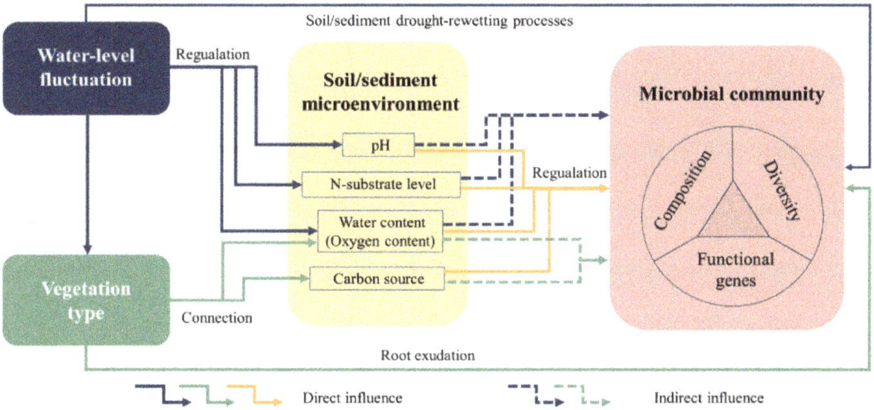

Figure 7. The topological analysis for the main effect mechanism of water level fluctuation and vegetation type on the microbial community.

5. Conclusions

In this study, the effects of water level and vegetation cover on the microbial community and functional genes were investigated by comparing and analyzing the environmental factors, microbial diversity, community composition, and functional genes under nine typical vegetation cover types in three vertical spatial zones (i.e., the landward zone, the interface zone, and the waterward zone) in the littoral zone of Lake Erhai. The results showed that water level fluctuations significantly and directly affected water content (WC) and oxidized and reduced nitrogen compounds due to the oxic–anoxic alternations and intensive material exchange conditions in the interface zone. The vegetation cover influenced the organic matter and total nitrogen content, as plant residues and root exudation supplied exogenous carbon and nitrogen. The microbial diversity in the interface sediments was the highest due to the alternating aerobic–anaerobic environments to which the dominant phyla and genera were more adaptable. The abundances of *amoA* AOA, *nirS*, and *amx* were higher than those of the other genes and were strongly related to flooding days and water content. In conclusion, water level fluctuation and vegetation type jointly affect the microbial community structure and nitrogen-related functional genes. This study provides insight into the impact of water level fluctuations in lakes on the nitrogen removal capacity of littoral zones.

Supplementary Materials: The following supporting information can be downloaded at: https://www.mdpi.com/article/10.3390/pr12061276/s1. Figure S1. The typical photographs of the sampling sites in the lakeshore. Figure S2. PLS-DA of microorganism on OUT level of the various soil/sediments of lakeshore. Figure S3. The Relative abundances of nitrogen-related functional genera in various samples of lakeshore. Figure S4. Heatmap of the top 50 genera in various soil/sediments of lakeshore. Figure S5. Heatmap of Spearman rank correlation coefficients between environmental variables and microbial communities and functional genes. Figure S6. Relationship between environmental variables, microbial community, and functional genes. Table S1. Basic information for various sites of lakeshore. Table S2. Primers of target genes used in the qPCR analysis [6,7,10,33–37]. Table S3. Sequencing statistics for various soil/sediment samples of lakeshore. Table S4. Microbial diversity indices for various soil/sediment samples of lakeshore.

Author Contributions: Conceptualization, J.Y. and J.C.; methodology, J.Y. and J.C.; investigation, W.L., F.Z. and Z.H.; software, J.Y. and W.L.; formal analysis, J.C. and W.L.; writing—original draft preparation, J.Y. and J.C.; writing—review and editing, J.Y. and Z.C.; visualization, J.Y.; supervision, Z.C. All authors have read and agreed to the published version of the manuscript.

Funding: This research was funded by the National Key Research and Development Project (No. 2021YFC3201003).

Data Availability Statement: The original contributions presented in the study are included in the article/Supplementary Material, further inquiries can be directed to the corresponding authors.

Acknowledgments: We thank the National Cryosphere Desert Data Center for providing the SRTM DEM data set of China. (http://www.ncdc.ac.cn, accessed on 13 June 2024). We thank Shenglin Yang for drawing the picture of the sampling sites on the lakeshore.

Conflicts of Interest: The authors declare no conflicts of interest.

References

1. Wang, W.; Liu, W.; Wu, D.; Wang, X.; Zhu, G. Differentiation of nitrogen and microbial community in the littoral and limnetic sediments of a large shallow eutrophic lake (Chaohu Lake, China). *J. Soils Sediments* **2019**, *19*, 1005–1016. [CrossRef]
2. Jeppesen, E.; Sorensen, P.; Johansson, L.; Sondergaard, M.; Lauridsen, T.; Nielsen, A.; Mejlhede, P. Recovery of lakes from eutrophication: Changes in nitrogen retention capacity and the role of nitrogen legacy in 10 Danish lakes studied over 30 years. *Hydrobiologia* **2024**. [CrossRef]
3. Yuan, J.; Wang, B.; Hou, Z.; Peng, J.; Li, D.; Chu, Z. Response of Nitrogen Removal Performance and Microbial Distribution to Seasonal Shock Nutrients Load in a Lakeshore Multicell Constructed Wetland. *Processes* **2023**, *11*, 2781. [CrossRef]
4. Zhu, G.; Wang, S.; Wang, W.; Zhou, L.; Jiang, B.; Op Den Camp, H.; Risgaard-Petersen, N.; Schwark, L.; Peng, Y.; Hefting, M.; et al. Hotspots of anaerobic ammonium oxidation at land–freshwater interfaces. *Nat. Geosci.* **2013**, *6*, 103–107. [CrossRef]
5. Yang, F.; Zhang, D.; Wu, J.; Chen, Q.; Long, C.; Li, Y.; Cheng, X. Anti-seasonal submergence dominates the structure and composition of prokaryotic communities in the riparian zone of the Three Gorges Reservoir, China. *Sci. Total. Environ.* **2019**, *663*, 662–672. [CrossRef] [PubMed]
6. Zhang, H.; Sun, L.; Li, Y.; Zhang, W.; Niu, L.; Wang, L. The bacterial community structure and N-cycling gene abundance in response to dam construction in a riparian zone. *Environ. Res.* **2021**, *194*, 110717. [CrossRef] [PubMed]
7. Zhang, M.; O'Connor, P.; Zhang, J.; Ye, X. Linking soil nutrient cycling and microbial community with vegetation cover in riparian zone. *Geoderma* **2021**, *384*, 114801. [CrossRef]
8. Middleton, J.A.; de Sosa, L.L.; Martin, B.C.; Jones, D.L.; Gleeson, D.B. Soil microbes of an urban remnant riparian zone have greater potential for N removal than a degraded riparian zone. *Environ. Microbiol.* **2020**, *22*, 3302–3314. [CrossRef] [PubMed]
9. Wang, S.; Wang, W.; Liu, L.; Zhuang, L.; Zhao, S.; Su, Y.; Li, Y.; Wang, M.; Wang, C.; Xu, L.; et al. Microbial Nitrogen Cycle Hotspots in the Plant-Bed/Ditch System of a Constructed Wetland with N_2O Mitigation. *Environ. Sci. Technol.* **2018**, *52*, 6226–6236. [CrossRef]
10. Zhang, Y.; Ji, G.; Wang, C.; Zhang, X.; Xu, M. Importance of denitrification driven by the relative abundances of microbial communities in coastal wetlands. *Environ. Pollut.* **2019**, *244*, 47–54. [CrossRef]
11. Liu, X.; Yuan, S.; Wang, H. Periodic inundations drive community assembly of amphibious plants in floodplain lakes. *Hydrobiologia* **2020**, *847*, 4207–4217. [CrossRef]
12. Li, T.; Wang, X.; Wang, X.; Huang, J.; Shen, L. Mechanisms Driving the Distribution and Activity of Mineralization and Nitrification in the Reservoir Riparian Zone. *Microb. Ecol.* **2023**, *6*, 1829–1846. [CrossRef] [PubMed]
13. Liu, X.; Chen, J.; Tang, B.; He, L.; Xu, Y.; Yang, C. Eco-Environmental Changes Due to Human Activities in the Erhai Lake Basin from 1990 to 2020. *Sci. Rep.* **2024**, *14*, 8646. [CrossRef] [PubMed]
14. Wen, Z.; Ma, Y.; Wang, H.; Cao, Y.; Yuan, C.; Ren, W.; Ni, L.; Cai, Q.; Xiao, W.; Fu, H.; et al. Water Level Regulation for Eco-social Services Under Climate Change in Erhai Lake Over the Past 68 years in China. *Front. Environ. Sci.* **2021**, *9*, 697694. [CrossRef]
15. Ji, Y.; Shi, W.; Qin, B. An indispensable role of overlying water in nitrogen removal in shallow lakes. *Sci. Total. Environ.* **2024**, *923*, 171487. [CrossRef] [PubMed]
16. Xiong, Z.; Guo, L.; Zhang, Q.; Liu, G.; Liu, W. Edaphic Conditions Regulate Denitrification Directly and Indirectly by Altering Denitrifier Abundance in Wetlands along the Han River, China. *Environ. Sci. Technol.* **2017**, *51*, 5483–5491. [CrossRef] [PubMed]
17. Zhang, Z.; Huang, X.; Zhou, Y.; Zhang, J.; Zhang, X. Discrepancies in Karst Soil Organic Carbon in Southwest China for Different Land Use Patterns: A Case Study of Guizhou Province. *Int. J. Environ. Res. Public Health* **2019**, *16*, 4199. [CrossRef] [PubMed]
18. Ding, L.; Zhou, J.; Li, Q.; Tang, J.; Chen, X. Effects of Land-Use Type and Flooding on the Soil Microbial Community and Functional Genes in Reservoir Riparian Zones. *Microb. Ecol.* **2022**, *83*, 393–407. [CrossRef] [PubMed]
19. Zhang, D.; Li, M.; Yang, Y.; Yu, H.; Xiao, F.; Mao, C.; Huang, J.; Yu, Y.; Wang, Y.; Wu, B.; et al. Nitrite and nitrate reduction drive sediment microbial nitrogen cycling in a eutrophic lake. *Water. Res.* **2022**, *220*, 118637. [CrossRef]
20. Li, D.; Chu, Z.; Zeng, Z.; Sima, M.; Huang, M.; Zheng, B. Effects of design parameters, microbial community and nitrogen removal on the field-scale multi-pond constructed wetlands. *Sci. Total Environ.* **2021**, *797*, 148989. [CrossRef]
21. Geng, S.; Cao, W.; Yuan, J.; Wang, Y.; Guo, Y.; Ding, A.; Zhu, Y.; Dou, J. Microbial diversity and co-occurrence patterns in deep soils contaminated by polycyclic aromatic hydrocarbons (PAHs). *Ecotoxicol. Environ. Saf.* **2020**, *203*, 110931. [CrossRef] [PubMed]
22. Feng, W.; Santonja, M.; Bragazza, L.; Buttler, A. Shift in plant-soil interactions along a lakeshore hydrological gradient. *Sci. Total. Environ.* **2020**, *742*, 140254. [CrossRef] [PubMed]
23. Seitzinger, S.; Harrison, J.A.; Böhlke, J.K.; Bouwman, A.F.; Lowrance, R.; Peterson, B.; Tobias, C.; Van Drecht, G. Denitrification across Landscapes and Waterscapes: A Synthesis. *Ecol. Appl.* **2006**, *16*, 2064–2090. [CrossRef] [PubMed]

24. Pucciariello, C.; Perata, P. New insights into reactive oxygen species and nitric oxide signalling under low oxygen in plants. *Plant. Cell Environ.* **2017**, *40*, 473–482. [CrossRef] [PubMed]
25. Penton, C.R.; Deenik, J.L.; Popp, B.N.; Bruland, G.L.; Engstrom, P.; St Louis, D.; Tiedje, J. Importance of sub-surface rhizosphere-mediated coupled nitrification–denitrification in a flooded agroecosystem in Hawaii. *Soil. Biol. Biochem.* **2013**, *57*, 362–373. [CrossRef]
26. Shi, B.; Cheng, X.; Jiang, S.; Pan, J.; Zhu, D.; Lu, Z.; Jiang, Y.; Liu, C.; Guo, H.; Xie, J. Unveiling the power of COD/N on constructed wetlands in a short-term experiment: Exploring microbiota co-occurrence patterns and assembly dynamics. *Sci. Total Environ.* **2024**, *912*, 169568. [CrossRef] [PubMed]
27. Wang, T.; Xiao, L.; Lu, H.; Lu, S.; Li, J.; Guo, X.; Zhao, X. Nitrogen removal from summer to winter in a field pilot-scale multistage constructed wetland-pond system. *J. Environ. Sci.* **2022**, *111*, 249–262. [CrossRef] [PubMed]
28. Jiang, X.; Liu, C.; Cai, J.; Hu, Y.; Shao, K.; Tang, X.; Gong, Y.; Yao, X.; Xu, Q.; Gao, G. Relationships between environmental factors and N-cycling microbes reveal the indirect effect of further eutrophication on denitrification and DNRA in shallow lakes. *Water. Res.* **2023**, *245*, 120572. [CrossRef] [PubMed]
29. Zhao, Y.; Wang, H.; Dong, W.; Chang, Y.; Yan, G.; Chu, Z.; Ling, Y.; Wang, Z.; Fan, T.; Li, C. Nitrogen removal and microbial community for the treatment of rural domestic sewage with low C/N ratio by A/O biofilter with Arundo donax as carbon source and filter media. *J. Water Process. Eng.* **2020**, *37*, 101509. [CrossRef]
30. Yang, Z.; Zhou, Q.; Sun, H.; Jia, L.; Zhao, L.; Wu, W. Metagenomic analyses of microbial structure and metabolic pathway in solid-phase denitrification systems for advanced nitrogen removal of wastewater treatment plant effluent: A pilot-scale study. *Water. Res.* **2021**, *196*, 117067. [CrossRef]
31. Zhang, L.; Huang, X.; Zhou, J.; Ju, F. Active predation, phylogenetic diversity, and global prevalence of myxobacteria in wastewater treatment plants. *ISME J.* **2023**, *17*, 671–681. [CrossRef] [PubMed]
32. Zhang, H.; Ma, B.; Huang, T.; Yang, W.; Liu, X.; Niu, L. Nitrogen removal from low carbon/nitrogen polluted water is enhanced by a novel synthetic micro-ecosystem under aerobic conditions: Novel insight into abundance of denitrification genes and community interactions. *Bioresour. Technol.* **2022**, *351*, 127013. [CrossRef]
33. Dionisi, H.M.; Layton, A.C.; Harms, G.; Gregory, I.R.; Robinson, K.G.; Sayler, G.S. Quantification of *Nitrosomonas oligotropha*-like ammonia-oxidizing bacteria and *nitrospira* spp. from full-scale wastewater treatment plants by competitive PCR. *Appl. Environ. Microbiol.* **2002**, *68*, 245–253. [CrossRef]
34. López-Gutiérrez, J.C.; Henry, S.; Hallet, S.; Martin-Laurent, F.; Philippot, L. Quantification of a novel group of nitrate-reducing bacteria in the environment by real-time PCR. *J. Microbiol. Methods* **2004**, *57*, 399–407. [CrossRef] [PubMed]
35. Bru, D.; Sarr, A.; Philippot, L. Relative abundances of proteobacterial membrane-bound and periplasmic nitrate reductases in selected environments. *Appl. Environ. Microbiol.* **2007**, *73*, 5971–5974. [CrossRef] [PubMed]
36. Kandeler, E.; Deiglmayr, K.; Tscherko, D.; Bru, D.; Philippot, L. Abundance of *narG*, *nirS*, *nirK*, and *nosZ* genes of denitrifying bacteria during primary successions of a glacier foreland. *Appl. Environ. Microbiol.* **2006**, *72*, 5957–5962. [CrossRef]
37. Scala, D.J.; Kerkhof, L.J. Nitrous oxide reductase (*nosZ*) gene-specific PCR primers for detection of denitrifiers and three *nosZ* genes from marine sediments. *FEMS Microbiol. Lett.* **1998**, *162*, 61–68. [CrossRef]

processes

MDPI

Article

Controlling Eutrophication of Aquaculture Production Water Using Biochar: Correlation of Molecular Composition with Adsorption Characteristics as Revealed by FT-ICR Mass Spectrometry

William F. Rance Bare [1], Ethan Struhs [2], Amin Mirkouei [1,3,*], Kenneth Overturf [4], Martha L. Chacón-Patiño [5,6], Amy M. McKenna [4,7], Huan Chen [5]

1 Department of Biological Engineering, University of Idaho, Idaho Falls, ID 83402, USA; bare9199@vandals.uidaho.edu
2 Department of Mechanical Engineering, University of Idaho, Idaho Falls, ID 83402, USA; stru4589@vandals.uidaho.edu
3 Department of Nuclear Engineering and Industrial Management, University of Idaho, Idaho Falls, ID 83402, USA; ksraja@uidaho.edu
4 Agricultural Research Service, United States Department of Agriculture, Hagerman, ID 83332, USA; ken.overturf@usda.gov (K.O.); mckenna@magnet.fsu.edu (A.M.M.)
5 National High Magnetic Field Laboratory, Ion Cyclotron Resonance Facility, Florida State University, Tallahassee, FL 32310, USA; chacon@magnet.fsu.edu (M.L.C.-P.); huan.chen@magnet.fsu.edu (H.C.)
6 International Joint Laboratory for Complex Matrices Molecular Characterization, iC2MC, TRTG, 76700 Harfleur, France
7 Department of Soil and Crop Sciences, Colorado State University, Fort Collins, CO 80523, USA
* Correspondence: amirkouei@uidaho.edu

Abstract: This study aims to construct a novel and sustainable approach for remediating aquaculture-generated water contamination using various engineered biochars. Particularly, this study focuses on capturing nitrogen and phosphorus from downstream water of commercial fish farms in Magic Valley, Idaho, containing approximately 2.26 mg/L of nitrogen and 0.15 mg/L of phosphorous. The results indicate that the proposed approach can improve downstream waters by adsorbing micronutrients (e.g., nitrogen-ammonia, nitrate-n + nitrite-n, and total phosphorus). Water treatment time and biochar pH are two key parameters strongly associated with adsorbed compounds. Molecular-level characterization of solvent-extracted organics from biochar materials (before and after water treatment) suggests increased levels of highly oxygenated molecules as a function of increasing water treatment time. Also, the results show the enrichment in organic species with higher molecular weight and increased double bond equivalents, with a compositional range similar to that of dissolved organic matter. Upon water treatment, extracted organics revealed higher abundances of compounds with higher H/C and O/C ratios. The engineered biochars, after water treatment, can be reused as nutrient-rich fertilizers. This study concluded that the engineered biochars could sequester more nitrogen and phosphorous over time. Also, the proposed approach can simultaneously increase fish production capacity and support the aquaculture industry in different regions by improving water quality and enabling aquaculture expansion.

Keywords: aquaculture production; engineered biochar; eutrophication; FT-ICR mass spectrometry; water treatment

1. Introduction

1.1. Challenges and Needs

The aquaculture industry produces aquatic organisms and seafood (e.g., fish, shellfish, and crustaceans) that can help to meet the growing consumer demand for healthy protein using natural, domestic resources [1]. According to the National Oceanic and Atmospheric

Administration (NOAA), the global seafood market reached $406 billion, and the United States (U.S.) imported over $21 billion (over six billion pounds) in seafood in 2020 [2]. Therefore, a sustainable aquaculture industry can provide an affordable and safe supply of seafood products, as well as train the workforce and create jobs in rural communities. However, the aquaculture industry is one of the prominent consumers of global water resources and one of the leading causes of eutrophication and water degradation [3].

Eutrophication is the result of oversupplied nutrients (e.g., nitrogen and phosphorus), pollutants, and wastes (e.g., uneaten fish food, fish feces, and drugs) from various operations (e.g., agriculture and aquaculture) to water bodies. Eutrophication changes water pH, oxygen, and temperature, as well as reduces water quality due to structural changes. For example, eutrophication can increase algae and marine plants' growth, which can threaten the survival of native fish species. Currently, there is no affordable method for addressing eutrophication and water degradation in large water bodies, such as rivers.

Aquaculture is a critical industry to the state of Idaho in the U.S., with most of the fish production located in southern Idaho (Hagerman area), approximately 40 km (around 25 miles) west of Twin Falls. Idaho permits around 65 aquaculture operations in the Hagerman area to discharge their downstream water into the Snake River. Currently, these fish farms produce 70–75% of all rainbow trout (over 30,000 tons annually) that are grown in the U.S. with over $100 million in annual revenue that has a huge effect on the southern Idaho economy and the employment of almost 1000 low skilled workers, directly or indirectly [4]. However, the U.S. Environmental Protection Agency (EPA) has determined that the discharge of water from aquaculture operations into the mid-reach of the Snake River increases the eutrophication risks, especially dense algal blooms and low water oxygen content that can cause changes to the ecosystem. Therefore, a cost-effective, environmentally friendly method to remove nutrients and pollutants prior to discharge to water bodies can address these challenges and consequently allow for increased production, reduced eutrophication, and economic stability in the region.

1.2. Background

Biochar Use for Water Remediation. Biochar has the potential to effectively improve eutrophic water by adsorbing micronutrients and contaminants due to its physicochemical properties. Heating organic materials or wastes (e.g., biomass feedstocks) in the absence of oxygen can thermally decompose their compounds to solid materials (biochar), liquid materials (bio-oil), and gases. These biomaterials have applications in various domains, such as addressing sustainability challenges across food-energy-water systems [5]. Pyrolysis is the main technology for producing biochar from organic materials using high-temperature processes without the presence of oxygen. Earlier studies showed that biochar (similar to charcoal) has the potential to remove micronutrients and pollutants (e.g., arsenic, cadmium, ammonium, phosphorus, and nitrate) from water and effluent streams [6,7]. Biochar is an adsorbent that can hold adsorbates (tiny particles, pollutants, and organic substances) in water due to its unique properties (e.g., surface area, architecture, porosity, and textual features). Recent studies investigated biochar adsorption capacities for immobilizing contaminants and mitigating water degradation [8]. For example, Yin et al. (2018) explored the application of modified biochar to remove ammonium, nitrate, and phosphate from eutrophic water [7]. They used modified biochar and Al-modified biochar from soybean straw, and their results show that Mg-/Al-modified biochar is an excellent material for eutrophic water treatment [7]. Novais et al. (2018) also investigated modified biochar from sugarcane straw and poultry manure for removing phosphorus from eutrophic water [9]. They concluded that biochar could be a safe solution for adsorbing phosphorus from water and reusing it in agriculture practices [9]. Shao et al. (2020) investigated nutrient removal and bacterial immobilization from water bodies in recirculating aquaculture systems using biochar from maize straw [10]. Their results show that biochar can improve the water quality and crab production yield [10]. Lin et al. (2021) applied biochar from spent mushroom substrate to improve water quality in aquaculture ponds, particularly in red claw crayfish farms [11]. Their results show that spent mushroom biochar can increase the water pH and reduce ammonia-N and sulfite levels [11]. Silva et al. (2021) explored biochar from waste-based materials for removing antibiotics from aquaculture effluents [12]. The latest studies by Chen et al. (2022) investigated biochar to immobilize heavy metals and polycyclic aromatic hydrocarbons in pond sediment and aquaculture products [13]. They concluded that biochar could remediate pollution in aquaculture environments while safeguarding the quality of aquatic products [13]. Also, Hu et al. (2022) and Hung et al. (2022) explored the use of biochar for removing contaminants (e.g., drugs) from aquaculture water bodies and ponds [14,15].

Inherent Organic Compounds in Biochar. Fast pyrolysis of biomass produces carbon residues, biochar, whose composition is shaped by the sequestration of recondensed pyrolysis liquids, consisting of ultra-complex mixtures of organic compounds, predominantly with high oxygen content. Previous reports have used solvent extractions to isolate organic species from biochar and characterize them via Fourier transform ion cyclotron resonance mass spectrometry (FT-ICR MS) [16], which offers the highest resolving power and mass accuracy among all available MS techniques [17,18]. Analysis of organic extracts from biochar has revealed thousands of compositions. In a typical FT-ICR mass spectrum, with tens of thousands of ions, each peak gets assigned a unique molecular formula based on mass accuracy and isotopic composition. Molecular formulae are then sorted into compound classes. For example, compounds with only C and H make up the HC class, whereas species with two N atoms and five O atoms, in addition to C and H, comprise the N_2O_5 class. Graphical representation of the molecular composition can be visualized by plotting three-dimensional plots of double bond equivalents (DBE), the number of rings plus double bonds to carbon, versus carbon number, with a color scale highlighting the relative abundance contributions of each compound within that heteroatom class. In addition, van Krevelen diagrams plot atomic H/C vs. O/C ratios [19,20]. Previously, McKenna et al. (2021) used solvent fractionation to extract organics from biochar, followed by subsequent characterization by atmospheric pressure photoionization coupled with 21 T FT-ICR MS [16] and reported functionalized and nonfunctionalized polycyclic aromatic hydrocarbons with aromatic structural motifs [16,21].

1.3. Study Focus and Objectives

The main objective of this study is to explore the remediation of aquaculture-generated water by a novel water treatment approach based on engineered biochars from woody feedstocks that have the potential to be recycled and reused as nutrient-rich fertilizers. Our earlier studies show that pinewood can effectively adsorb micronutrients (P and N) from water [22]. Thus, this study focuses on biochars from different pinewoods (e.g., lodgepole or mixed ponderosa pine and spruce trees). The novelty of this study lies in addressing eutrophication risks, using a sustainable method to remove contaminants from aquaculture effluents before discharging them to water bodies. The applied approach could promote sustainable aquaculture practices and create an entirely new industry, employing used biochar after water treatment as nutrient-rich soil conditioners or fertilizers due to its high carbon and nutrient content. We seek to address several research questions related to the adsorption capacity of engineered biochars (especially N and P), the chemistry of adsorbed and released materials after water treatment, and the overall environmental performance of the designed water treatment method. Also, this study explores solutions to address some of the issues associated with the complex molecular nature of biochar filters and soluble organic species that can harbor in their intricate microstructure. Characterization studies via ultra-high-resolution mass spectrometry can provide valuable information on extractable organics and components at the molecular level before and after using biochar for water treatment. The investigation of such a complex mixture is central for potential applications in water treatment. The overall characterization of the materials at the molecular level will enable more accurate modeling and scale-up of valorization processes that will, in turn, allow more efficient utilization of water and food resources, increasing their economic value and helping the nation's natural supplies.

2. Materials and Methods

2.1. Materials

Biochar. This study uses two different biochars: (1) from older growth ponderosa pine and spruce trees, with little to no lodgepole in them, and from the slash piles of timbering operations in Medford, Oregon, labeled as OB (Oregon Biochar), and (2) from lodgepole pine in southeast Idaho, labeled as IB (Idaho Biochar). Biochar production processes are slightly different for Oregon and Idaho Biochars; all samples were dry sieved for plus 100 microns before water treatment.

Water. One of the prime prerequisites for establishing an aquaculture operation is a representative water source. In this study, the water is provided from Thousand Springs (the outlet for the east Snake Plain aquifer), which flows more than 5.66 m^3/s (around 200 cubic feet per second) from several springs in the canyon wall. The Thousand Springs water is ideal for growing rainbow trout due to the temperature range of 13.5–15.5 °C (57–60 °F) and its quality, such as pH range (Figure 1a).

Figure 1. (**a**) Thousand Springs water for growing rainbow trout, (**b**) testing tanks, utilizing fourth-use waters from the main rearing ponds that contain relatively higher levels of nitrogen and phosphorus, and (**c**) large size water filter bag with 300 g biochar for 30 min treatment and 450 g biochar for 12–48 h treatment, attached to the water pipe with ties.

2.2. Methods

Biochar Production Method. Biochar samples are produced from woody biomass feedstocks (e.g., pinewood) under different process configurations (e.g., conversion temperature and residence time). The IB production process includes size reduction (grinding), drying, and slow pyrolysis (at 400–450 °C under 10–15 psi for 30 min) using the built-in-house slow pyrolysis reactors [23]. The OB was provided by Oregon Biochar Solutions [24].

Water Treatment Method. Field tests were conducted at the USDA facility equipped with water from production fish farms, located north of Buhl, Idaho, adjacent to the Snake River, utilizing fourth-use waters from their main rearing ponds, known to contain relatively higher levels of nitrogen and phosphorus (Figure 1b). The utilized field test area has been developed by the U.S. Department of Agriculture (USDA) in cooperation with the University of Idaho aquaculture research facilities in Hagerman for many field studies and research projects. For the field experiments, each testing tank can control the water flow into the tank. Flow rates for this study were set at 7.57 L/min. Time intervals were designed to establish the saturation time for different sample sizes. Particularly, 24 testing tanks were used, including six experiments (3 for OB and 3 for IB) using 300 g biochar for 30 min;

and 18 experiments (9 for OB and 9 for IB) using 450 g biochar for 12, 24, and 48 h, each time-period in triplicate. Each biochar sample was placed in a large water filter bag, and the bag was attached to the water pipe with zip ties (Figure 1c). The used water filter bags have a pore size of 100 microns. Water treatment experiments were run in triplicate to evaluate the removal efficiency of nitrogen and phosphorus from the aquaculture production water using different biochar samples.

Characterization Methods. Several analytical laboratories assisted this study by analyzing biochar and water before and after treatment. For biochar nitrogen-ammonium analysis, a KCI extractable (ASA 33-3.2) was used for preparation, followed by the colorimetric method (ASA 33-7.3) [25,26]. Similarly, for biochar nitrogen-nitrate + nitrate analysis, a KCI extractable (ASA 33-3.2) was used for preparation, followed by the colorimetric method (ASA 33-8.3) [25,26]. For total phosphorus, biochar was digested in 30% nitric acid and analyzed by Inductively Coupled Plasma Optical Emission Spectrometry (ICP-OES). For water analysis, EPA protocols 365.1 and 300.1 were used to assess total phosphorus and nitrogen, respectively [27,28]. For characterization of the extractable organic compounds, 30 g of biochar were placed in cellulose thimbles and extracted in a Soxhlet apparatus with separate solvent mixtures during 24 h of acetone/cyclohexane 1/1 and toluene/tetrahydrofuran/methanol 1/1/1. The extracted species in solution were desolvated under nitrogen, weighed, and stored in the dark prior to FT-ICR MS analysis.

Molecular Characterization by 21 T FT-ICR MS. 21-tesla FT-ICR MS coupled with negative-ion electrospray ionization (-ESI) was used to (a) compare the extractable organics from different biochar samples before and after water treatment, (b) to understand the effect of biochar production and water treatment conditions, and (c) to identify biochar production conditions that produce materials with high efficiency for the removal of specific water contaminants (e.g., nitrogen and phosphorus containing compounds).

Extracted organics were ionized by negative-ion electrospray ionization, extensively used to characterize polar, oxygen-containing compounds through deprotonation that occurs in solution [29,30]. Fractions were dissolved in 1:1:1 toluene:tetrahydrofuran: methanol at a 150 μg/mL concentration, infused at 0.5 μL/min, and ionized with a needle voltage of −3.0 kV, with typical source conditions, 3. Samples were diluted in Ions and analyzed with a custom-built 21T FT-ICR mass spectrometer, in which 1×10^6 charges were accumulated for 1–3 ms in an external ion trap using automatic gain control (AGC) [31,32]. Ions were transferred to the dynamically harmonized ICR via a decreasing auxiliary radio frequency. Excitation and detection were achieved on the same pair of electrodes of the ICR cell, operated at 6 V. 100 time-domain transients of 3.1 s were collected and averaged with Predator Software Corev1.2.3 for each sample. Mass spectra were internally calibrated with abundant homologous series based on the "walking" calibration method [33]. PetroOrg© Software NS 18.0.6 [34] assigned elemental composition assignment and data visualization (e.g., DBE vs. carbon number and van Krevelen diagrams). Only classes with a combined %R.A. above 0.15% total RA will be considered.

3. Results and Discussion

3.1. Results

Biochar Analysis. Both IB and OB experiments appeared to capture (sequester) nitrogen-ammonia, nitrate-n + nitrite-n, and total phosphorus over time. IB experiments show significant nutrient (N and P) adsorption, especially nitrogen-ammonia and nitrate-n + nitrite-n. The results show that IB phosphorus increased from an average of ~240, 380, to 430 μg/g after 12, 24, and 48 h water treatment, respectively (Figure 2a). IB and OB initially released phosphorous into the water but gradually sequestered phosphorous over time. Also, IB and OB samples adsorbed micronutrients from downstream water over time but never sequestered phosphorous within the time frame of the experiments (0.5–48 h) at a level higher than the unused (before treatment) biochars. Figure 2 provides insight into relationships between the experimental parameters of interest. Nitrate, nitrite, and ammonium adsorption efficiency of biochar increased with longer residence times for

IB. Contrary to our previous results of short water treatment (residence) time experiments (5–60 min) for the static water treatment system, there is a visible correlation between the adsorption of phosphorus and nitrogen compounds seen in IB samples [22]. Key parameters (e.g., residence time and biochar pH) had relatively strong associations with adsorbed compounds. OB, compared to IB, has high total phosphorus and pH, which seems to have led to the desorption of phosphorus and a significant decrease in pH with prolonged residence times. Overall, IB results appear more promising as a solution to reduce eutrophication (oversupplied nutrients) from fish farm effluent. Longer residence times should be applied to identify the optimal treatment time for the adsorption of phosphorus and nitrogen compounds. Table S1 in Supplementary Materials provides IB and OB results before and after water treatment.

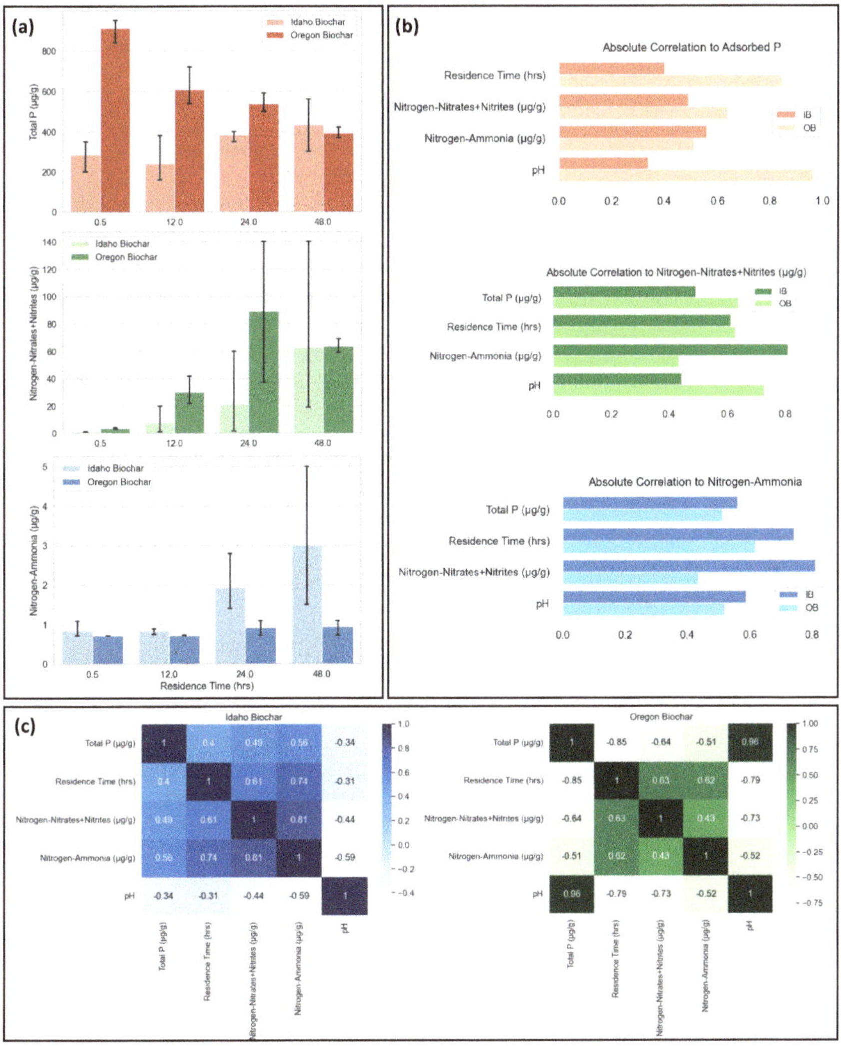

Figure 2. Field results and comparison: (**a**) biochar adsorption of phosphorus, nitrates, nitrites, and ammonia after specified residence times; (**b**) absolute Pearson coefficients relative to adsorbed compound; and (**c**) correlation of key experimental parameters by biochar source.

Water Results. During the water tests, water was run through 300 g of biochar for approximately 0.5 h; both IB and OB released phosphorous into the water, but the amount was significantly larger for OB experiments. The water running through the testing tanks used in this study contained approximately 0.15 mg/L of total phosphorous and 2.26 mg/L of nitrate. Table S2 in Supplementary Materials presents water results before and after 0.5 h water treatment, using IB and OB.

Molecular Characterization Results. Biochars, before and after being used in water treatment, were extracted with two separate solvent mixtures. First, acetone/cyclohexane extracts non-polar/moderately polar species entrained within the char complex microstructure. Next, acetone and cyclohexane target the extraction of less polar/highly aromatic compounds [35,36]. The second solvent mixture, toluene/tetrahydrofuran/methanol, assisted in the isolation of more aromatic/polarizable compounds. Each fraction was quantified gravimetrically. Table 1 shows the mass yields and indicates the IB leached higher amounts of organics (around 5-fold) compared to OB, and in general, more material was extracted with toluene/THF/MeOH, suggesting the dominance of polar organics within IB's structure.

Table 1. Weight (wt.%) percentage of extracted organics relative to the amount of biochar used in each extraction.

Sample	Acetone/Cyclohexane	Tol/THF/MeOH	Treatment (h)
Untreated Idaho	1.9	2.8	0
IB 1	2.4	3.5	12
IB 2	3.1	4.6	24
IB 3	3.6	4.7	48
Untreated Oregon	0.1	0.3	0
OB 1	0.1	0.6	12
OB 2	0.4	0.7	24
OB 3	0.3	0.9	48

The separate fractions were analyzed via negative-ion electrospray ionization (-ESI) coupled with 21-tesla FT-ICR MS. Solvent fractionation with different solvent chemistry extends the compositional space for detected species in complex mixtures via mass spectrometry, as previously shown for fossil fuels [17,37]. Tens of thousands of peaks were assigned in each sample based on mass accuracy and isotopic structure. Then, the elemental compositions were sorted based on the number and type of heteroatom (i.e., nitrogen, oxygen, sulfur) classes. Figure 3 presents the heteroatom class distribution for O_x classes detected in the acetone/cyclohexane and toluene/THF/MeOH fractions for IB and OB biochars. Species that contain one oxygen correspond to O_1 (phenol, alcohol) in negative-ESI and O_2 (carboxylic acid), whereas polyfunctional oxygen moieties comprise classes with more than two oxygens (O_{3+} per molecule). The acetone fraction from IB revealed an increase in oxygen numbers as a function of increasing water treatment time. For instance, the untreated biochar featured abundant compounds with six or fewer oxygens and dominance of O_4-O_6 classes. Conversely, compounds were detected after water treatment with much higher oxygen numbers (up to O_{17}). This effect is less pronounced in the Tol/THF/MeOH extracted organics. Still, for untreated biochars, low-order oxygen-containing classes, such as O_2 and O_3, are more abundant compared to the extracted organics after water treatment. Compositional trends in the class distribution are more difficult to capture for OB. However, the bar graphs indicate that OB2, after 24 h of water treatment, contains abundant compounds with ultra-high oxygen content (e.g., O_7, O_8, and O_9 classes).

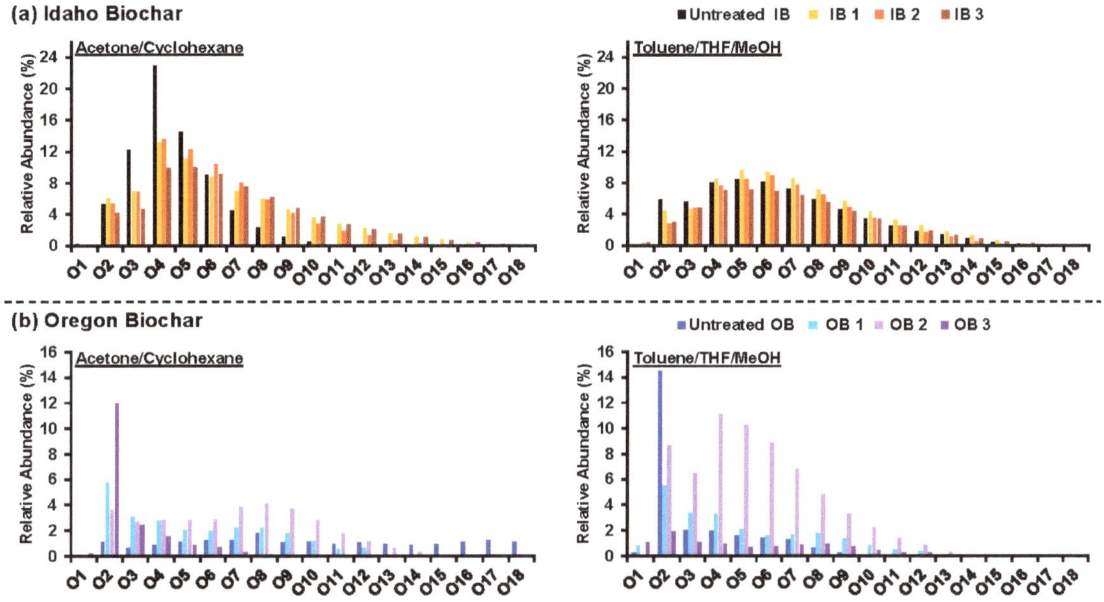

Figure 3. Heteroatom class distribution for oxygen classes derived from negative-ion FT ICR Ms at 21tesla for biochar solvent fractions. Ox (O_1, O_2, . . ., O_{18}) were identified for the acetone/cyclohexane (left) and toluene/THF/MeOH (right) fractions for (**a**) IB and (**b**) OB samples.

Molecular formulae are represented in compositional plots of DBE vs. carbon numbers for acetone/cyclohexane and Tol/THF/MeOH fractions (Figure 4). The extracted organics from OB feature a more limited compositional range, in terms of both DBE and carbon numbers, with abundant detection of compounds with DBE < 20, afar from the red dotted line. This line is known as the polycyclic aromatic hydrocarbon limit and is a compositional boundary for planar organic molecules. Compositions close to the polycyclic aromatic hydrocarbon limit are highly aromatic/pericondensed and alkyl depleted. The results show that IB samples feature more species with compositions close to the polycyclic aromatic hydrocarbon line, with DBE up to 30. This suggests abundant content of highly aromatic, oxygen-containing molecules, likely similar to oxyPAHs [38], which, to some extent, resemble the composition of natural organic matter [39]. The compositional range for the extracted organics with Tol/THF/MeOH is much more extended and is highly enriched in compositions with DBE values above 20, in which relative abundance increases as a function of increasing water treatment.

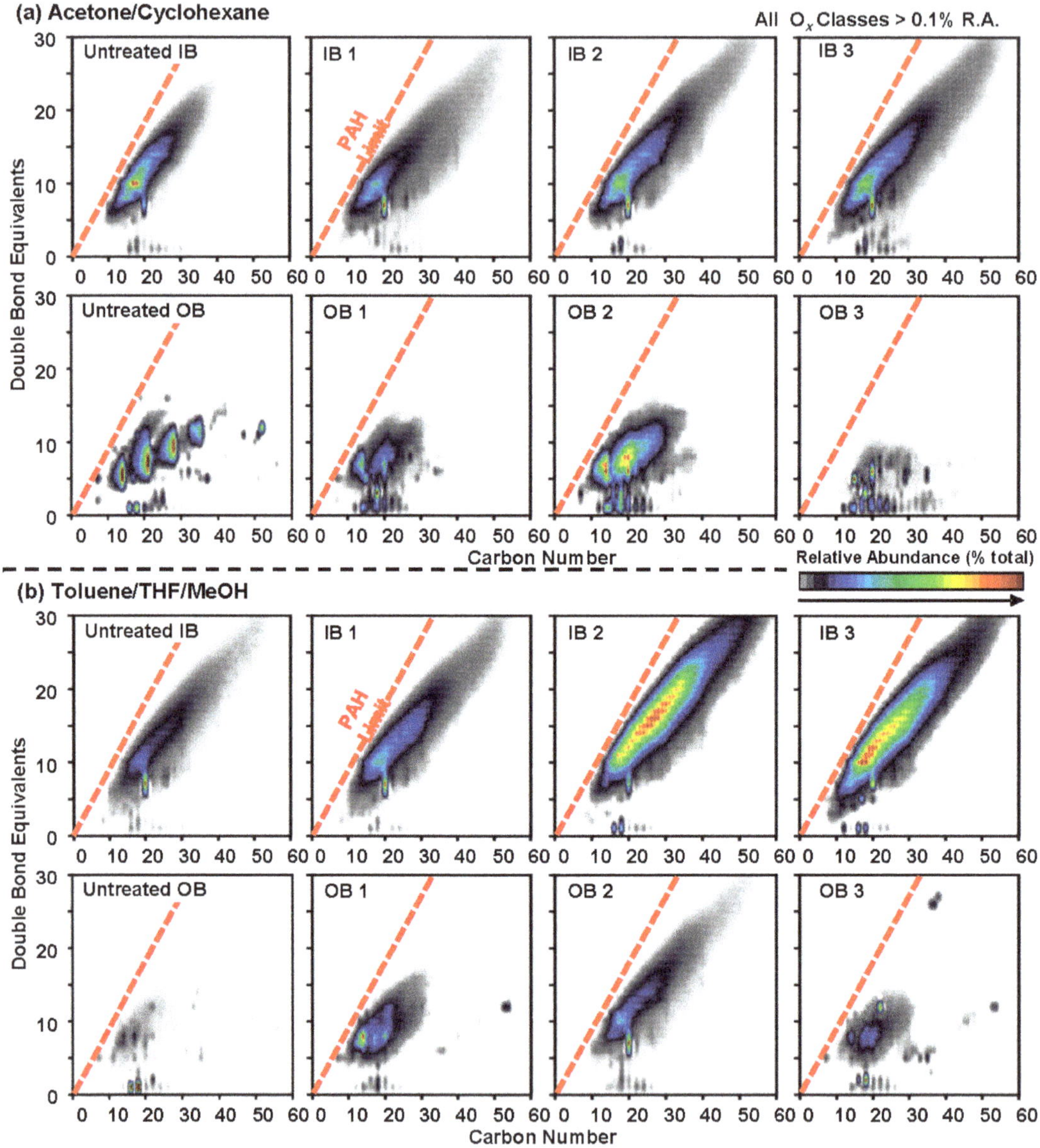

Figure 4. Compositional ranges of DBE vs. carbon numbers for (**a**) acetone/cyclohexane and (**b**) Tol/THF/MeOH fractions.

Figure 5 presents the compositional plots of H/C vs. O/C ratios via van Krevelen diagrams. The color scale represents the aromaticity index: non-aromatic (black, AI $\leq$ 0.5), aromatic (gold, 0.67 > AI > 0.5), and condensed aromatic (red, AI $\geq$ 0.6 [40]). The results indicate that after 48 h of water treatment, the amount of compositions (each data point contains at least one molecular formula) increases in each of the "aromaticity zones." In other words, after water treatment, the plots contain more data points. Particularly, the acetone fraction for IB reveals dramatic changes in composition.

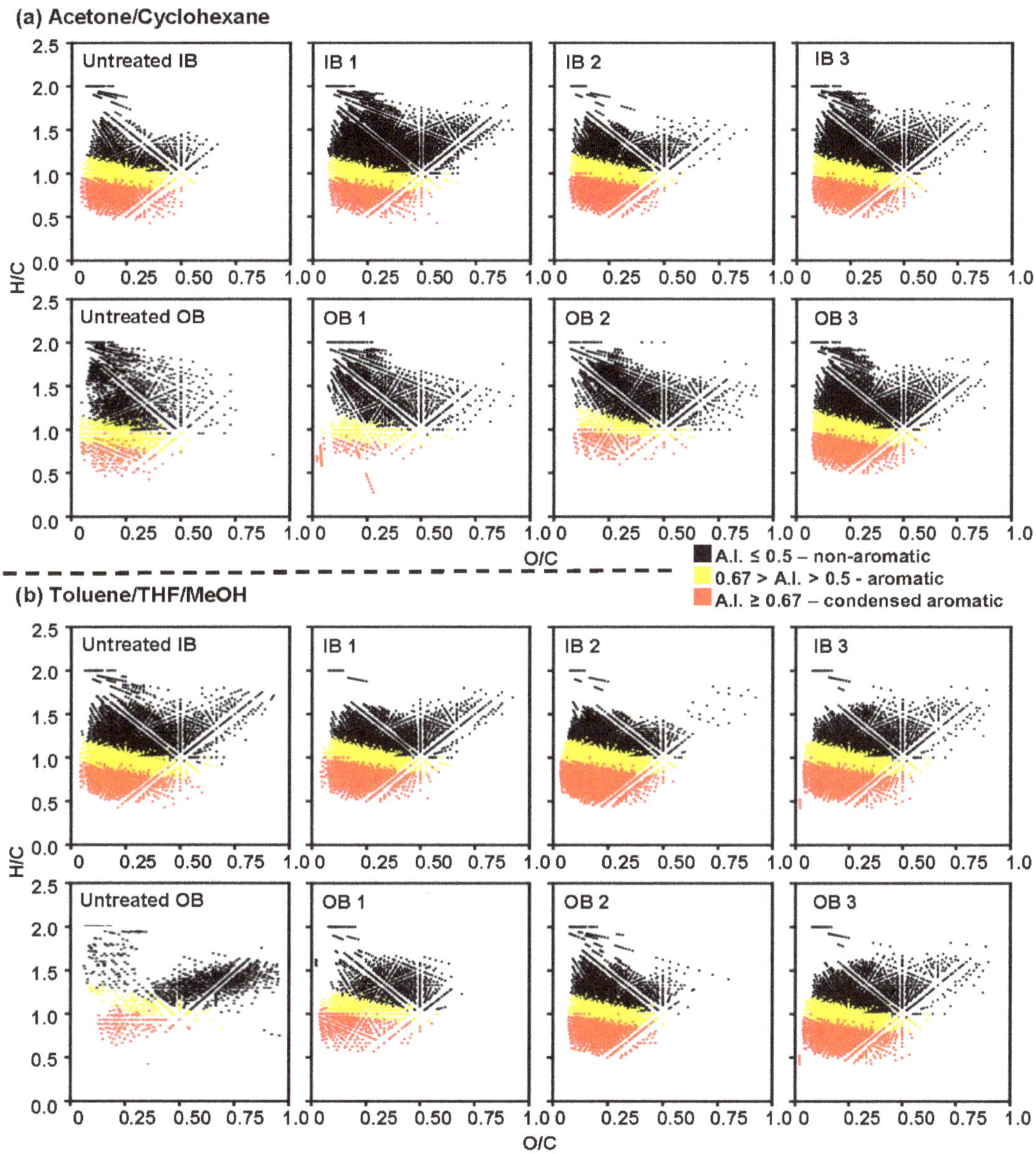

Figure 5. Compositional ranges of H/C vs. O/C ratios via van Krevelen diagrams for (**a**) acetone/cyclohexane and (**b**) Tol/THF/MeOH fractions.

Figure 6 presents expanded van Krevelen diagrams for the acetone fractions for IB before and after 48 h of water treatment. The color scale is the relative abundance as detected by MS and normalized within all the plotted compositions. "New" compositions are detected after 48 h of water treatment, as highlighted in red. Such species are likely captured from the water, over time, feature much higher oxygen content (O/C > 0.5), and are likely water soluble.

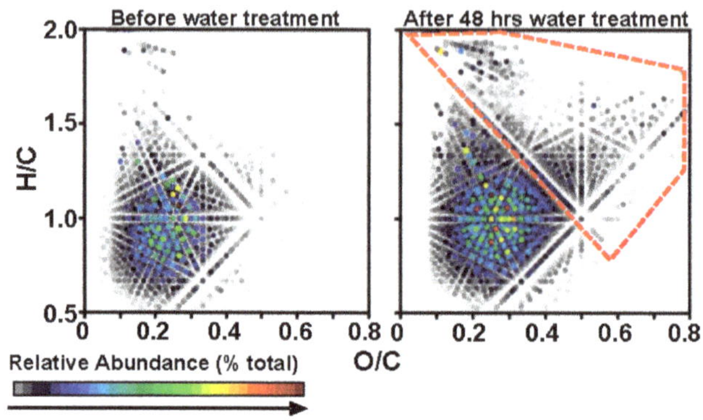

Figure 6. Expanded van Krevelen diagrams of acetone/cyclohexane fractions for IB before and after 48 h of water treatment. New compositions are detected after 48 h of water treatment, as shown in the red dashed box.

3.2. Discussion

The seafood market has become well-established over the past 40 years and contributes to addressing food supply and local job creation. However, the effluents from the downstream water of these farms include a wide array of contaminants, which result in eutrophication. Earlier studies in 2009 reported that eutrophication cost in the U.S. due to various reasons, such as global warming and water pollution from excess nutrients, is around $2.2 billion annually [41]. Western States (e.g., Idaho and California) have strong aquaculture industries, and the current limitation for fish farms in Idaho for releasing total phosphorus has been set at 0.075 mg/L. These permits were issued in 2012 (or later) and are currently under review for reissue due to high eutrophication risks. Lowering the nutrient limits by EPA and the Idaho Department of Environmental Quality (DEQ) could have devastating effects on the current operations, such as either closing fish farms or reducing their production rate, which can subsequently create significant economic hardship for the entire region.

Sustainable solutions using renewable materials (e.g., biochar) can improve water quality and the performance of the aquaculture industry. The renewable materials sector has frequently been seeking in-depth knowledge about the nature and intrinsic properties of biomaterials. The challenges in resolving the complex composition of thermochemically-derived biochar from biomass feedstocks has precluded its appropriate utilization in renewable applications, leaving open new opportunities for the development of alternative usages, such as water treatment and soil-plant health improvement.

The best result was observed with IB 3.1 (48 h treatment time), as presented in Table S1 (in Supplementary Materials), where the total nitrogen sequestration was 155 µg/g, and phosphorus sequestration was about 60 µg/g. The best result among the OB samples was OB2.2 (24 h treatment time) with a total nitrogen sequestration of 140.93 µg/g. The OB showed no capacity for phosphorus sequestration. In fact, the OB had a higher initial P concentration of 1200 µg/g, and the total P content decreased with the water treatment. The adsorption of nitrogen and phosphorus species by biochar can be better understood by correlating their molecular structures with the adsorption capacity. The adsorption capacity is attributed to chemisorption via electron donor–acceptor (π–π) interactions, interaction with O-containing functional groups forming hydrogen bonds, weaker physisorption (due to van der Waals forces), and ion exchange [42]. Biochar can adsorb N and P simultaneously. However, the inhibiting effect of adsorption among similarly charged anions has been reported due to competition for adsorption sites between NO_3^- and PO_4^{3-} and competition for exchange reactions [43]. The P removal capacity of the biochar is high when the surface is modified by ions of alkali metals, such as Ca and Mg [44]. The lower phosphorus sequestration levels of OB and IB could be attributed to the absence of alkali metal or Al and Fe cations in the biochar.

The degree of aromatic compounds present in biochar is influenced by the pyrolysis condition. Increasing the temperature, resident time, and air level during pyrolysis would result in a higher level of aromatic compounds in biochar [45]. Very high pyrolysis temperature results in the loss of a large number of polar functional groups, such as -CH, C=C, and C=O, resulting in enhanced hydrophobicity of the biochar surface and poor adsorption characteristics. Slow pyrolyzed biochar reveals predominantly fused aromatic C rings with H bonds due to surface carboxylate or phenolate groups [46]. An increase in the surface negative charge by substituting the edge of the C-H fraction of aromatic rings with COO^- groups under oxidation conditions increases the cation exchange capacity of the biochar. This helps with the adsorption of NH_4^+ but not with anions, such as NO_3^- or PO_4^{3-}.

The point of zero charge of the pine-derived biochar was reported to be at pH 8.47 [47]. In this study, the pH of untreated water was in the range of 7.61–7.67, which kept the biochar surface in an acidic condition and promoted the adsorption of anions, such as NO_3^-, NO_2^-, $H_2PO_4^-$, HPO_4^{2-} and PO_4^{3-}. It is also noted that the pH varied after the water treatment. The change in the pH (ΔpH) was in the range of −0.7 to 3.1, depending on the experimental conditions. A pH condition higher than 8.5 would render the biochar surface negatively charged and, therefore, limit the adsorption of anions. At higher pH conditions, the adsorption of nitrogen in the form of NH_4^+ could be promoted.

In this investigation, no intentional metal functional groups were introduced. Both the untreated IB and OB had a nitrogen content of <1.72 μg/g in the starting material, which indicated that both the biochar materials had very limited nitrogen functional groups for effective adsorption. Therefore, the non-metal functional group initially available for adsorption reaction could be the oxygen functional group, such as—COOH, -C=O, -COC-, -COO, -CHO, and -OH [48]. However, after exposure to nitrogen-rich water, the surface of the biochar could be modified with nitrogen species, and nitrogen functional groups could be present in the form of NH_4-N, NO_2-N, and NO_3-N. Furthermore, the initial higher concentrations of phosphorus (500 and 1200 μg/g in IB and OB, respectively) indicated that these materials possibly contained phosphorus functional groups, such as -P=O, -P-O-C, -P=OOH, and P-O-P chains [49]. The nitrogen functional groups and phosphorus functional groups are considered electroactive functional groups that can donate or accept electrons for binding the adsorbates.

The characterization of pine-based biochars derived from Idaho and Orgon sources, using FT-ICR MS with -ESI, indicated that the untreated IB contained compounds with six or fewer oxygens and dominance of O_4-O_6 classes (extracted by acetone, and therefore, these are less polar/highly aromatic compounds), as shown in Figure 3a. On the other hand, the untreated OB showed a higher content of highly aromatic/more polarizable compounds of the 2-oxygen class, as shown in Figure 3b (Toluene/THF/MeOH extraction). After water treatments, compounds with much higher oxygen numbers were detected in both types of biochar. This observation implied that high oxidizing conditions prevailed during the water treatment. The IB contained a large amount of highly aromatic, oxygen-containing molecules that are possibly alkyl depleted (as shown In Figure 4) as the abundance points were closer to the polycyclic aromatic hydrocarbon limit line. The compounds lying below the polycyclic aromatic hydrocarbon limit line will have planar organic molecules, while compounds lying above the line will have "buckybowl"-type structures (convex−concave pi-conjugated surfaces in hemispherical structures) [50]. On the other hand, the OB showed a limited compositional range in terms of the double bond equivalent (DBE = c − h/2 + n/2 +1 for the compound $C_cH_hN_nO_oS_s$) less than 20, and the abundance points lying well below the polycyclic aromatic hydrocarbon limit line. The van Krevelen diagrams (Figure 5) indicated that the contents of aromatic and condensed aromatic compounds were higher in untreated IB samples than in untreated OB samples. The abundance of aromaticity index (AI) increased with the water treatment.

Analysis of the nitrogen and phosphorus adsorption results and correlation of the results with the high-resolution mass spectroscopy data indicated that the adsorption mechanisms operating on the biochar surface are complex. Several mechanisms could synergistically interact, such as physisorption, chemisorption, electrostatic interaction, and ion exchange. Other researchers have reported similar observations [7,9,51]. Dai et al. (2020) investigated the utilization of biochar for nitrogen and phosphorus removal due to the high removal rate and other environmental benefits [52]. Earlier studies achieved a maximum capturing rate of around 0.70 mg/g for ammonium (NH_4^+), 40.63 mg/g for nitrate (NO_3^-), and 74.47 mg/g for phosphate (PO_4^{3-}), using different engineered biochars [7]. Higher adsorption characteristics of IB than OB could be attributed to the high concentrations of alkyl-depleted highly aromatic oxygen-containing molecules. The limited sequestration of phosphorus by IB could be due to the site-specific competition between nitrogen and phosphorus ions. Modification of the biochar with Ca^{2+}, Mg^{2+}, Fe^{2+}, or Al^{3+} cations could have improved the phosphorus sequestration capacity.

4. Conclusions

This study focuses on removing water contaminants from the effluent of aquaculture facilities, using woody-based biochars that can reduce eutrophication (oversupplied nutrients) from fish farm effluent. Particularly, this study investigates (a) the nitrogen-ammonia, nitrate-n + nitrite-n, and total phosphorus adsorption rate and amount, as well as saturation time at 3.7–7.5 L/min downstream water flow, (b) the adsorbing and releasing of materials after water treatment, using different pinewood biochars, and (c) the effectiveness level of the proposed water treatment method to remove a significant amount of contaminant, especially nutrients (nitrogen and phosphorus). The results indicate that the proposed approach can improve downstream fish farm waters by adsorbing micronutrients (e.g., nitrogen-ammonia, nitrate-n + nitrite-n, and total phosphorus). Water treatment time and biochar pH are the key parameters that had relatively strong associations with adsorbed compounds. Overall, the results appear more promising as a solution to reduce micronutrients from fish farm effluent. Longer residence times should be applied to identify the optimal treatment time for the adsorption of phosphorus and nitrogen compounds.

Molecular-level characterization of solvent-extracted organics from biochar materials suggests increased levels of highly oxygenated molecules as a function of increasing water-treatment time. Also, the results show the enrichment in organic species with higher molecular weight (higher carbon number) and increased double bond equivalents, with a compositional range similar to that of dissolved organic matter. Upon water treatment, extracted organics revealed higher abundances of compounds with higher H/C and O/C ratios. The engineered biochars can be reused as nutrient-rich soil conditioners or fertilizers after water treatment. This study concluded that the engineered biochars could sequester more nitrogen and phosphorous over more time. Also, the proposed approach can simultaneously increase fish production capacity and support the aquaculture industry in the western U.S. and other regions by improving water quality and enabling aquaculture expansion.

This study has identified the following areas that need further investigation to determine engineered biochar potential to effectively remove nitrogen and phosphorus from fish farm effluents:

- Investigation of different woody-based biochars with greater nutrient removal capacity from aquaculture effluents.
- Investigation of various engineered biochar production with different physicochemical properties to identify a material with higher micronutrient adsorption rates.
- Investigation of modified biochars with other chemical materials, such as magnesium chloride ($Mgcl_2$) or aluminum chloride ($Alcl_3$), to increase nutrient adsorption rate from eutrophic waters.
- Investigation of released materials or pollutants from biochar into downstream receiving waters.

- Investigation of used biochar after water treatment for soil improvement.

Supplementary Materials: The following supporting information can be downloaded at https://www.mdpi.com/article/10.3390/pr11102883/s1, Table S1: Biochar results before and after water treatment, Table S2: Water results before and after 0.5 h treatment with IB and OB.

Author Contributions: W.F.R.B., E.S., A.M., K.O., M.L.C.-P. and K.S.R. contributed to the conceptualization, methodology, data collection, analyses, and writing the manuscript. W.F.R.B. obtained Oregon biochar, E.S. produced Idaho biochar, and K.O. performed the field experiments. M.L.C.-P., A.M.M. and H.C. performed the characterization of extractable organic compounds 21 T FT-ICR MS. All authors contributed to writing the manuscript. All authors have read and agreed to the published version of the manuscript.

Funding: This research was funded by the United States Geological Survey, 104b grant.

Data Availability Statement: All FT-ICR MS data files are publicly available via the Open Science Framework (https://osf.io/2yxuk/ (1 October 2022)) at https://doi.org/10.17605/OSF.IO/2YXUK, accessed on 1 October 2022. All data needed to evaluate the conclusions are present in the manuscript and/or the Supplementary Materials.

Acknowledgments: The authors wish to acknowledge the United States Geological Survey (USGS) 104b grants for funding this project. The authors also wish to acknowledge inputs, resources, and support from the United States Department of Agriculture, Agricultural Research Service (USDA-ARS), University of Idaho Aquaculture Research Institute (UI-ARI), Riverence Provisions LLC, Idaho Water Resources Research Institute (IWRRI), National High Magnetic Field Laboratory (MagLab) supported by the National Science Foundation Division of Materials Research and Division of Chemistry through DMR 1644779, the State of Florida, and Oregon Biochar Solutions. Publication of this article was funded by the University of Idaho - Open Access Publishing Fund.

Conflicts of Interest: The authors declare no conflict of interest.

References

1. USDA Aquaculture. Available online: https://www.usda.gov/topics/farming/aquaculture (accessed on 5 December 2022).
2. Miller, A. The U.S. Is Not Harvesting as Many Fish as It Could, Driving Up Imports. Available online: https://www.cnbc.com/2022/10/05/us-underfishing-drives-up-imports.html (accessed on 5 December 2022).
3. Mirkouei, A. A Renewable Solution For Polluted Waters: Biochar Explained. Forbes. Available online: https://www.forbes.com/sites/aminmirkouei/2021/06/28/a-renewable-solution-for-polluted-waters-biochar-explained/ (accessed on 15 January 2022).
4. University of Idaho Aquaculture Research and Extension Activities Provided by UI Extension, Twin Falls County. Available online: https://www.uidaho.edu/extension/county/twin-falls/aquaculture (accessed on 5 December 2022).
5. Hersh, B.; Mirkouei, A.; Sessions, J.; Rezaie, B.; You, Y. A Review and Future Directions on Enhancing Sustainability Benefits across Food-Energy-Water Systems: The Potential Role of Biochar-Derived Products. *AIMS Environ. Sci.* **2019**, *6*, 379. [CrossRef]
6. Yoon, K.; Cho, D.-W.; Tsang, D.C.; Bolan, N.; Rinklebe, J.; Song, H. Fabrication of Engineered Biochar from Paper Mill Sludge and Its Application into Removal of Arsenic and Cadmium in Acidic Water. *Bioresour. Technol.* **2017**, *246*, 69–75. [CrossRef] [PubMed]
7. Yin, Q.; Wang, R.; Zhao, Z. Application of Mg–Al-Modified Biochar for Simultaneous Removal of Ammonium, Nitrate, and Phosphate from Eutrophic Water. *J. Clean. Prod.* **2018**, *176*, 230–240. [CrossRef]
8. Yin, Q.; Liu, M.; Ren, H. Removal of Ammonium and Phosphate from Water by Mg-Modified Biochar: Influence of Mg Pretreatment and Pyrolysis Temperature. *BioResources* **2019**, *14*, 6203–6218. [CrossRef]
9. Novais, S.V.; Zenero, M.D.O.; Barreto, M.S.C.; Montes, C.R.; Cerri, C.E.P. Phosphorus Removal from Eutrophic Water Using Modified Biochar. *Sci. Total Environ.* **2018**, *633*, 825–835. [CrossRef]
10. Shao, Y.; Zhong, H.; Mao, X.; Zhang, H. Biochar-Immobilized *Sphingomonas* sp. and *Acinetobacter* sp. Isolates to Enhance Nutrient Removal: Potential Application in Crab Aquaculture. *Aquac. Environ. Interact.* **2020**, *12*, 251–262. [CrossRef]
11. Lin, J.-C.; Cheng, A.-C.; Shiu, Y.-L.; Wong, Y.-C.; Yeh, S.-P.; Simangunsong, T.; Liu, C.-H. Using the Biochar Produced from Spend Mushroom Substrate to Improve the Environmental Condition of Aquaculture Pond. *Aquac. Res.* **2021**, *52*, 3532–3539. [CrossRef]
12. Silva, C.P.; Pereira, D.; Calisto, V.; Martins, M.A.; Otero, M.; Esteves, V.I.; Lima, D.L. Biochar-TiO_2 Magnetic Nanocomposites for Photocatalytic Solar-Driven Removal of Antibiotics from Aquaculture Effluents. *J. Environ. Manag.* **2021**, *294*, 112937. [CrossRef]
13. Chen, Y.; Su, J.; Zhao, H.; Li, J.-Y.; Wang, J.; Wang, Q.; Yin, J.; Jin, L. In-Situ Biochar Amendment Mitigates Dietary Risks of Heavy Metals and PAHs in Aquaculture Products. *Environ. Pollut.* **2022**, *308*, 119615. [CrossRef]
14. Hu, F.; Zhang, T.; Liang, J.; Xiao, J.; Liu, Z.; Dahlgren, R.A. Impact of Biochar on Persistence and Diffusion of Antibiotic Resistance Genes in Sediment from an Aquaculture Pond. *Env. Sci. Pollut. Res.* **2022**, *29*, 57918–57930. [CrossRef]

15. Hung, C.-M.; Chen, C.-W.; Huang, C.-P.; Dong, C.-D. N-Doped Metal-Free Biochar Activation of Peroxymonosulfate for Enhancing the Degradation of Antibiotics Sulfadiazine from Aquaculture Water and Its Associated Bacterial Community Composition. *J. Environ. Chem. Eng.* **2022**, *10*, 107172. [CrossRef]
16. McKenna, A.M.; Chacón-Patiño, M.L.; Chen, H.; Blakney, G.T.; Mentink-Vigier, F.; Young, R.B.; Ippolito, J.A.; Borch, T. Expanding the Analytical Window for Biochar Speciation: Molecular Comparison of Solvent Extraction and Water-Soluble Fractions of Biochar by FT-ICR Mass Spectrometry. *Anal. Chem.* **2021**, *93*, 15365–15372. [CrossRef]
17. Chacón-Patiño, M.L.; Rowland, S.M.; Rodgers, R.P. Advances in Asphaltene Petroleomics. Part 1: Asphaltenes Are Composed of Abundant Island and Archipelago Structural Motifs. *Energy Fuels* **2017**, *31*, 13509–13518. [CrossRef]
18. Marshall, A.G.; Hendrickson, C.L.; Jackson, G.S. Fourier Transform Ion Cyclotron Resonance Mass Spectrometry: A Primer. *Mass Spectrom. Rev.* **1998**, *17*, 1–35. [CrossRef]
19. Van Krevelen, D.W. Graphical-Statistical Method for the Study of Structure and Reaction Processes of Coal. *Fuel* **1950**, *29*, 269–284.
20. Kim, S.; Kramer, R.W.; Hatcher, P.G. Graphical Method for Analysis of Ultrahigh-Resolution Broadband Mass Spectra of Natural Organic Matter, the van Krevelen Diagram. *Anal. Chem.* **2003**, *75*, 5336–5344. [CrossRef]
21. Chacón-Patiño, M.L.; Rowland, S.M.; Rodgers, R.P. The Compositional and Structural Continuum of Petroleum from Light Distillates to Asphaltenes: The Boduszynski Continuum Theory As Revealedby FT-ICR Mass Spectrometry. In *The Boduszynski Continuum: Contributions to the Understanding of the Molecular Composition of Petroleum*; ACS Symposium Series; American Chemical Society: Washington, DC, USA, 2018; Volume 1282, pp. 113–171. ISBN 978-0-8412-3294-5.
22. Bare, R.; Struhs, E.; Mirkouei, A.; Overturf, K.; Small, B. Engineered Biomaterials for Reducing Phosphorus and Nitrogen Levels from Downstream Water of Aquaculture Facilities. *Processes* **2023**, *1*, 1029. [CrossRef]
23. Struhs, E.; Mirkouei, A.; You, Y.; Mohajeri, A. Techno-Economic and Environmental Assessments for Nutrient-Rich Biochar Production from Cattle Manure: A Case Study in Idaho, USA. *Appl. Energy* **2020**, *279*, 115782. [CrossRef]
24. Oregon Biochar Solutions Ultra-Pure Biochar from Forest Wood in the Pacific Northwest. Available online: https://www.chardirect.com (accessed on 5 December 2022).
25. Maynard, D.G.; Kalra, Y.P.; Crumbaugh, J.A. *Nitrate and Exchangeable Ammonium Nitrogen (Chapter 6)*; Taylor and Francis Group: Boca Raton, FL, USA, 2007; ISBN 978-0-8493-3586-0.
26. EPA. *Methods for Chemical Analysis of Water and Wastes, "Nitrogen, Nitrate-Nitrite" Method 353.2 (Colorimetric Automated, Cadmium Reduction)*; US Environmental Protection Agency: Cincinnati, OH, USA, 1984.
27. EPA. *Method 300.1, Determination of Inorganic Anions in Drinking Water by Ion Chromatography Rev. 1.0*; US Environmental Protection Agency: Cincinnati, OH, USA, 1993.
28. EPA. *Method 365.1, Revision 2.0: Determination of Phosphorus by Semi-Automated Colorimetry*; US Environmental Protection Agency: Cincinnati, OH, USA, 1993.
29. Bahureksa, W.; Borch, T.; Young, R.B.; Weisbrod, C.R.; Blakney, G.T.; McKenna, A.M. Improved Dynamic Range, Resolving Power, and Sensitivity Achievable with FT-ICR Mass Spectrometry at 21 T Reveals the Hidden Complexity of Natural Organic Matter. *Anal. Chem.* **2022**, *94*, 11382–11389. [CrossRef]
30. Niles, S.F.; Chacón-Patiño, M.L.; Chen, H.; McKenna, A.M.; Blakney, G.T.; Rodgers, R.P.; Marshall, A.G. Molecular-Level Characterization of Oil-Soluble Ketone/Aldehyde Photo-Oxidation Products by Fourier Transform Ion Cyclotron Resonance Mass Spectrometry Reveals Similarity Between Microcosm and Field Samples. *Environ. Sci. Technol.* **2019**, *53*, 6887–6894. [CrossRef]
31. Chacón-Patiño, M.L.; Moulian, R.; Barrère-Mangote, C.; Putman, J.C.; Weisbrod, C.R.; Blakney, G.T.; Bouyssiere, B.; Rodgers, R.P.; Giusti, P. Compositional Trends for Total Vanadium Content and Vanadyl Porphyrins in Gel Permeation Chromatography Fractions Reveal Correlations between Asphaltene Aggregation and Ion Production Efficiency in Atmospheric Pressure Photoionization. *Energy Fuels* **2020**, *34*, 16158–16172. [CrossRef]
32. Smith, D.F.; Podgorski, D.C.; Rodgers, R.P.; Blakney, G.T.; Hendrickson, C.L. 21 Tesla FT-ICR Mass Spectrometer for Ultrahigh-Resolution Analysis of Complex Organic Mixtures. *Anal. Chem.* **2018**, *90*, 2041–2047. [CrossRef]
33. Savory, J.J.; Kaiser, N.K.; McKenna, A.M.; Xian, F.; Blakney, G.T.; Rodgers, R.P.; Hendrickson, C.L.; Marshall, A.G. Parts-Per-Billion Fourier Transform Ion Cyclotron Resonance Mass Measurement Accuracy with a "Walking" Calibration Equation. *Anal. Chem.* **2011**, *83*, 1732–1736. [CrossRef]
34. Florida State University PetroOrg Software. Available online: https://nationalmaglab.org/user-facilities/icr/software/ (accessed on 2 September 2022).
35. Chacón-Patiño, M.L.; Rowland, S.M.; Rodgers, R.P. Advances in Asphaltene Petroleomics. Part 2: Selective Separation Method That Reveals Fractions Enriched in Island and Archipelago Structural Motifs by Mass Spectrometry. *Energy Fuels* **2018**, *32*, 314–328. [CrossRef]
36. Chacón-Patiño, M.L.; Smith, D.F.; Hendrickson, C.L.; Marshall, A.G.; Rodgers, R.P. Advances in Asphaltene Petroleomics. Part 4. Compositional Trends of Solubility Subfractions Reveal That Polyfunctional Oxygen-Containing Compounds Drive Asphaltene Chemistry. *Energy Fuels* **2020**, *34*, 3013–3030. [CrossRef]
37. Rodgers, R.P.; Mapolelo, M.M.; Robbins, W.K.; Chacón-Patiño, M.L.; Putman, J.C.; Niles, S.F.; Rowland, S.M.; Marshall, A.G. Combating Selective Ionization in the High Resolution Mass Spectral Characterization of Complex Mixtures. *Faraday Discuss.* **2019**, *218*, 29–51. [CrossRef]

38. Glattke, T.J.; Chacón-Patiño, M.L.; Hoque, S.S.; Ennis, T.E.; Greason, S.; Marshall, A.G.; Rodgers, R.P. Complex Mixture Analysis of Emerging Contaminants Generated from Coal Tar- and Petroleum-Derived Pavement Sealants: Molecular Compositions and Correlations with Toxicity Revealed by Fourier Transform Ion Cyclotron Resonance Mass Spectrometry. *Environ. Sci. Technol.* **2022**, *56*, 12988–12998. [CrossRef]
39. Qi, Y.; Xie, Q.; Wang, J.-J.; He, D.; Bao, H.; Fu, Q.-L.; Su, S.; Sheng, M.; Li, S.-L.; Volmer, D.A.; et al. Deciphering Dissolved Organic Matter by Fourier Transform Ion Cyclotron Resonance Mass Spectrometry (FT-ICR MS): From Bulk to Fractions and Individuals. *Carbon Res.* **2022**, *1*, 3. [CrossRef]
40. Koch, B.P.; Witt, M.; Engbrodt, R.; Dittmar, T.; Kattner, G. Molecular Formulae of Marine and Terrigenous Dissolved Organic Matter Detected by Electrospray Ionization Fourier Transform Ion Cyclotron Resonance Mass Spectrometry. *Geochim. Cosmochim. Acta* **2005**, *69*, 3299–3308. [CrossRef]
41. Environmental Protection Agency (EPA). Nutrient Indicators Dataset. Available online: https://www.epa.gov/nutrient-policy-data/nutrient-indicators-dataset (accessed on 5 December 2022).
42. Zhu, D.; Pignatello, J.J. Characterization of Aromatic Compound Sorptive Interactions with Black Carbon (Charcoal) Assisted by Graphite as a Model. *Environ. Sci. Technol.* **2005**, *39*, 2033–2041. [CrossRef]
43. Liu, H.; Dong, Y.; Wang, H.; Liu, Y. Ammonium Adsorption from Aqueous Solutions by Strawberry Leaf Powder: Equilibrium, Kinetics and Effects of Coexisting Ions. *Desalination* **2010**, *263*, 70–75. [CrossRef]
44. Feng, Y.; Zhao, D.; Qiu, S.; He, Q.; Luo, Y.; Zhang, K.; Shen, S.; Wang, F. Adsorption of Phosphate in Aqueous Phase by Biochar Prepared from Sheep Manure and Modified by Oyster Shells. *ACS Omega* **2021**, *6*, 33046–33056. [CrossRef] [PubMed]
45. Shafizadeh, F.; Sekiguchi, Y. Development of Aromaticity in Cellulosic Chars. *Carbon* **1983**, *21*, 511–516. [CrossRef]
46. Liu, Z.; Quek, A.; Kent Hoekman, S.; Balasubramanian, R. Production of Solid Biochar Fuel from Waste Biomass by Hydrothermal Carbonization. *Fuel* **2013**, *103*, 943–949. [CrossRef]
47. Santos, L.B.; Striebeck, M.V.; Crespi, M.S.; Ribeiro, C.A.; De Julio, M. Characterization of Biochar of Pine Pellet. *J. Therm. Anal. Calorim.* **2015**, *122*, 21–32. [CrossRef]
48. Zhang, C.; Zhang, L.; Gao, J.; Zhang, S.; Liu, Q.; Duan, P.; Hu, X. Evolution of the Functional Groups/Structures of Biochar and Heteroatoms during the Pyrolysis of Seaweed. *Algal Res.* **2020**, *48*, 101900. [CrossRef]
49. Yang, Y.; Piao, Y.; Wang, R.; Su, Y.; Liu, N.; Lei, Y. Nonmetal Function Groups of Biochar for Pollutants Removal: A Review. *J. Hazard. Mater. Adv.* **2022**, *8*, 100171. [CrossRef]
50. Kanao, E.; Kubo, T.; Naito, T.; Matsumoto, T.; Sano, T.; Yan, M.; Otsuka, K. Differentiating π Interactions by Constructing Concave/Convex Surfaces Using a Bucky Bowl Molecule, Corannulene in Liquid Chromatography. *Anal. Chem.* **2019**, *91*, 2439–2446. [CrossRef]
51. Zhang, Z.; Yan, L.; Yu, H.; Yan, T.; Li, X. Adsorption of Phosphate from Aqueous Solution by Vegetable Biochar/Layered Double Oxides: Fast Removal and Mechanistic Studies. *Bioresour. Technol.* **2019**, *284*, 65–71. [CrossRef]
52. Dai, Y.; Wang, W.; Lu, L.; Yan, L.; Yu, D. Utilization of Biochar for the Removal of Nitrogen and Phosphorus. *J. Clean. Prod.* **2020**, *257*, 120573. [CrossRef]

processes

MDPI

Article

Response of Nitrogen Removal Performance and Microbial Distribution to Seasonal Shock Nutrients Load in a Lakeshore Multicell Constructed Wetland

Jing Yuan [1,2], Bin Wang [1,2], Zeying Hou [1,2], Jiayu Peng [1,2], Dan Li [3] and Zhaosheng Chu [1,2,*]

1 National Engineering Laboratory for Lake Pollution Control and Ecological Restoration, Chinese Research Academy of Environmental Sciences, Beijing 100012, China; yuan.jing@craes.org.cn (J.Y.)
2 State Environmental Protection Key Laboratory for Lake Pollution Control, Chinese Research Academy of Environmental Sciences, Beijing 100012, China
3 College of Urban Construction, Nanjing Tech University, Nanjing 211816, China
* Correspondence: chuzs@craes.org.cn

Abstract: Multicell constructed wetlands (MCWs) on lakeshores are a prospective treatment technique. However, the factors affecting the nutrient removal performance of lakeshore MCWs at the field scale are unclear. This study chose a field-scale lakeshore MCW with the highest mass removal efficiency (approximately 49,175.12 mg m^{-2} day^{-1}) for total nitrogen removal in the wet season to investigate the response of nitrogen removal and microbial distribution to seasonal shock nutrients load. The mass loading rates in the wet season were as high as 43~72 times over those in the dry season. Hence, a storage pond (SP), as a forebay retention cell, was necessary to mitigate the shock loads of the influent, which is beneficial to nitrogen removal of the MCW system. The two major genera in the sediments are heterotrophic nitrification–aerobic denitrification bacteria, and the abundance and species of the nitrogen-related functional genera were higher in the wet season than the dry season. According to the results of redundancy analysis, the hydraulic residence time (29.4%, F = 2.2, $p < 0.1$) and hydraulic loading rate (85.9, F = 36.5, $p < 0.05$) were the major factors explaining microbial community variation, instead of environmental factors (temperature, pH, and dissolved oxygen). The shock loads of influent and the periodic saturation in sediments contributed to a complicated oxygen and nitrogen nutrient exchange environment resulting in higher abundance and species of nitrogen-related microbes, which is beneficial to nitrogen removal in lakeshore MCWs. The results provided a scientific basis for the optimal design of constructed wetlands on lakeshores.

Keywords: lakeshore wetlands; nitrogen removal; nonpoint source pollution; hydraulic characters; microbial community; spatial and temporal regulation

1. Introduction

Nonpoint source (NPS) pollution is one of the crucial sources resulting in surface water pollution, especially agricultural runoff, as the primary form in rural nonpoint sources and contains excessive nitrogenous and phosphorus compounds that cause eutrophication in recipient water bodies, in which the total nitrogen concentration could be as high as nearly 50 mg L^{-1} [1–3]. Lakeshores are transition zones between terrestrial and aquatic environments and are vital to mitigate diffuse NPS pollution before entering a lake [4]. To control NPS pollution, constructed wetlands (CWs) in lakeshores have become a popular eco-friendly and cost-effective method to capture and purify parts of nutrient-rich runoff water before they reach lakes, reducing nutrient loads discharging into lakes [5].

In contrast to common CWs, multicell constructed wetlands (MCWs) in lakeshores increase the nutrient purification process through biodegradation, adsorption, and assimilation and consist of various kinds of treatment cells, such as vertical flow, horizontal flow, and eco-floating beds [3,6,7]. Studies have shown that the combination of vertical and horizontal flow increases the removal of nitrogen, which is applied to purify NPS pollution [8]. In addition, the greater water depths in the cells of MCWs contribute to storing and purifying wastewater because there is more space without increasing the land source of MCW systems [9]. Li et al. [10] studied 63 MCWs grouped into four kinds depending on various water depths and treatment processes, and the maximum nitrogen removal efficiency was 63.7%. Wang et al. [11] constructed two pilot-scale MCW systems to treat micro-polluted river water with a low C/N ratio, and the total nitrogen removal rates were between 19.56% and 34.84%. The design and operational parameters were vital factors in contaminant removal efficiency, including hydraulic residence time (HRT), hydraulic loading rate (HLR), mass loading rate (MLR), water depth, influent contamination, etc. [12,13]. However, in previous studies and reviews, most empirical design and operational parameters were highly related to lab-scale/pilot-scale experiments [11,14]. Moreover, those at the field scale have rarely been reported, which are easily affected by environmental factors, such as climate, rainfall, and season, which are uncontrollable but crucial to the operational performance of MCWs in engineering practice [15]. Therefore, the hydraulic character of the MCW system, characterized by the seasonal variation in the design and operational parameters, is one of the critical factors affecting nutrient removal.

Numerous studies have demonstrated that assimilation and storage by plants and microbes serve as the primary nitrogen removal pathways in MCWs [5,16–18]. The microbial nitrogen cycle within CWs is driven by nitrogen-related microorganisms, such as heterotrophic denitrification genera, facultative autotrophic denitrification genera, autotrophic denitrification genera, anaerobic ammonium oxidation genera, and nitrification genera [9,19]. Environmental conditions in MCW systems influence nitrogen conversions, while the wet and dry seasons, with variations in precipitation volume, are essential in the microbial nitrogen cycle. For example, surface runoff caused by rainfall brings more contaminants into MCWs, and nitrogen-converting microbes fluctuate in microbial communities during the wet and dry seasons along with changes in nitrogen contents [20]. Lakeshore zones are regarded as biochemical hotspots of nitrogen cycling because of the fluctuating hydrological regime and intensive material exchange between inland areas and water, particularly oxygen and nitrogen nutrients [21]. However, our knowledge about nitrogen contaminants carried by NPS pollution and the effect of design parameters in lakeshore MCWs on microbial community distribution is limited.

In this study, we aimed to investigate the response of nitrogen removal and microbial community structure to the design and operational parameters (HRT, HLR, and MLR) in a field-scale lakeshore MCW. We hypothesized that the microbial community related to the fluctuating hydrology and nitrogen nutrients of the influent in the wet and dry seasons was beneficial to nitrogen removal and purification of the lakeshore wetlands. Therefore, the primary objectives of this study were (1) to characterize the hydraulic conditions of the MCW system depending on the seasonal variation of the design and operational parameters, (2) to identify the spatial and temporal characteristics of the microbial structure, and (3) to disclose the effects of seasonal parameters on the removal performance and the microbial community distribution.

2. Materials and Methods

2.1. Sites and MCW System Description

Lake Erhai is a subtropical plateau lake in the city of Dali, Yunnan Province, Southwest China. The mean annual precipitation of Lake Erhai is approximately 1000 mm, and more than 50% of the yearly precipitation concentrated in the summer months is attributed to the monsoon climate. It is evident in Figure S1 that most of the precipitation in 2021 was concentrated from May to October, which was regarded as the wet season in this study, while the other months were the dry season, with precipitation being less than 50 mm. The maximum and minimum temperatures of 2021 were 23.65 °C (in June) and 3.26 °C (in January), respectively, and the average annual temperature was 16.14 °C.

Lake Erhai is in the mesotrophic status and has been at risk of changing to the eutrophic stage in recent years due to NPS pollution [22]. To control the water quality, many MCWs are applied to detain stormwater runoff and purify the discharge before entering Lake Erhai. The Lake Erhai Ecological Restoration and Constructed Wetlands Project located on the lakeshore, which is 129 km long and approximately 900 hectares surrounding the lake, broke ground in 2019 and was completed recently in succession. The MCW at Dazhuang Village was chosen as the research site because it is one of the surrounding-lake MCWs with typical design and structures (Figure 1a). The influent of the MCW system was from nonpoint source pollution, whose effluent was mixed with that of other MCWs and eventually discharged into Lake Erhai. Figure 1b shows an overview diagram, and Table S1 shows the basic information of this MCW system. The chosen MCWs consisted of four cells, including a storage pond (SP), followed by an ecological oxidation pond (EP), and then a horizontal subsurface flow constructed wetland (HSCW) and surface flow constructed wetland (SFCW). The treatment process is shown in Figure 1c, and photographs of the four cells are shown in Figure S2.

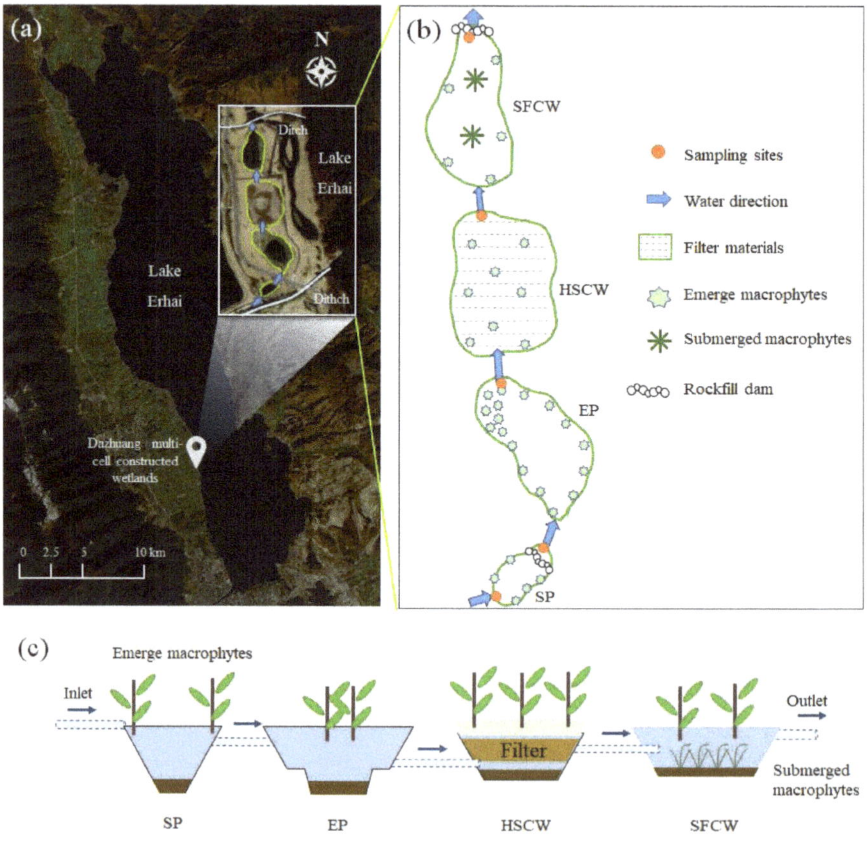

Figure 1. Basic information on the MCW system: (**a**) location of Lake Erhhai and the MCW system; (**b**) the layout of MCW system; (**c**) the treatment process. SP: storage pod; EP: ecological oxidation pond; HSCW: horizontal subsurface flow constructed wetland; SFCW: surface flow constructed wetland.

2.2. Sample Collection and Analyses

Water samples were collected with sample bottles monthly from January to December 2021 in the inlets and outlets of the MCW system (from five sites and repeated three times) at a certain depth range (0–10 cm) for further analysis. All samples were analyzed immediately after sampling. The environmental parameters, namely the temperature (T), pH, and dissolved oxygen (DO) content, were analyzed with specific portable meters (Mettler Company, Columbus, OH, USA).

To investigate the microbial community in the MCW system, surface sediment samples (0–10 cm) were collected using the sediment column sampler device at four cells. Three parallel samples in the device were mixed thoroughly as one composite sample and repeated at two random points beside the sample site per cell. The sediment samples were collected in the dry season in January 2021 and in the wet season in June 2021. All the fresh samples were kept in polyethylene plastic zip lock bags and refrigerated at −60 °C.

Total nitrogen (TN), nitrate (NO_3^--N), and ammonium (NH_4^+-N) concentrations were measured according to standard methods [23]. TN concentrations were determined by the alkaline potassium persulfate digestion UV spectrophotometric method. NO_3^--N concentrations were detected with an ultraviolet–visible spectrophotometer at a wavelength of 410 nm. NH_4^+-N concentrations were assayed using Nessler's reagent colorimetry method.

2.3. Microbial Analyses

The DNA from the MCW system's sediment samples was extracted using a PowerSoil™ DNA extraction kit (MO BIO Laboratories, Carlsbad, CA, USA). A NanoDrop™ 2000 spectrophotometer (Thermo Fisher Scientific, Waltham, MA, USA) was used to quantify and evaluate the genomic DNA. Universal primers 515FmodF (5′-GTGYCAGCMGCCGCGGTAA-3′) and 806RmodR (5′-GGAC TACNVGGGTWTCTAAT-3′) were used for polymerase chain reaction (PCR) [9]. PCRs were performed as described in a previous study [9]. The purified amplicons were pooled in equimolar and paired-end (2 × 250 bp) sequencing and conducted using an Illumina MiSeq PE300 platform (Illu-mina Inc., San Diego, CA, USA) at Majorbio Bio-Pharm Technology Co., Ltd. (Shanghai, China). The raw data were processed and analyzed following the guidelines described by Li et al. [9]. The process results were conducted to further analyze the OUT-level species richness and alpha diversity index values (Shannon, ACE, Chao1, and Simpson indices).

2.4. Statistical Analyses

The mass loading rate per square meter (MLR, mg m^{-2} day^{-1}), hydraulic loading rate (HLR, m day^{-1}), and hydraulic residence time (HRT, day) were calculated using Equations (1)–(3), respectively. Equations (4)–(6) were used to determine the removal efficiency (RE, %), mass removal rate per square meter (MRR, mg m^{-2} day^{-1}), and mass removal rate per cubic meter (MRR_V, mg m^{-3} day^{-1}), respectively.

$$MLR = C_{in} \times Q_{in}/A \tag{1}$$

$$HLR = Q_{in}/A \tag{2}$$

$$HRT = V/Q_{in} \tag{3}$$

$$RE = (C_{in} \times Q_{in} - C_{out} \times Q_{out})/(C_{in} \times Q_{in}) \times 100\% \tag{4}$$

$$MRR = (C_{in} \times Q_{in} - C_{out} \times Q_{out})/A \times 1000 \tag{5}$$

$$MRR_V = (C_{in} \times Q_{in} - C_{out} \times Q_{out})/V \times 100 \tag{6}$$

Here, C_{in} and C_{out} represent the nitrogen concentrations in influent and effluent (mg L^{-1}), respectively; Q_{in} and Q_{out} represent the daily flow water quantity of influent and effluent (m^3 day^{-1}), respectively; V is the water capacity of each cell (m^3); and A is the water surface area of each cell (m^2).

The constrained ordination of redundancy analysis (RDA) was performed by Canoco 5.0. Furthermore, linear and nonlinear regression simulations and Spearman's rank correlation coefficients were calculated and performed using Origin 2021 and Microsoft Excel 365.

3. Results and Discussion

3.1. Characteristics of Water Quantity and Nitrogen Concentrations in Influent and Effluent

3.1.1. Water Quantity

The water quantity of the influent varied greatly between the dry and wet seasons. The amount of influent in the dry season was below 10×10^3 m^3 $month^{-1}$, except January (18.66×10^3 m^3 $month^{-1}$), while most of that in the wet season was more than 100×10^3 m^3 $month^{-1}$, and the highest water quantity was in August, which was as high as 643×10^3 m^3 $month^{-1}$. The amounts of influent water in the wet season were approximately 20–60 times greater than those in the dry season, attributed to the concentrated precipitation in summer, and the rainfall amount decreased sharply from November. The characteristics of the water quantity of the influent were consistent with the local rainfall patterns, as shown in Figure S1. The influent water from the lakeshore wetlands' ditches mostly consisted of irrigation drainage and wastewater from the catchment. In the wet season, the amount of the influent rises gradually along with the increasing surface runoff for rainy reasons and agricultural runoff during the growing season [3].

3.1.2. Nitrogen Concentration

The nitrogen concentrations of the influent and effluent are shown in Figure 2, and the seasonal variation in water quality is presented in Table 1. Compared to the influent, the nitrogen concentrations in the effluent decreased significantly, suggesting effective nitrogen interception in the lakeshore MCW. Overall, the average TN, NO_3^--N, and NH_4^+-N concentrations in the influent over the year were 21.47 ± 12.23 mg L^{-1}, 18.58 ± 12.04 mg L^{-1}, and 0.45 ± 0.39 mg L^{-1}, respectively. The average nitrogen concentrations in the wet season were higher than in the dry season, indicating that nutrients carried by rain runoff were the primary reason for the water quality fluctuations. In the wet season, the TN concentrations were 14.75–52.90 mg L^{-1}, while those in the dry season were all below 24.05 mg L^{-1}. Thus, the lakeshore MCW needs to cope with the high nutrient load of stormwater runoff, especially in the flood season. Considering the nitrogen composition in the influent, the higher concentration of NO_3^--N (2.40–47.57 mg L^{-1}) than that of NH_4^+-N (0.08–1.53 mg L^{-1}) resulted from the high percentage of agricultural lands (65.07%) in the catchment (Table S2), and NO_3^--N was the primary form of nitrogen fertilizer in the Lake Erhai basin, which is evidence that agricultural nonpoint sources were the primary pollution source for the lakeshore MCW [3].

Table 1. Nitrogen concentration and removal performance of the lakeshore MCW.

		c_{in} (mg L^{-1})	c_{out} (mg L^{-1})	RE (%)	MRR (mg m^{-2} day^{-1})	MRR_v (mg m^{-3} day^{-1})	MLR (mg m^{-2} day^{-1})	HLR (m day^{-1})	HRT (day)
Wet (n = 6)	TN	27.30 ± 14.36	10.11 ± 8.83	65.95 ± 25.01	15,741.45 ± 17,419.21	21680.63 ± 23991.40	846.65 ± 1173.06	0.84 ± 0.90	2.42 ± 3.20
	NO_3^--N	24.43 ± 13.89	8.54 ± 8.43	68.61 ± 26.41	15,088.87 ± 17,337.83	20,781.84 ± 23,879.31	791.85 ± 1135.34		
	NH_4^+-N	0.62 ± 0.49	0.21 ± 0.19	53.64 ± 30.57	295.09 ± 339.01	406.43 ± 466.91	13.53 ± 14.57		
Dry (n = 6)	TN	15.64 ± 4.99	9.21 ± 7.63	46.22 ± 38.60	180.03 ± 108.41	247.96 ± 149.32	15.60 ± 7.29	0.04 ± 0.02	24.53 ± 9.71
	NO_3^--N	12.72 ± 5.31	6.86 ± 7.36	49.66 ± 42.61	154.41 ± 127.25	212.67 ± 175.26	11.30 ± 5.35		
	NH_4^+-N	0.29 ± 0.10	0.15 ± 0.07	45.11 ± 25.16	5.77 ± 6.33	7.95 ± 8.72	0.33 ± 0.31		
Year (n = 12)	TN	21.47 ± 12.23	9.66 ± 8.26	56.09 ± 32.68	7960.74 ± 14,281.83	10,964.30 ± 19,670.30	431.12 ± 902.14	0.44 ± 0.74	13.47 ± 13.45
	NO_3^--N	18.58 ± 12.04	7.70 ± 7.96	59.13 ± 35.22	7621.64 ± 14,052.49	10,497.25 ± 19,354.43	401.57 ± 867.22		
	NH_4^+-N	0.45 ± 0.39	0.18 ± 0.15	49.37 ± 27.06	150.43 ± 274.02	207.19 ± 377.40	6.93 ± 12.00		

Data are presented as means ± standard deviation (S.D.) with n = 3 or n = 6.

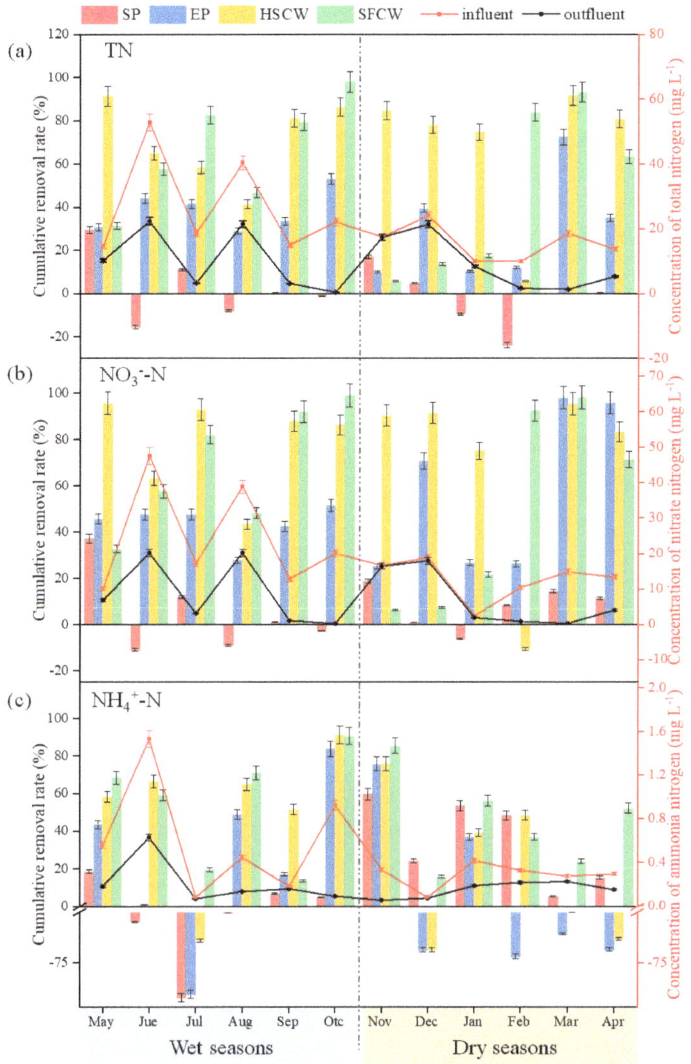

Figure 2. Removal rates of the lakeshore MCW and nitrogen concentrations in the influent and effluent. (**a**) TN; (**b**) NO_3^--N; (**c**) NH_4^+-N. SP: storage pod; EP: ecological oxidation pond; HSCW: horizontal subsurface flow constructed wetland; SFCW: surface flow constructed wetland.

3.2. Seasonal Variation in Design and Operational Parameters

The significant differences in water quantity and quality of the influent between various seasons can greatly influence the operational parameters of the lakeshore MCW, such as the mass loading rate (MLR), hydraulic loading rate (HLR), and hydraulic residence time (HRT) [13]. Seasonal variations in MLR, HLR, and HRT were investigated as follows, and monthly data are provided in Table S3.

3.2.1. MLR

MLR is a comprehensive result of the water quantity and quality of the influent. As shown in Table 1, the average MLRs of TN, NO_3^--N, and NH_4^+-N in the wet season were 846.65 ± 1173.06, 791.85 ± 1135.34, and 13.53 ± 14.57 mg·m^{-2} day^{-1}, respectively, which were higher than those in the dry season (15.60 $\pm$ 7.29, 11.30 $\pm$ 5.35, and 0.33 $\pm$ 0.31 mg m^{-2} day^{-1}, respectively). The higher MLRs in the wet season were attributed to the large water quantity and high nitrogen concentration caused by concentrated rainfall. It is a common feature of the surface runoff in the basin of Lake Erhai that the nutrient loading fluctuates seasonally [9,10]. The seasonal variation was more evident in the lakeshore MCW for the average nitrogen MLRs in the wet season, which were as high as 43~72 times over those in the dry season, while the storing multipond CWs further than 10 km away from the lake were only 2~3 times in early research [24]. Therefore, the seasonal shock loads of the lakeshore MCW are characterized by intermittent and impulse-type discharges into the receiving waters. Interestingly, the average nitrogen concentration in the wet season was only 1.75–2.16 times that in the dry season, so the water quantity of the influent was affected more than the nitrogen concentration on the MLRs in the wet season, while the opposite trend was observed in the dry season. This is another piece of evidence illustrating that the large water quantity caused by concentrated rainfall was one of the critical factors for fluctuating loads in the wet season. In addition, intermittent desiccations of EP were observed in the lakeshore MCW in the dry season, as shown in Figure S3, which were attributed to less rainfall and rapid evaporation in the dry season.

3.2.2. HLR

HLR is one of the vital parameters related to the storage capacity or amount of influent [13,24]. As shown in Table 1, the HLRs of the lakeshore MCW were usually below 0.05 m day^{-1} in the dry season, while those in the wet season were almost beyond 0.05 m day^{-1}, especially the HLR as high as 2.60 m day^{-1} in August. As shown in Figure S1, the precipitation in August was the highest over the year, and the evaporation was lower than that of other months in the wet season, such as July and September. The influent amount was in accordance with the precipitation trend, which was highly related to HLR, so HLRs in August were highest over the year. Because of the high response between the precipitation and HLR, the rainfall fluctuations in both the wet and dry seasons contributed to the fluctuating HLR. Ruan et al. [25] investigated the influent factors on the purification of 75 field-scale CWs in rural and urban areas, and the results showed that the HLRs of 45 CWs were below 0.5 m day^{-1}, while those of only 5 CWs exceeded 1 m day^{-1}. Overall, the HLRs of the lakeshore MCW were slightly higher than those of other constructed wetlands. Comparing the HLRs in various MCW cells, as shown in Figure 3, the HLRs of the SP cell in the wet season (1.93–61.67 m day^{-1}) were significantly higher than those of other cells. The HLR sharply decreased after flowing through the SP cell, and the highest HLR of the next cell (the EP) in the wet season was only 10.72 m day^{-1}, far from that of SP (61.67 m day^{-1}). The value gap between the two cells was the biggest in August, and the HLR decreased by as much as 82.50% after flowing through the SP. Therefore, it was such an efficient way to mitigate the HLR fluctuation and shock loads in the wet season that the storage pond was set as the first cell of the lakeshore MCW.

3.2.3. HRT

HRT dominates the reaction time, directly affects the biochemical process, and leads to fluctuations in the removal efficiency [13]. The HRTs of the lakeshore MCW were also different between the wet and dry seasons, as shown in Table 1. The HRTs in the wet season were shorter (0.28–8.87 days) than those in the dry season (9.63–38.32 days), attributed to the concentrated rainfall, which increased the water quantity in the influent. It was reported that the HRT and HLR were highly related in MCW purification, and the HRT was negatively related to the HLR of the influent [13], as shown in Figure 3. The unusual decrease in the HRT in January may be related to the manual cleaning of sludge of the MCWs during the low water level. The HRTs for the four MCW cells are shown in Figure 3, and the shortest HRT (the annual average value was approximately 0.79 days) in all the MCW cells appeared in the SP, which was related to the highest HLR in the SP among the four MCW cells. The relatively long HRTs in the other three MCW cells benefited from adequate storage in the SP, which provided sufficient reaction time for biochemical remediation during nitrogen purification in the MCW system. This was evidence that the forebay SP was essential in the lakeshore MCW to cope with the water quantity and quality fluctuations before entering the lake.

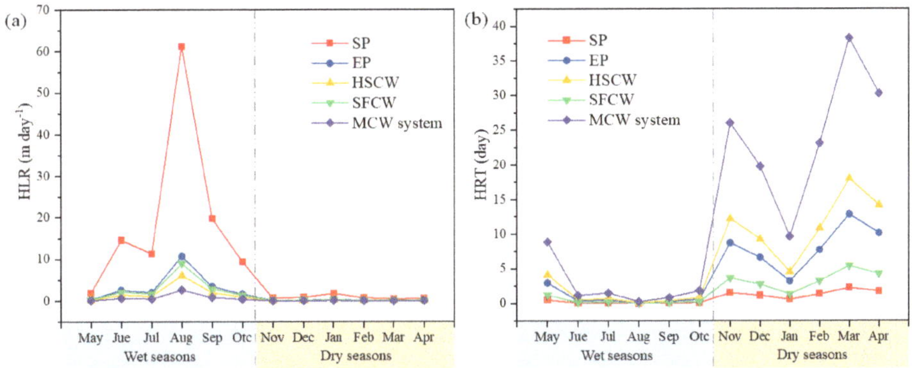

Figure 3. HLRs (**a**) and HRTs (**b**) of various cells of the lakeshore MCW system. SP: storage pod; EP: ecological oxidation pond; HSCW: horizontal subsurface flow constructed wetland; SFCW: surface flow constructed wetland.

3.3. Comparison of Nitrogen Removal Performances

3.3.1. Seasonal nitrogen removal performance of the lakeshore MCW system

The REs of the wetland in various seasons are shown in Table 1 and Table S3. The differences in N removal observed under various seasons and treatment processes are statistically significant ($p < 0.05$), as shown by one-way ANOVA in Table S4. Overall, the nitrogen REs of the wetland in the wet season were higher than those in the dry season because the average TN, NH_4^+-N, and NO_3^--N REs of the wetland in the wet season were 65.95 ± 25.01%, 53.64 ± 30.57%, and 68.61 ± 26.41%, respectively. In comparison, those in the dry season were 46.22 ± 38.60%, 45.11 ± 25.16%, and 49.66 ± 42.61%, respectively. Similarly, the MRRs and MRR_vs in the wet season were higher than those in the dry season. A higher nitrogen concentration and water quantity of the influent were positively related to a higher nitrogen removal effect in the wet season, illustrating a high potential purification and operation effectiveness of the lakeshore MCW in the wet season. Compared to TN and NO_3^--N, the lower NH_4^+-N REs were related to the relatively low NH_4^+-N concentrations in the influent. The average value over the year was 0.45 ± 0.39 mg L^{-1} (as shown in Table 1), which was far from the surface water quality standard class III (1 mg L^{-1}) and the first grade A standard of the municipal sewage discharge standard (5 mg L^{-1}) in China [26]. This was also the main reason for the significantly lower MRR and MRR_v of NH_4^+-N than those of TN and NO_3^--N.

3.3.2. Nitrogen Removal Performances of the Various Cells

The nitrogen concentrations of the influent along the flowing route in the lakeshore MCW are shown in Figure S4, directly showing the purification effectiveness. The nitrogen concentrations of the influent were significantly decreased after flowing through each wetland cell in general, although those of certain cells in several months slightly fluctuated. To further analyze the contribution of the RE of each cell to those of the MCW system, the cumulative REs from the influent of the first cell (EP) to the outlet of the other three cells were sequentially calculated, as shown in Figure 2, that is, the cumulative RE of the SFCW precisely showed the nitrogen purification ability of the MCW system.

The SP was the first cell in the lakeshore MCW, and the maximum design water depth (1 m) was higher than that in the other cells, as shown in Table S1. As the above results showed, it functioned as a pre-retention pond and was essential in water quantity regulation. The REs of nitrogen in the SP were not as satisfactory as those in other cells, and the nitrogen concentrations in the outlet of the SP were even slightly increased, especially in June and August. It was a sacrifice for purification effectiveness caused by nitrogen resuspension carried by the particles in the influent. In the EP, the RE of NO_3^--N was the best (the average value was 46.42 ± 28.00%), and that of TN was the second best (the average value was 32.60 ± 20.93%). This was attributed to the extensive plant coverage and varying water depth in the EP, contributing to the creation of different aerobic, facultative, and anaerobic oxygen zones in the sediments, which had an advantage for denitrification by microorganisms. For the HSCW, the REs of TN and NH_4^+-N were the highest (the average values were 55.46 ± 28.01% and 29.43 ± 22.61%, respectively) among the four cells. The differences between the REs in the two seasons of the HSCW were not as noticeable, indicating that the primary mechanism for nitrogen retention of the HSCW was pollution adsorption by the lava filters, which were barely affected by the seasons. The high nitrogen retention performance was attributed to the vertical flow path in the filtration bed of the HSCW, which provides adequate contact between the wastewater and lava fillers and is barely affected by the seasons. The adsorption mechanism for nitrogen retention in the HSCW was quicker and had more retention ability compared to the biological treatment by the plants and microbes in the wetland. In the SFCW, the nitrogen concentrations in the influent were lower than those in the effluent in other seasons, except that the NH_4^+-N REs of the SFCW in the dry season were approximately 13.32%. Nevertheless, the performance of the SFCW in nitrogen removal has potential because the nitrogen concentrations of the influent were lower than those in other kinds of MCWs [7,11].

3.4. Effects of Seasonal Parameters on the Nitrogen Removal

To further analyze the coupling effects of the seasonal variation in parameters and nitrogen removal properties, Spearman's rank correlation coefficients between the nitrogen removal performance (RE and MRR) and the design and operational parameters in the two seasons were analyzed (Figure 4). The corresponding relations in the wet season differed from those in the dry season. The RE was negatively related to the MLR and HLR in the wet season while being positively associated with the HRT. A lower MLR or HLR and a longer HRT benefited settling down the particles in the influent and for adequate biochemical degradation of nitrogen pollutants, contributing to a satisfactory removal rate [27]. However, the relationships between MRR and the design and operational parameters were reversed, and a high mass loading of pollutants in the wastewater increased the MRR because the MRR was positively correlated with the MLR or HLR. Surprisingly, the MRR was negatively related to the HRT, meaning that the longer the wastewater retention was, the lower the efficiency of contaminant mass intercepted. Therefore, to balance RE and MRR to achieve a satisfactory purification performance of wetlands, the HRT should be limited to a specific range that is not very high [24]. In addition, a higher nitrogen concentration in the influent was related to a higher RE and MRR, meaning that the potential nitrogen retention capacity was not achieved in CWs in the wet season.

In the dry season, REs of TN and NO_3^--N were slightly negatively related to HLR, MLR, nitrogen concentrations, water quantity, and MRR, which were significantly different from those in the wet season and inconsistent with the previously reported storing MCW [24]. This was attributed to the excessively shallow water depth of lakeshore MCWs in the dry season, and even the bottoms of certain ponds were exposed (Figure S3). Hence, the biochemical purification process of the MCW was not typical, which was also why the nitrogen REs were lower than those in the wet season. The nitrogen concentrations and water quality were lower in the dry season than the wet season, so the purification demand was lower. The primary ecosystem service functions of the lakeshore MCWs were not retaining and purifying the contaminants in this period.

Season		MLR(TN)	MLR(NO_3^--N)	MLR(NH_4^+-N)	HLR	HRT	c_{in}(TN)	c_{in}(NO_3^--N)	c_{in}(NH_4^+-N)	Q_{in}	Q_{out}
Wet season	RE(TN)	-0.1428	-0.1429	-0.2021	-0.0857	0.0862	0.1433	0.1502	-0.1429	-0.0878	-0.0863
	RE(NO_3^--N)	-0.0857	-0.0854	-0.0880	-0.0286	0.0299	0.0857	0.0871	-0.0866	-0.0289	-0.0285
	RE(NH_4^+-N)	-0.1431	-0.1427	0.4857	-0.2571	0.2571	0.3143	0.3189	0.5429	-0.2571	-0.2573
	MRR(TN)	0.9428 **	0.9429 **	0.8859 *	0.8862 *	-0.8902 *	0.7714	0.7723	0.2016	0.8861 *	0.8857 *
	MRR(NO_3^--N)	0.9431 **	0.9435 **	0.8849 *	0.8901 *	-0.8867 *	0.7719	0.7733	0.1987	0.8842 *	0.8860 *
	MRR(NH_4^+-N)	0.5429	0.5442	0.9444 **	0.3714	-0.3720	0.7143	0.7149	0.6571	0.3714	0.3721
Dry season	RE(TN)	-0.6572	-0.5430	-0.0289	-0.4890	0.4866	-0.2573	-0.4283	-0.2572	-0.4862	-0.4869
	RE(NO_3^--N)	-0.6581	-0.5429	-0.0291	-0.4857	0.4849	-0.2568	-0.4291	-0.2565	-0.4856	-0.4870
	RE(NH_4^+-N)	-0.0861	-0.2571	0.7143	0.0857	-0.0848	-0.4857	-0.3714	0.8857 *	0.0860	0.0857
	MRR(TN)	-0.6566	-0.5433	-0.0286	-0.4857	0.4863	-0.2575	-0.4286	-0.2571	-0.4870	-0.4857
	MRR(NO_3^--N)	-0.7143	-0.1429	-0.1433	-0.3725	0.3721	-0.2580	-0.2571	-0.4286	-0.3721	-0.3714
	MRR(NH_4^+-N)	-0.0871	-0.4279	0.9429 **	0.3714	-0.3714	-0.7143	-0.6571	0.9988 ***	0.3715	0.3721

-1.00 -0.75 -0.50 -0.25 0.00 0.25 0.50 0.75 1.00

Figure 4. Heatmap of Spearman's rank correlation coefficients. * $0.01 < p \leq 0.05$, ** $0.001 < p \leq 0.01$, *** $p \leq 0.001$.

3.5. Spatial and Temporal Variations in Microbial Community

The functions and behaviors of bacteria are one of the major factors affecting the performance of lakeshore MCWs. The response of the microbial community in various cells to fluctuating nitrogen loads was investigated to promote the possible reason for the seasonal variability in the nitrogen removal performance of lakeshore MCWs. The Illumina sequencing statistics of different sediment samples from the four cells in the two seasons are compared in Tables S5 and S6. To ensure sequencing uniformity [28], the species richness and alpha diversity index values were calculated based on 5845 operational taxonomic units (OTUs) (ranging from 1065 to 5334 per sample). The ACE and Chao1 indices reflect the microbial community richness, and Sobs is the observed richness. The Shannon and Simpson indices indicate bacterial community diversity. Therefore, the microbial community richness and diversity in the sediments of the EP and SFCW were relatively higher than those in the other cells, while those of the HSCW were the lowest. This was attributed to the high-density planting in the SFCW and EP, which was essential in rhizosphere bacterial activity, community richness, diversity, and function [7,15]. A complicated oxygen environment with different aerobic, facultative, and anaerobic oxygen zones easily occurs in the rhizoplane and water for the oxygen released from the root, which is beneficial for increasing the microbial community richness [29]. As shown in Table S1, the HSCW and SFCW had a high plant coverage (more than 90%), followed by EP. The amphibious plants (*Iris pseudacorus* L. and *Cyperus involucratus*) in the HSCW, as shown in Figure S2, covered the surface soil layer upon the filter layer through which the water fluid passes. The sediment moisture and oxygen concentrations in the HSCW differed from the root microenvironments underwater in other cells, resulting in significant variations in the microbial community richness [30]. Similarly, the microbial richness and diversity in the sediment were lower than those in the dry season. Because of the deeper depth of the ponds in the wet season for the concentrated precipitation, the oxygen in the rhizoplane and water was lower in the wet season than in the dry season, which was not friendly to the aerobic microorganisms.

To investigate the difference in the microbial community between the two seasons in the four cells, PCA based on Bray–Curtis dissimilarity was performed, and the result is shown in Figure 5a. Microbial communities are grouped by cells resulting from water quality and rhizosphere environmental changes in the four cells. For the SFCW samples, the differences within groups were more significant than those between groups, as shown in Figure S5. This was attributed to the substantial changes in sediment moisture between the two seasons, and the oxygen content and substrate availability were consequently changed after SFCW rewetting [31]. Interestingly, the samples in the wet season clustered together while those in the dry season could be separated, indicating that the treatment technologies in each cell and spatial heterogeneity had more significant potential impacts on microbial community distribution in the dry season than in the wet season.

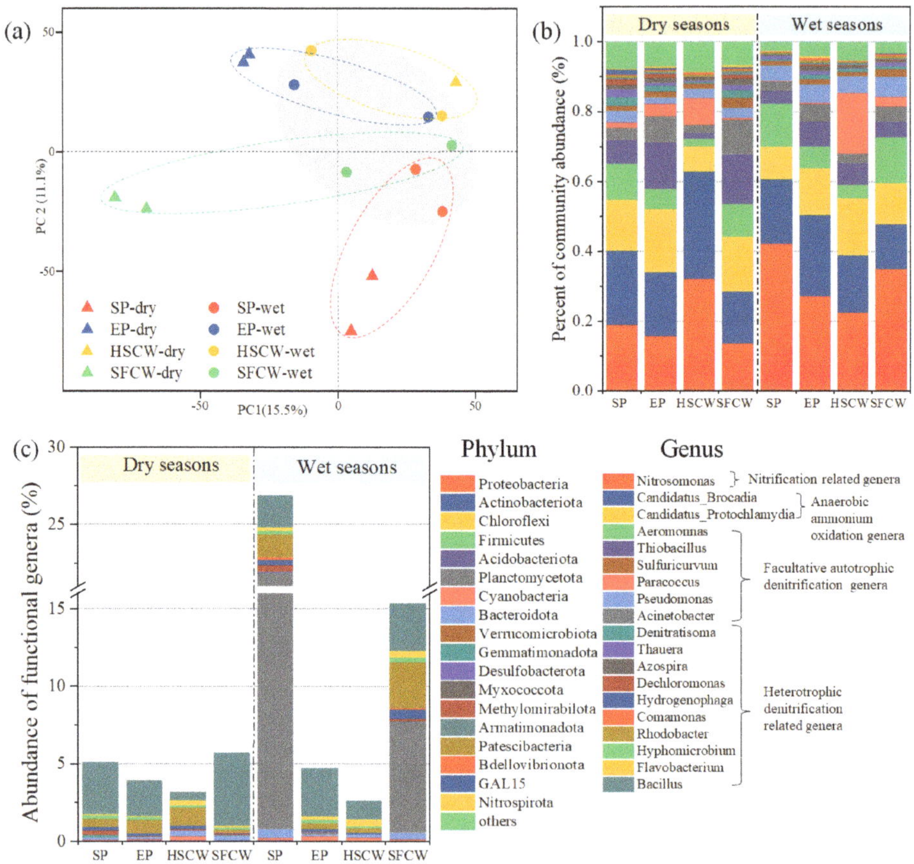

Figure 5. PCA and relative abundance of bacteria at the phylum and genus levels. (**a**) PCA analysis of bacterial communities from different sediments of four cells in various seasons (R = 0.1338, $p < 0.05$); (**b**) relative abundances of bacteria at the phylum level; (**c**) relative abundances of nitrogen-related functional genera. SP: storage pod; EP: ecological oxidation pond; HSCW: horizontal subsurface flow constructed wetland; SFCW: surface flow constructed wetland.

The microbial community composition and bacterial relative abundances at the phylum level are shown in Figure 5b. The most abundant phylum detected in the lakeshore MCW was *Proteobacteria*, ranging from 13.75% to 42.21%, followed by *Actinobacteriota* (12.73–30.67%), *Chloroflexi* (7.37–18.18%), *Firmicutes* (2.12–13.12%), *Acidobacteriota* (1.64–13.35%), *Planctomycetota* (2.12–9.71%), and *Cyanobacteria* (0.17–17.52%). *Proteobacteria* is identified as the most abundant phylum in the sediments of constructed wetlands for wastewater treatment [32]. Most of the nitrifying bacteria belong to *Proteobacteria* and *Nitrospirota* [33]. The high relative abundance of *Proteobacteria* but incredibly low *Nitrospirota* indicated that *Proteobacteria* executed NH_4^+-N conversion in the lakeshore MCWs. In the wet season, the total relative abundance of denitrificans, such as *Proteobacteria*, *Chloroflexi*, *Firmicutes*, and *Nitrospirota*, was higher than that in the dry season, which was the result of differences in the water temperatures and the operating parameters of the lakeshore CWs between the two seasons [34].

The structural evolution of the known bacterial communities at the genus level is shown in Figure S6, and nitrogen-related functional genera were abundant in the lakeshore MCW, as shown in Figure 5c, which demonstrated the complex biological nitrogen removal pathways in the lakeshore MCW, such as nitrification, denitrification, and anammox. *Bacillus* and *Rhodobacter* are the two major genera in the lakeshore MCW, and both are heterotrophic nitrification–aerobic denitrification (HN-AD) bacteria. The decrease in *Bacillus* was accompanied by the increase in *Rhodobacter*, which may be attributed to the competitive relationship. Research on HN-AD bacteria has been popular at home and abroad for altering the traditional theory suggesting that nitrification can be performed only by autotrophic bacteria and denitrification can be carried out only under anaerobic conditions, which is evidence for simultaneous nitrification and denitrification (SND), and *Pseudomonas* was one of the typical SND bacteria [35]. The greater number of nitrifying bacteria in the wet season explained the higher nitrogen removal in the lakeshore MCW. The abundance and species of the nitrogen-related functional genera were higher in the wet season than the dry season. This was the reason for the better nitrogen removal of the lakeshore MCWs in the wet season. Interestingly, the relative abundance of *Acinetobacter* in the SP and SFCW in the wet season was higher than that in the dry season. *Acinetobacter* is the primary bacterium related to nitrogen and phosphorus in enhanced biological phosphorus removal (EBPR) sludge, for which the high temperature of 22.5 °C during the wet season was most conducive to growth and reproduction [36,37]. In addition, a high phosphate concentration and an extensive phosphorus loading of the influent in the wet season resulted in phosphorus-accumulating bacteria growth. The high relative abundance of *Acinetobacter* showed potential phosphorus removal of the lakeshore MCW, which needs further investigation.

3.6. Effects of Seasonal Parameters on the Microbial Community

The activity of microorganisms was directly affected by environmental conditions such as pH, DO, and temperature, and the nitrogen cycle in sediments of the lakeshore MCW was affected by the nitrogen concentrations and the design and operational parameters, such as the MLR, HLR, and HRT [9,20,38]. To explore the effects of seasonal shock nutrient loads on microbial communities, RDA was used to identify the relationships between the microbial community (based on abundances at the phylum and genus levels), design and operational parameters (based on nitrogen concentrations, MLR, HLR, and HRT), and environmental conditions (based on pH, DO, and temperature). The interactive-forward-selection analysis method was chosen to select the best subset of the variables to summarize the variation in species composition. Figure 6 displays the structural evolutions of the phylum and nitrogen-related genera in various cells. The RDA results showed that the major variables to the variation in species composition were HRT, HLR, MLR, TN and NO_3^--N concentrations, DO, and pH.

As shown in Figure 6a, the RDA between phyla distributions in each MCW cell and seasonal factors indicated that the HRT (29.4%, F = 2.2, $p < 0.1$) was the main factor explaining variation in the microbial community composition at the phylum level (Figure 6a). A longer HRT value supports adequate reaction time for biotic processes, which affects the microbial community [24] and explains the excellent nitrogen removal rate in the lakeshore MCW. In addition, the design and operational parameters related to the shock loadings (such as MLR and HLR) also affected the microbial community, especially the MLR of NO_3^--N, because nitrate level was one of the key factors that control the structure of denitrifying communities [39]. In addition, the microbial communities at the phylum level were clustered by season in the SP, EP, and SFCW, demonstrating that the influence of seasonal changes in surface runoff on the biochemical purification of the lakeshore MCW is significant. However, the microbial communities of the HSCW in the two seasons were similar and separated from the other cells for the various nitrogen retention mechanisms dependent on the adsorption of the filler, which is the reason for the higher nitrogen RE in the HSCW.

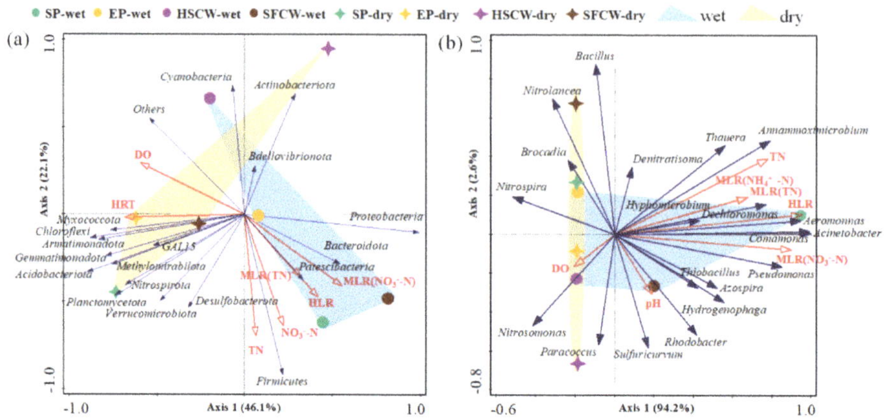

Figure 6. RDA between microbial community, operational parameters, and environemntal conditions in the lakeshore MCW: (**a**) at the phylum level; (**b**) at the nitrogen-related functional guera level. SP: storage pod; EP: ecological oxidation pond; HSCW: horizontal subsurface flow constructed wetland; SFCW: surface flow constructed wetland.

As shown in Figure 6b, HLR, the design and operational parameter, was the significant factor explaining variation in the nitrogen-related functional genera composition (85.9, $F = 36.5$, $p < 0.05$). The typical character of shock loads carried by the surface runoff in the Lake Erhai basin by intermittent and impulse-type discharges into the wetlands resulted in large HLRs, bringing enormous disturbances and water exchange to enrich oxygen in the water column by changing hydrological and hydraulic performance [9], which was beneficial to create complicated environmental conditions for the biochemical process. In addition, anaerobic ammonium oxidation genera and facultative autotrophic denitrification-related genera (such as *Aeromonnas*, *Acinetobacter*, and *Pseudomonas*) were positively correlated with nitrogen concentration, and heterotrophic denitrification related genera (such as *Comamonas*, *Hydrogenophaga*, *Azospira*, and *Rhodobacter*) were positively correlated with NO_3^--N MLR. A high nitrogen concentration in the influent was conducive to the domestication and aggregation of nitrogen-related functional genera. Additionally, nitrification-related genera (such as *Nitrosomonas* and *Nitrospira*) were positively correlated with DO. Periodic saturation of wetland sediments due to significant changes in seasonal streamflow would create fluctuating oxic/anoxic conditions for ammonium nitrification by *Nitrosomonas* and subsequent nitrite oxidation by *Nitrospira*, which is vital for nitrogen retention [40]. Therefore, the seasonal shock loads contributed to a unique community of the nitrogen-related microbes, which benefited the favorable nitrogen removal in the lakeshore MCW.

3.7. Regulation Suggestion for Lakeshore Multicell Constructed Wetlands

The lakeshore MCW is a prospective technique to control the nonpoint sources of water pollution before the surface runoff enters the lake. Compared to the traditional CWs, the seasonal shock loads were a key factor affecting the nitrogen removal performance. So, a forebay retention cell was necessary to mitigate the shock loads. To improve the efficiency of the practice engineering program, a combination of the storing pond/wetland project with the adjacent eco-purification CW project was a good idea for the best management practice (BMP). In addition, the methods contributed to a complicated oxygen environment, which is friendly to the rhizosphere bacterial community diversity and nitrogen-related microbe richness, which were suggested to improve the nitrogen removal performance. For example, various treatment processes, including the horizontal subsurface flow constructed wetland with filters and the surface flow constructed wetland, showed good nitrogen retention properties in this study. Finally, this study allowed us to gain more insight into how nitrogen and phosphorus collaborative purification of lakeshore MCWs will be necessary in the future.

4. Conclusions

In summary, this study investigated the response of nitrogen removal and microbial community structure to the design and operational parameters (HRT, HLR, and MLR) in a field-scale lakeshore MCW. The surface runoff by intermittent and impulse-type discharges into the wetlands resulted in the seasonal shock loads of the influent. The fluctuating hydrology and nitrogen nutrients of the influent in the wet and dry seasons were so evident that the MLRs in the wet season were as high as 43~72 times over those in the dry season. The high MLRs in the wet season were one of the major factors affecting the nitrogen retention rates of the lakeshore wetland, which resulted in higher average removal efficiencies of TN, NH_4^+-N, and NO_3^--N in the wet season than in the dry season. The sufficient effective storage in SP contributed to the adequate HRTs in the MCW system under the shock loads. Therefore, a forebay retention cell is crucial to alleviate the shock loads in the lakeshore MCW system. High-density planting in the SFCW and EP contributed to a complicated oxygen environment with different aerobic, facultative, and anaerobic oxygen zones that easily occur in the rhizoplane and water, which played an essential role in rhizosphere bacterial community richness and diversity. The HRT and HLR were the major factors explaining the phylum and nitrogen-related genera variations. More nitrifying and denitrifying bacteria appeared in the wet season, and alternating aerobic and anaerobic conditions occur in the water quantity fluctuation in the wet and dry seasons, resulting in coupled nitrification and denitrification, which explains the higher nitrogen removal in the lakeshore CWs. The results provide a scientific basis for the optimal design of constructed wetlands in lakeshores.

Supplementary Materials: The following supporting information can be downloaded at: https://www.mdpi.com/article/10.3390/pr11092781/s1. Figure S1: Rainfall, evaporation and influent amount in 2021; Figure S2: Photographs of the cells in the lakeshore of MCW; Figure S3: Comparisons of EP in wet and dry seasons; Figure S4: The nitrogen concentrations of the influent along the flowing rate. (**a**) TN; (**b**) NO_3^--N; (**c**) NH_4^+-N; Figure S5: The visualization of beta diversity and analysis of similarities (ANOSIM); Figure S6: Relative abundances of the top 100 genera in various sediments of cells; Table S1: Basic information of the multi-cell constructed wetlands; Table S2: Land use in the catchment around the lakeshore MCW; Table S3: Monthly nitrogen concentration and removal performance of the lakeshore MCW; Table S4: Sequencing statistics for the sediment samples of lakeshore MCW; Table S5: Microbial diversity indices for the cells of lakeshore MCW; Table S6: Microbial diversity indices for the cells of lakeshore MCW.

Author Contributions: Conceptualization, J.Y. and D.L.; methodology, J.Y. and Z.H.; investigation, B.W.; software, J.Y. and J.P.; formal analysis, J.Y. and B.W.; writing—original draft preparation, J.Y.; writing—review and editing, J.Y. and Z.C.; visualization, J.Y.; supervision, Z.C. All authors have read and agreed to the published version of the manuscript.

Funding: This research was funded by the National Key Research and Development Project of the Ministry of Science and Technology of China (No. 2021YFC3201003).

Data Availability Statement: Not applicable.

Conflicts of Interest: The authors declare no conflict of interest.

References

1. Qin, B.; Zhang, Y.; Zhu, G.; Gao, G. Eutrophication control of large shallow lakes in China. *Sci. Total Environ.* **2023**, *881*, 163494. [CrossRef] [PubMed]
2. Wang, J.; Lu, J.; Zhang, Z.; Han, X.; Zhang, C.; Chen, X. Agricultural non-point sources and their effects on chlorophyll-a in a eutrophic lake over three decades (1985–2020). *Environ. Sci. Pollut. Res.* **2022**, *29*, 46634–46648. [CrossRef] [PubMed]

3. Li, D.; Ye, B.; Hou, Z.; Chu, Z.; Zheng, B. Long-term performance and microbial distribution of a filed-scale storing multi-pond constructed wetland with Ottelia acuminata for the treatment of non-point source pollution. *J. Clean. Prod.* **2020**, *262*, 121367. [CrossRef]
4. Cole, L.J.; Stockan, J.; Helliwell, R. Managing riparian buffer strips to optimise ecosystem services: A review. *Agric. Ecosyst. Environ.* **2020**, *296*, 106891. [CrossRef]
5. Walton, C.R.; Zak, D.; Audet, J.; Petersen, R.J.; Lange, J.; Oehmke, C.; Wichtmann, W.; Kreyling, J.; Grygoruk, M.; Jabłońska, E.; et al. Wetland buffer zones for nitrogen and phosphorus retention: Impacts of soil type, hydrology and vegetation. *Sci. Total Environ.* **2020**, *727*, 138709. [CrossRef] [PubMed]
6. Fernandez-Fernandez, M.I.; Vega, P.T.M.D.; Jaramillo-Morán, M.A.; Garrido, M. Hybrid constructed wetland to improve organic matter and nutrient removal. *Water* **2020**, *12*, 2023. [CrossRef]
7. Wang, T.; Xiao, L.; Lu, H.; Lu, S.; Li, J.; Guo, X.; Zhao, X. Nitrogen removal from summer to winter in a field pilot-scale multistage constructed wetland-pond system. *J. Environ. Sci.* **2022**, *111*, 249–262. [CrossRef]
8. Byeon, C.; Nam, B.E. An assessment of the ecological functions of a sustainable structured wetland biotope (SSB). *Ecol. Eng.* **2020**, *145*, 105723. [CrossRef]
9. Li, D.; Chu, Z.; Zeng, Z.; Sima, M.; Huang, M.; Zheng, B. Effects of design parameters, microbial community and nitrogen removal on the field-scale multi-pond constructed wetlands. *Sci. Total Environ.* **2021**, *797*, 148989. [CrossRef]
10. Li, D.; Chu, Z.; Huang, M.; Zheng, B. Multiphasic assessment of effects of design configuration on nutrient removal in storing multiple-pond constructed wetlands. *Bioresour. Technol.* **2019**, *290*, 121748. [CrossRef]
11. Wang, T.; Xiao, L.; Lu, H.; Lu, S.; Zhao, X.; Liu, F. Effect of the influent substrate concentration on nitrogen removal from summer to winter in field pilot-scale multistage constructed wetland–pond systems for treating low-C/N river water. *Sustainability* **2021**, *13*, 12456. [CrossRef]
12. Mendes, L.R. Nitrogen Removal from Agricultural Subsurface Drainage by Surface-Flow Wetlands: Variability. *Processes* **2021**, *9*, 156. [CrossRef]
13. Li, D.; Zheng, B.; Liu, Y.; Chu, Z.; He, Y.; Huang, M. Use of multiple water surface flow constructed wetlands for non-point source water pollution control. *Appl. Microbiol. Biotechnol.* **2018**, *102*, 5335–5368. [CrossRef] [PubMed]
14. Zhao, Y.; Wang, H.; Dong, W.; Chang, Y.; Yan, G.; Chu, Z.; Ling, Y.; Wang, Z.; Fan, T.; Li, C. Nitrogen removal and microbial community for the treatment of rural domestic sewage with low C/N ratio by A/O biofilter with Arundo donax as carbon source and filter media. *J. Water Process. Eng.* **2020**, *37*, 101509. [CrossRef]
15. Chen, D.; Gu, X.; Zhu, W.; He, S.; Wu, F.; Huang, J.; Zhou, W. Denitrification- and anammox-dominant simultaneous nitrification, anammox and denitrification (SNAD) process in subsurface flow constructed wetlands. *Bioresour. Technol.* **2019**, *271*, 298–305. [CrossRef]
16. Song, S.; Wang, B.; Yang, T.; Gu, Y.; Sheng, S.; Zhao, D.; An, S.; Li, A. Performance and Bacteria Communities of a Full-Scale Constructed Wetland Treating the Secondary Effluent after Multi-Years' Operation. *Processes* **2023**, *11*, 1469. [CrossRef]
17. Erler, D.; Eyre, B.; Davison, L. The contribution of anammox and denitrification to sediment N2 production in a surface flow constructed wetland. *Environ. Sci. Technol.* **2008**, *42*, 9144–9150. [CrossRef]
18. Kabuba, J.; Lephallo, J.; Rutto, H. Comparison of various technologies used to eliminate nitrogen from wastewater: A review. *J. Water Process. Eng.* **2022**, *48*, 102885. [CrossRef]
19. He, K.; Qin, H.; Wang, F.; Ding, W.; Yin, Y. Importance of the Submerged Zone during Dry Periods to Nitrogen Removal in a Bioretention System. *Water* **2020**, *12*, 876. [CrossRef]
20. Lu, S.; Sun, Y.; Lu, B.; Zheng, D.; Xu, S. Change of abundance and correlation of Nitrospira inopinata-like comammox and populations in nitrogen cycle during different seasons. *Chemosphere* **2020**, *241*, 125098. [CrossRef]
21. Zhu, G.; Wang, S.; Wang, W.; Wang, Y.; Zhou, L.; Jiang, B.; Op Den Camp, H.J.M.; Risgaard-Petersen, N.; Schwark, L.; Peng, Y.; et al. Hotspots of anaerobic ammonium oxidation at land–freshwater interfaces. *Nat. Geosci.* **2013**, *6*, 103–107. [CrossRef]
22. Lin, S.; Shen, S.; Zhou, A.; Lyu, H. Sustainable development and environmental restoration in Lake Erhai, China. *J. Clean. Prod.* **2020**, *258*, 120758. [CrossRef]
23. APHA. *Standard Methods for the Examination of Water and Wastewater*, 22nd ed.; American Public Health Association: Washington, DC, USA, 2012.
24. Li, D.; Zheng, B.; Chu, Z.; Liu, Y.; Huang, M. Seasonal variations of performance and operation in field-scale storing multipond constructed wetlands for nonpoint source pollution mitigation in a plateau lake basin. *Bioresour. Technol.* **2019**, *280*, 295–302. [CrossRef] [PubMed]
25. Ruan, W.; Cai, H.; Xu, X.; Man, Y.; Wang, R.; Tai, Y.; Chen, Z.; Vymazal, J.; Chen, J.; Yang, Y.; et al. Efficiency and plant indication of nitrogen and phosphorus removal in constructed wetlands: A field-scale study in a frost-free area. *Sci. Total Environ.* **2021**, *799*, 149301. [CrossRef]
26. Ren, Z.; Fu, X.; Zhang, G.; Li, Y.; Qin, Y.; Wang, P.; Liu, X.; Lv, L. Study on performance and mechanism of enhanced low-concentration ammonia nitrogen removal from low-temperature wastewater by iron-loaded biological activated carbon filter. *J. Environ. Manag.* **2022**, *301*, 113859. [CrossRef] [PubMed]
27. Wu, H.; Zhang, J.; Ngo, H.H.; Guo, W.; Hu, Z.; Liang, S.; Fan, J.; Liu, H. A review on the sustainability of constructed wetlands for wastewater treatment: Design and operation. *Bioresour. Technol.* **2015**, *175*, 594–601. [CrossRef] [PubMed]

28. Wu, Y.; Han, R.; Yang, X.; Fang, X.; Chen, X.; Yang, D.; Zhang, R. Correlating microbial community with physicochemical indices and structures of a full-scale integrated constructed wetland system. *Appl. Microbiol. Biotechnol.* **2016**, *100*, 6917–6926. [CrossRef] [PubMed]
29. Sun, H.; Xu, S.; Wu, S.; Wang, R.; Zhuang, G.; Bai, Z.; Deng, Y.; Zhuang, X. Enhancement of facultative anaerobic denitrifying communities by oxygen release from roots of the macrophyte in constructed wetlands. *J. Environ. Manag.* **2019**, *246*, 157–163. [CrossRef] [PubMed]
30. Vinther, F.P. Measured and simulated denitrification activity in a cropped sandy and loamy soil. *Biol. Fertil. Soils* **1992**, *14*, 43–48. [CrossRef]
31. Xue, L.; Ren, H.; Li, S.; Leng, X.; Yao, X. Soil bacterial community structure and co-occurrence pattern during vegetation restoration in karst rocky desertification area. *Front. Microbiol.* **2017**, *8*, 2377. [CrossRef] [PubMed]
32. Verduzo Garibay, M.; Fernández Del Castillo, A.; de Anda, J.; Senés-Guerrero, C.; Gradilla-Hernández, M.S. Structure and activity of microbial communities in response to environmental, operational, and design factors in constructed wetlands. *Int. J. Environ. Sci. Technol.* **2022**, *19*, 11587–11612. [CrossRef]
33. Mobarry, B.K.; Wagner, M.; Urbain, V.; Rittmann, B.E.; Stahl, D.A. Phylogenetic probes for analyzing abundance and spatial organization of nitrifying bacteria. *Appl. Environ. Microbiol.* **1996**, *62*, 2156–2162. [CrossRef] [PubMed]
34. Zhang, J.; Yang, Y.; Zhao, L.; Li, Y.; Xie, S.; Liu, Y. Distribution of sediment bacterial and archaeal communities in plateau freshwater lakes. *Appl. Microbiol. Biotechnol.* **2015**, *99*, 3291–3302. [CrossRef] [PubMed]
35. Pochana, K.; Keller, J.; Lant, P. Model development for simultaneous nitrification and denitrification. *Water Sci. Technol.* **1999**, *39*, 235–243. [CrossRef]
36. Martín, H.G.; Ivanova, N.; Kunin, V.; Warnecke, F.; Barry, K.W.; Mchardy, A.C.; Yeates, C.; He, S.; Salamov, A.A.; Szeto, E.; et al. Metagenomic analysis of two enhanced biological phosphorus removal (EBPR) sludge communities. *Nat. Biotechnol.* **2006**, *24*, 1263–1269. [CrossRef]
37. Wentzel, M.C.; Lotter, L.H.; Loewenthal, R.E. Metabolic behaviour of Acinetobacter spp. in enhanced biological phosphorus removal—A biochemical model. *Water SA* **1986**, *12*, 209–224.
38. Yan, L.; Xie, C.; Xu, X.; Che, S. Effects of revetment type on the spatial distribution of soil nitrification and denitrification in adjacent tidal urban riparian zones. *Ecol. Eng.* **2019**, *132*, 65–74. [CrossRef]
39. Liu, X.; Tiquia, S.M.; Holguin, G.; Wu, L.; Nold, S.C.; Devol, A.H.; Luo, K.; Palumbo, A.V.; Tiedje, J.M.; Zhou, J. Molecular Diversity of Denitrifying Genes in Continental Margin Sediments within the Oxygen-Deficient Zone off the Pacific Coast of Mexico. *Appl. Environ. Microbiol.* **2003**, *69*, 3549–3560. [CrossRef]
40. Middleton, J.A.; de Sosa, L.L.; Martin, B.C.; Jones, D.L.; Gleeson, D.B. Soil microbes of an urban remnant riparian zone have greater potential for N removal than a degraded riparian zone. *Environ. Microbiol.* **2020**, *22*, 3302–3314. [CrossRef] [PubMed]

processes

Article

Surface Runoff and Diffuse Nitrogen Loss Dynamics in a Mixed Land Use Watershed with a Subtropical Monsoon Climate

Congyu Hou [1,2], Zhifeng Yang [1,2] and Wei Ouyang [1,2,*]

[1] Advanced Interdisciplinary Institute of Environment and Ecology, Beijing Normal University, Zhuhai 519087, China; congyuhou@bnu.edu.cn (C.H.); zfyang@bnu.edu.cn (Z.Y.)
[2] School of Environment, Beijing Normal University, Beijing 100875, China
* Correspondence: wei@bnu.edu.cn

Abstract: The surface hydrology and diffuse pollution loading is hypothesized based on the unique characteristics under different rainfall types in watersheds with a subtropical monsoon climate. This study evaluated the effects of different rainfall event intensity on surface runoff and diffuse nitrogen loss in a subtropical watershed located in the latitude range between 22.5° north and 25° north in southern China by adapting the hydrological model SWAT (Soil and Water Assessment Tool). The rainfall events were classified into four categories including light rainfall (LR), moderate rainfall (MR), intense rainfall (IR), and extreme rainfall (ER). Our results showed that the ER events had the lowest occurrences of 7 days per year, but they contributed the most to total surface runoff (54.29%) and TN losses (57.63%) in the subtropical watershed researched. Although forests and pastures had a better effect on reducing diffuse pollution, their surface runoff and nitrogen loss were still proved to be serious under ER events. As for the nitrogen loss form, water-soluble nitrogen oxides nitrogen (ON-N) loss was the dominant form of total nitrogen (TN) loss, which accounted for 90% of the total loss in the simulated watershed in the subtropical monsoon climate region. Under LR events, however, nitrate nitrogen (NO_3-N) accounted for 74.54% and 55.59% of TN losses from paddy fields and dry lands, respectively. The effects of antecedent soil water content (ASW) on surface runoff and TN loss were higher under ER events. The surface runoff and TN loss in the high-ASW condition were 1.17–3.86 times and 1.34–1.76 times higher, respectively, than in the low-ASW condition. These findings highlighted the important influence of ER events on diffuse pollution and have significant implications for controlling diffuse pollution under different rainfall types. This conclusion proved that extra attention and efforts are required in handling non-point-source pollution in subtropical monsoon watersheds than the watersheds in cooler regions because the natural systems including forest and pastures are no longer effective enough in controlling surface runoff and TN loss during ER.

Keywords: surface runoff; diffuse pollution; nitrogen loss; rainfall types; land use

1. Introduction

Watershed diffuse pollution has become an important cause of water quality deterioration and it is also the direct consequence of hydrological processes under rainfall events [1]. Nitrogen (N), as a primary component of the diffuse pollution, is sensitive to different climatic conditions [2]. The climatic condition contributes to 6.12–26.20% change in total nitrogen (TN) loss [3]. In particular, the climatic conditions such as extreme rainfall events and heavy annual rainfall are reported to be critical in influencing nitrogen loss through hydrological features, litter decomposition, and soil nitrogen transformation [4–6]. Moreover, the runoff and TN loss are expected to be serious in watersheds with both high annual rainfall and temperature [7]. Their loss coefficients in temperate climate regions were usually 8 to 12 times higher than in cold climates [8]. Therefore, the extreme rainfall events during rainy seasons have become major concerns regarding diffuse nitrogen pollution [9]. What is worse, climate change, which has been accelerating during the last century, has led to higher annual rainfall and more extreme rainfall events globally, especially to the regions with subtropical monsoon climate [10,11]. Therefore, it is of vital importance to assess the impacts and understand the mechanisms by which changes in climatic conditions affect the watershed diffuse pollution under subtropical monsoon climate.

Except for the temporal and spatial distribution tendencies of the watershed diffuse pollution, the soil N biochemical cycle can also be affected by climate factors; hence, the forms of nitrogen loss from terrestrial ecosystems could vary significantly throughout different climate zones due to the different soil N transformation processes [6,12]. For instance, dissolved organic nitrogen is the main form of diffuse nitrogen in the watersheds with subtropical acidic forest soil, while nitrate nitrogen (NO_3-N) is dominate in watersheds with temperate acidic forest soils [12]. Among the impacting factors, rainfall characteristics are one of the most critical in affecting the nitrogen form in the watershed diffuse pollution, such as adsorbed organic nitrogen (ON-N) and dissolved NO_3-N. The NO_3-N was reported to be relatively dominant under light rainfall (LR) conditions, while the ON-N was dominant under heavy rainfall conditions [13]. For example, the percentage of ON-N was expected to increase from 44.87% to 61.92% of the TN loss when the rainfall intensity changed from 45 mm/h to 120 mm/h on bare loess soil and the percentage of NO_3-N decreased from 48.40% to 29.40% of the TN loss when the storm return period changed from less than 1 year to more than 10 years in a sub-watershed of the Mahantango Creek, USA [14,15]. However, the watersheds under subtropical monsoon climate face different conditions. Diffuse pollution has been proved to be serious and critical in these watersheds. What is worse, the mechanisms by which changes in climatic conditions affect the temporal and spatial distribution behaviors and nitrogen loss forms remained unclear. Hence, it is important to assess surface runoff and nitrogen loss characteristics in watersheds with subtropical monsoon climate.

Another critical factor affecting watershed diffuse nitrogen pollution is the soil parameters and land use changes within the watershed, which could explain more than 70% of the total nitrogen in the river under some extreme cases [16,17]. Among the soil parameters that could lead to watershed diffuse nitrogen pollution changes, the antecedent soil water content (ASW) is one of the most critical factors that affects soil water storage and infiltration capacity, thereby influencing diffuse pollution mechanisms [18]. Land use changes such as changes in crop types, land management practices, land use distributions, and riverbank buffers could affect watershed diffuse nitrogen pollution distribution behaviors [19–23]. Hou et al. found that a decrease of nearly 10% in the nitrogen load in surface runoff could be achieved by redistributing land uses within the experimental watershed without any changes in any types of land management practices [24]. The modelling of nine management practice scenarios indicated that pasture installation resulted in a 97% reduction in NO_3-N loss in a watershed of central Iowa [25]. Meanwhile, different land use types can affect the forms of watershed nitrogen loss significantly. For instance, in stream water of catchments dominated by tea plantations, the total ON-N and NO_3-N concentrations were 3.27 and 5.40 times higher than the concentrations in forested catchments, respectively [26]. However, as watersheds are complex and robust ecosystems that are sensitive to both natural and artificial factors, the dynamics of how changes in land use and climate change impact the watershed diffuse nitrogen pollution forms and distribution behaviors under subtropical monsoon climate have remained unclear. Additionally, the interactions between rainfall patterns and land use changes have introduced extra uncertainties in the watershed nitrogen pollution behaviors. Therefore, it is critical and essential for us to understand the dynamics of the watershed diffuse nitrogen pollution processes that are related to land use changes and different rainfall patterns in the subtropical monsoon climate regions.

The Soil and Water Assessment Tool (SWAT), a useful model for the assessment of watershed hydrological processes and nutrient loading in mixed land use watersheds [27], was employed in this study to estimate the overall influence of rainfall patterns on watershed diffuse nitrogen pollution in a typical watershed under subtropical monsoon climate. Specifically, the main objectives of this study were to (1) assess the characteristics of surface runoff and TN loss under different rainfall types; (2) investigate the forms of nitrogen loss under different rainfall types; (3) clarify the effects of soil hydrological factors, especially ASW, on surface runoff and TN loss in a mixed land use watershed with subtropical monsoon climate.

2. Materials and Methods

2.1. Study Area Description and Rainfall Patterns

The Dongjiang watershed, located in southern China (Figure 1), experiences a subtropical monsoon climate with an average annual temperature of 21 °C and rainfall of 1800 mm. The elevation ranged from −14 m to 1491 m in the Dongjiang watershed with six weather stations and three hydrological stations used in this research, which made the watershed a suitable watershed for setting up hydraulic models for assessing non-point-source pollution and surface runoff. More than 72% of the total area is covered by forests and more than 13% by agricultural lands. Reports and data from local governments have shown that the excessive application of fertilizers has increased the diffuse nitrogen pollution to water bodies, which accelerated water quality degradation in the watershed. Some of the possible reasons that could lead to the extensive use of fertilizer could be the need for higher crop production from local farmers and lack of regulations in fertilizer application processes in this region. Up to now, no detailed reports have been revealed concerning the reasons that were directly related with this excessive application of fertilizers.

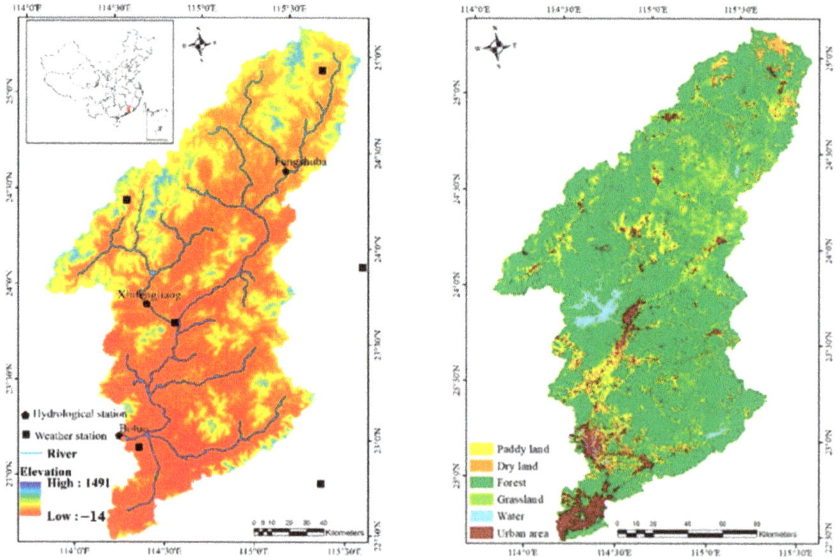

Figure 1. Location, topographic distribution, and land use (2015) of Dongjiang Watershed.

The rainfall in this watershed showed significantly temporal variations with a distinctive wet season from April to September. According to annual rainfall, the study years were classified into three scenarios: dry (1989, 1991, 1996, 2003, 2004, 2007, 2009, 2012, 2013), normal (1988, 1990, 1992, 1993, 1994, 1995, 1998, 1999, 2002, 2005, 2014, 2017), and wet years (1997, 2001, 2006, 2008, 2011, 2015, 2016). The wet years had an annual rainfall from 1981.62 mm to 2487.47 mm, while normal years were from 1587.91 mm to 1973.40 mm, and dry years were from 1287.52 mm to 1582.84 mm, as shown in Table 1. By adapting the above classification numbers, the non-parametric Least Significant Difference (LSD) test showed a significant difference between groups at a significance level of 0.05 (Table 1), which proved that differences existed between groups. All the statistics tests were prepared in R (version 4.2.0).

Table 1. Characteristics of dry, normal, and wet years based on the yearly rainfall.

Classification	Mean mm/y	Max mm/y	Min mm/y	STDEV mm/y	LSD Test *
Wet	2176.45	2487.47	1981.62	307.97	a
Normal	1766.82	1973.40	1587.91	100.52	b
Dry	1422.16	1582.84	1287.52	109.25	c

Note: * The significant difference is at the level of 0.05 and different letters show that significant differences existed between groups.

The daily precipitation was classified into four groups with the percentile threshold method, which is commonly used with a percentile value (95th percentile) as the extreme threshold during 1988 to 2017 [28]. The rainfall threshold of the whole watershed was calculated using the mean precipitation of each weather station weighted by covering area within the watershed. The definitions of LR and MR were based on the classification from the China Meteorological Administration. The daily rainfall amount (R_i) was divided into four types: 1 mm $\leq$ LR< 10 mm; 10 mm $\leq$ MR < 25 mm; 25 mm $\leq$ intense rainfall (IR) < 55.9 mm; and extreme rainfall (ER) $\geq$ 55.9 mm. The LR had the highest annual occurrence (81 days) and the lowest rainfall amount (358 mm). The annual occurrence was 38 days for MR and 21 days for IR, representing 26.55% and 32.93% of the annual rainfall, respectively. Although ER had the lowest occurrence frequency (7 days on average), it contributed 24.96% of the annual rainfall amount. The rainfall amount of ER (575 mm) was almost equal to MR (611 mm). The main difference between wet, normal, and dry years was rainfall occurrences and amount of IR and ER (Table 2). The rainfall occurrence of IR and ER in wet years was 7 and 3 days higher than normal years, respectively, and 12 and 5 days higher than dry years. The rainfall amount of ER in wet years was 47.90% and 104.24% higher than in normal and dry years, respectively.

Table 2. Statistical features of rainfall pattern in dry, normal, and wet years from 1988 to 2017.

	Rainfall Occurrence								Rainfall Amount (mm)			
	LR		MR		IR		ER		LR	MR	IR	ER
	Day	%	Day	%	Day	%	Day	%				
Wet	84 ± 13 a	52.17	41 ± 7 a	25.47	27 ± 6 a	16.77	9 ± 3 a	5.59	379 ± 65 a	665 ± 124 a	998 ± 232 a	799 ± 312 a
Normal	84 ± 12 a	56.38	39 ± 5 a	26.17	20 ± 4 a	13.42	6 ± 2 a	4.03	374 ± 54 a	627 ± 82 a	737 ± 157 a	540 ± 231 b
Dry	74 ± 10 a	58.27	34 ± 5 a	26.77	15 ± 5 b	11.81	4 ± 2 b	3.15	320 ± 51 a	541 ± 92 a	539 ± 180 b	385 ± 175 b
Average	81 ± 12	55.10	38 ± 5	25.85	21 ± 5	14.29	7 ± 3	4.76	358 ± 51	611 ± 99	758 ± 190	575 ± 239

Note: LR, MR, IR, and ER represent light rainfall, moderate rainfall, intense rainfall, and extreme rainfall, respectively. Different letters in the same column indicate that different hydrological years reach significant differences at the 0.05 level.

2.2. SWAT Model Setup

The SWAT model (version 2012) was used to simulate the surface runoff and nitrogen loss in the watershed. The model databases included Digital Elevation Model (DEM), land use, soil properties, meteorological data, and agricultural management information. The DEM was obtained from the Geospatial Data Cloud (http://www.gscloud.cn, accessed on 1 July 2020). Land use properties of 1990, 2000, 2010, and 2015 were obtained from the Resource and Environment Data Cloud Platform, Chinese Academy of Sciences (http://www.resdc.cn/, accessed on 23 May 2020) and were used in the model simulation to represent the land use of 1985–1994, 1995–2004, 2005–2011, 2012–2017 (Table 3). The database of soil properties was obtained from the Institute of Soil Science, Chinese academy of Science (access date 21 November 2019). The three major soil types in Dongjiang watershed are red soil, latosol, and paddy soil, which occupy 39.39%, 36.95%, and 16.93% of the total area, respectively. Daily meteorological data (rainfall, maximum and minimum temperature, relative humidity, wind speed, and daylight hours) from 1985 to 2017 were collected from the National Meteorological Information Centre, China Meteorological Administration (http://data.cma.cn, accessed on 12 March 2020). The solar radiation data were calculated by daylight hours through the method proposed by Neitsch, and further details can be found in the model's theoretical documentation [29]. The dominant crops for paddy field and dryland were rice and peanut, respectively. They were set as an annual rotation tillage system. The date of sowing, fertilization, and harvesting, as well as the amount of fertilizer, were imported into the model. According to the local agricultural survey, the average amount of nitrogen fertilizer (element N) applied in each crop season was 200 kg/ha and 187 kg/ha for rice and peanuts, respectively. The model run from 1985 to 2017 and the data from 1985 to 1987 was used for model warm-up. All the GIS data and SWAT were prepared in ArcGIS (version 10.2).

Table 3. Land use composition dynamics for Dongjiang watershed from 1990 to 2015 (%).

Land Use	1990	2000	2010	2015
Paddy field	11.40	11.28	10.64	4.98
Dry land	5.70	5.48	5.15	8.77
Forest	75.97	75.84	75.68	72.54
Pasture	3.79	3.84	3.46	8.00
Urban area	1.30	1.72	2.62	2.04
Water	1.84	1.84	2.45	3.67

2.3. Model Calibration and Validation

The daily stream flow data obtained from three hydrological stations for the 2011–2014 period (Fengshuba, Xinfengjiang, and Boluo) and the monthly TN load obtained from the Boluo station from 2013 to 2014 were used for model calibration and validation. As one of the most widely used algorithms for model calibration and validation, the Sequential Uncertainty Fitting (SUFI2) algorithm provides the researchers the ability to assess the fitness between two sequential time series with uncertainty. As a result, calibration and validation analyses were conducted using the SUFI2 algorithm in the SWAT-CUP (SWAT-CUP version 5.1.6.2), which was developed by the Neprash Corporation and Texas A&M University. Satisfactory model performance of daily stream flow and monthly TN load were achieved during calibration and validation periods (Figure 2). Coefficients of determination values (R^2) were greater than 0.6 and the Nash–Sutcliffe efficiencies (E_{NS}) were greater than 0.5. More importantly, the SWAT model set up could precisely simulate the impacts from extreme weather events such as long-term dry periods and extreme weather events, both of which could play critical roles in generating TN loss and surface runoff. Hence, the SWAT model was reliable for simulating the surface runoff and nitrogen loss in the watershed.

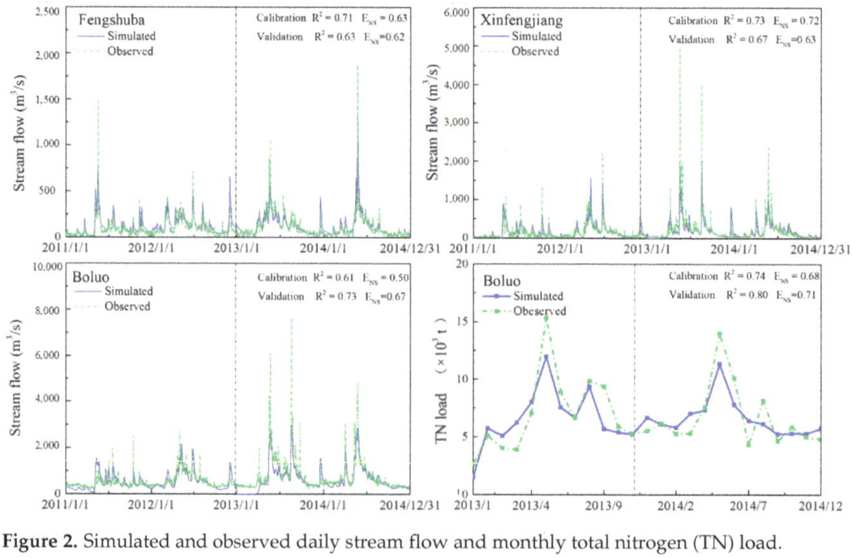

Figure 2. Simulated and observed daily stream flow and monthly total nitrogen (TN) load.

2.4. Data Analysis

To investigate the effects of ASW on surface runoff and nitrogen loss under different rainfall types, the soil water conditions were divided into three groups based on soil ASW values. The soil moisture condition was defined as 'low' when the ASW value was lower than the 25th percentile of all watershed ASW values, as 'high' when the value was higher than the 75th percentile, and as 'medium' when the value was between the 25th and 75th percentiles.

The differences between rainfall occurrence and amount of each rainfall type were assessed using the one-way analysis of variance (ANOVA). To explore the impacts of hydrological factors (rainfall, soil water, percolation water, and ASW) on watershed surface runoff and nitrogen loss, the simulation years were classified into three categories including dry, normal, and wet years based on annual rainfall amount. Based on the results, the outputs of 2015, 2016, and 2017 were classified as typical normal, wet, and dry years, respectively. The ternary plots were used to present the dynamic effects of rainfall, soil water, and percolation water on surface runoff and nitrogen loss. Each factor was rescaled from 0 to 1, and the plots represented the relative values.

3. Results

3.1. Daily Dynamics of Diffuse TN Loading under Rainfall

To understand the dynamics of how diffuse nitrogen loss is influenced by daily rainfall, the relationship between daily streamflow at the watershed outlet and watershed mean daily rainfall was evaluated. The watershed monitored daily rainfall, simulated stream flow, and nitrogen loading in 2015 are shown in Figure 3. The rainy period mainly occurred from May to August with more extreme rainfall events and higher base value. The rainfall amount of LR, MR, IR, and ER was 358, 611, 758, and 575 mm, with the corresponding occurrence rates at 55.10%, 25.85%, 14.29%, and 4.76%, respectively. Daily stream flow and TN loading change showed a similar trend to rainfall, which meant rainfall was the direct influencing factor for these changes. Additionally, the changes in TN loss and stream flow were higher than precipitation, which proved that TN loss and surface runoff were expected to magnify the increasing level of precipitation and result in more serious ecological concerns and risks.

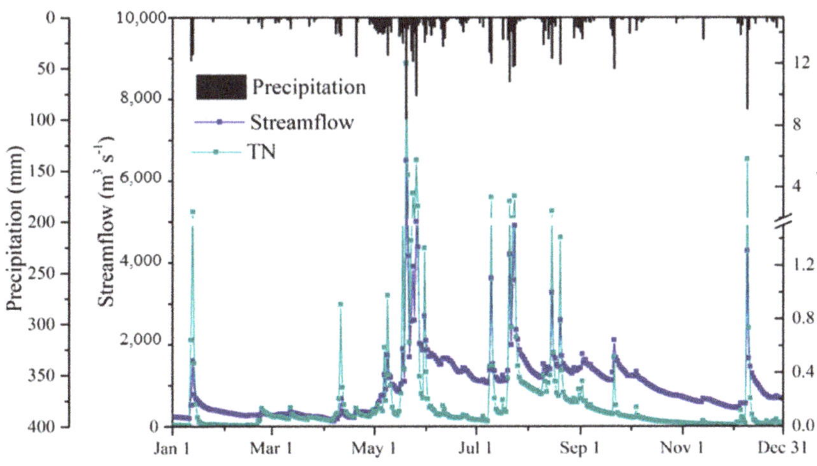

Figure 3. Watershed daily rainfall, stream flow at outlet, and diffuse TN loading of watershed.

The scatterplot of the response of TN losses to rainfall in this watershed is shown in Figure 4a, in which rainfall intensity was divided into four levels according to the daily rainfall amount, as detailed in Section 2.1. The average TN losses under ER (658.2*t*) were over 38 times more than LR (15.2*t*) and over 18 times more than IR (32.8*t*), which meant the TN loss was sensitive to the rainfall. One-way ANOVA showed TN losses under different rainfall levels were significantly different ($p < 0.01$). It was also found that TN losses had significant strong Pearson correlations with daily rainfall amount. The TN losses depended exponentially on rainfall amount as fitted in Figure 4a, while residual error also meant that rainfall was not the only major influencing factor on TN losses. Meanwhile, the ON-N losses showed a similarly exponential relationship with rainfall amount as TN losses, as shown in Figure 4b. While NO_3-N showed weak Pearson correlation with rainfall amount ($r = 0.272$), there were no significant differences between the NO_3-N values of ER and IR or ER and MR. This finding indicated that the effect of rainfall intensity on TN loss change was that heavy rainfall was expected to accelerate ON-N loss more than NO_3-N.

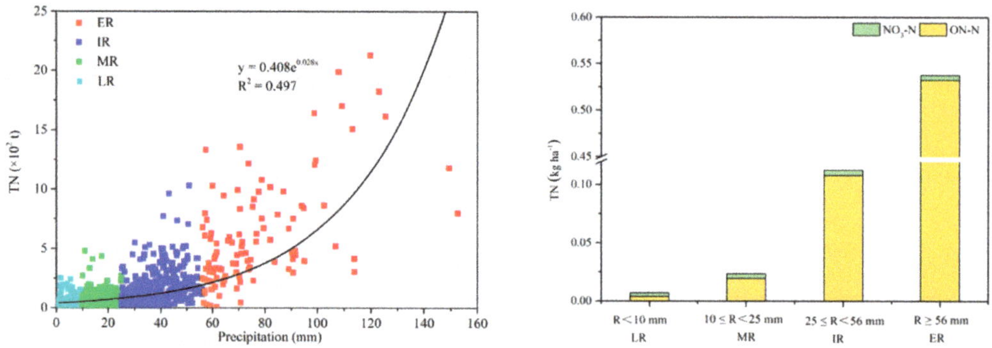

(**a**) (**b**)

Figure 4. Watershed TN (**a**) and ON-N (**b**) losses at outlet under four rainfall types (LR, light rainfall; MR, moderate rainfall; IR, intense rainfall; ER, extreme rainfall).

3.2. Watershed Diffuse TN Loading from Different Land Uses and Rainfall Conditions

In order to highlight the impacts of land use on TN losses, the watershed daily diffuse TN losses from four main types of land uses were compared. As shown in Figure 5, TN loading increased exponentially with daily rainfall intensity, increasing for all the four types of land use, while the rising tendency varied among different land use types. The average TN losses from all four land uses were lowest under LR events by 0.0015, 0.0050, 0.0363, and 0.0491 kg ha^{-1} for forest, pasture, paddy fields, and dry lands, respectively. Furthermore, the values increased significantly, up to 0.2873, 1.5479, 1.4371, and 2.2674 kg ha^{-1} under ER events. It was obvious that the average TN losses were minimal for forest land, especially under ER events. Moreover, IR events contributed to 34.34, 35.07, 38.32, and 39.60% of TN losses for forest, pasture, paddy fields, and dry lands, respectively, while ER events contributed to 26.59, 31.08, 44.26, and 46.15% of TN losses from forest, pasture, paddy fields, and dry lands. As a result, IR and ER were the dominant rainfall types that could cause diffuse TN pollution in this watershed in the subtropical monsoon climate region. The order of sensitivity of TN loss to daily rainfall intensity of the four types of land uses from high to low was dry lands, paddy field, pasture, and forest.

The proportion of diffused nitrogen forms is shown in Figure 5b. Like Figure 4b, ON-N losses from each land use type increased with daily rainfall exponentially; both TN and ON-N losses showed a significant difference between the four rainfall levels for all the land use types ($p < 0.01$), while NO_3-N losses increased gradually from LR to IR and then decreased slightly under ER except for paddy fields, for which NO_3-N loss increased from LR to ER. The differences in NO_3-N losses between MR, IR, and ER were not always statistically significant, as shown in Figure 5b. Moreover, although NO_3-N losses changed slightly with rainfall level, LR events contributed to most of the NO_3-N losses, as it occurred more frequently than any other kind of rainfall; its contribution to NO_3-N losses was 34.88, 42.80, 50.47, and 56.63% from forest, pasture, paddy fields, and dry lands, respectively. By contrast, ER events contributed to only 2.99, 4.34, 3.57, and 2.97% of NO_3-N loss from forest, pasture, paddy fields, and dry lands, respectively. Hence, LR was the dominant rain condition that caused NO_3-N losses in this watershed, but IR and ER were the main type of rainfall causing ON-N and TN losses.

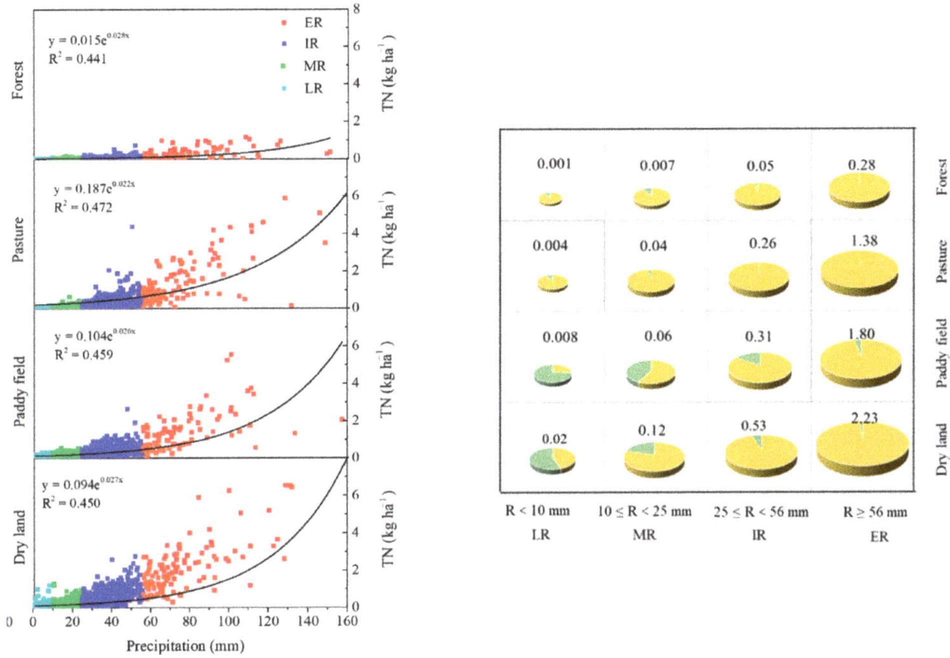

(a) (b)

Figure 5. (**a**) Streamflow and (**b**) TN loss from different land uses under four rainfall types. Green area referred to NO_3-N and yellow area referred to ON-N.

Moreover, paddy fields and dry lands contributed most to TN losses, at 28.80 and 47.52%, and contributed the most to NO_3-N losses, at 54.46 and 46.94%, while dry lands and pastures contributed most to ON-N losses, at 51.59% and 29.66%. Dry lands were the most critical diffused nitrogen source in both forms of TN losses, while forests were the least critical source.

3.3. *Impact of ASW on Watershed Surface Runoff and TN Loss*

After analyzing the impacts of rainfall and land use types on TN loss, the effects of ASW on TN loss under different rainfall types were further compared for detailed dynamics (Figure 6). Our results showed that high ASW would accelerate TN losses under rainfall to a limited degree and was not statistically significant ($p > 0.05$). Under the ER type of rainfall, TN loss under high-ASW conditions was 1.18, 1.05, 1.54, and 1.59 times the values under low-ASW conditions for forests, dry lands, paddy fields, and pastures, respectively. In dry lands, surface runoff was 33.59 mm under low-ASW conditions, but increased to 37.08 mm under high-ASW conditions. The latter was about 1.18, 1.64, and 1.97 times the values in paddy fields, pastures, and forests in the high-ASW condition. TN loss in pastures was 2.27 kg·ha^{-1} under high-ASW conditions, and the same values for dry lands were 1.36 and 7.83 times the values in paddy fields and forests, respectively.

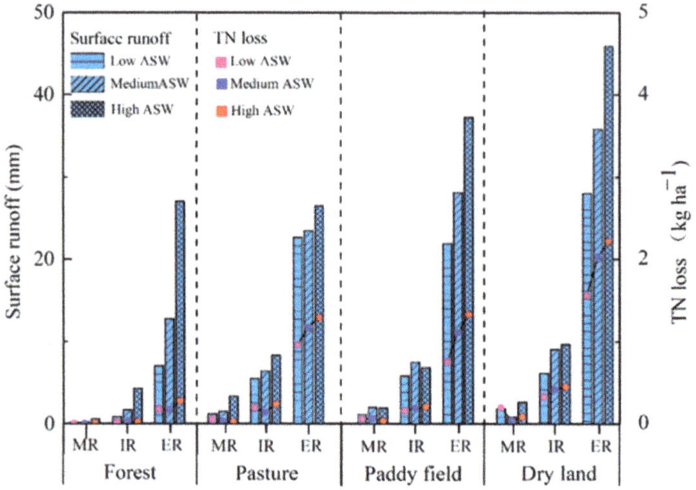

Figure 6. Impact of ASW on surface runoff and TN loss under different rainfall intensities and land use conditions.

4. Discussion

4.1. Surface Runoff and Diffuse TN Loss under Various Rainfall Types

The relation between surface runoff and diffuse pollution loss is directly influenced by the rainfall characteristics [30]. ER, which is characterized by high rainfall intensity and low occurrences, plays an important role in causing severe runoff and nutrient losses [31]. In the subtropical watershed researched in this paper, although ER events had the lowest occurrences of 7 days per year, they contributed the most to total surface runoff (54.29%) and TN losses (57.63%). Additionally, low soil infiltration capacity promoted nitrogen loss since it increased soil erosion by accelerating surface runoff [32].

Land use was another critical factor that could influence surface runoff and nitrogen loss. Based on the simulated results, the surface runoff and TN loss from farmlands were higher than that from forests and pastures under all types of rainfall researched in this paper (LR, MR, IR, and ER), which proved that natural lands were advantageous in reducing surface runoff and nitrogen loss compared with artificial agricultural lands [33]. During this process, a higher percentage of rainfall was absorbed by soil or penetrated to the deeper layers of soil columns in forests and pastures as soil water or percolation water instead of flowing out as runoff [34]. Agricultural lands, however, usually have the lower soil infiltration capacity and generate more surface runoff and nitrogen loss [35]. TN loss in dry lands was about 1.7 times higher than paddy fields due to the fact that the surface layers of the dry soil were more vulnerable to precipitation compared with the paddy field. As a result, all types of precipitation could penetrate deeper into the dry soil surface and resulted in increased loss in both TN and sediments. One major soil property that could reflect this difference is that the paddy field usually has lower soil infiltration capacity than dry lands [4]. Our results also showed that the surface runoff and TN loss from forests and pastures were more sensitive to rainfall types than paddy fields and dry lands. The percentage of surface runoff and TN loss under ER events in forests and pastures (58.41–70.67%) was higher than in paddy fields and dry lands (48.60–50.60%), despite their lower values (Figures 3 and 4). One possible reason for this finding was that the relatively high ASW of the soil in forests and pastures was expected to significantly reduce the soil interception capacity and the soil with decreased interception capacity would generate more surface runoff and TN loss during ER events [35]. This finding proved that forests and pastures were effective in controlling surface runoff and TN loss during LR, MR, and IR events, but not as effective during ER events. For agricultural lands, frequent fertilizer applications, compacted soil surface, and low soil infiltration capacity were all expected to result in serious surface runoff and nutrient loss under any type of rainfall event. However, the agricultural lands under ER events should still be carefully managed, as the surface runoff and TN loss were still the most serious and critical among all types of rainfall.

In conclusion, surface runoff and TN loss were proved to be serious under ER for forests and pastures and under all types of rainfall for the agricultural lands. Meanwhile, ER was found to be the type of rainfall with the most surface runoff and TN loss for all types of land uses.

4.2. Different Forms of Nitrogen Loss under a Subtropical Monsoon Climate

The previous research by Deelstra et al. has found that nitrogen loss amount was relatively high in the watersheds with high annual rainfall and temperature [36]. This is likely due to the acceleration of nitrogen-releasing processes from biomass residues as a result of the increased activity of soil microorganisms and fauna under high soil moisture and temperature conditions [7,37]. Thus, more nitrogen diffuses to water bodies during the rainfall–runoff process in watersheds with high rainfall and temperature. The findings in our research are consistent with previous research in which runoff and TN were significantly higher in a temperate climate region compared with a cold climate region [8].

Furthermore, the dominant form of nitrogen loss among different climate regions was ON-N, implying that soil erosion-induced pollution was critical for the diffuse pollution of these regions [38]. The fraction of ON-N became larger as temperature and rainfall increased, and the ON-N percentage of total nitrogen loss was the highest among all the nitrogen loss forms in watersheds with subtropical monsoon climate (90–95%) in our simulation results. A possible reason for this conclusion was that higher soil moisture and more frequent rainfall events were expected to promote the loss of ON-N that absorbed on soil particles via soil erosion processes. The NO_3-N loss ratio, however, would decrease as annual temperature increased even if the annual rainfall amounts were similar [15,39]. This could be attributed to the accelerated soil nitrogen mineralization process under higher temperature [40]. This finding was also consistent with the previous research pointing out that the increasing winter temperature mobilized the large stores of ON-N and increased nitrate runoff [16]. The component of nitrogen could originate from the fact that biomass residues in forests and pastures could be easily decomposed in a relatively short period of time due to high temperature and humidity in subtropical regions [7].

The ON-N is adsorbed by soil particles and transported with sediment during soil erosion in the nitrogen loss process [41]. Generally, increased rainfall was expected to result in increased ON-N loss via soil erosion [14,15]. The same patterns could also be observed in this study, as the percentage of ON-N loss increased as rainfall increased. Under ER events, more than 97.22% of the TN loss was in the form of ON-N, the majority of which is generated via soil erosion.

However, the dominant nitrogen loss forms showed critical variance under different rainfall and land use types. For forests and pastures, the ON-N loss accounted for more than 90% of TN loss regardless of rainfall type, which implied that nitrogen loss from these two land use types was mainly related to soil erosion in the simulated watershed. For agricultural lands, NO_3-N, stored in the upper soil layers from excessive fertilizer application, was highly soluble in water and could be easily washed away via the runoff process [42]. This could partially explain the reason that the percentages of NO_3-N loss in the paddy field and dry lands were higher than in forests and pastures. The fact that the ON-N loss ratios were lower in paddy fields than in dry lands might result from the decreased soil erosion and nitrogen loss in paddy fields with higher field ridge and water storage capacity [4,43].

4.3. Implications for Subtropical Watershed Management

The ER events were extremely critical in subtropical watersheds, especially under high antecedent moisture conditions and in wet years. Compared with forests and pastures in the same region, agricultural lands were more vulnerable to more serious rainfall events because significant changes in surface runoff and nitrogen loss ratio took place when rainfall events moved from LR to ER. Therefore, ecological engineering structures and buffer strips should be implemented along agricultural lands to naturalize the negative impacts on surface runoff and nitrogen loss from frequent and unpredictable extreme rainfall in subtropical regions [44].

Meanwhile, even though forests and pastures were proved to be sources of TN loss and surface runoff, the recommendations for those two types of land use were not applicable due to the fact that they belong to the native local ecosystems. As part of the local ecosystems, the pollution and surface runoff should be a natural part of the processes within a reasonable range. However, as climate change has become more and more significant and critical, the increase in TN loss and surface runoff has become more and more concerning for decision-makers and should be taken into consideration in future.

Furthermore, soil management practices such as conservation tillage, mulching, and hedgerow systems could also be implemented in the watershed in subtropical monsoon regions [31]. Agricultural diffuse pollution should be further minimized by reducing fertilizer application, increasing fertilizer use efficiency, and adopting no-tillage management [45]. Moreover, increasing the ridge height of paddy fields is an effective way to enhance field water storage and reduce surface runoff and nitrogen loss [4]. For forest and pasture lands, more care needs to be taken to prevent them from generating more TN loss and surface runoff and becoming a critical source of pollution and ecological concerns under on-going climate change.

5. Conclusions

The characteristics of surface runoff and diffuse nitrogen loss under different rainfall types in a typical subtropical monsoon watershed were investigated using the SWAT model. The results demonstrated that surface runoff and TN loss were expected to become more serious as rainfall amount increased. ER events had the lowest occurrences, but it contributed the most to surface runoff and TN loss. Although forests and pastures had a better effect on reducing diffuse pollution, their surface runoff and nitrogen loss were still proved to be serious under ER events. ON-N loss was the dominant form of TN loss, accounting for 90% of the total loss in the simulated watershed in the subtropical monsoon climate region. The percentage of ON-N loss increased as rainfall events changed from LR to ER and nearly all the TN loss was in the form of ON-N under ER events. ON-N was the dominant form of nitrogen loss generated from forests and pastures regardless of rainfall type, while NO_3-N loss was dominant under LR events for paddy fields and dry lands. The effects of ASW on surface runoff and TN loss were higher under ER events. The surface runoff and TN loss in the high-ASW condition were up to 4 times and about 1.5 times higher, respectively, than in the low-ASW condition.

The findings in this research have proved that ER events should be considered the priority period in reducing nitrogen loss and surface runoff generated from the watersheds in subtropical monsoon region. Notably, due to the extremely high annual rainfall amount and daily rainfall intensity in this climate region, forests and pastures, which were usually considered to be effective in controlling nitrogen loss and surface runoff, would still produce a considerable amount of TN loss and surface runoff via soil erosion. This conclusion proved that extra attention and efforts were required in handling non-point-source pollution in subtropical monsoon watersheds than the watersheds in cooler regions because the natural systems including forest and pastures were no longer effective enough in controlling surface runoff and TN loss during ER. However, as our paper only analyzed the changes brought to surface runoff and TN loss without assessing the ecological risks that could be triggered by these changes, some of the future research could be applied to the area assessing the harm and risk of the local ecosystems, especially under climate changes. Additionally, some of the non-point-source pollution could be solved by civil engineering processes designed using bio-chemical processes, especially for the extensive amount of

TN loss in the surface runoff during ER. It would be beneficial both to local ecosystems as well as farmers if the TN loss via agricultural land could be further utilized.

As a result, it is of vital importance for the decision-makers to make necessary plans against ecological disasters triggered by extreme weather events, especially for the watersheds in the subtropical monsoon region, such as the Dongjiang watershed researched in this paper. Additionally, possible land management such as raising paddy fields' ridge heights and growing protective crops during rainy seasons should also be applied to the agricultural land in this region as well.

Author Contributions: Writing—original draft, C.H.; Writing—review & editing, W.O.; Supervision, Z.Y. All authors have read and agreed to the published version of the manuscript.

Funding: The research presented here benefits from financial support from the National Key Research and Development Program of China (2022YFC3202202).

Institutional Review Board Statement: Not applicable.

Informed Consent Statement: Not applicable.

Data Availability Statement: Not applicable.

Conflicts of Interest: The authors declare no conflict of interest. The authors declare that they have no known competing financial interests or personal relationships that could have appeared to influence the work reported in this paper.

References

1. Kumwimba, M.N.; Meng, F.; Iseyemi, O.; Moore, M.T.; Bo, Z.; Tao, W.; Liang, T.J.; Ilunga, L. Removal of non-point source pollutants from domestic sewage and agricultural runoff by vegetated drainage ditches (VDDs): Design, mechanism, management strategies, and future directions. *Sci. Total Environ.* **2018**, *639*, 742–759. [CrossRef] [PubMed]
2. Oygarden, L.; Deelstra, J.; Lagzdins, A.; Bechmann, M.; Greipsland, I.; Kyllmar, K.; Povilaitis, A.; Iital, A. Climate change and the potential effects on runoff and nitrogen losses in the Nordic–Baltic region. *Agr. Ecosyst. Environ.* **2014**, *198*, 114–126. [CrossRef]
3. Nguyen, H.H.; Recknagel, F.; Meyer, W. Effects of projected urbanization and climate change on flow and nutrient loads of a Mediterranean catchment in South Australia. *Ecohydrol. Hydrobiol.* **2018**, *19*, 279–288. [CrossRef]
4. Gao, X.; Ouyang, W.; Hao, Z.; Shi, Y.; Wei, P.; Hao, F. Farmland-atmosphere feedbacks amplify decreases in diffuse nitrogen pollution in a freeze-thaw agricultural area under climate warming conditions. *Sci. Total Environ.* **2017**, *579*, 484–494. [CrossRef] [PubMed]
5. Ng Kee Kwong, K.F.; Bholah, A.; Volcy, L.; Pynee, K. Nitrogen and phosphorus transport by surface runoff from a silty clay loam soil under sugarcane in the humid tropical environment of Mauritius. *Agri. Ecosyst. Environ.* **2002**, *91*, 147–157. [CrossRef]
6. Philben, M.; Ziegler, S.E.; Edwards, K.A.; Kahler, R.; Benner, R. Soil organic nitrogen cycling increases with temperature and precipitation along a boreal forest latitudinal transect. *Biogeochemistry* **2016**, *127*, 397–410. [CrossRef]
7. Peña-Peña, K.; Irmler, U. Nitrogen and carbon losses from decomposing litter in natural and agroecosystems of two different climate regions of Brazil. *Eur. J. Soil Biol.* **2018**, *86*, 26–33. [CrossRef]
8. Zhang, Y.; Zhou, Y.; Shao, Q.; Liu, H.; Lei, Q.; Zhai, X.; Wang, X. Diffuse nutrient losses and the impact factors determining their regional differences in four catchments from North to South China. *J. Hydro.* **2016**, *543*, 577–594. [CrossRef]
9. Huang, J.C.; Lee, T.Y.; Kao, S.J.; Hsu, S.C.; Lin, H.J.; Peng, T.R. Land use effect and hydrological control on nitrate yield in subtropical mountainous watersheds. *Hydrol. Earth Syst. Sci.* **2012**, *16*, 699–714. [CrossRef]
10. Domroes, M.; Schaefer, D. Recent climate change affecting rainstorm occurrences: A case study in East China. *Clim. Past.* **2008**, *4*, 303–309. [CrossRef]
11. The National Aeronautics and Space Administration (NASA). Climate Change: Vital Signs of the Planet. 2022. Available online: https://climate.nasa.gov/ (accessed on 5 October 2022).
12. Zhang, J.; Tian, P.; Tang, J.; Yuan, L.; Ke, Y.; Cai, Z.; Zhu, B.; Müller, C. The characteristics of soil N transformations regulate the composition of hydrologic N export from terrestrial ecosystem. *J. Geophys. Res. Biogeosci.* **2016**, *121*, 1409–1419. [CrossRef]
13. Drewry, J.J.; Newham, L.T.H.; Croke, B.F.W. Suspended sediment, nitrogen and phosphorus concentrations and exports during storm-events to the Tuross estuary, Australia. *J. Environ. Manag.* **2009**, *90*, 879–887. [CrossRef] [PubMed]
14. Wu, L.; Peng, M.; Qiao, S.; Ma, X. Assessing impacts of rainfall intensity and slope on dissolved and adsorbed nitrogen loss under bare loessial soil by simulated rainfalls. *Catena* **2018**, *170*, 51–63. [CrossRef]
15. Zhu, Q.; Schmidt, J.P.; Buda, A.R.; Bryant, R.B.; Folmar, G.J. Nitrogen loss from a mixed land use watershed as influenced by hydrology and seasons. *J. Hydrol.* **2011**, *405*, 307–315. [CrossRef]
16. Hessen, D.O.; Hindar, A.; Holtan, G. The significance of nitrogen runoff for eutrophication of freshwater and marine recipients. *Oceanology* **1997**, *26*, 312–320.

17. Li, W.; Zhai, L.; Lei, Q.; Wollheim, W.M.; Liu, J.; Liu, H.; Hu, W.; Ren, T.; Wang, H.; Liu, S. Influences of agricultural land use composition and distribution on nitrogen export from a subtropical watershed in China. *Sci. Total Environ.* **2018**, *642*, 21–32. [CrossRef] [PubMed]
18. Hardie, M.A.; Cotching, W.E.; Doyle, R.B.; Holz, G.; Lisson, S.; Mattern, K. Effect of antecedent soil moisture on preferential flow in a texture-contrast soil. *J. Hydrol.* **2011**, *398*, 191–201. [CrossRef]
19. Acero Triana, J.S.; Chu, M.L.; Guzman, J.A.; Moriasi, D.N.; Steiner, J.L. Evaluating the risks of groundwater extraction in an agricultural landscape under different climate projections. *Water* **2020**, *12*, 400. [CrossRef]
20. Botero-Acosta, A.; Chu, M.L.; Huang, C. Impacts of environmental stressors on nonpoint source pollution in intensively managed hydrologic systems. *J. Hydrol.* **2019**, *579*, 124056. [CrossRef]
21. Han, J.; Guzman, J.A.; Chu, M.L. Gully erosion susceptibility considering spatiotemporal environmental variables: Midwest US region. *J. Hydrol. Reg.* **2022**, *43*, 101196. [CrossRef]
22. Hou, C.; Chu, M.L.; Botero-Acosta, A.; Guzman, J.A. Modeling field scale nitrogen non-point source pollution (NPS) fate and transport: Influences from land management practices and climate. *Sci. Total Environ.* **2021**, *759*, 143502. [CrossRef] [PubMed]
23. Triana, J.S.A.; Chu, M.L.; Shipley, N.J.; Van Riper, C.J.; Stewart, W.P.; Suski, C.D. A decision-making framework for evaluating environmental tradeoffs in enhancing ecosystem services across complex agricultural landscapes. *J. Environ. Manag.* **2022**, *314*, 115077. [CrossRef] [PubMed]
24. Hou, C.; Chu, M.L.; Guzman, J.A.; Triana, J.S.A.; Moriasi, D.N.; Steiner, J.L. Field scale nitrogen load in surface runoff: Impacts of management practices and changing climate. *J. Environ. Manag.* **2019**, *249*, 109327. [CrossRef]
25. Chaplot, V.; Saleh, A.; Jaynes, D.B.; Arnold, J. Predicting water, sediment and NO_3-N loads under scenarios of land-use and management practices in a flat watershed. *Water Air Soil Pollut.* **2004**, *154*, 271–293. [CrossRef]
26. Jacobs, S.R.; Breuer, L.; Butterbach-Bahl, K.; Pelster, D.E.; Rufino, M.C. Land use affects total dissolved nitrogen and nitrate concentrations in tropical montane streams in Kenya. *Sci. Total Environ.* **2017**, *603–604*, 519–532. [CrossRef] [PubMed]
27. Ouyang, W.; Gao, X.; Wei, P.; Gao, B.; Lin, C.; Hao, F. A review of diffuse pollution modeling and associated implications for watershed management in China. *J. Soils Sediments* **2017**, *17*, 1527–1536. [CrossRef]
28. Toride, K.; Cawthorne, D.L.; Ishida, K.; Kavvas, M.L.; Anderson, M.L. Long-term trend analysis on total and extreme precipitation over Shasta Dam watershed. *Sci. Total Environ.* **2018**, *626*, 244–254. [CrossRef]
29. Neitsch, S.L.; Arnold, J.G.; Kiniry, J.R.; Williams, J.R.; King, K.W. *Soil and Water Assessment Tool Theoretical Documentation: Version 2005*; Texas Water Resources Institute: College Station, TX, USA, 2005.
30. Truman, C.; Strickland, T.; Potter, T.; Franklin, D.; Bosch, D.; Bednarz, C. Variable rainfall intensity and tillage effects on runoff, sediment, and carbon losses from a loamy sand under simulated rainfall. *J. Environ. Qual.* **2007**, *36*, 1495–1502. [CrossRef] [PubMed]
31. Dai, C.; Liu, Y.; Wang, T.; Li, Z.; Zhou, Y. Exploring optimal measures to reduce soil erosion and nutrient losses in southern China. *Agric. Water Manag.* **2018**, *210*, 41–48. [CrossRef]
32. Assouline, S. Rainfall-induced soil surface sealing: A critical review of observations, conceptual models, and solutions. *Vadose Zone J.* **2004**, *3*, 570–591. [CrossRef]
33. Hao, F.; Zhang, X.; Wang, X.; Ouyang, W. Assessing the relationship between landscape patterns and nonpoint-source pollution in the Danjiangkou Reservoir Basin in China. *J. Am. Water Resour. Assoc.* **2012**, *48*, 1162–1177. [CrossRef]
34. Qiu, G.Y.; Yin, J.; Tian, F.; Geng, S. Effects of the "Conversion of Cropland to Forest and Grassland Program" on the water budget of the Jinghe River catchment in China. *J. Environ. Qual.* **2011**, *40*, 1745–1755. [CrossRef] [PubMed]
35. Brown, A.E.; Zhang, L.; Mcmahon, T.A.; Western, A.W.; Vertessy, R.A. A review of paired catchment studies for determining changes in water yield resulting from alterations in vegetation. *J. Hydrol.* **2005**, *310*, 28–61. [CrossRef]
36. Deelstra, J.; Iital, A.; Povilaitis, A.; Kyllamr, K.; Greipsland, I.; Bliche-Mathiesen, G.; Jansons, V.; Koskiaho, J.; Lagzdins, A. Reprint of "Hydrological pathways and nitrogen runoff in agricultural dominated catchments in Nordic and Baltic countries". *Agric. Ecosyst. Environ.* **2014**, *198*, 65–73. [CrossRef]
37. Mao, R.; Zhang, X.; Song, C.; Wang, X.; Finnegan, P. Plant functional group controls litter decomposition rate and its temperature sensitivity: An incubation experiment on litters from a boreal peatland in northeast China. *Sci. Total Environ.* **2018**, *626*, 678–683. [CrossRef]
38. Ouyang, W.; Xu, X.; Hao, Z.; Gao, X. Effects of soil moisture content on upland nitrogen loss. *J. Hydrol.* **2017**, *546*, 71–80. [CrossRef]
39. Huang, H. Impacts of Land-Use, Soil Property and Climate Change on NPS Pollution Loss under Long-Term Agricultural Activities. Ph.D. Thesis, Beijing Normal University, Beijing, China, 2016. (In Chinese).
40. Fan, M.; Shibata, H. Simulation of watershed hydrology and stream water quality under land use and climate change scenarios in Teshio River watershed, northern Japan. *Ecol. Indic.* **2015**, *50*, 79–89. [CrossRef]
41. Hao, C.; Yan, D.; Xiao, W.; Shi, M.; He, D.; Sun, Z. Impacts of typical rainfall processes on nitrogen in typical rainfieldss of black soil region in Northeast China. *Arab. J. Geosci.* **2015**, *8*, 6745–6757. [CrossRef]
42. Dekun, S. Characteristics of nitrogen loss on slope under simulated rainfall. *Bull. Soil Water Conserv.* **2009**, *29*, 98–101.
43. Ouyang, W.; Wu, Y.; Hao, Z.; Zhang, Q.; Bu, Q.; Gao, X. Combined impacts of land use and soil property changes on soil erosion in a mollisol area under long-term agricultural development. *Sci. Total Environ.* **2018**, *613–614*, 798–809. [CrossRef] [PubMed]

44. Cristan, R.; Aust, W.M.; Bolding, M.C.; Barrett, S.M.; Munsell, J.F.; Schilling, E. Effectiveness of Forestry best management practices in the United States: Literature review. *For. Ecol. Manag.* **2016**, *360*, 133–151. [CrossRef]
45. Wei, P.; Ouyang, W.; Gao, X.; Hao, F.; Hao, Z.; Liu, B. Modified control strategies for critical source area of nitrogen (CSAN) in a typical freeze-thaw watershed. *J. Hydrol.* **2017**, *551*, 518–531. [CrossRef]

processes

MDPI

Review

Cutting-Edge Solutions for Soil and Sediment Remediation in Shipyard Environments

Jae Ho Jung [1,2,†], Md Akhte Khirul [1,†], Dohyoung Kang [1], Hobin Jee [1], Chanwoo Park [1], Yudam Jung [1], Seunghyun Song [1] and Euntae Yang [1,*]

1 Department of Marine Environmental Engineering, Gyeongsang National University, Tongyoung 53064, Gyeongsangnam-do, Republic of Korea
2 IDEA Environment Service Co., Ltd., Gwangju 12767, Gyeonggi-do, Republic of Korea
* Correspondence: yet83@gnu.ac.kr
† These authors contributed equally to this work.

Abstract: Shipyards are significant industrial sources of environmental pollution, releasing substantial amounts of heavy metals, petroleum hydrocarbons, and organic solvents into soil and groundwater during shipbuilding and maintenance operations. Such contamination not only affects the shipyard premises but also poses serious environmental threats to nearby communities, raising concerns about the long-term sustainability of the shipbuilding industry. Given the increasing global emphasis on sustainable industrial practices, addressing shipyard-related pollution has become a critical environmental challenge. This review aims to provide a comprehensive understanding of the pollution issues associated with shipyards and explore effective remediation strategies. It focuses on contamination in both soil and groundwater, and covers pollution generated throughout the shipbuilding and maintenance lifecycle. First, it examines previous studies to identify the major contaminants and pollution sources typically found at shipyard sites. Next, the paper reviews recent advances in soil and groundwater remediation technologies, including physical, chemical, and biological methods tailored to the unique challenges of shipyard environments. Finally, the review discusses current limitations in remediation practices and outlines potential directions for future research and technological development.

Keywords: heavy metal; persistent organic compounds; shipyard; soil remediation

1. Introduction

The shipbuilding industry has long played a pivotal role in the national economy and has contributed significantly to various sectors, including maritime transportation and defense [1]. This is not merely about constructing vessels; it is a complex, integrated sector that brings together technologies, materials, and human resources from a wide range of industries. Due to this multidisciplinary nature, various synthetic chemical products are extensively used at shipyard sites, resulting in the release of substantial amounts of pollutants [2] (Figure 1). Despite the significant discharge of pollutants from shipyards, the proper treatment of these contaminants has been insufficient. Although environmental regulations were strengthened in the 1990s, leading to the introduction of wastewater treatment facilities and hazardous substance management systems, large quantities of pollutants have continued to enter the soil, groundwater, and marine environments. This ongoing contamination is evidenced by persistent reports of soil, groundwater, and marine pollution in areas surrounding shipyards [2,3].

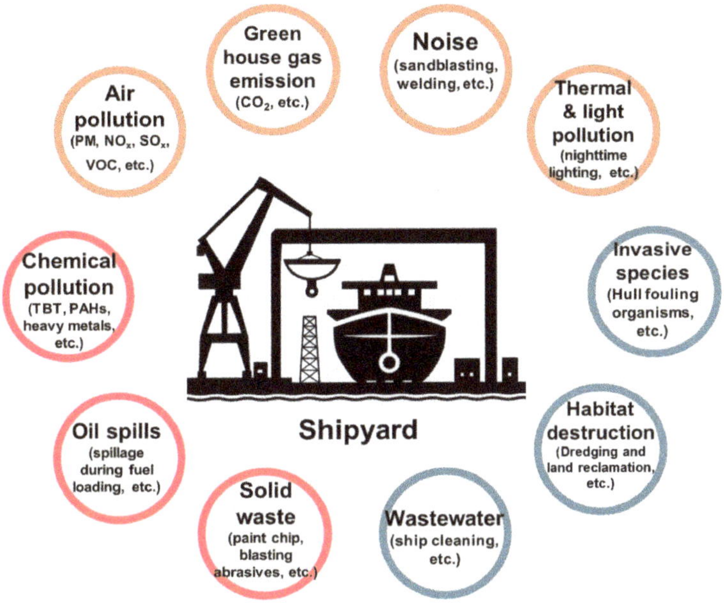

Figure 1. Environmental impacts of shipyards.

This poses potential environmental risks to the surrounding ecosystems and local communities and may significantly undermine the long-term sustainability of the shipbuilding industry, ultimately affecting future generations [4]. Therefore, to address these issues, continuous monitoring of the environment around shipyards, the adoption of pollution prevention technologies, and the remediation of contaminated environments are essential [5,6]. Furthermore, the remediation technologies must be appropriately tailored to the specific characteristics of the shipbuilding industry.

This review is based on a wide survey of shipyard contamination cases reported across multiple global sites, and focuses on remediation technologies considered applicable to shipyard soils. To this end, we selectively reviewed key studies published over the past five years, with literature retrieved from major databases including Scopus and Web of Science. Keywords such as "shipyard soil contamination," "remediation technologies," "heavy metals," and "persistent organic pollutants" were used. Recent research trends were also considered to reflect the evolving focus on technology integration and field applicability.

In this review, we explore the remediation technologies applicable to shipyard environments that have already been contaminated. To this end, we review previous studies that analyze the contamination of shipyard sites to identify common pollution trends and sources. We then examine both conventional and cutting-edge remediation techniques for restoring contaminated soil and groundwater. Real-world case studies of remediation projects in shipyard contexts are also discussed to assess practical applicability. Furthermore, we analyze the challenges and limitations of current technologies and conclude by presenting future perspectives and potential directions for overcoming these barriers.

Unlike previous reviews that addressed soil remediation in general industrial contexts, this work specifically focuses on the complex contamination conditions unique to shipyard sites and systematically compares practical case studies with emerging remediation technologies.

2. Contaminants Profile in Shipyards

2.1. Key Pollutants and Their Origins

Shipbuilding and dismantling activities have long been recognized as significant sources of hazardous substances released into surrounding soils and marine sediments [2]. These pollutants are not randomly distributed but are closely associated with specific industrial operations, such as metal cutting, surface coating, and equipment dismantling [7]. This section reviews the key contaminants commonly found in soils and sediments near shipyards, based on findings from prior studies, and explores their respective sources.

A summary of the major contaminants detected in shipyard soils, along with their primary sources, maximum concentrations, and exceedance of regulatory standards, is provided in Table 1. Polycyclic aromatic hydrocarbons (PAHs) are among the most prevalent organic pollutants detected in areas adjacent to shipyards [2,8]. These compounds typically originate from high-temperature processes such as fuel combustion and metal cutting. Numerous studies conducted near decommissioned shipyards have reported elevated concentrations of PAHs in both soil and nearby sediments, often exceeding international regulatory thresholds [2,8]. In some cases, PAH congener ratios confirmed a thermal origin, indicating a direct link to dismantling activities involving intense heat [8]. For instance, benzo[a]pyrene and fluoranthene have been detected at concentrations up to 2400 μg/kg in sediment samples, significantly exceeding U.S. EPA and Dutch sediment quality standards [2,8].

Polychlorinated biphenyls (PCBs), once widely used in industrial paints and insulating fluids, have been detected in marine sediments near ship dismantling sites [9]. Due to their chemical stability and lipophilic properties, PCBs persist in sediments over long periods. In certain regions, their concentrations have reached levels associated with potential ecological harm. Concentrations as high as 847 μg/kg have been reported, surpassing the Effects Range Median (ERM) for marine ecosystems, while even lower levels (124 ng/g) still suggest moderate risk [10].

DDT and its degradation products, despite being banned or strictly regulated in many countries, are still found in areas where antifouling paints were historically used. Soil and sediment analyses from a former shipyard in southern China revealed significant accumulations of total DDT, with concentrations far exceeding both national and international safety guidelines [11]. Specifically, levels of up to 235.6 mg/kg have been recorded in sediments, exceeding thresholds set by both Chinese and Dutch regulatory frameworks [11]. These findings underscore the long-term environmental persistence of organochlorine-based coatings.

Tributyltin (TBT), a highly toxic biocide historically used in antifouling paints, continues to be detected in sediments near former shipyard facilities, despite its global ban under the International Maritime Organization (IMO). In some locations, TBT concentrations exceed ecological threshold values, and the compound is known to cause severe endocrine disruption in marine organisms. Measured TBT levels have exceeded 36,000 ng Sn/g in some sediment samples, far above the ERM guideline, with known impacts such as imposex in gastropods and endocrine disruption in higher organisms [12]. Other organotin compounds, such as mono- and dibutyltin, have also been detected at concentrations above 5000 ng/g, exceeding WHO aquatic toxicity thresholds [13,14].

In addition to these persistent organic pollutants (POPs), several heavy metals have also been reported in shipyard environments [2,5,15]. Copper (Cu), commonly found in marine paints and electrical components, has been detected at elevated levels in soils and leachates surrounding inactive shipyards. In some cases, the concentrations exceeded the U.S. EPA's Universal Treatment Standards (UTS) [16]. Similarly, high Cu levels have been observed in soils near shipyards in South Korea, posing risks to soil quality and ecosystem health. Cu concentrations up to 1321 mg/kg have been reported in soil samples, clearly surpassing ERM values and representing a major ecological hazard [15].

Lead (Pb) remains a key contaminant, particularly in areas where painted surfaces and electrical systems have been dismantled [15]. Pb has been found not only in the soil but also in food crops cultivated near shipbreaking yards, raising concerns about human exposure through the food chain [17]. Its strong binding affinity allows it to persist in soil over long periods, which is consistent with its historical use in paints and electronic materials [18]. Pb concentrations have reached up to 312 mg/kg in shipyard soil, exceeding both ERM and USEPA UTS guidelines [15]. Zinc (Zn), associated with galvanization and surface coatings, has been detected at concentrations exceeding ecological guidelines in multiple shipyard areas [19]. Maximum Zn levels reported include 2147 mg/kg in soil, indicating severe enrichment in contaminated zones [15]. Like Cu and Pb, Zn often co-occurs with other heavy metals, amplifying the cumulative ecological risks to both terrestrial and aquatic environments [20]. Cadmium (Cd), though typically found at lower concentrations, poses a unique risk due to its high mobility and bioaccumulative potential [21]. Soil and crop analyses near shipbreaking yards in Bangladesh revealed the presence of Cd in both soil and leafy vegetables, indicating a direct route for human exposure through diet [16]. Moreover, the slightly acidic to neutral pH of these soils may enhance cadmium mobility, further increasing its environmental impact [16,22]. While absolute Cd concentrations are often modest, their uptake by crops has led to exceedances of WHO/FAO food safety limits, especially for leafy vegetables consumed locally [21].

Table 1. Major contaminants in shipyards with regulatory standards.

Pollutant	Primary Source	Maximum Detected Concentration	Exceedance of Standards	Refs.
PAHs	Fuel combustion, high-temperature cutting	2400 µg/kg (sediment)	Exceeds U.S. EPA and Dutch standards	[2]
DDT	Antifouling paints	235.6 mg/kg (sediment)	Exceeds Chinese and Dutch soil quality standards	[11]
PCBs	Electrical equipment, industrial coatings	124.33 ng/g (sediment)	Exceeds moderate to heavy contamination level	[23]
		847 µg/kg	Exceeds effects range median (ERM)	[10]
OCPs (DDT derivatives)	Residues from antifouling agents	–	Exceeds ERM and effect range low (ERL) standards	[10]
TBT	Antifouling paints, ship surface coatings	36,292 ng Sn/g	Banned under IMO; exceeds ERM	[12]
Cu	Paint coatings, dismantling of electrical cables	1321 mg/kg	Exceeds ERM	[15]
Pb	Fuel residues, corrosion of painted surfaces	312 mg/kg	Exceeds ERM and U.S. EPA Universal Treatment Standards (UTS)	[15]
Zn	Metal corrosion, antifouling paints	2147 mg/kg	Exceeds ERM	[15]

Table 1. *Cont.*

Pollutant	Primary Source	Maximum Detected Concentration	Exceedance of Standards	Refs.
Cd	Welding, antifouling agents	High plant transfer factor	Exceeds WHO/FAO plant intake standards	[21]
Other Organotin Compounds	Antifouling paints, paint waste	5380 ng/g (sediment)	Exceeds ERM and WHO aquatic toxicity thresholds	[13,14]

2.2. Ecotoxicity and Human Health Risk

As previously discussed, a diverse array of contaminants, including PAHs, heavy metals, organochlorine pesticides, and organotin compounds, coexist in shipyard environments. While each of these substances poses significant individual toxicity, there is increasing concern about their potential to exert synergistic effects when present together [24,25].

Table 2 summarizes the ecological and human health effects of major contaminants commonly detected at shipyard sites. To begin with PAHs, high-molecular-weight compounds such as benzo[a]pyrene and benz[a]anthracene are well-documented for their carcinogenic and mutagenic properties. In a study conducted in Hong Kong, Chiu et al. [2] found that concentrations of these PAHs exceeded the U.S. EPA guidelines as well as the Soil Screening Levels established for Superfund Region III. Furthermore, the study reported a BOD_5/COD ratio of 0.021 in leachate, indicating the presence of POPs with low biodegradability, thereby limiting the self-purification capacity of the soil. Guo et al. [11] conducted a risk assessment under a residential exposure scenario and found that DDT exposure via soil ingestion and dermal contact resulted in estimated cancer risks exceeding the threshold of 10^{-6} to 10^{-5}. These findings highlight the long-term health implications of chronic exposure to organochlorine pesticides. Yılmaz et al. [10] identified high ecotoxicity in the Aliağa region of Turkey, with sediment concentrations of PAHs and PCBs surpassing the ERM. Additionally, total DDT levels exceeded the Effects Range Low (ERL) at all sampled locations, suggesting widespread ecological stress.

Organotin compounds, particularly TBT, are regarded as among the most persistent and toxic pollutants found in shipyard sediments [12]. Despite a global ban imposed by the IMO, Kim et al. [12] reported TBT concentrations exceeding 580 ng/g, which are known to induce imposex in gastropods and endocrine disruption in higher organisms. According to Kotrikla and Staniszewska et al. [13,14], other organotin compounds similarly exhibit severe neuroendocrine and immune system toxicity.

Heavy metals such as Cu, Pb, Zn, and Cd have also consistently exceeded ERM benchmarks in various studies [2,17,19,21]; Chowdhury and Rasid [21] observed significant accumulation of these metals, particularly Cd, in the edible parts of leafy vegetables, with levels surpassing the WHO/FAO food safety guidelines. Given their high bioavailability and mobility, these metals pose long-term risks to food safety and are associated with chronic toxicities, including renal impairment and skeletal disorders.

The interactive effects of these pollutants are further exemplified in a study by Choi and Jeon [15], who investigated a shipyard site in Busan, South Korea. They reported severe contamination (Enrichment Factor > 10) and observed combined organic and inorganic pollution. This mixed pollution scenario was found to exert significant ecotoxicological pressure on benthic organisms and to impair overall ecosystem functioning.

Therefore, minimizing contaminant emissions is essential, but comprehensive site remediation is equally critical to reduce long-term risks and support the redevelopment of former shipyard sites while minimizing adverse impacts on the surrounding environment.

Table 2. Ecotoxicity and human health risk of the pollutants detected in shipbuilding yards.

Pollutant	Primary Risk Type	Ecological Impact	Human Health Impact	Refs.
PAHs	Carcinogenic, mutagenic	Toxic to marine life and benthic sediments	Skin contact/inhalation → Increased cancer risk	[2]
DDT Derivatives	Endocrine disruption, carcinogenic	Reduced survival rate of benthic organisms	Soil ingestion → Cancer and reproductive abnormalities	[11]
PCBs	Persistent toxicity, immune suppression	Reproductive issues and genetic mutations in marine species	Immune system impairment, potential carcinogenicity	[10]
Tributyltin (TBT)	Hormonal disruption, reproductive toxicity	Induces imposex in gastropods	Affects liver, thyroid, and reproductive organs	[12]
Other Organotin Compounds	Induction of imposex, toxicity to crustaceans	Sediment toxicity, decreased biodiversity	Neuroendocrine disruption from long-term exposure	[13,14]
Copper, Lead, Zinc	Neurotoxicity, bioaccumulation	Inhibits microbial activity and impairs plant growth	Developmental issues, skin disorders	[15]
Cd	Nephrotoxicity, skeletal damage	High accumulation and transfer in plants	Dietary intake → Kidney damage and bone abnormalities	[21]

3. Recent Advancement of Remediation Technologies

Various technologies are currently being applied to soil remediation [26–28]. In this section, soil remediation methods applicable for shipyard remediation are categorized into physical, chemical, and biological treatments. The fundamental principles of each approach and representative studies conducted to date are reviewed (Figure 2).

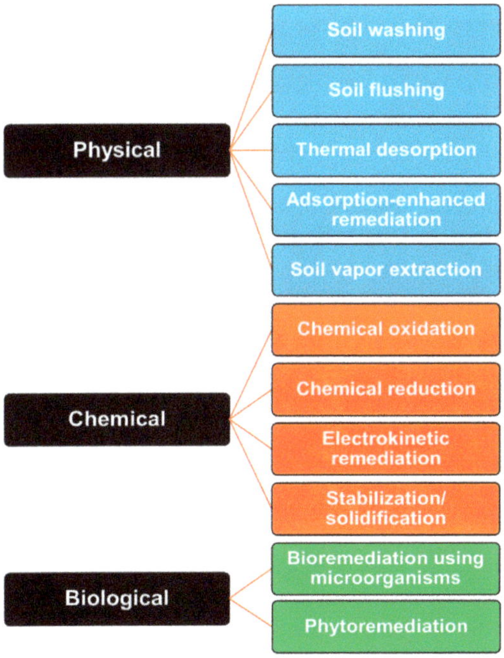

Figure 2. Classification of remediation technologies.

3.1. Physical Remediation

Physical remediation technologies aim to remove or isolate contaminants from polluted soil using mechanical or physical means. These methods are particularly effective in complexly contaminated environments such as shipyard sites. Key approaches include soil washing, thermal desorption, adsorption-based treatments, and soil vapor extraction (SVE). The underlying principles of each technique, along with recent representative studies, are summarized in Figure 3 and Table 3.

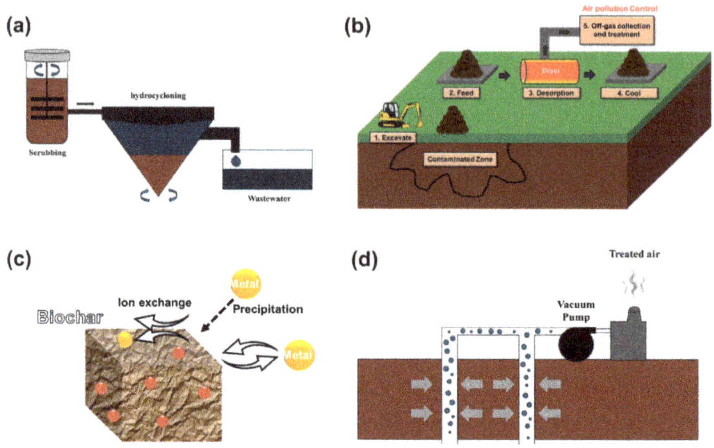

Figure 3. Illustration of representative physical remediation techniques: (**a**) soil washing, (**b**) thermal desorption, (**c**) adsorption-enhanced remediation, and (**d**) soil vapor extraction [26–29].

3.1.1. Soil Washing

Soil washing, a representative physical remediation technique, is increasingly recognized as an effective and relatively fast solution for treating complex contamination typically found at industrial sites like shipyards [30]. The process involves excavating contaminated soil and applying washing solutions, such as acids, chelating agents, or surfactants, to separate and remove pollutants that are bound to soil particles through physical and chemical interactions (Figure 3a) [31]. This method has proven especially efficient in removing heavy metals [31,32]. In addition, it shows promising results for certain organic pollutants such as Total petroleum hydrocarbon (TPH) and PAHs when surfactant-based agents are used [33]. Given the diversity of shipyard activities, including painting, metal cutting, and machinery dismantling, contamination often includes both metals and POPs. Therefore, soil washing is particularly well-suited for sites where pollutants are concentrated in surface soil and excavation is feasible.

Soil washing has long been recognized as an effective remediation strategy for metal-contaminated soils [31,32]; however, traditional methods are often limited by high wash water demand and poor selectivity for co-existing organic pollutants [31]. Recent studies have addressed these limitations by incorporating nanomaterials and biodegradable chelating agents, thereby improving contaminant removal while minimizing secondary pollution. For instance, Grimison et al. [34] demonstrated field-scale application of soil washing for PFAS-contaminated firefighting sites, achieving an average PFOS reduction of ~90% using a recycled washing solution. Similarly, Yuan et al. [35] applied citric acid-based washing with nZVI-supported biochar for Cr(VI) removal and showed effective recovery of the washing agent, achieving 79.9% removal from soil and 98.9% from effluent. Addressing PAH contamination, Zhang et al. [33] used surfactant-enhanced washing on coking plant soil, reaching up to 50.2% removal under optimized multi-step conditions. Further, Rabieian and Taghavijeloudar [36] integrated advanced oxidation (UV/O_3) into a continuous soil washing system for phenanthrene remediation, achieving over 98% removal from soil and complete effluent detoxification.

Innovative techniques have also been proposed for metal recovery from wash water, such as the electrochemical treatment of EDTA complexes by Huang et al. [37], enabling up to 95% heavy metal removal with 96.8% EDTA recovery. For aged PAH-contaminated soils, Xing et al. [38] employed thiacalix-4-arene tetrasulfonate, achieving notable removal efficiencies (up to 55.1%) while elucidating interaction mechanisms via spectroscopic analyses. Corres et al. [39] explored electrostatic separation to optimize washing conditions for ten toxic elements, revealing variable recovery efficiencies based on element type and operating voltage. Finally, Keochanh and Phenrat [40] showed that magnet-assisted nZVI significantly enhanced chlorpyrifos removal (>99.95%), demonstrating the synergy of magnetic separation with nanoscale reactive media. These recent developments collectively reflect a shift toward more selective, resource-efficient, and integrative soil washing systems, particularly relevant to industrial sites such as shipyards, where mixed contaminant profiles pose additional remediation challenges.

3.1.2. Thermal Desorption

Thermal desorption is a physical remediation technology that removes contaminants from soil by heating it to volatilize or decompose organic pollutants, and in some cases, certain heavy metals such as Cd [41,42] (Figure 3b). This method is particularly advantageous for industrial sites like shipyards where complex contamination is prevalent, due to its rapid treatment capability and ease of application [13].

Recent research has focused on improving both efficiency and ecological compatibility through optimized processes and integrated treatment approaches. For example, Liu et al. [43] investigated the thermal behavior of PAH-contaminated soil, emphasizing pollutant volatilization patterns and changes in soil functionality after treatment. Similarly, Sang et al. [44] demonstrated the effectiveness of low-temperature indirect thermal desorption for lubricating oil-contaminated soil, reporting significant contaminant removal alongside microbial community recovery and improved soil properties.

Wang et al. [45] applied thermal desorption to heavy petroleum-contaminated soil, achieving effective contaminant reduction and ecological risk mitigation while highlighting the potential for reuse of treated residues. Bykova et al. [46] evaluated treatment performance across diverse landscapes (tundra, taiga, forest-steppe), underscoring the need for site-specific operational strategies.

In terms of optimization, Li et al. [42] developed a sustainability-based evaluation tool to guide the operational design of thermal desorption systems, enhancing overall removal performance. Hu et al. [27] combined low-temperature thermal desorption with secure landfilling for oil-based drill cuttings, demonstrating pollutant control, reduced health risk, and economic feasibility.

Complementary strategies have also emerged. Ilyas et al. [47] compared the effects of pyrochar-assisted phytoremediation and thermal desorption for crude oil-contaminated soil, revealing synergistic benefits in soil recovery and contaminant reduction. Zhang et al. [48] proposed an optimized numerical model to guide thermal desorption for pyrene-contaminated soils, offering new insights into temperature–time optimization. Xue et al. [49] explored electromagnetic induction-based thermal desorption and demonstrated effective remediation of petroleum hydrocarbons at lower energy inputs. Finally, Yue et al. [50] applied biomass co-pyrolysis to Cd-contaminated soil, achieving Cd removal efficiencies exceeding 96% in laboratory soil and 92% in field soil at 550 °C under a nitrogen atmosphere. Collectively, these studies demonstrate a clear evolution of thermal desorption from simple high-temperature treatment toward more integrated and precise remediation technologies that can effectively address mixed contaminants.

3.1.3. Adsorption-Enhanced Remediation

Adsorption-enhanced remediation is a physicochemical technique that selectively removes or immobilizes heavy metals and certain organic pollutants from contaminated soils using solid-phase sorbents (Figure 3c). In recent years, significant advances have been made in the development of bio-based sorbents, many of which offer improved performance and compatibility with other remediation technologies. Zhang et al. [51] demonstrated that MnO_2-coated rice husk biomass effectively removed cadmium from both soil and aqueous phases, highlighting its potential as a high-performance sorbent prepared via a simple synthesis route. Similarly, Chen et al. [52] applied carboxylated cellulose nanocrystals embedded with α-FeOOH for the simultaneous removal and immobilization of Pb(II), Cd(II), and As(V), identifying synergistic effects of adsorption and precipitation processes. Lin et al. [53] reported that biochar/g-C_3N_4 composites exhibited strong adsorption and degradation capacities for petroleum hydrocarbons, indicating their potential for addressing organic contaminants.

In the context of mixed contamination, Qi et al. [54] utilized biochar loaded with mixed bacteria to enhance the immobilization of uranium and cadmium in co-contaminated soils. Li et al. [55] developed polyaniline-embedded polyvinyl alcohol/sodium alginate hydrogel beads for Cr(VI) removal, extending their application from water treatment to soil remediation. Qi et al. [56] further improved adsorption efficiency for Cu^{2+}, Cd^{2+}, and Pb^{2+} using MgO-loaded nitrogen and phosphorus self-doped biochar, while Zhang et al. [57] employed ball-milled, phosphorus-loaded biochar to remediate alkaline soils contaminated with Pb and Cd.

Lastly, Li et al. [58] integrated β-cyclodextrin/hydrothermal biochar derived from waste cotton into an electrokinetic-permeable reactive barrier (EK-PRB) system, achieving both high metal immobilization efficiency and resource circularity. These studies collectively demonstrate a trend toward multifunctional, sustainable sorbents that not only enhance remediation performance but also address complex contaminant mixtures characteristic of industrial sites.

3.1.4. Soil Vapor Extraction

SVE is an in situ remediation technology that removes VOCs and aromatic hydrocarbons (e.g., BTEX) by extracting contaminated vapors from the unsaturated zone (Figure 3d) [59]. This method is particularly effective in non-cohesive soils with high porosity and is widely applied for the early-stage treatment of volatile contaminants due to its low energy demand and ease of implementation [60].

Recently, Zhang et al. [29] developed a machine learning-based predictive model to evaluate the efficiency of SVE systems, identifying contaminant concentration, soil moisture content, and temperature as critical influencing parameters. Liu et al. [61] combined pneumatic fracturing with SVE to enhance VOC removal in low-permeability clay soils, effectively overcoming mass transfer limitations caused by soil structure.

Additionally, Lyu et al. [59] proposed a sequential treatment system that integrated SVE with an internal combustion engine to utilize extracted vapors as a fuel source, thereby minimizing secondary emissions. Xu et al. [62] improved SVE efficiency in low-permeability soils by establishing pressure gradients through air sparging, which enhanced the desorption and mass transfer of volatile pollutants.

Meanwhile, de Melo Henrique et al. [63] employed an electrokinetic soil flushing-enhanced SVE system to significantly improve the removal of hexachlorocyclohexanes, highlighting its potential for treating persistent halogenated compounds. These advancements collectively indicate a shift toward hybridized SVE processes that are tailored for complex site conditions and contaminant profiles.

Table 3. Previous studies of physical remediation techniques.

Method	Soil Type	Contaminants	Experimental Conditions	Efficiency	Refs.
Soil washing	Clay-rich firefighting site	PFASs (PFOS and PFHxS, etc.)	Field-scale soil washing with agent recycling	Avg. ~90% removal (PFOS116→11 μg/L)	[34]
	Cr-spiked agricultural	Cr	Citric acid wash + nZVI-BC for effluent	79.9% (soil), 98.9% (effluent)	[35]
	Coking plant soil, low TOC	16 PAHs (e.g., FLT, PYR, and BaP)	Single/multi-step surfactant washing	~50.2% (best, TX100 multi-wash)	[33]
	Phenanthrene (PHE)	Phenanthrene-spiked loam	Ultrasound + surfactant wash + UV/O_3	98% (soil), 100% (effluent)	[36]
	Soil washing effluent (not direct soil)	Cu, Pb, and Cd (as EDTA complexes)	Electrochemical AC process with graphite electrodes	49–95% metal removal, 96.8% EDTA recovery	[37]
	Fine sandy clay loam, artificially spiked with PAHs (Nap, Flua, Pyr); aged 1 year	Naphthalene (Nap), Fluoranthene (Flua), and Pyrene (Pyr)	Batch soil washing using thiacalix-4-arene tetrasulfonate (TCAS) under controlled pH and temperature; interaction mechanisms studied via FTIR, NMR, UV-vis, and computational modeling	Nap: 55.1%, Flua: 47.7%, Pyr: 50.0% at 0.7% TCAS concentration	[38]
	Soil from a former lead mining district in Linares, Spain; granulometric fraction (1000–2000 μm)	10 potentially toxic elements (PTEs): As, Cd, Cr, Cu, Hg, Mo, Ni, Pb, Sb, and Zn	Peak recoveries at 40 kV: Zn (83.3%), Mo (81.0%), Cu (62.2%), Ni (57.3%) Lowest for Hg (13.2%)	Electrostatic separation at 12 voltages (20–41.5 kV); optimized via a new PPAA-U (Penalized Attributive Analysis) index that balances yield and recovery while penalizing variance	[39]
	Silt loam from Thailand, artificially spiked with 40 mg/kg Chlorpyrifos	Chlorpyrifos (CPF)	Magnetic separation + nanoscale zero-valent iron (nZVI) assisted by activation and reductive/oxidative degradation processes	>99.95% CPF removal using activated AZ or AZ + H_2O_2	[40]
	soil washing effluent (SWE) containing PHE, not the soil itself	Phenanthrene (PHE), one of the U.S. EPA priority PAHs	Ultrasound-activated persulfate (US/PS) oxidation with APG surfactant; reaction influenced by pH, PS dose, temperature, anions, and organic matter; mechanism confirmed by EPR, DFT, GC-MS	Up to 79.8% degradation in SWE (initial PHE = 6 mg/L, optimized US/PS conditions)	[64]
	Real uranium-contaminated soil with different Fe/U ratios (5:1, 2.5:1, 1.5:1); up to 160 Bq/g U(VI)	U(VI), Fe, Al, Si, Ca, Mg, K, and Na	3-cycle soil washing with H_2SO_4 followed by pH-adjusted neutralization using hydrazine (N_2H_4) for co-precipitation; major metal behavior analyzed by ICP-OES and XRF	≥99.4% U(VI) removal with 3-step washing using ≥0.05 M H_2SO_4 Clearance levels achieved: <81 ppm (1 Bq/g)	[65]

Table 3. *Cont.*

Method	Soil Type	Contaminants	Experimental Conditions	Efficiency	Refs.
Thermal desorption	Loess soil (Gansu, China)	Phenanthrene, Pyrene, and Benzo(a)pyrene (PAHs)	50–800 °C, 2 h, argon atmosphere, tube furnace	>99% removal at 400 °C; <5% below 200 °C	[43]
	Loam, clay, sand (China)	Waste lube oil	200–500 °C, 60 min, indirect TD under N_2 atmosphere	93% at 350 °C	[44]
	Artificially contaminated clay loam soil (Shenyang, China)	Heavy petroleum (alkanes, aromatics, resins, and asphaltenes)	100–600 °C, 5–60 min, bench-scale furnace (TD system)	87.26–98.69% at 500–600 °C after 60 min	[45]
	Gleysols, Podzols, Phaeozems (Russia)	Petroleum hydrocarbons	Ex situ thermal desorption, 250 °C	Reduced from up to 24,850 mg/kg to ≤410 mg/kg	[46]
	Artificially diesel-contaminated soil (Baoding, China)	Diesel hydrocarbons (C10–C40)	Muffle furnace; 150 °C, 350 °C, 550 °C; treated until ≤826 mg/kg	350 °C: Residual ≈ 280 mg/kg; 550 °C: Residual ≈ 5 mg/kg	[42]
	Oil-based drill cuttings (Hubei, China)	PAHs, Heavy metals (As, Cr, etc.)	Pilot rotary reactor; 250–300 °C, 10–30 min	PAHs: up to 96.27% at 300 °C / 30 min; Heavy metals: low	[27]
	Crude oil-spiked soil (Rawalpindi, Pakistan)	Crude oil (10%, 20%)	Thermal desorption at 500 °C, 60 min, lab scale	19.7% degradation at 20% contamination level	[47]
	Sandy soil (Suqian, China)	Pyrene (PAH)	Bench-scale TD; 100–400 °C, 2–60 min; optimized temp: 200 °C	99.1% at 400 °C/10 min, 94.9% at 200 °C/40 min	[48]
	Diesel-spiked sandy soil (Wuhan, China)	TPH	Electromagnetic induction LTTD; 200–300 °C, 20–60 min	93.44–97.65% (optimum at 216 °C, 21 min)	[49]
	Cd-contaminated soil (spiked + field soil, Hunan, China)	Cd	Biomass co-pyrolysis (WS/soil = 0.4), 550 °C, 1 h, N_2 atmosphere	Up to 96.5% (lab soil); >92% (4 field soils)	[50]
Adsorption-enhanced remediation	Petroleum-contaminated soil	Cd	MnO_2 modified biochar	115.04 mg g^{-1} (Cd)	[51]
	Contaminated soil	Pb, Cd, and As	Fe/cellulose nanocrystals	126.06 (Pb), 53.07 (Cd), and 15.80 mg g^{-1} (As)	[52]
	Petroleum-contaminated soil	TPH	Biochar/g-C_3N_4	80 mg g^{-1} (TPH)	[53]
	U and Cd co-contaminated soil	U and Cd	Mixed bacteria 9	69% (U) and 56% (Cd)	[54]
	Artificially contaminated soil	Cr	Polyaniline/polyvinyl alcohol/sodium alginate composite gel bead	83.1 mg g^{-1} (Cr)	[55]
	Heavy metal-contaminated soils	Cu, Cd, and Pb	MgO-loaded fish scale biochar	505.8 (Cu), 327.2 (Cd), and 661.2 mg g^{-1} (Pb)	[56]
	Farmland soil	Cd and Pb	Ball-milled P-loaded biochar materials	18.7 (Cd) and 126 mg g^{-1} (Pb)	[57]
	Heavy metal-contaminated soils	Pb and Cd	Novel β-CD/hydrothermal biochar	50.44 (Pb) and 33.77 mg g^{-1} (Cd)	[58]

Table 3. *Cont.*

Method	Soil Type	Contaminants	Experimental Conditions	Efficiency	Refs.
Soil vapor extraction (SVE)	Sandy, silty, clay	Organic pollutants including VOCs (e.g., BTEX), SVOCs, and chlorinated hydrocarbons	Machine learning (XGB, SVM, ANN, RF) trained on 11 variables (e.g., time, temperature, soil structure)	Predicted with R^2 = 0.9799 using XGB model; actual removal varies by conditions	[29]
	Low-permeability clay artificially contaminated with ammonia gas (NH_3); lab-scale soil column and 2D plate models	Ammonia gas	Pneumatic fracturing combined with SVE; tested air pressure (0.1–0.5 MPa), fracturing point spacing (10–25 cm), and extraction modes (continuous, intermittent, stepwise, gas-assisted)	Up to 98% removal under continuous extraction + bottom gas injection at 20 kPa; optimal vapor flow rate = 9 L/min	[61]
	Loam soil spiked with toluene (0.25–1 g/kg); water content adjusted (1–10%)	Toluene	Soil Vapor Extraction (SVE) + Internal Combustion Engine (ICE) for VOC off-gas destruction; optimized via Response Surface Methodology (RSM); thermal reuse of ICE exhaust applied to SVE	SVE: 100% DRE (Destruction Removal Efficiency) via ICE Desorption time optimized to 615 min at 0.3 g/kg, 10 L/min air flow, 9% moisture Pulsed operation improved efficiency and saved 1.104 kWh/kg	[59]
	Low-permeability soil (clayey), spiked with MTBE	MTBE (a common gasoline additive VOC)	Air sparging	Up to 28% MTBE removal at 100 kPa with enhanced air sparging; improvement over conventional by +23.7%	[62]
	Synthetic compacted silty loam	γ-HCH, ε-HCH, α-HCH, and δ-HCH	Up to 92.9% removal in 15 days with SVE + EKSF at 3.0 V/cm (vs. 70% with SVE only)	Combined Soil Vapor Extraction (SVE) and Electrokinetic Soil Flushing (EKSF)	[63]

3.2. Chemical Remediation

Chemical remediation encompasses a range of in situ treatment techniques that leverage redox reactions to transform or degrade contaminants directly within the subsurface [66]. It is particularly effective for managing a broad spectrum of pollutants, including organic compounds and toxic metals, which are commonly encountered in industrial sites [30]. Recent advances focus on enhancing the selectivity, sustainability, and field applicability of chemical oxidants and reductants through the integration of nanomaterials, catalysts, and bio-based activators. These technologies can be broadly categorized into three main approaches such as chemical oxidation and reduction, electrokinetic remediation, and stabilization/solidification. Each method targets specific contaminant profiles and geochemical conditions, as illustrated in Figure 4. Recent studies demonstrating their efficiency and innovation across diverse contaminated sites are systematically summarized in Table 3.

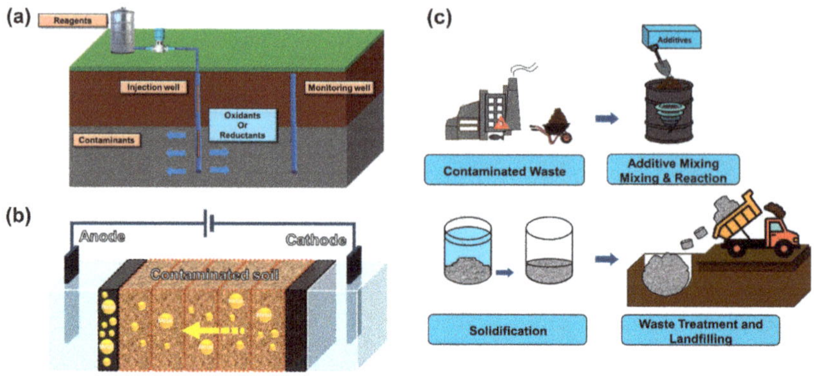

Figure 4. Illustrations of representative chemical remediation techniques: (**a**) Chemical oxidation and reduction, (**b**) electrokinetic remediation, and (**c**) stabilization/solidification [67–69].

3.2.1. Chemical Oxidation and Reduction

Chemical oxidation and chemical reduction are widely applied in situ remediation techniques designed to treat a broad range of contaminants, including VOCs, PAHs, chlorinated solvents (such as trichloroethylene and perchloroethylene), and toxic metals (Figure 4a). These methods are particularly relevant for industrial sites, where co-contamination is common and rapid, and large-scale treatment is often necessary [70].

Recent developments have focused on combining nanomaterials, biodegradable activators, and biochar to improve treatment efficiency, sustainability, and field applicability. The evolution of these methods reflects a shift toward integrated remediation approaches that minimize secondary pollution and adapt to site-specific conditions.

For example, Zeng et al. [71] developed a pilot-scale system using biochar-supported nanoscale zero-valent iron to activate persulfate for treating soil contaminated with aromatic hydrocarbons. This approach significantly improved degradation efficiency under real field conditions.

Song et al. [72] demonstrated the effectiveness of sulfidated nanoscale zero-valent iron in treating cadmium-contaminated groundwater at a zinc smelter site. Their study confirmed high reduction efficiency and operational stability in highly acidic environments.

Wang et al. [73] introduced a novel oxidation strategy using polymetallic-doped graphitic carbon nitride to generate hydrogen peroxide directly within the contaminated zone. This in situ oxidant generation method reduces the need for external chemical inputs, lowering costs and environmental risks.

In the treatment of persistent chlorinated organics, Liu et al. [74] combined chemical reduction with enhanced bioremediation to accelerate trichloroethylene degradation in groundwater. Their integrated system showed superior performance in both speed and long-term stability. Reginatto et al. [75] explored the application of nanoscale zero-valent iron in clay-rich soils contaminated with hexavalent chromium. Laboratory-scale column tests confirmed the feasibility of applying this technology to low-permeability soils typical of shipyard environments.

Recent research has also emphasized the ecological impact of these chemical processes. Geng et al. [76] used high-resolution mass spectrometry to assess the structural transformation of soil organic matter after oxidation treatments. Meanwhile, Hong et al. [77] investigated how different oxidation methods affect microbial community structure in petroleum-contaminated soils. Rodrigues et al. [78] provided a thorough overview of field applications of chemical reduction methods, particularly focusing on the use of zero-valent metals in permeable reactive barriers to remediate chlorinated organic pollutants in complex contaminated sites such as shipyards. Lastly, Mouvet et al. [79] reported on a field-scale chemical reduction project targeting chlordecone-contaminated soils in the French Caribbean. The study not only demonstrated successful contaminant immobilization but also evaluated soil recovery and agronomic potential post-treatment.

3.2.2. Electrokinetic Remediation

Electrokinetic remediation is an in situ technique that employs low-intensity direct current to mobilize and extract contaminants from fine-grained or low-permeability soils such as clay and silt [30]. As illustrated in Figure 3b, an electric potential is applied between embedded electrodes (anode and cathode) in the contaminated soil, inducing electroosmotic flow, electromigration, and electrophoresis [80]. These mechanisms facilitate the transport of charged contaminants, such as metal ions, toward the electrodes, where they can be collected or treated [81].

Recent research has focused on optimizing electrode configurations, enhancing electrokinetic transport using surfactants or biodegradable complexing agents, and integrating electrokinetic systems with biological or chemical treatment processes to increase remediation efficiency; Kim et al. [81] compared one-dimensional and two-dimensional electrode arrangements and reported that the two-dimensional system significantly improved the removal efficiency of Pb and Zn from mine-impacted soils. Shi et al. [82] combined electrokinetics with bioremediation to treat trichloroethylene (TCE) in low-permeability soils, observing enhanced microbial activity and dehalogenation near the electrode zones. Similarly, Narenkumar et al. [83] applied a biosurfactant-assisted bio-electrokinetic system, achieving effective mobilization and removal of Cu and chromium (Cr), while maintaining microbial compatibility. Sun et al. [84] demonstrated that electrode placement plays a critical role in soft clay environments, where a vertical anode–cathode arrangement enhances electroosmotic flow. In another study, Wang et al. [85] used biodegradable complexing agents to chelate Pb and Cu, which significantly improved their mobility and removal efficiency, achieving over 75% removal.

3.2.3. Stabilization/Solidification

Stabilization/solidification is a passive in situ or ex situ remediation technique that physically immobilizes contaminants by incorporating them into a stable solid matrix, thereby reducing their leachability and environmental mobility [86]. As shown in Table 4 and illustrated in Figure 4c, Recent advances in stabilization/solidification technologies focus on low-carbon binders, industrial waste reuse, and long-term durability under environmental stress. Several studies have reported excellent performance improvements in both contaminant immobilization and mechanical integrity. Fan et al. [86] compared coal fly ash-based geopolymer with ordinary Portland cement for solidifying MSWI fly ash. The geopolymer achieved over 95% reduction in Pb leachability and demonstrated higher unconfined compressive strength (10.4 MPa) than cement (6.2 MPa), meeting both environmental and structural performance thresholds. Wang et al. [87] developed a low-alkalinity, clay-based binder to stabilize Pb-contaminated soils without cement. The treatment reduced Pb leaching by approximately 90% and maintained compressive strength of ~2 MPa after 28 days—suitable for capping or subgrade applications. Chen et al. [88] applied a composite of red mud and blast furnace slag under alkali activation to Cd-contaminated soils. Cd leachability was suppressed by over 96%, and compressive strength improved to 5.2 MPa with a 10% binder dosage. Xia et al. [89] evaluated a hydroxyapatite-based binder in field-scale applications at a Pb-Zn smelter site. The system retained strength above 3.5 MPa over 90 days and reduced Pb concentrations to <5 mg/L, remaining within regulatory leaching limits. Liu et al. [90] investigated the durability of a Zn-contaminated soil solidified with a cement-soda residue mix under freeze–thaw cycling. After six cycles, Zn leaching remained below 2 mg/L, and compressive strength was maintained at ~4.1 MPa. Zhang et al. [91] compared calcium oxide (CaO) and magnesium oxide (MgO) activated GGBS binders for Zn-contaminated clay slurry. MgO-activated binders outperformed CaO in both Zn immobilization (88%) and unconfined strength (~6.5 MPa), indicating better stability and lower long-term risk.

3.3. Biological Remediation

Biological remediation refers to the use of microorganisms, plants, or their enzymatic systems to degrade, transform, or immobilize environmental contaminants under natural or engineered conditions [30]. Compared to physicochemical approaches, biological methods are generally more sustainable and better suited for the long-term treatment of organic pollutants and selected heavy metals. These techniques are applicable to shipyard soils where mixed contamination is prevalent.

As illustrated in Figure 5 and summarized in Table 3, biological remediation can be broadly categorized into microbial and phytoremediation strategies. Each has been applied either individually or in combination with other techniques to improve remediation efficiency and ecological compatibility.

Table 4. Previous studies of chemical remediation.

Method	Soil Type	Contaminants	Experimental Conditions	Efficiency	Refs.
Chemical oxidation	Aquifer soil	Hydrocarbons (e.g., benzene, toluene, ethylbenzene, and xylene)	Experiment with 0.2 and 50 mM concentrations of H_2O_2 and PMS in a microenvironment	Maximum geochemical change observed with 50 mM PMS treatment	[92]
	Technosol	Aromatic hydrocarbons (e.g., benzene, toluene, and xylene)	Combination of nano zero-valent iron (nZVI) and biochar to activate PS for site-scale remediation of aromatic hydrocarbons	99% removal of contaminants	[71]
	Aqueous media	Tinidazole (an antibiotic)	Tinidazole removal from aqueous media by heat-activated PS	95% removal of tinidazole	[93]
	Aqueous media	Trichloroethylene (TCE)	Combination of PS and dithionite for the removal of TCE	85% TCE removal with the PS/dithionite system	[94]
	Aqueous media	Various organic contaminants	In situ generation of H_2O_2 for contaminant removal from aqueous media	90% removal of contaminants	[73]
	Organic contaminated soil	Various organic contaminants	Comparison of natural oxidant demand (NOD) of different oxidants in soil	Comparison of NOD effects of each oxidant	[95]

Table 4. *Cont.*

Method	Soil Type	Contaminants	Experimental Conditions	Efficiency	Refs.
Chemical oxidation	Aqueous media	Fluoroquinolone antibiotics (e.g., ciprofloxacin)	Use of sludge biochar combined with PDS to degrade fluoroquinolones in soil	85% fluoroquinolone removal	[96]
	Diesel-contaminated soil	Diesel contaminants	Combined use of O_3 and H_2O_2 for diesel-contamination removal	90% diesel-contamination removal	[97]
	Aqueous media	Organic contaminants (e.g., phenol and chlorophenols)	Organic degradation with PMS in the presence of metal oxides	80% organic degradation rate	[98]
	Diesel-contaminated soil	Diesel-contaminated soil	Comparison of Fenton and base-activated persulfate treatments	Enhanced cation exchange capacity and germination index post-treatment	[76]
	Total petroleum hydrocarbon (TPHs) contaminated soil	TPH	Application of 5% H_2O_2; monitoring over 11 days	64.8% TPH reduction; increased soil respiration rate	[77]
Chemical reduction	Contaminated groundwater	Cd, Zn, Ni, and Al	Field application at a zinc smelter; S-nZVI injected into acidic groundwater	Cd removal increased from 7.2% (nZVI) to 99.8% (S-nZVI); pH neutralized to >6	[72]
	Contaminated nitisol	Chlordecone (CLD)	Field trials with 2.8% and 4.0% ZVI (dw/dw) in 0–40 cm soil layer; treatment durations of 37 and 94 days	CLD concentration reduced by up to 74% (2.8% ZVI) and 69% (4.0% ZVI); 75% of the decrease achieved within 21–24 days	[79]
	Contaminated groundwater	Trichloroethylene (TCE)	Field application combining EZVI injection with biostimulation; monitoring over several months	Significant reduction in TCE concentrations; enhanced dechlorination rates observed	[74]
	Contaminated groundwater	1,2,3-Trichloropropane (TCP)	Pilot-scale injections at Marine Corps Base Camp Pendleton; monitoring over four years	CP reductions exceeding 90% maintained for at least four years post-injection	[99]

Table 4. *Cont.*

Method	Soil Type	Contaminants	Experimental Conditions	Efficiency	Refs.
Chemical reduction	Contaminated clayey soil	Cr	Batch and column tests with varying nZVI/Cr(VI) ratios and injection pressures	98% Cr(VI) reduction at optimal conditions	[75]
	Chromium-contaminated soil	Cr	Batch experiments assessing Cr(VI) removal efficiency	97.96% Cr(VI) removal after 21 days	[100]
	Kaolinite	Phenanthrene	Permeable reactive barrier installation	Effective in situ remediation	[101]
	Copper-contaminated soil	Cu	Two-step soil washing technique using $NH_2OH{\cdot}HCl$ and humic substances	Achieved up to 53% Cu removal efficiency	[102]
	Aqueous media	Chlorinated organic compounds (e.g., TCE)	Synthesis and application of S-nZVI for enhanced dechlorination	S-nZVI demonstrated higher reactivity towards TCE compared to traditional nZVI	[78]
	Aqueous media	Arsenic	Use of modified nZVI for arsenic removal	Enhanced removal efficiency due to surface modification	[103]
Electrokinetic remediation	Smelter soil	Cd, Cu, Pb, Ni, and Zn	2D electrode array, 8 days, 1 V/cm, 20–75 mA	79.8% (Cd), 82.6% (Cu), 83.7% (Ni), 34.3% (Pb), and 81.1% (Zn)	[81]
	Low-permeability clay	Trichloroethylene (TCE)	~1 V/cm, ~25 mA	73% (TCE)	[82]
	Heavy-metal-contaminated soft clay	Cu	20–60 V, 10–30 mA	42.2% (Cu)	[84]
	Heavy metal-contaminated soil	Cr	7 days, 30 V	75%	[83]
	Artificially contaminated soil	Pb, Na, and Cl	1–3 days, ~1 V/cm	41% (Pb), 82% (Na), and 69% (Cl)	[104]
	Real mine tailings	Pb and Zn	9 days, 2 V/cm, 20–70 mA	51% (Pb) and 38% (Zn)	[105]
	Soft soils	Cu and Pb	2 V/cm, 160 mA	30% (Cu) and 10% (Pb),	[68]
	Heavy metals contaminated soil	Cu, Pb, and Zn	5 days, 110.2 mA	95.69% (Cu), 94.42% (Pb) and 95.05% (Zn)	[85]

Table 4. *Cont.*

Method	Soil Type	Contaminants	Experimental Conditions	Efficiency	Refs.
Solidification/stabilization (S/S)	MSWI fly ash	Heavy metals (Pb, Zn, and Cd)	Comparison between coal fly ash-based geopolymer and Portland cement; curing time 28 days; leaching test; compressive strength test; chemical speciation analysis	Immobilization efficiency >99% for Pb, Zn, Cd using geopolymer; compressive strength up to 22.4 MPa.	[86]
	Lead-zinc smelting slag (LZSS)	Heavy metals (Pb, Zn, Cu, and Cr)	Composite geopolymer using PFA and BFS; optimized alkali-activator ratios; single and orthogonal design; 28-day curing; leaching tests (TCLP, CN-HJ 299-2007).	Max compressive strength: 47.39 MPa; up to 80% LZSS solidified with < toxic leaching levels; suitable for landfill and construction.	[89]
	High-Pb-contaminated shooting range	Pb	PC and GGBS binary cement with KDP, ISSA, and PG additives; curing for 7 and 28 days; TCLP and semi-dynamic leaching tests; compressive strength tests; XRD, TGA, SEM-EDX analysis.	TCLP Pb reduction >99.9% with BC; compressive strength up to 40 MPa; 69.4% reduction in diffusion coefficient of Pb with KDP.	[106]
	Antimony-contaminated soil (Harbin, China, simulated contamination)	Sb	Six Fe/Fe^{2+}-FA-CA composite materials tested; 40 freeze-thaw cycles; TCLP and CT leaching tests; UCS and SEM/XRD analysis; 7 and 28-day curing.	Best composite (FFC) showed 92.92% SE (TCLP) and 93.37% (CT); UCS maintained > 0.815 MPa after 40 F-T cycles; Sb leaching < 0.06 mg/L.	[107]
	Pb-Zn smelter contaminated site	Pb, Zn, Cd, and organics (COD)	8% SPC (superphosphate + CaO, 3:1); 33 and 256 days curing; field trial; TCLP and HJT 557-2010 leaching; COD, ANC, SEP, XRD, DCP tests.	>95% reduction in Pb, Zn, Cd leachability; COD below 40 mg/L by 33 days; increased strength; ANC improvement; formation of phosphate precipitates.	[108]

Table 4. *Cont.*

Method	Soil Type	Contaminants	Experimental Conditions	Efficiency	Refs.
Solidification/ stabilization (S/S)	Zn-contaminated clay slurry (simulated dredged sediment)	Zn	Binders: OPC, CaO-GGBS, MgO-GGBS; Zn = 2000/4000 mg/kg; GGBS: 10%, MgO 0–3%, CaO 0–1.5%; Curing: 56 days; pH, UCS, leaching tests; XRD	MgO-GGBS yielded highest UCS and lowest Zn leaching (0.01–0.1 mg/kg); CaO-GGBS showed poor performance due to $Ca(Zn_2(OH)_6)\cdot 2H_2O$ formation; MgO-GGBS suitable as OPC alternative.	[91]
	Zn-contaminated clayey soil (artificially prepared)	Zn	Cement and soda residue mixed at various ratios (C3SR7, C4SR6, C5SR5); Zn^{2+} concentrations 0–10,000 mg/kg; 90-day curing; 0–10 freeze–thaw cycles; UCS, SEM, EDS, MIP tests for strength, crack, and pore analysis.	Max UCS at 2000 mg/kg Zn; UCS decreased 34.9–56% after 10 F–T cycles; higher cement ratio improved F–T resistance; micro- and macro-pore changes linked to strength loss.	[90]
	Alkali-activated fiber-reinforced composites (AAFRCs) using incineration fly ash (IFA)	Heavy metals (Cd, Pb, Zn, Cu, and Cr) from municipal solid waste incineration fly ash	IFA partially replaced GGBS and SF at 0–60%; mechanical tests (tensile, compressive, flexural); SEM/EDS; fiber pullout test; leaching test (TCLP); optimized at ≤40% IFA; 28-day curing at 20 ± 1 °C after 80 °C heat curing.	Up to 40% IFA replacement achieved 73.78 MPa compressive strength and 5.19% tensile strain; effective immobilization of heavy metals (Pb, Cd, Zn); IFA-AAMs considered safe for civil use.	[109]
	Pb-contaminated kaolin clay	Pb	GGBS-MgO (9:1) binder; binder contents of 12%, 15%, 18%; Pb = 2% wt.; Simulated acid rain (pH 2, 4, 7); 28-day curing; semi-dynamic leaching, ANC, MIP, XRD; leachability, diffusion coefficients, pore structure, and hydration phases studied.	CFL of Pb decreased with higher binder content and higher SAR pH; Dobs as low as 3.77×10^{-18} m^2/s at pH 7.0; Pb immobilized as hydrocerussite; higher binder content improved acid buffering and pore compactness.	[110]

Table 4. *Cont.*

Method	Soil Type	Contaminants	Experimental Conditions	Efficiency	Refs.
Solidification/ stabilization (S/S)	Artificially prepared Cd-contaminated soil (Taiyuan, China)	Cd	RAS binder composed of RM:GGBS:CCR: PG = 5:10:2:3; Cd = 0–2500 mg/kg; RAS dosage = 0–20%; Curing = 7–28 days; pH, UCS, leaching toxicity (CN-GB 5085.3), BCR, MIP, XRD, SEM-EDS, FTIR.	Compressive strength 4–15× untreated soil, all >1 MPa; Cd leaching reduced by >96%; Residual Cd form increased; Total leaching $\leq$ 0.01 mg/L after 28 d; Cd stabilized via precipitation, C–S–H, AFt formation, ion exchange, and complexation.	[88]
	Contaminated soils	As and Pb	pH adjustment (lime)	96.2% and 98.8%	[87]

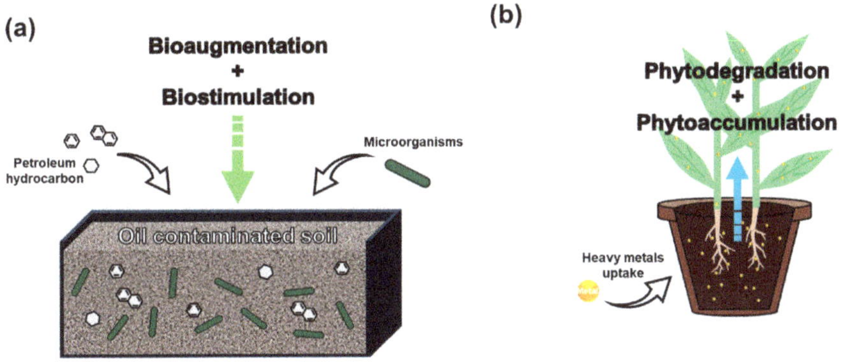

Figure 5. Schematic illustration of (**a**) microbial remediation and (**b**) phytoremediation [111,112].

3.3.1. Microbial Remediation

Microbial remediation involves the use of indigenous or introduced microorganisms to degrade, transform, or immobilize contaminants in polluted soils. This strategy primarily includes biostimulation (enhancing native microbial activity through nutrient or surfactant addition) and bioaugmentation (introducing specific strains with high degradation capacity) [30]. As shown in Table 5, this approach has been widely used in hydrocarbon-contaminated sites. For example, Sánchez Mata et al. [113] applied a mixture of enzymes, surfactants, and vermicompost to stimulate microbial activity in automotive oil-contaminated soils, achieving a total petroleum hydrocarbon (TPH) reduction of over 70% within 28 days. Similarly, Wu et al. [114] showed that the addition of nutrients significantly accelerated hydrocarbon degradation and improved microbial diversity in oil-impacted soils. Bioaugmentation techniques have also shown high potential, especially when combined with biostimulation. In a comparative study, Yaman [111] found that combined bioaugmentation-biostimulation outperformed natural attenuation by achieving over 85% crude oil removal efficiency in 40 days. Haleyur et al. [115] revealed that bioaugmentation with PAH-degrading bacteria significantly enhanced degradation rates in aged soils and led to a shift in the microbial community structure toward hydrocarbon degraders.

Recent developments also highlight hybrid microbial systems, integrating microbial remediation with biochar or electrokinetics to overcome challenges like low pollutant bioavailability. For instance, Wang et al. [116] used Stenotrophomonas acidaminiphila supported on agricultural biochar to remediate Cr(VI)-contaminated soil, achieving over 90% reduction in Cr(VI) via coupled bioreduction and adsorption. Furthermore, Barba et al. [117] compared biostimulation and bioaugmentation under electrokinetic fields and found the hybrid system yielded faster degradation of 2,4-D herbicides and improved microbial viability around electrodes. These studies demonstrate that microbial remediation, particularly when combined with supportive amendments or physical assistance (e.g., biochar, electric fields), can offer effective, environmentally friendly solutions for legacy pollution in shipyard soils and similar industrial environments.

3.3.2. Phytoremediation

Phytoremediation refers to the use of plants and their associated rhizosphere microorganisms to remove, stabilize, or transform contaminants in soil and water [118]. It is particularly attractive for long-term and sustainable remediation at marginally contaminated or shallowly polluted sites due to its low cost, minimal invasiveness, and ecological co-benefits [119]. As illustrated in Figure 5 and summarized in Table 5, recent research emphasizes a range of mechanisms, including phytoextraction, phytostabilization, and rhizodegradation, often enhanced through soil amendments or microbial inoculation. Wiafe et al. [120] applied both organic and inorganic amendments to enhance phytoavailability and recovery of heavy metals, achieving over 50% reduction in metal mobility. Hou et al. [121] demonstrated the synergistic effect of plant-microbe systems in degrading PAHs in petroleum-contaminated soils, with degradation rates exceeding 60%. Similarly, Chen et al. [122] conducted a meta-analysis showing that biochar amendments significantly increased plant uptake efficiency of metals such as Cd, Pb, and Zn by over 30% in many cases.

To improve chelation-assisted uptake, Li et al. [123] used GLDA, a biodegradable chelating agent, with Tagetes patula, reporting high cadmium removal without phytotoxic effects. Chigbo and Batty [124] reported over 70% Cu and pyrene removal using Zea mays with EDDS, confirming the viability of dual-contaminant phytoremediation. Direct phytoextraction was also explored by Shehata et al. [125], showing meaningful reductions in Pb, Cd, and Ni levels in soil.

More advanced strategies involve microbial facilitation. Liu et al. [126] demonstrated enhanced uptake of heavy metals via co-cropping techniques using plant growth-promoting rhizobacteria. Integrated approaches were validated by Lai et al. [127], who achieved over 65% reduction in TPH and heavy metals by combining phytoremediation with microbial inoculants. Moreover, crop rotation strategies were evaluated by Yang et al. [112], revealing significant variation in phytoextraction efficiency based on annual fluxes of Cd. Finally, Bhat et al. (2022) [128] provided a comprehensive review highlighting the scalability, eco-compatibility, and future potential of phytoremediation as a green technology.

Table 5. Previous studies on biological remediation.

Method	Soil Type	Contaminants	Experimental Conditions	Efficiency	Refs.
Bioremediation	Sandy loam	Automotive residual oil	pH 4.5; 37 °C; enzyme addition + 3–5% vermicompost; 49 days	99.9% removal efficiency	[113]
	Clay	Petroleum Hydrocarbons	Temperature-controlled environment with nutrient amendments	88%	[129]
	Contaminated soils	PAHs (Polycyclic Aromatic Hydrocarbons)	Inoculation with PAH-degrading microbes	99%	[115]
	Clay	2,4-dichlorophe noxyacetic acid	Temperature-controlled environment with applied electric field 1.0 V cm^{-1}	75~85%	[117]
	Contaminated soils	TPH	open and closed microcosms systems, Humidity-controlled environment with nutrient amendments	84.1% and 78.7%	[130]
	Crude Oil-Contaminated Soils	TPH	microbial inoculum and nutrient addition	35~66%	[111]
	Contaminated Soils	Cr	Humidity-controlled environment with microbial inoculum	74.94%	[116]
	Petroleum-contaminated soil	TPH	Room temperature environment with nutrient addition	60% and 34%	[114]
	Petroleum-contaminated soil	TPH	room temperature of 20–25 °C (autumn) and 10 °C(winter).	59.89% and 51.38%	[131]
Phytoremediation	Agricultural soil	As, Hg, Cd, and Pb	Field experiment using *T. capensis* and *H. psittacorum* with foliar application of fulvic acid, auxin, compost, coconut coir, and sulfur	Enhanced uptake and translocation of heavy metals due to amendments altering soil pH and microbial activity	[120]
	Petroleum-contaminated soil	Polycyclic aromatic hydrocarbons (PAHs)	Use of microorganism-assisted phytoremediation in heavily polluted soils	Significant degradation of PAHs observed, indicating enhanced remediation efficiency	[121]

Table 5. *Cont.*

Method	Soil Type	Contaminants	Experimental Conditions	Efficiency	Refs.
Phytoremediation	Contaminated soil	Cd, Pb, Cu, and Zn	Application of biochar derived from various sources to contaminated soils	Reduction in bioavailable Cd by 52%, Pb by 46%, Cu by 29%, and Zn by 36%	[122]
	Agricultural soil	Cd	Application of tetrasodium glutamate diacetate (GLDA) with *Tagetes patula* L.	Removal of 12.9% of Cd from contaminated agricultural land	[123]
	Co-contaminated soil	Heavy metals and PAHs	Application of EDTA and citric acid to enhance phytoextraction	Enhanced plant growth by 41% and increased phytoextraction of Cu; pyrene dissipation improved by up to 90.4%	[124]
	Contaminated soil	Cr, Co, Cd, and Mn	Use of kenaf and flax plants with sulfur and gibberellin treatments over two growing seasons	Double-season phytoremediation efficiencies: Cr (66.87%), Mn (65.63%), Co (54.66%), Cd (23.40%)	[125]
	Contaminated soil	Heavy metals	Co-cropping of *Sedum alfredii* with alfalfa, associated with plant growth-promoting rhizobacteria (PGPR)	Enhanced phytoextraction of heavy metals due to synergistic plant-microbe interactions	[126]
	Contaminated soil	TPH and heavy metals	Use of *Populus bonatii* and *Crotalaria juncea* with earthworms and petroleum-degrading bacteria over 20 months	Higher TPH removal efficacy at 30–60 cm soil depth; more Zn than Cr absorbed and accumulated by plants	[127]
	Heavy metal-contaminated soil	Cd	Evaluation of phytoremediation potential based on the balance between annual input and output fluxes of Cd in farmland soil	Demonstrated potential for Cd removal through plant uptake	[112]
	Contaminated soil	Heavy metals	Use of hyperaccumulator plant species in various soil types	Effective removal of heavy metals from polluted areas	[128]

4. Practical Examples of Remediation and Redevelopment in Shipyard Environments

Table 6 summarizes representative examples of shipyard remediation and redevelopment projects from various regions. Although multiple advanced technologies such as soil washing, in situ chemical oxidation, and bioremediation have been introduced in the literature, the actual remediation practices at shipyards have largely relied on conventional physical approaches such as excavation and dredging. This trend can be attributed to multiple practical factors, including the urgency of site reuse, economic constraints, and the complex heterogeneity of contamination at such industrial sites. The following cases highlight specific contamination profiles, applied remediation methods, their performance, and the scale of intervention. At the Hunters Point Naval Shipyard in San Francisco (~3.5 million m^2), the site was impacted by petroleum hydrocarbons, VOCs (e.g., benzene, TCE), PCBs (<0.74 mg/kg), heavy metals (e.g., Pb up to 287 mg/kg), and radionuclides (e.g., Cs-137, Sr-90) [132–134]. The remediation involved the excavation of approximately 1 million tons of contaminated soil, the installation of landfill capping systems, vapor extraction, and the removal of radiologically contaminated structures. These interventions achieved more than 95% removal of PCB and TPH concentrations, with radionuclide levels reduced to near-background conditions. Advanced in situ chemical or biological treatments were not actively utilized despite the complexity of contamination. While the site presented a complex mixture of pollutants, advanced in situ chemical or biological treatments were not predominantly used.

Similarly, At the Mare Island Naval Shipyard (~22.7 million m^2), the remediation strategy was similarly excavation-focused, involving the removal of over 300 underground storage tanks (USTs), excavation of hotspots contaminated with PCBs (11–14 mg/kg), lead (up to 8300 mg/kg), zinc (11,000 mg/kg), TPH (82 mg/L), and various explosives [134–136]. A pilot-scale thermal desorption test was conducted, but large-scale implementation was not pursued. The project achieved over 90% reduction in key contaminants and enabled the site to be certified for commercial reuse, with regulatory monitoring scheduled until 2050.

The Philadelphia Navy Yard (~5.9 million m^2) was remediated through a combination of asbestos abatement, soil excavation targeting PCB (>50 mg/kg), lead (>2000 mg/kg), petroleum hydrocarbons, and UST removal [137,138]. Groundwater contaminated with VOCs was treated through pump-and-treat systems. Post-remediation assessments indicated PCB levels below 1 mg/kg, and asbestos was fully removed, enabling large-scale redevelopment that currently hosts over 150 businesses [137].

At the Charlestown Navy Yard in Boston (~526,000 m^2), the site contained PCBs (~12 mg/kg), PAHs (benzo[a]pyrene up to 21 mg/kg), dioxins (~0.046 mg/kg), and lead (~7 mg/kg). Remediation activities included excavation of approximately 917 m^3 of contaminated soil and sediment, PCB equipment removal, and capping of residual contamination zones [139–141]. The site has since been redeveloped as a National Historical Park [139].

In Varvsstaden (Kockums Shipyard), Malmö (~180,000 m^2), remediation targeted petroleum hydrocarbons, chlorinated solvents (e.g., TCE, PCE), and heavy metals. The excavation was conducted in grid-based units (~400 m^2), with high-contamination soils (~20,000 m^3) removed for off-site disposal. Soil washing was used for partial reuse of lower-risk materials, and the site now meets Swedish residential soil standards [142,143].

Table 6. Projects of shipyard remediation and redevelopment.

Project (Location)	Primary Contaminants	Remediation Technologies Used	Efficiency	Refs.
Hunters Point Naval Shipyard (San Francisco, CA, USA; ~3,504,581 m^2)	Volatiles-Halogenated: 1,1,1-TCA (25,300:g/kg); PCE (4360:g/kg); and TCE (60,100:g/kg); heavy metals (lead ~287 mg/kg); PCBs (<0.74 mg/kg); radionuclides (Cs-137, Sr-90, etc.);	Excavation (~1 million tons), soil vapor extraction, landfill capping, radiological cleanup	95%+ PCB/TPH reduction, radionuclide levels near-background, redevelopment-ready, ongoing monitoring	[132–134]
Mare Island Naval Shipyard (Vallejo, CA, USA; ~22,662,416 m^2)	PCBs (11–14 mg/kg), lead (up to 8300 mg/kg), zinc (11,000 mg/kg), TPH (82 mg/L), mercury, explosives, asbestos, low-level radionuclides	300+ USTs removed, hotspot excavation, pilot thermal desorption, $130M radiological cleanup	>90% PCB/TPH reduction, all land certified clean, commercial reuse enabled, monitored to 2050	[134–136]
Philadelphia Navy Yard (Philadelphia, PA, USA; ~5,888, 181 m^2)	Asbestos, PCBs (>50 mg/kg), lead (>2000 mg/kg), PAHs, petroleum hydrocarbons, VOCs	Asbestos abatement, PCB/lead soil excavation, UST removal, pump-and-treat for solvents	Full reuse; <1 mg/kg PCBs, asbestos removed, 150+ businesses by 2025	[137,138]
Charlestown Navy Yard (Boston, MA, USA; ~526,092 m^2)	PCBs (~12 mg/kg), PAHs (benzo[a]pyrene ~21 mg/kg), dioxins (~0.046 mg/kg), heavy metals (lead ~7 mg/kg), asbestos	~917 m^3 soil excavation, PCB machinery removal, lead/asbestos abatement, capping	>90% PAH/PCB reduction, National Historical Park reuse, residual PCBs capped	[139–141]
Varvsstaden (Kockums Shipyard) (Malmö, Sweden; ~180,000 m^2)	PAHs, mineral oils, heavy metals (Pb, Cu, Zn), chlorinated solvents (TCE/PCE), no TBT	Grid excavation (~400 m^2 cells), soil washing, 20,000 m^3 hazardous soil disposal, clean fill	Swedish residential standard met, >80% material reuse, urban redevelopment ongoing	[142,143]
Oslo Harbor Cleanup (Oslo, Norway)	Heavy metals (Hg ~0.8 mg/kg, Pb ~94 mg/kg), PAH16 (~3.6 mg/kg), PCBs (~28 µg/kg), TBT (~112 µg/kg)	700,000 m^3 dredging, CAD, 0.5 m clay cap, active monitoring	TBT reduced to 2.4 µg/kg, metals to near-background, fishing bans lifted	[144–147]
Howards Bay (Fraser Shipyard) (Superior, WI, USA)	Lead (>83 mg/kg), mercury, PAHs, TBT (>1.6 µg/kg), shipyard debris	~64,742 m^3 redging, cement stabilization, sand/topsoil capping	99%+ TBT reduction, bay reopened, dredge reused in green space	[148–151]
Tongyoung (Shina Shipyard) (Gyeongsangnam-do, South Korea; ~180,000 m^2)	Heavy metals (Cu, Cd, As, Pb, Zn) above KR Region 1 standards, TBT, TPH, down to 7 m depth	Grid excavation up to 7 m, high-pressure soil washing, reuse of clean soil, slurry wall installation	99% metal cleanup efficiency, the cleanup targets Korea's Region 1 soil standards, enabling future redevelopment into a marina and a mixed-use district	[152]

The Oslo Harbor Cleanup (Norway) primarily addressed sediment contamination with heavy metals (e.g., Hg ~0.8 mg/kg, Pb ~94 mg/kg), PAH16 (~3.6 mg/kg), PCBs (~28 μg/kg), and TBT (~112 μg/kg). Approximately 700,000 m^3 of sediments were dredged, confined in a designated CAD system, and capped with 0.5 m of clay. TBT levels were reduced to 2.4 μg/kg, and heavy metals were decreased to near-background levels, allowing the reactivation of fishing and harbor activities [144–147].

In Howards Bay (Fraser Shipyard), Wisconsin, roughly 64,742 m^3 of contaminated sediments were dredged and stabilized using cement-based solidification. These materials were then disposed of in licensed landfills and capped with clean sand and topsoil. TBT was reduced by over 99%, and portions of the treated sediment were reused for landscaping and green infrastructure development [148–151].

Finally, the Tongyeong (Shina Shipyard) project in South Korea (~180,000 m^2) is one of the few examples that actively adopted advanced soil treatment methods. Heavy metals (e.g., Cu, Cd, As, Pb, Zn), TPH, and TBT were detected at concentrations exceeding Korea's Region 1 residential standards, reaching depths of up to 7 m. A combination of deep grid excavation and high-pressure soil washing was applied. Recovered clean soils were reused, and a slurry wall was installed for hydraulic containment. The remediation achieved over 99% removal efficiency for heavy metals, making the site suitable for future redevelopment into a mixed-use and marina district [152].

Collectively, these case studies demonstrate that physical removal techniques such as excavation and dredging have remained the primary remediation strategies at shipyard sites, not merely because of their operational simplicity, but also because they offered reliable, large-scale contaminant removal under complex and heterogeneous conditions. These methods provided a practical response to urgent redevelopment timelines, regulatory certainty, and site-specific constraints. However, it is also important to situate these choices within their historical and technological context. Most remediation projects reviewed here were implemented between the early 1990s and 2010s, a period when many advanced in situ technologies, such as in situ chemical oxidation, electrokinetic remediation, or large-scale bioremediation, were either still in the pilot phase or lacked the regulatory frameworks necessary for full-scale deployment. At that time, their performance under field conditions remained uncertain, and their operational risks were perceived to be high. In that sense, the dominance of conventional physical methods does not necessarily reflect a disregard for innovation, but rather a rational response to the technological readiness levels, economic constraints, and risk-averse institutional decision-making frameworks of the time. As remediation technologies continue to evolve, addressing issues of cost, efficiency, and regulatory compatibility, there is strong potential for a more integrated and adaptive use of advanced techniques in future shipyard redevelopment projects.

5. Evaluation of Soil Remediation Technologies for Shipyards

The previous section emphasized that remediation projects at shipyard sites have predominantly relied on conventional physical methods, such as excavation and dredging. This tendency stems from site-specific constraints and technological limitations, as well as the operational simplicity and predictability of these approaches. Only a few soil remediation techniques discussed in this review, such as soil washing and soil vapor extraction, have occasionally been applied. However, the increasing complexity of contamination, rising demand for sustainable remediation strategies, stricter environmental regulations, and growing expectations for post-remediation land reuse suggest that broader and more flexible remediation approaches will be required in the future. Accordingly, this section examines the technical advantages and limitations, economic and environmental considerations, and practical applicability of selected alternative remediation methods in shipyard settings. Based on this multifaceted analysis, the study provides a comprehensive evaluation of the potential integration of these technologies into future shipyard soil remediation strategies.

5.1. Technological Advantages and Limitations

A variety of remediation technologies have been developed and applied to address the diverse nature of soil contamination. In shipyard settings, where contamination typically includes heavy metals, petroleum hydrocarbons, and POPs, the selection of an appropriate remediation strategy requires careful consideration of each technology's technical performance, applicability to site conditions, and potential constraints.

Table 7 summarizes the advantages and challenges of the selected soil remediation technologies [30]. Soil washing separates contaminants from soil particles using water or chemical solutions, often aided by physical agitation. It is effective for soils contaminated with heavy metals and hydrophobic organic pollutants. However, the process generates large volumes of wastewater and is generally more suitable for coarse-grained soils, as fine particles reduce separation efficiency. According to the U.S. EPA [153], the soil washing process typically consumes between approximately 640 L/m^3 and 3960 L/m^3 of water, depending on the specific process configuration and soil characteristics.

Soil flushing involves injecting chemical solutions into the subsurface to mobilize and extract contaminants in situ. It minimizes surface disruption and is technically applicable to sensitive areas. Its performance, however, is highly dependent on soil permeability and hydraulic control. In poorly permeable soils, flushing efficiency declines, and there is a risk of uncontrolled migration of mobilized contaminants; in a full-scale field application at a Korean military site, modified in situ soil flushing combined with air sparging achieved TPH removal rates of about 53–62% over depths of 6–9 m and increased the soil's hydraulic conductivity nearly fivefold, demonstrating that flushing efficiency is highly dependent on soil permeability, with lower-permeability layers showing reduced treatment effectiveness [154].

Thermal desorption heats soil to volatilize and remove organic contaminants. It is particularly effective for persistent pollutants such as PAHs, PCBs, and TBT. Technically, it provides high removal rates, but it requires significant energy input and uniform heating, making it less suitable for large or deeply contaminated areas. Previously reported field data indicate thermal desorption systems typically require substantial energy inputs, with electrical resistance heating consuming approximately 200–400 kWh/m^3, and gas-fired thermal conduction systems using between 620 and 890 kWh/m^3, depending on site-specific conditions and heating configuration [155].

Adsorption-enhanced remediation utilizes materials such as activated carbon or biochar to immobilize pollutants within the soil matrix. This method offers versatility for both organic and inorganic contaminants. Nevertheless, its success depends on the correct selection and dosage of the adsorbent material, and it may produce secondary solid-phase waste that requires further handling.

Soil vapor extraction removes volatile organic compounds from unsaturated zones by applying a vacuum to draw contaminated vapors out of the soil. It is effective in permeable, dry soils and can be combined with other technologies. Limitations include the need for off-gas treatment systems and the inapplicability of saturated or fine-grained soils. Off-gas treatment systems, which are essential to managing extracted vapor and preventing environmental release, can account for up to 50% of total soil vapor extraction remediation project costs [156].

Chemical oxidation introduces oxidants such as hydrogen peroxide, permanganate, or persulfate to break down organic contaminants in place. It provides rapid degradation and can be applied across heterogeneous soils. However, incomplete reactions, secondary by-product formation, and challenges in oxidant distribution reduce its reliability in complex field conditions [157]. For example, field-scale in situ chemical oxidation treatment has limitations in fully mineralizing petroleum hydrocarbons into CO_2 and H_2O, often resulting in the formation of various low-molecular-weight hydrophilic intermediates. In particular, BTEX and MTBE generate complex mixtures of aromatic and aliphatic intermediates, and the complete removal of these by-products is essential to prevent secondary contamination [158,159].

Table 7. Benefits and challenges of soil remediation technologies [30].

Technology	Advantages	Challenges
Soil washing	• Effective for heavy metals and hydrophobic organics • Physically separates contaminants	• Large volumes of wastewater • May require further treatment of effluents
Soil flushing	• In situ application possible • Lower soil disturbance	• Potential spreading of contamination • Efficiency depends on soil permeability
Thermal desorption	• Broad-spectrum pollutant removal • High treatment efficiency	• Energy-intensive • Difficult to scale for deep contamination
Adsorption-enhanced remediation	• Enhances contaminant binding and immobilization • Applicable to a wide range of contaminants	• Requires careful selection of adsorbents • May produce secondary waste
Soil vapor extraction	• Effective for VOCs • Can be combined with other methods	• Only applicable to unsaturated, permeable soils • Requires off-gas treatment
Chemical oxidation	• Rapid degradation of organic pollutants • Applicable in heterogeneous soils	• May lead to the formation of reactive or toxic by-products • Requires careful dosing
Chemical reduction	• Effective for heavy metals and chlorinated organics • Permanent immobilization possible	• Limited by heterogeneous soil conditions • Requires precise reagent distribution
Electrokinetic remediation	• Effective for fine-grained soils • Simultaneous removal of metals and organics	• High energy consumption • Requires stable electrolyte control and system monitoring
Stabilization/solidification	• Improves physical stability and immobilizes contaminants	• Volume increase in treated material • Potential leaching over time
Microbial bioremediation	• Capable of degrading specific organic pollutants biologically	• Slow kinetics • Strongly influenced by temperature, pH, and nutrient levels
Phytoremediation	• Applicable to surface-level contamination across wide areas	• Limited to shallow zones • Long remediation durations required

Chemical reduction targets redox-sensitive contaminants like chlorinated solvents and certain metals by chemically reducing them to less toxic or immobilized forms. It offers long-term stabilization potential but requires precise control of redox conditions and effective reagent delivery, particularly in heterogeneous subsurface environments.

Chemical reduction efficiency is limited in low-permeability soils such as clays and tills, primarily due to the restricted injection rate of treatment solutions [160].

Electrokinetic remediation uses a low-voltage electric current to mobilize charged contaminants across electrodes embedded in the soil. It is one of the few technologies suitable for fine-grained, low-permeability soils, allowing for the removal of both organics and metals. However, it is energy-intensive and demands strict control of electrolyte composition and system stability over extended durations. Studies on periodic-power electrokinetic remediation have demonstrated energy consumption ranging from 45 to 3669 kWh/m^3, depending on soil chemistry and applied power regimes, underlining the need for careful energy optimization in field applications [161].

Stabilization/solidification involves mixing binding agents with contaminated soil to immobilize pollutants and improve structural integrity. It is technically effective for preventing contaminant leaching. However, it increases the volume of treated material and poses long-term monitoring challenges due to possible matrix degradation [162].

Microbial bioremediation relies on microbial activity to degrade biodegradable organic contaminants, especially petroleum hydrocarbons. It is applicable in situ with minimal disturbance, but it is slow, sensitive to environmental factors (e.g., temperature, pH), and ineffective for recalcitrant compounds [163].

Phytoremediation employs plants to extract, degrade, or stabilize contaminants in shallow soils. It is technically feasible over large surface areas but limited by the depth of root systems, the slow pace of remediation, and seasonal dependencies. Phytoremediation is generally limited to the effective rooting depth of plants, which typically ranges from 0.3 m for grasses to 5–6 m for woody species like poplars under field conditions. Depending on contaminant type and site characteristics, remediation may require multiple growing seasons, with typical treatment durations spanning 2–5 years for hydrocarbons and significantly longer for heavy metals [164].

In summary, while each technology offers distinct advantages under specific conditions, their field-scale applicability in shipyard environments depends heavily on contaminant characteristics, soil properties, and operational constraints. As shipyard sites increasingly face complex and diverse contamination profiles, a nuanced understanding of these technologies' technical trade-offs is essential to inform appropriate remediation design.

5.2. Economic and Life Cycle Assessment

While the technical efficacy of remediation technologies is a primary consideration in site selection and engineering design, their full-scale implementation depends equally on cost efficiency and environmental sustainability. This is particularly critical for shipyard sites, where large-scale remediation must address diverse contaminants, including heavy metals, petroleum hydrocarbons, PAHs, and persistent organic pollutants, under complex spatial and regulatory constraints. Cost analysis and life cycle assessment (LCA) approaches have quantified these differences, providing an evidence-based foundation for remediation planning [165–168].

Among the soil remediation methods, soil washing is effective for removing heavy metals and PAHs but entails substantial secondary burdens from sludge and wash water treatment. It also requires high energy input. Typical treatment costs range from approximately $80 to $200 per ton, and its CO_2 footprint is lower than thermal and desorption or off-site landfilling [169,170]. Soil flushing, which injects surfactants or chelating agents into the subsurface, has a similar cost profile—generally $75–$210 per ton—though its field applicability may vary based on soil permeability and contaminant distribution [169].

In contrast, thermal desorption allows rapid remediation and is highly effective for VOCs and other volatile contaminants. However, this method demands substantial capital investment and energy input, leading to the highest global warming potential (GWP) among the reviewed technologies—typically 300–400 kg CO_2-eq per ton. Costs can also reach up to \$300 per ton [171,172]. While some LCA models suggest long-term environmental benefits from thorough source removal and reduced plume migration, these gains may only offset initial impacts over extended timeframes [165].

Adsorption-enhanced remediation represents a more flexible and low-cost alternative. However, its environmental footprint is highly dependent on the production route of the adsorbents. Cradle-to-gate LCA studies show that coal-based GAC produces the highest emissions (5.2–5.5 kg CO_2-eq/kg), while biomass-derived GACs—from coconut or wood—have significantly lower GWPs (0.9–1.1 kg CO_2-eq/kg). Reactivated GAC, with a GWP of 2.1 kg CO_2-eq/kg, also offers a promising, more sustainable option [173].

Soil vapor extraction is simple to operate and highly effective for VOC removal. While its overall cost is moderate, the treatment of off-gases can incur significant additional expenses [174]. Typical operations range around \$20–\$220 per ton of soil for extraction and treatment of off-gases [174,175]. In contrast, despite its rapid action, chemical oxidation performs poorly in most LCA categories due to the use of strong reagents like Fenton's reagent or persulfate, which contribute significantly to ecotoxicity and human toxicity [176]. Chemical reduction using zero-valent iron (ZVI) exhibits intermediate GWP, typically ranging from 70 to 180 kg CO_2-eq per ton of soil treated, depending on synthesis methods [176]. Among production methods, sodium borohydride-based synthesis shows the lowest environmental burden, whereas hydrogen gas reduction incurs higher impacts due to energy intensity. Key contributors include electricity usage, $NaBH_4$ manufacturing, and reagent emissions. Cleaner energy sources and optimized production routes can mitigate these effects.

Electrokinetic remediation, especially when combined with permeable reactive barriers, demonstrates the potential for both organic and inorganic contaminants. Field-scale costs are approximately \$200 per m^3 [177]. In terms of carbon footprint, electrokinetic remediation shows 0.36–0.49 kg CO_2-eq/kg soil, with energy consumption contributing 88–93% of the overall impact. Atrazine-contaminated soils required the highest energy, while Chlorsulfuron needed the least. Renewable energy integration (e.g., solar) could cut water footprint by up to 90% [178].

Among biological approaches, a recent project in Southeast Asia reported \$310/m^3 (~\$174 material costs) for combined bioaugmentation–biostimulation [179]. A recent LCA-based study on petroleum-contaminated saline soils demonstrated that bioremediation using inorganic nutrients (e.g., urea) can result in significant environmental burdens, including 84.2 kg CO_2-eq per ton and high impacts on human health and ecosystem quality [180]. In contrast, organic waste-based bioremediation (e.g., molasses with bacterial inoculants) reduced the overall environmental footprint by up to 90%. These findings highlight that nutrient selection critically affects the sustainability of bioremediation processes. Integrating waste valorization within bioremediation offers both ecological and circular economy benefits.

Lastly, phytoremediation is recognized as a sustainable option for heavy metal-contaminated soils, with LCA studies showing 30–100% lower environmental impacts compared to excavation or soil washing [181]. However, long treatment durations (18 months to 16 years) and biomass disposal remain key challenges. TEA indicates that soil preparation and biomass management drive most of the costs, while chelator-assisted phytoextraction can boost metal uptake by up to 200-fold, enhancing efficiency. Integrating

LCA and techno-economic analysis supports informed, site-specific decision-making for sustainable remediation.

5.3. Integrated Assessment and Site-Specific Considerations

The previous two sections discussed the strengths and limitations of soil remediation technologies, with a particular focus on their economic and environmental aspects. Based on these considerations and previous studies, this section identifies six key evaluation criteria and conducts an integrated assessment of each technology [30]. These criteria—efficiency, cost, eco-friendliness, ease of application, speed, and treatment scale—provide a comprehensive framework for assessing practical performance.

Figure 6 summarizes how each representative technology performs across these dimensions. Thermal desorption and chemical oxidation show strong performance in terms of efficiency and speed, making them highly effective for the rapid removal of persistent contaminants. However, their high energy requirements and associated emissions lead to lower scores in cost and environmental sustainability. These characteristics make them suitable for hotspot or time-sensitive remediation but less viable for long-term, large-scale applications.

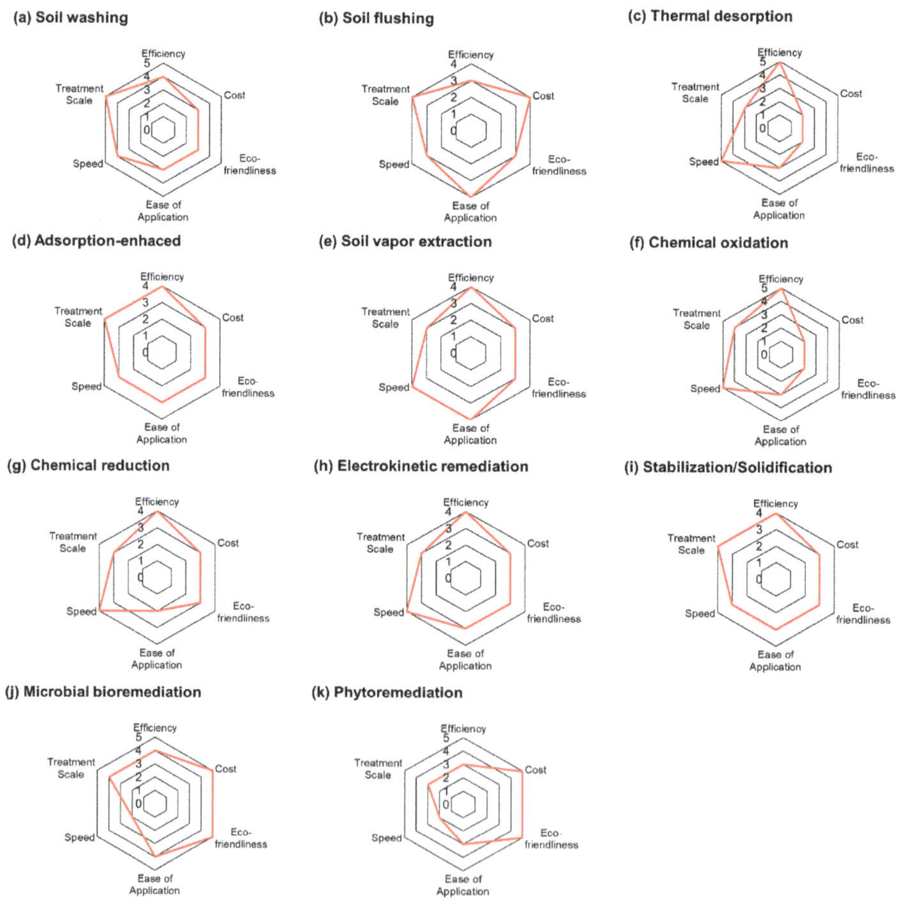

Figure 6. Performance evaluation of representative soil remediation technologies (rated from 1 = poor to 5 = excellent) [28].

In contrast, phytoremediation and microbial bioremediation stand out for their environmental and economic advantages. They are associated with the lowest GWP emissions and offer additional benefits such as landscape restoration and potential biomass utilization. That said, their slow treatment rates and limited applicability to deep or complex contamination reduce their potential as standalone solutions. Instead, they are better positioned as follow-up or polishing stages within a multi-step remediation process.

Soil washing demonstrates relatively balanced performance across all criteria, particularly excelling in efficiency and treatment capacity. Its flexibility and compatibility with diverse contamination profiles highlight its value in shipyard contexts, especially when integrated into phased or combined strategies. Stabilization and solidification also offer consistent results across various pollutant types, though their long-term effectiveness may be influenced by environmental changes and material durability.

Electrokinetic remediation and soil flushing show variable performance depending heavily on soil characteristics, contaminant type, and energy availability. While they can be highly effective under specific conditions, especially in fine-grained or low-permeability soils, their broader applicability remains limited. Similarly, technologies like adsorption-enhanced remediation and soil vapor extraction offer clear advantages for targeted use (e.g., VOC removal or shallow zone treatment) but fall short in scalability and speed.

Overall, the comparative results reinforce the idea that no single technology satisfies all remediation goals. Optimal outcomes will depend on selecting and combining technologies based on the dominant site constraints, contamination profiles, and long-term land use plans. Hybrid approaches, where high-impact, front-end methods such as thermal treatment or washing are followed by sustainable polishing techniques like phytoremediation or solidification, can offer more balanced solutions. To translate this understanding into practical strategies, it is essential to apply structured decision-making frameworks that integrate environmental, technical, and socio-economic dimensions.

5.4. Toward Practical Applicability of Selected Soil Remediation Technologies for Shipyard Remediation

Table 8 highlights differences in the range of pollutants that each technology can effectively target. This table displays the applicability of each technology to major contaminants commonly found in shipyard environments. Soil washing shows the broadest spectrum of applicability, capable of addressing PAHs, PCBs, heavy metals, and organotin compounds. Its ability to physically separate contaminants from the soil matrix allows it to respond flexibly to diverse contamination profiles, making it a highly promising primary remediation technology for complex shipyard sites. Thermal desorption similarly demonstrated high applicability across almost all major contaminants, particularly excelling in treating persistent pollutants such as high concentrations of PAHs, PCBs, and organotin compounds. However, as discussed earlier, its high energy demands, and operational costs must be considered when applying it at large scales. Chemical oxidation proved effective mainly for organic pollutants such as PAHs, PCBs, TBT, and other organotin compounds, indicating its potential for hotspot remediation in shipyard sites where organic contamination predominates. Chemical reduction exhibited strong treatment capabilities for heavy metals and some organotin compounds, although its overall applicable contaminant range was more limited. Given its selective reaction characteristics, careful site-specific design is required for its effective application.

Phytoremediation and bioremediation showed applicability primarily to petroleum hydrocarbons and low concentrations of PAHs but were less effective against PCBs and heavy metals. Therefore, these technologies are more suitable as complementary strategies to stabilize residual contamination or support long-term ecological restoration rather than serving as primary remediation methods for shipyard sites. However, recent LCA-based assessments suggest that phytoremediation, despite its limitations, offers significant advantages in terms of low greenhouse gas emissions, land aesthetics, and cost-effectiveness when coupled with biomass valorization strategies such as anaerobic digestion or energy recovery. In scenarios involving moderate contamination and long-term site redevelopment goals, phytoremediation may serve as a viable secondary or finishing-stage option.

Table 8. Summary of pollutant types remediated by each technology (○ = applicable, △ = partially applicable, × = not applicable).

Technology	PAHs	PCBs	TBT	Heavy Metals	OCPs	Other Organotin Compounds
Soil washing	○	○	○	○	○	○
Soil flushing	△	×	×	○	×	×
Thermal desorption	○	○	○	○	○	○
Adsorption-enhanced	○	○	○	○	○	○
Soil vapor extraction	○	○	×	×	×	×
Chemical oxidation	○	○	○	×	○	○
Chemical reduction	×	×	○	○	×	○
Electrokinetic remediation	×	×	×	○	×	×
Stabilization/solidification	△	○	○	○	○	○
Bioremediation	○	×	×	×	×	×
Phytoremediation	○	×	×	△	×	×

Soil flushing and electrokinetic remediation were primarily applicable to heavy metals (e.g., Pb, Cd, Zn, Cu) and are particularly promising for fine-grained soils or low-permeability formations. However, their applicability to organic contaminants remains limited. Furthermore, electrokinetic remediation's high electricity consumption and potential for secondary waste streams necessitate consideration of energy sources and long-term environmental trade-offs. Stabilization and solidification techniques show consistent performance in immobilizing a broad range of pollutants but raise concerns regarding future leachability and monitoring requirements under changing environmental conditions, such as those driven by climate change.

Site heterogeneity presents a significant obstacle to the universal deployment of any single technology. The co-occurrence of diverse contaminants at varying depths and concentrations often requires a hierarchical or hybrid strategy. For example, the initial removal of high-concentration contaminants through soil washing or thermal desorption could be followed by bioremediation or phytoremediation to address residual contamination in a sustainable manner. Importantly, site-specific geostatistical characterization and modeling of contaminant mobility should guide remediation design to prevent cross-media transfer and uncontrolled migration, especially in coastal sites vulnerable to sea-level rise or groundwater intrusion.

The practicality of advanced remediation techniques also varies significantly depending on regional socio-economic contexts. In developing countries, barriers such as limited financial resources, weak regulatory enforcement, and lack of technical capacity can hinder adoption. In such cases, cost-effective and ecologically adaptive approaches such as phytoremediation or low-tech bioremediation, supported by local biomass reuse strategies, may offer more feasible options.

Future shipyard site restoration projects must carefully balance economic feasibility, environmental sustainability, and long-term resilience by tailoring remediation strategies to specific site conditions and contamination profiles. The integration of LCA and TEA into remediation planning processes allows stakeholders to evaluate trade-offs more systematically and select combinations of technologies that align with both short-term cleanup goals and long-term land use visions.

6. Conclusions

In this review, soil contamination issues at shipyard sites, applicable soil remediation technologies, practical cases of soil and sediment remediation, and the advantages and limitations of various remediation approaches were comprehensively discussed. Particular attention was given to evaluating future directions for the development and application of soil remediation technologies in shipyard environments. To date, compared to remediation efforts for other contaminated sites, documented remediation cases specifically targeting shipyard soils have been relatively limited. Moreover, even in the available examples, the remediation methods employed were predominantly simple physical techniques such as excavation and dredging, rather than the use of advanced or integrated treatment technologies. Given the unique characteristics of shipyard environments, such as the co-occurrence of petroleum hydrocarbons, heavy metals, PCBs, and organotin compounds, as well as site-specific challenges like coastal proximity and subsurface heterogeneity, future remediation strategies must evolve beyond single-process approaches. A more effective restoration of shipyard soils will require the integration of multiple remediation technologies, carefully selected and combined based on site-specific contamination profiles, environmental conditions, and project goals. Specifically, a hierarchical or hybrid treatment framework should be pursued, wherein high-concentration contaminants are initially removed using high-efficiency methods such as soil washing or thermal desorption, followed by secondary or complementary treatments such as bioremediation or phytoremediation for residual contaminants. Additionally, careful attention must be paid to secondary environmental issues, particularly the generation and management of contaminated wastewater produced during soil remediation processes. Efficient handling, treatment, and disposal of remediation by-products are essential to ensure that secondary pollution does not compromise the overall effectiveness and environmental integrity of remediation efforts. Finally, considerations of economic feasibility, environmental sustainability, and long-term resilience must be balanced when designing shipyard soil remediation projects. Advancing the application of combined and adaptive remediation technologies will be critical to achieving more efficient, sustainable, and cost-effective restoration of complexly contaminated shipyard sites.

Author Contributions: Conceptualization, Methodology, Data curation, validation, D.K., H.J., C.P., Y.J., and S.S.; Writing—review, and editing, Writing—original draft, Visualization, Funding acquisition, Formal analysis, Data curation, Conceptualization, E.Y.; Writing—original draft, Visualization, Methodology, Data curation, Conceptualization, J.H.J.; Writing—original draft, Visualization, Methodology, Formal analysis, Data curation, Conceptualization, M.A.K. All authors have read and agreed to the published version of the manuscript.

Funding: This work was supported by the Research Resurgence under the Glocal University 30 Project at Gyeongsang National University in 2024.

Data Availability Statement: All data are presented in the article.

Conflicts of Interest: Jae Ho Jung was employed by the company IDEA Environment Service Co., Ltd. The remaining authors declare that the research was conducted in the absence of any commercial or financial relationships that could be construed as a potential conflict of interest.

Abbreviations

The following abbreviations are used in this manuscript:

BOD	Biochemical Oxygen Demand
BTX	Benzene, Toluene, and Xylene
COD	Chemical Oxygen Demand
Cd	Cadmium
Cr	Chromium
Cu	Copper
DDT	Dichlorodiphenyltrichloroethane
EDTA	Ethylenediaminetetraacetic Acid
EDX	Energy-Dispersive X-Ray Spectroscopy
EK-PRB	Electrokinetic-Permeable Reactive Barrier
ERL	Effects Range Low
ERM	Effects Range Median
FAO	Food And Agriculture Organization
GLDA	Tetrasodium Glutamate Diacetate
IMO	International Maritime Organization
LZSS	Lead-Zinc Smelting Slag LZSS
MSWI	Municipal Solid Waste Incinerator
NZVI	Nanoscale Zero-Valent Iron
OCP	Octacalcium Phosphate
OCPs	Organochlorine Pesticides
PAHs	Polycyclic Aromatic Hydrocarbons
PCBs	Polychlorinated Biphenyls
PFAS	Per-And Poly-Fluoroalkyl Substances
PFOS	Perfluorooctanesulfonic Acid
POPs	Persistent Organic Pollutants
Pb	Lead
SEM	Scanning Electron Microscopy
SVE	Soil Vapor Extraction
TBT	Tributyltin
TBT	Tributyltin
TCAS	Tetrasulfonate
TCE	Trichloroethylene
TGA	Thermogravimetric Analysis
TOC	Total Organic Carbon
TPH	Total Petroleum Hydrocarbons
USEPA	U.S. Environmental Protection Agency
UTS	Universal Treatment Standards
VOCs	Volatile Organic Compounds
WHO	World Health Organization
XRD	X-Ray Crystallography
Zn	Zinc

References

1. Kwak, S.-J.; Yoo, S.-H.; Chang, J.-I. The role of the maritime industry in the Korean national economy: An input–output analysis. *Mar. Policy* **2005**, *29*, 371–383. [CrossRef]
2. Chiu, S.; Ho, K.; Chan, S.; So, O.; Lai, K. Characterization of contamination in and toxicities of a shipyard area in Hong Kong. *Environ. Pollut.* **2006**, *142*, 512–520. [CrossRef] [PubMed]
3. Makri, P.; Hermides, D.; Kontakiotis, G.; Zarkogiannis, S.D.; Besiou, E.; Janjuhah, H.T.; Antonarakou, A. Integrated Ecological Assessment of Heavily Polluted Sedimentary Basin within the Broader Industrialized Area of Thriassion Plain (Western Attica, Greece). *Water* **2022**, *14*, 382. [CrossRef]

4. Karakas, S.; Acar, A.Z.; Kirmizi, M. Maritime Sustainability: Navigating Complex Challenges and Ecological Footprints. In *Sustainable Development Seen Through the Lenses of Ethnoeconomics and the Circular Economy*; Springer: Berlin/Heidelberg, Germany, 2024; pp. 19–47.
5. Eklund, B.; Eklund, D. Pleasure Boatyard Soils are Often Highly Contaminated. *Environ. Manag.* **2014**, *53*, 930–946. [CrossRef]
6. Paipa-Sanabria, E.; González-Montoya, D.; Coronado-Hernández, J.R. Green Technologies for Environmental Air and Water Impact Reduction in Ships: A Systematic Literature Review. *J. Mar. Sci. Eng.* **2025**, *13*, 839. [CrossRef]
7. Hayman, B.; Dogliani, M.; Kvale, I.; Fet, A.M. *Technologies for Reduced Environmental Impact from Ships-Ship Building, Maintenance and Dismantling Aspects*; ENSUS-2000: Newcastle upon Tyne, UK, 2000.
8. Li, P.; Dsikowitzky, L.; Diao, X.; Yang, F.; Li, Q.X.; Schwarzbauer, J. Unusual tin organics, DDX and PAHs as specific pollutants from dockyard work in an industrialized port area in China. *Chemosphere* **2020**, *243*, 125284. [CrossRef]
9. Hong, S.H.; Yim, U.H.; Shim, W.J.; Oh, J.R. Congener-Specific Survey for Polychlorinated Biphenlys in Sediments of Industrialized Bays in Korea: Regional Characteristics and Pollution Sources. *Environ. Sci. Technol.* **2005**, *39*, 7380–7388. [CrossRef]
10. Yılmaz, A.; Karacık, B.; Yakan, S.D.; Henkelmann, B.; Schramm, K.-W.; Okay, O.S. Organic and heavy metal pollution in shipbreaking yards. *Ocean Eng.* **2016**, *123*, 452–457. [CrossRef]
11. Guo, G.; Zhang, C.; Wu, G.; Ding, Q.; Wang, S.; Li, F. Health and ecological risk-based characterization of soil and sediment contamination in shipyard with long-term use of DDT-containing antifouling paint. *Sci. Total Environ.* **2013**, *450–451*, 223–229. [CrossRef]
12. Kim, N.S.; Shim, W.J.; Yim, U.H.; Ha, S.Y.; An, J.G.; Shin, K.H. Three decades of TBT contamination in sediments around a large scale shipyard. *J. Hazard. Mater.* **2011**, *192*, 634–642. [CrossRef]
13. Kotrikla, A. Environmental management aspects for TBT antifouling wastes from the shipyards. *J. Environ. Manag.* **2009**, *90*, S77–S85. [CrossRef] [PubMed]
14. Staniszewska, M.; Radke, B.; Namieśnik, J.; Bolałek, J. Analytical methods and problems related to the determination of organotin compounds in marine sediments. *Int. J. Environ. Anal. Chem.* **2008**, *88*, 747–774. [CrossRef]
15. Choi, J.-S.; Jeon, S.-K. A Geo-statistical Assessment of Heavy Metal Pollution in the Soil Around a Ship Building Yard in Busan, Korea. *Korean Soc. Mar. Environ. Saf.* **2018**, *24*, 907–915. [CrossRef]
16. Islam, M.N.; Ganguli, S.; Saha, N.; Mamun Huda, M.; Hoque, M.A.; Peng, C.; Ng, J.C. Uncovering the impact of mega-scale shipbreaking yards on soil and crop quality in Bangladesh: A spatiotemporal dynamics and associated health risks of metal/loid contamination. *J. Hazard. Mater.* **2024**, *464*, 132931. [CrossRef]
17. Hasan, A.B.; Reza, A.H.M.S.; Kabir, S.; Siddique, M.A.B.; Ahsan, M.A.; Akbor, M.A. Accumulation and distribution of heavy metals in soil and food crops around the ship breaking area in southern Bangladesh and associated health risk assessment. *SN Appl. Sci.* **2020**, *2*, 155. [CrossRef]
18. Alam, I.; Barua, S.; Ishii, K.; Mizutani, S.; Hossain, M.M.; Rahman, I.M.M.; Hasegawa, H. Assessment of health risks associated with potentially toxic element contamination of soil by end-of-life ship dismantling in Bangladesh. *Environ. Sci. Pollut. Res.* **2019**, *26*, 24162–24175. [CrossRef]
19. Ali, M.M.; Islam, M.S.; Islam, A.R.M.T.; Bhuyan, M.S.; Ahmed, A.S.S.; Rahman, M.Z.; Rahman, M.M. Toxic metal pollution and ecological risk assessment in water and sediment at ship breaking sites in the Bay of Bengal Coast, Bangladesh. *Mar. Pollut. Bull.* **2022**, *175*, 113274. [CrossRef]
20. Rahman, M.S.; Rahman, M.; Jolly, Y.N.; Hossain, M.K.; Semme, S.A.; Paray, B.A.; Arai, T.; Yu, J.; Hossain, M.B. Heavy metals in afforested mangrove sediment from the world's largest delta: Distributional mapping, contamination status, risk assessment and source tracing. *Mar. Pollut. Bull.* **2024**, *203*, 116429. [CrossRef]
21. Chowdhury, N.; Rasid, M.M. Heavy metal contamination of soil and vegetation in ambient locality of ship breaking yards in Chittagong, Bangladesh. *J. Environ. Sci. Toxicol. Food Technol.* **2016**, *10*, 20–27.
22. Rakib, M.R.J.; Rahman, M.A.; Onyena, A.P.; Kumar, R.; Sarker, A.; Hossain, M.B.; Islam, A.R.M.T.; Islam, M.S.; Rahman, M.M.; Jolly, Y.N.; et al. A comprehensive review of heavy metal pollution in the coastal areas of Bangladesh: Abundance, bioaccumulation, health implications, and challenges. *Environ. Sci. Pollut. Res.* **2022**, *29*, 67532–67558. [CrossRef]
23. Luo, Y.; Tong, G.; Song, Q.; Tao, P.; Jin, M.; Gu, N.; Zheng, M.; Yu, X.; Yu, X. Impacts of shipyard oil leakage on the PAHs and PCBs occurrence in Xiangshan Bay, China. *Mar. Environ. Res.* **2023**, *190*, 106057. [CrossRef] [PubMed]
24. Biswas, B.; Sarkar, B.; Mandal, A.; Naidu, R. Heavy metal-immobilizing organoclay facilitates polycyclic aromatic hydrocarbon biodegradation in mixed-contaminated soil. *J. Hazard. Mater.* **2015**, *298*, 129–137. [CrossRef] [PubMed]
25. Gauthier, P.T.; Norwood, W.P.; Prepas, E.E.; Pyle, G.G. Metal–PAH mixtures in the aquatic environment: A review of co-toxic mechanisms leading to more-than-additive outcomes. *Aquat. Toxicol.* **2014**, *154*, 253–269. [CrossRef] [PubMed]
26. Dermont, G.; Bergeron, M.; Mercier, G.; Richer-Laflèche, M. Soil washing for metal removal: A review of physical/chemical technologies and field applications. *J. Hazard. Mater.* **2008**, *152*, 1–31. [CrossRef]

27. Hu, G.; Liu, H.; Chen, C.; Hou, H.; Li, J.; Hewage, K.; Sadiq, R. Low-temperature thermal desorption and secure landfill for oil-based drill cuttings management: Pollution control, human health risk, and probabilistic cost assessment. *J. Hazard. Mater.* **2021**, *410*, 124570. [CrossRef]
28. Ambika, S.; Kumar, M.; Pisharody, L.; Malhotra, M.; Kumar, G.; Sreedharan, V.; Singh, L.; Nidheesh, P.; Bhatnagar, A. Modified biochar as a green adsorbent for removal of hexavalent chromium from various environmental matrices: Mechanisms, methods, and prospects. *Chem. Eng. J.* **2022**, *439*, 135716. [CrossRef]
29. Zhang, S.; Zhao, J.; Zhu, L. Predicting removal efficiency of organic pollutants by soil vapor extraction based on an optimized machine learning method. *Sci. Total Environ.* **2024**, *927*, 172438. [CrossRef]
30. Aparicio, J.D.; Raimondo, E.E.; Saez, J.M.; Costa-Gutierrez, S.B.; Álvarez, A.; Benimeli, C.S.; Polti, M.A. The current approach to soil remediation: A review of physicochemical and biological technologies, and the potential of their strategic combination. *J. Environ. Chem. Eng.* **2022**, *10*, 107141. [CrossRef]
31. Liu, J.; Zhao, L.; Liu, Q.; Li, J.; Qiao, Z.; Sun, P.; Yang, Y. A critical review on soil washing during soil remediation for heavy metals and organic pollutants. *Int. J. Environ. Sci. Technol.* **2021**, *19*, 601–624. [CrossRef]
32. Chiu, J.C.; Gani, P. Soil Washing Methods for Effective Removal of Heavy Metal Contaminants. *Ind. Domest. Waste Manag.* **2024**, *4*, 56–71. [CrossRef]
33. Zhang, M.; Chen, W.; Chuan, X.; Guo, X.; Shen, X.; Zhang, H.; Wu, F.; Hu, J.; Wu, Z.; Wang, X. Remediation of heavily PAHs-contaminated soil with high mineral content from a coking plant using surfactant-enhanced soil washing. *Sci. Total Environ.* **2024**, *909*, 168499. [CrossRef] [PubMed]
34. Grimison, C.; Knight, E.R.; Nguyen, T.M.H.; Nagle, N.; Kabiri, S.; Bräunig, J.; Navarro, D.A.; Kookana, R.S.; Higgins, C.P.; McLaughlin, M.J. The efficacy of soil washing for the remediation of per-and poly-fluoroalkyl substances (PFASs) in the field. *J. Hazard. Mater.* **2023**, *445*, 130441. [CrossRef] [PubMed]
35. Yuan, Z.; Peng, A.; Chu, Z.; Zhang, X.; Huang, H.; Mi, Y.; Xia, D.; Wu, X.; Ye, Z.; Tao, Y. Sustainable remediation of Cr (VI)-contaminated soil by soil washing and subsequent recovery of washing agents using biochar supported nanoscale zero-valent iron. *Sci. Total Environ.* **2024**, *921*, 171107. [CrossRef] [PubMed]
36. Rabieian, M.; Taghavijeloudar, M. Simultaneously removal of PAHs from contaminated soil and effluent by integrating soil washing and advanced oxidation processes in a continuous system: Water saving, optimization and scale up modeling. *Water Res.* **2024**, *256*, 121563. [CrossRef]
37. Huang, J.; Xiao, X.; Chen, B. Insight into the electrochemical process of EDTA-assisted soil washing effluent under alternating current. *J. Hazard. Mater.* **2024**, *470*, 134115. [CrossRef]
38. Xing, L.-Y.; Meng, G.-H.; Yang, J.-X.; Xu, M.-H.; Xu, Y.-M.; Xie, H.-X.; Wang, A.-J.; Sun, Y.-L. PAHs removal by soil washing with thiacalix [4] arene tetrasulfonate. *Environ. Sci. Ecotechnol.* **2024**, *21*, 100422. [CrossRef]
39. Corres, X.; Sierra, C.; Diez-Mestas, A.; Gallego, J.; Baragaño, D. A novel heuristic tool for selecting the best upgrading conditions for the removal of potentially toxic elements by soil washing. *J. Hazard. Mater.* **2024**, *466*, 133529. [CrossRef]
40. Keochanh, D.; Phenrat, T. Magnet-assisted nanoscale zerovalent iron soil washing in chlorpyrifos-contaminated soil remediation: Proof of concept. *Process Saf. Environ. Prot.* **2024**, *191*, 723–735. [CrossRef]
41. Sikdar, S.K.; Grosse, D.; Rogut, I. Membrane technologies for remediating contaminated soils: A critical review. *J. Membr. Sci.* **1998**, *151*, 75–85. [CrossRef]
42. Li, Y.; Wei, M.; Yu, B.; Liu, L.; Xue, Q. Thermal desorption optimization for the remediation of hydrocarbon-contaminated soils by a self-built sustainability evaluation tool. *J. Hazard. Mater.* **2022**, *436*, 129156. [CrossRef]
43. Liu, C.; Shi, H.; Wang, C.; Fei, Y.; Han, Z. Thermal remediation of Soil contaminated with polycyclic aromatic hydrocarbons: Pollutant removal process and influence on Soil Functionality. *Toxics* **2022**, *10*, 474. [CrossRef] [PubMed]
44. Sang, Y.; Yu, W.; He, L.; Wang, Z.; Ma, F.; Jiao, W.; Gu, Q. Sustainable remediation of lube oil-contaminated soil by low temperature indirect thermal desorption: Removal behaviors of contaminants, physicochemical properties change and microbial community recolonization in soils. *Environ. Pollut.* **2021**, *287*, 117599. [CrossRef] [PubMed]
45. Wang, S.; Cheng, F.; Shao, Z.; Wu, B.; Guo, S. Effects of thermal desorption on ecotoxicological characteristics of heavy petroleum-contaminated soil. *Sci. Total Environ.* **2023**, *857*, 159405. [CrossRef] [PubMed]
46. Bykova, M.V.; Alekseenko, A.V.; Pashkevich, M.A.; Drebenstedt, C. Thermal desorption treatment of petroleum hydrocarbon-contaminated soils of tundra, taiga, and forest steppe landscapes. *Environ. Geochem. Health* **2021**, *43*, 2331–2346. [CrossRef]
47. Ilyas, N.; Shoukat, U.; Saeed, M.; Akhtar, N.; Yasmin, H.; Khan, W.; Iqbal, S. Comparison of plant growth and remediation potential of pyrochar and thermal desorption for crude oil-contaminated soils. *Sci. Rep.* **2021**, *11*, 2817. [CrossRef]
48. Zhang, S.; Zhao, J.; Zhu, L. New insights into thermal desorption remediation of pyrene-contaminated soil based on an optimized numerical model. *J. Hazard. Mater.* **2024**, *461*, 132687. [CrossRef]
49. Xue, Y.; Chen, L.; Xiang, L.; Zhou, Y.; Wang, T. Experimental investigation on electromagnetic induction thermal desorption for remediation of petroleum hydrocarbons contaminated soil. *J. Environ. Manag.* **2023**, *328*, 117200. [CrossRef]

50. Yue, R.; Zhang, X.; Zhong, Y.; Chen, Z.; Zhao, Y.; Wang, D.; Wang, Z.; Mao, X. Thermal reduction-desorption of cadmium from contaminated soil by a biomass co-pyrolysis process. *J. Hazard. Mater.* **2022**, *423*, 126937. [CrossRef]
51. Zhang, Y.; Li, A.; Liu, L.; Duan, X.; Ge, W.; Liu, C.; Qiu, G. Enhanced remediation of cadmium-polluted soil and water using facilely prepared MnO_2-coated rice husk biomass. *Chem. Eng. J.* **2023**, *457*, 141311. [CrossRef]
52. Chen, Y.; Fan, J.; Ma, R.; Xue, Y.; Ma, Q.; Chen, X.; Yuan, S.; Zheng, H.; Qiu, Y.; Fu, R. Enhanced Remediation of Pb (Ii), Cd (Ii) and As (V) Contaminated Water and Soil by A-Feooh Incorporated Carboxylated Cellulose Nanocrystals: Synergistic Effect and Immobilization Mechanism. *SSRN*. Available online: https://papers.ssrn.com/sol3/papers.cfm?abstract_id=4091491 (accessed on 15 May 2025). [CrossRef]
53. Lin, H.; Yang, Y.; Shang, Z.; Li, Q.; Niu, X.; Ma, Y.; Liu, A. Study on the enhanced remediation of petroleum-contaminated soil by biochar/g-C_3N_4 composites. *Int. J. Environ. Res. Public Health* **2022**, *19*, 8290. [CrossRef]
54. Qi, X.; Gou, J.; Chen, X.; Xiao, S.; Ali, I.; Shang, R.; Wang, D.; Wu, Y.; Han, M.; Luo, X. Application of mixed bacteria-loaded biochar to enhance uranium and cadmium immobilization in a co-contaminated soil. *J. Hazard. Mater.* **2021**, *401*, 123823. [CrossRef] [PubMed]
55. Li, Y.; Wen, J.; Xue, Z.; Yin, X.; Yuan, L.; Yang, C. Removal of Cr (VI) by polyaniline embedded polyvinyl alcohol/sodium alginate beads— Extension from water treatment to soil remediation. *J. Hazard. Mater.* **2022**, *426*, 127809. [CrossRef] [PubMed]
56. Qi, X.; Yin, H.; Zhu, M.; Yu, X.; Shao, P.; Dang, Z. MgO-loaded nitrogen and phosphorus self-doped biochar: High-efficient adsorption of aquatic Cu_{2+}, Cd_{2+}, and Pb_{2+} and its remediation efficiency on heavy metal contaminated soil. *Chemosphere* **2022**, *294*, 133733. [CrossRef] [PubMed]
57. Zhang, P.; Xue, B.; Jiao, L.; Meng, X.; Zhang, L.; Li, B.; Sun, H. Preparation of ball-milled phosphorus-loaded biochar and its highly effective remediation for Cd-and Pb-contaminated alkaline soil. *Sci. Total Environ.* **2022**, *813*, 152648. [CrossRef]
58. Li, Y.; Shao, M.; Huang, M.; Sang, W.; Zheng, S.; Jiang, N.; Gao, Y. Enhanced remediation of heavy metals contaminated soils with EK-PRB using β-CD/hydrothermal biochar by waste cotton as reactive barrier. *Chemosphere* **2022**, *286*, 131470. [CrossRef]
59. Lyu, Z.; Zhou, W.; Gao, X.; Zheng, H.; Jia, J. Remediation of toluene-contaminated soils by sequential treatment: Soil vapor extraction systems and internal combustion engine units. *J. Contam. Hydrol.* **2025**, *271*, 104532. [CrossRef]
60. Cao, W.; Zhang, L.; Miao, Y.; Qiu, L. Research progress in the enhancement technology of soil vapor extraction of volatile petroleum hydrocarbon pollutants. *Environ. Sci. Process. Impacts* **2021**, *23*, 1650–1662. [CrossRef]
61. Liu, Y.; Huang, F.; Zhou, D.; Wang, T.; Wang, G. Study of factors affecting the remediation of contaminated soil through pneumatic fracturing and soil vapor extraction. *J. Hazard. Mater.* **2024**, *467*, 133776. [CrossRef]
62. Xu, L.; Hu, X.; Zha, F.; Kang, H.; Fang, L.; Kang, B.; Chu, C.; Yang, C. Mass transfer enhancement of air sparging on VOCs contaminated low-permeability soil by establishing pressure gradient. *Chemosphere* **2023**, *313*, 137416. [CrossRef]
63. de Melo Henrique, J.M.; Isidro, J.; Sáez, C.; López-Vizcaíno, R.; Yustres, A.; Navarro, V.; Dos Santos, E.V.; Rodrigo, M.A. Enhancing soil vapor extraction with EKSF for the removal of HCHs. *Chemosphere* **2022**, *296*, 134052. [CrossRef]
64. Zhang, L.; Wang, T.; Zhang, M.; Wang, L.; Kang, G.; Wu, S.; She, Y. Ultrasonically activated persulfate process for the degradation of phenanthrene in soil-washing effluent: Experimental, DFT calculation and toxicity evaluation. *J. Environ. Chem. Eng.* **2024**, *12*, 113035. [CrossRef]
65. Jun, B.-M.; Oh, M.; Jung, J.-Y.; Eun, H.-C.; Yoon, Y. A feasibility study on clearance levels for real U (VI)-contaminated soil and the mechanisms involving metal ions after neutralization of effluents generated from the soil-washing process. *Nucl. Eng. Technol.* **2025**, *57*, 103518. [CrossRef]
66. Alazaiza, M.Y.; Albahnasawi, A.; Ali, G.A.; Bashir, M.J.; Copty, N.K.; Amr, S.S.A.; Abushammala, M.F.; Al Maskari, T. Recent advances of nanoremediation technologies for soil and groundwater remediation: A review. *Water* **2021**, *13*, 2186. [CrossRef]
67. Petrillo, A.; Fraternali, F.; Acampora, A.; Di Chiara, G.; Colangelo, F.; Farina, I. Innovative Solidification and Stabilization Techniques Using Industrial By-Products for Soil Remediation. *Appl. Sci.* **2025**, *15*, 4002. [CrossRef]
68. Hu, W.; Cheng, W.-C.; Wen, S. Investigating the effect of degree of compaction, initial water content, and electric field intensity on electrokinetic remediation of an artificially Cu-and Pb-contaminated loess. *Acta Geotech.* **2023**, *18*, 937–949. [CrossRef]
69. Siegrist, R.L.; Crimi, M.; Thomson, N.R.; Clayton, W.S.; Marley, M.C. In situ chemical oxidation. In *Chlorinated Solvent Source Zone Remediation*; Springer: Berlin/Heidelberg, Germany, 2014; pp. 253–305.
70. Nie, J.; Wang, Q.-m.; Han, L.-j.; Li, J.-s. Synergistic remediation strategies for soil contaminated with compound heavy metals and organic pollutants. *J. Environ. Chem. Eng.* **2024**, *12*, 113145. [CrossRef]
71. Zeng, Y.; Li, T.; Ding, Y.; Fang, G.; Wang, X.; Ye, B.; Ge, L.; Gao, J.; Wang, Y.; Zhou, D. Biochar-supported nano-scale zerovalent iron activated persulfate for remediation of aromatic hydrocarbon-contaminated soil: An in-situ pilot-scale study. *Biochar* **2022**, *4*, 64. [CrossRef]
72. Song, I.-G.; Kang, Y.-G.; Kim, J.-H.; Yoon, H.; Um, W.Y.; Chang, Y.-S. Assessment of sulfidated nanoscale zerovalent iron for in-situ remediation of cadmium-contaminated acidic groundwater at a zinc smelter. *J. Hazard. Mater.* **2023**, *441*, 129915. [CrossRef]
73. Wang, L.; Ma, J.; Guo, Q.; Liu, L.; Shou, J.; Sun, A.; Zhao, L. In situ generation of hydrogen peroxide using polymetallic-doped g-C_3N_4 for pollutant removal. *Appl. Sci.* **2021**, *11*, 10797. [CrossRef]

74. Liu, M.-H.; Hsiao, C.-M.; Lin, C.-E.; Leu, J. Application of combined in situ chemical reduction and enhanced bioremediation to accelerate TCE treatment in groundwater. *Appl. Sci.* **2021**, *11*, 8374. [CrossRef]
75. Reginatto, C.; Cecchin, I.; Heineck, K.S.; Thomé, A.; Reddy, K.R. Use of nanoscale zero-valent iron for remediation of clayey soil contaminated with hexavalent chromium: Batch and column tests. *Int. J. Environ. Res. Public Health* **2020**, *17*, 1001. [CrossRef] [PubMed]
76. Geng, W.; Wei, Y.; Ke, Y.; Qin, J.; Yu, X.; Guo, X.; Long, M. Unveiling molecular transformations of soil organic matter after remediation by chemical oxidation based on ESI-FT-ICR-MS analysis. *ACS EST Eng.* **2023**, *3*, 831–840. [CrossRef]
77. Hong, Y.K.; Kim, J.W.; Lee, D.J.; Kim, H.S.; Park, Y.-H.; Kim, S.C. Effect of In-Situ Oxidation Method on Soil Biological Properties in Total Petroleum Hydrocarbon Contaminated Soil. *J. Soil Sci. Fertil.* **2023**, *56*, 90–100. [CrossRef]
78. Rodrigues, R.; Betelu, S.; Colombano, S.; Tzedakis, T.; Masselot, G.; Ignatiadis, I. In situ chemical reduction of chlorinated organic compounds. *Environ. Soil Remediat. Rehabil. Exist. Innov. Solut.* **2020**, 283–398. [CrossRef]
79. Mouvet, C.; Collet, B.; Gaude, J.-M.; Rangon, L.; Bristeau, S.; Senergues, M.; Lesueur-Jannoyer, M.; Jestin, A.; Hellal, J.; Woignier, T. Physico-chemical and agronomic results of soil remediation by In Situ Chemical Reduction applied to a chlordecone-contaminated nitisol at plot scale in a French Caribbean banana plantation. *Environ. Sci. Pollut. Res.* **2020**, *27*, 41063–41092. [CrossRef]
80. Cameselle, C.; Reddy, K.R. Electrobioremediation: Combined Electrokinetics and Bioremediation Technology for Contaminated Site Remediation. *Indian Geotech. J.* **2022**, *52*, 1205–1225. [CrossRef]
81. Kim, S.-O.; Jeong, J.Y.; Lee, W.-C.; Yun, S.-T.; Jo, H.Y. Electrokinetic remediation of heavy metal-contaminated soils: Performance comparison between one-and two-dimensional electrode configurations. *J. Soils Sediments* **2021**, *21*, 2755–2769. [CrossRef]
82. Shi, C.; Tong, M.; Cai, Q.; Li, Z.; Li, P.; Lu, Y.; Cao, Z.; Liu, H.; Zhao, H.-P.; Yuan, S. Electrokinetic-enhanced bioremediation of trichloroethylene-contaminated low-permeability soils: Mechanistic insight from spatio-temporal variations of indigenous microbial community and biodehalogenation activity. *Environ. Sci. Technol.* **2023**, *57*, 5046–5055. [CrossRef]
83. Narenkumar, J.; Das, B.; Abilaji, S.; Sathishkumar, K.; AlSalhi, M.S.; Devanesan, S.; Rajasekar, A.; Malik, T. Biosurfactant-assisted bio-electrokinetic enhanced remediation of heavy metal-contaminated soil. *Front. Microbiol.* **2024**, *15*, 1458369. [CrossRef]
84. Sun, Z.; Xu, S.; Zhang, J.; Eugene, B.D.; Li, S. Effect of Electrode Positioning on Electrokinetic Remediation of Contaminated Soft Clay with Surface Electrolyte. *Toxics* **2024**, *12*, 758. [CrossRef]
85. Wang, Y.; Han, Z.; Li, A.; Cui, C. Enhanced electrokinetic remediation of heavy metals contaminated soil by biodegradable complexing agents. *Environ. Pollut.* **2021**, *283*, 117111. [CrossRef] [PubMed]
86. Fan, C.; Wang, B.; Ai, H.; Qi, Y.; Liu, Z. A comparative study on solidification/stabilization characteristics of coal fly ash-based geopolymer and Portland cement on heavy metals in MSWI fly ash. *J. Clean. Prod.* **2021**, *319*, 128790. [CrossRef]
87. Wang, L.; Cho, D.-W.; Tsang, D.C.; Cao, X.; Hou, D.; Shen, Z.; Alessi, D.S.; Ok, Y.S.; Poon, C.S. Green remediation of As and Pb contaminated soil using cement-free clay-based stabilization/solidification. *Environ. Int.* **2019**, *126*, 336–345. [CrossRef]
88. Chen, T.; Wang, L.; He, B.; Peng, X.; Nie, X.; Ma, F.; Han, P.; Bai, X. Study on the solidification/stabilization of cadmium-contaminated soil by red mud-assisted blast furnace slag under excitation conditions. *J. Clean. Prod.* **2024**, *435*, 140505. [CrossRef]
89. Xia, M.; Muhammad, F.; Zeng, L.; Li, S.; Huang, X.; Jiao, B.; Shiau, Y.; Li, D. Solidification/stabilization of lead-zinc smelting slag in composite based geopolymer. *J. Clean. Prod.* **2019**, *209*, 1206–1215. [CrossRef]
90. Liu, J.; Zha, F.; Xu, L.; Kang, B.; Yang, C.; Feng, Q.; Zhang, W.; Zhang, J. Strength and microstructure characteristics of cement-soda residue solidified/stabilized zinc contaminated soil subjected to freezing–thawing cycles. *Cold Reg. Sci. Technol.* **2020**, *172*, 102992. [CrossRef]
91. Zhang, Y.; Ong, Y.J.; Yi, Y. Comparison between CaO-and MgO-activated ground granulated blast-furnace slag (GGBS) for stabilization/solidification of Zn-contaminated clay slurry. *Chemosphere* **2022**, *286*, 131860. [CrossRef]
92. Kim, E.-J.; Park, S.; Adil, S.; Lee, S.; Cho, K. Biogeochemical alteration of an aquifer soil during in situ chemical oxidation by hydrogen peroxide and peroxymonosulfate. *Environ. Sci. Technol.* **2021**, *55*, 5301–5311. [CrossRef]
93. Zarei, A.A.; Bazrafshan, E.; Mosafer, J.; Foroughi, M.; Khaksefidi, R.; Boghsani, G.T.; Mohammadi, L.; Dargahi, A. In situ chemical oxidation of tinidazole in aqueous media by heat-activated persulfate: Kinetics, thermodynamic, and mineralization studies. *Appl. Water Sci.* **2024**, *14*, 71. [CrossRef]
94. Liu, Y.; Zhang, Y.; Zhou, A. A potential novel approach for in situ chemical oxidation based on the combination of persulfate and dithionite. *Sci. Total Environ.* **2019**, *693*, 133635. [CrossRef]
95. Dangi, M.B.; Urynowicz, M.A.; Schultz, C.L.; Budhathoki, S. A comparison of the soil natural oxidant demand exerted by permanganate, hydrogen peroxide, sodium persulfate, and sodium percarbonate. *Environ. Chall.* **2022**, *7*, 100456. [CrossRef]
96. Fang, Z.; Zhou, Z.; Xue, G.; Yu, Y.; Wang, Q.; Cheng, B.; Ge, Y.; Qian, Y. Application of sludge biochar combined with peroxydisulfate to degrade fluoroquinolones: Efficiency, mechanisms and implication for ISCO. *J. Hazard. Mater.* **2022**, *426*, 128081. [CrossRef] [PubMed]
97. Chen, W.-Y.; Pokhrel, P.; Wang, Y.-S.; Lin, S.-L.; Liu, M.-H. Combined application of ozone and hydrogen peroxide to degrade diesel contaminants in soil and groundwater. *Water* **2021**, *13*, 3404. [CrossRef]

98. Park, S.; Oh, S.; Kim, I. Role of Borate Buffer in Organic Degradation by Peroxymonosulfate in the Presence of Metal Oxides. *Water* **2021**, *13*, 2698. [CrossRef]
99. Lea, K.; Eric, S.; Rula, D. *Verification Monitoring for In Situ Chemical Reduction Using Zero-Valent Zinc, A Novel Technology for Remediation of Chlorinated Alkanes*. 2020. Available online: https://serdp-estcp.mil/projects/details/785bb4d4-9971-4722-a636-6ae8f895c636 (accessed on 15 May 2025).
100. Huang, T.; Li, Y.; Qian, J.; Liu, S.; Wang, J. Remediation of Cr (VI)-contaminated soil by double-modified nanoscale zero-valent iron: Performance and mechanism. *J. Soils Sediments* **2024**, *24*, 1724–1738. [CrossRef]
101. Galvão, Á.G.; Costa, L.G.; Costa, E.C.d.A.; da Silva, D.R.; Martínez-Huitle, C.A.; Vieira dos Santos, E. Application of a Zero-Valent Iron/Cork as Permeable Reactive Barrier for In Situ Remediation of Phenanthrene in Soil. *Catalysts* **2022**, *12*, 1591. [CrossRef]
102. Wang, L.; Wei, J.; Yang, L.; Chen, Y.; Wang, M.; Xiao, L.; Yuan, G. Enhancing soil remediation of copper-contaminated soil through washing with a soluble humic substance and chemical reductant. *Agronomy* **2023**, *13*, 1754. [CrossRef]
103. Singh, P.; Pal, P.; Mondal, P.; Saravanan, G.; Nagababu, P.; Majumdar, S.; Labhsetwar, N.; Bhowmick, S. Kinetics and mechanism of arsenic removal using sulfide-modified nanoscale zerovalent iron. *Chem. Eng. J.* **2021**, *412*, 128667. [CrossRef]
104. Hussain, A.A.; Kamran, K.; Ishaq, M.; Akram, A.; Hina, M. Optimization of electroosmotic flow to enhance the removal of contaminants from low-permeable soils. *J. Appl. Electrochem.* **2023**, *53*, 1245–1258. [CrossRef]
105. Asadollahfardi, G.; Sarmadi, M.S.; Rezaee, M.; Khodadadi-Darban, A.; Yazdani, M.; Paz-Garcia, J.M. Comparison of different extracting agents for the recovery of Pb and Zn through electrokinetic remediation of mine tailings. *J. Environ. Manag.* **2021**, *279*, 111728. [CrossRef]
106. Wang, L.; Yu, K.; Li, J.-S.; Tsang, D.C.; Poon, C.S.; Yoo, J.-C.; Baek, K.; Ding, S.; Hou, D.; Dai, J.-G. Low-carbon and low-alkalinity stabilization/solidification of high-Pb contaminated soil. *Chem. Eng. J.* **2018**, *351*, 418–427. [CrossRef]
107. Jiang, M.; Wang, K.; Xue, R.; Zhao, Q.; Wang, L.; Wang, W.; Ding, J. Remediation of antimony-contaminated soil with composite materials of Fe/Fe_{2+}-fly ash-concrete additive in seasonal freezing regions. *Process Saf. Environ. Prot.* **2022**, *167*, 247–256. [CrossRef]
108. Xia, W.-Y.; Du, Y.-J.; Li, F.-S.; Guo, G.-L.; Yan, X.-L.; Li, C.-P.; Arulrajah, A.; Wang, F.; Wang, S. Field evaluation of a new hydroxyapatite based binder for ex-situ solidification/stabilization of a heavy metal contaminated site soil around a Pb-Zn smelter. *Constr. Build. Mater.* **2019**, *210*, 278–288. [CrossRef]
109. Kan, L.; Shi, R.; Zhao, Y.; Duan, X.; Wu, M. Feasibility study on using incineration fly ash from municipal solid waste to develop high ductile alkali-activated composites. *J. Clean. Prod.* **2020**, *254*, 120168. [CrossRef]
110. Wu, H.-L.; Jin, F.; Bo, Y.-L.; Du, Y.-J.; Zheng, J.-X. Leaching and microstructural properties of lead contaminated kaolin stabilized by GGBS-MgO in semi-dynamic leaching tests. *Constr. Build. Mater.* **2018**, *172*, 626–634. [CrossRef]
111. Yaman, C. Performance and kinetics of bioaugmentation, biostimulation, and natural attenuation processes for bioremediation of crude oil-contaminated soils. *Processes* **2020**, *8*, 883. [CrossRef]
112. Yang, Y.; Xiao, C.; Wang, F.; Peng, L.; Zeng, Q.; Luo, S. Assessment of the potential for phytoremediation of cadmium polluted soils by various crop rotation patterns based on the annual input and output fluxes. *J. Hazard. Mater.* **2022**, *423*, 127183. [CrossRef]
113. Sánchez Mata, O.; Aguilera Flores, M.M.; Ureño García, B.G.; Ávila Vázquez, V.; Cabañas García, E.; Franco Villegas, E.A. Bioremediation of Automotive Residual Oil-Contaminated Soils by Biostimulation with Enzymes, Surfactant, and Vermicompost. *Int. J. Environ. Res. Public Health* **2023**, *20*, 6600. [CrossRef]
114. Wu, M.; Dick, W.A.; Li, W.; Wang, X.; Yang, Q.; Wang, T.; Xu, L.; Zhang, M.; Chen, L. Bioaugmentation and biostimulation of hydrocarbon degradation and the microbial community in a petroleum-contaminated soil. *Int. Biodeterior. Biodegrad.* **2016**, *107*, 158–164. [CrossRef]
115. Haleyur, N.; Shahsavari, E.; Jain, S.S.; Koshlaf, E.; Ravindran, V.B.; Morrison, P.D.; Osborn, A.M.; Ball, A.S. Influence of bioaugmentation and biostimulation on PAH degradation in aged contaminated soils: Response and dynamics of the bacterial community. *J. Environ. Manag.* **2019**, *238*, 49–58. [CrossRef]
116. Wang, X.; Zhao, R.; Wu, H.; Jia, X.; Liu, Y.; Zhou, G.; Chen, S.; Zhao, F.; Li, L.; Hu, S. Enhanced bioremediation of hexavalent chromium via Stenotrophomonas acidaminiphila 4–1 assisted with agricultural wastes-derived biochar. *Biochem. Eng. J.* **2024**, *208*, 109355. [CrossRef]
117. Barba, S.; Villaseñor, J.; Rodrigo, M.A.; Cañizares, P. Biostimulation versus bioaugmentation for the electro-bioremediation of 2,4-dichlorophenoxyacetic acid polluted soils. *J. Environ. Manag.* **2021**, *277*, 111424. [CrossRef] [PubMed]
118. Sharma, J.K.; Kumar, N.; Singh, N.P.; Santal, A.R. Phytoremediation technologies and their mechanism for removal of heavy metal from contaminated soil: An approach for a sustainable environment. *Front. Plant Sci.* **2023**, *14*, 1076876. [CrossRef] [PubMed]
119. Phang, L.-Y.; Mingyuan, L.; Mohammadi, M.; Tee, C.-S.; Yuswan, M.H.; Cheng, W.-H.; Lai, K.-S. Phytoremediation as a viable ecological and socioeconomic management strategy. *Environ. Sci. Pollut. Res.* **2024**, *31*, 50126–50141. [CrossRef]
120. Wiafe, S.; Boahen, C.; Bandoh, T. Enhanced Phytoremediation of Heavy Metals By Soil-applied Organic And Inorganic Amendments: Heavy Metal Phytoavailability, Accumulation, And Metal Recovery. *Soil Sediment Contam. Int. J.* **2024**, *34*, 634–658. [CrossRef]

121. Hou, L.; Liu, R.; Li, N.; Dai, Y.; Yan, J. Study on the efficiency of phytoremediation of soils heavily polluted with PAHs in petroleum-contaminated sites by microorganism. *Environ. Sci. Pollut. Res.* **2019**, *26*, 31401–31413. [CrossRef]
122. Chen, D.; Liu, X.; Bian, R.; Cheng, K.; Zhang, X.; Zheng, J.; Joseph, S.; Crowley, D.; Pan, G.; Li, L. Effects of biochar on availability and plant uptake of heavy metals–A meta-analysis. *J. Environ. Manag.* **2018**, *222*, 76–85. [CrossRef]
123. Li, H.; Jin, R.; Xu, Z.; Hu, H.; Kalkhajeh, Y.K.; Zhao, Y.; Zhan, L. Application of chelate GLDA for remediating Cd-contaminated farmlands using *Tagetes patula* L. *Environ. Sci. Pollut. Res.* **2023**, *30*, 3774–3782. [CrossRef]
124. Chigbo, C.; Batty, L. Chelate-assisted phytoremediation of Cu-pyrene-contaminated soil using *Z. mays*. *Water Air Soil Pollut.* **2015**, *226*, 74. [CrossRef]
125. Shehata, S.M.; Badawy, R.K.; Aboulsoud, Y.I. Phytoremediation of some heavy metals in contaminated soil. *Bull. Natl. Res. Cent.* **2019**, *43*, 189. [CrossRef]
126. Liu, Z.-F.; Ge, H.-G.; Li, C.; Zhao, Z.-P.; Song, F.-M.; Hu, S.-B. Enhanced phytoextraction of heavy metals from contaminated soil by plant co-cropping associated with PGPR. *Water Air Soil Pollut.* **2015**, *226*, 29. [CrossRef]
127. Lai, W.-L.; Lee, F.-Y.; Chen, C.S.; Hseu, Z.-Y.; Kuo, Y.-L. The removal efficacy of heavy metals and total petroleum hydrocarbons from contaminated soils by integrated bio-phytoremediation. *J. Soil Groundw. Environ.* **2014**, *19*, 35–44. [CrossRef]
128. Bhat, S.A.; Bashir, O.; Haq, S.A.U.; Amin, T.; Rafiq, A.; Ali, M.; Américo-Pinheiro, J.H.P.; Sher, F. Phytoremediation of heavy metals in soil and water: An eco-friendly, sustainable and multidisciplinary approach. *Chemosphere* **2022**, *303*, 134788. [CrossRef] [PubMed]
129. Kalhor, A.X.; Movafeghi, A.; Mohammadi-Nassab, A.D.; Abedi, E.; Bahrami, A. Potential of the green alga *Chlorella vulgaris* for biodegradation of crude oil hydrocarbons. *Mar. Pollut. Bull.* **2017**, *123*, 286–290. [CrossRef]
130. Cerqueira, V.S.; Maria do Carmo, R.P.; Camargo, F.A.; Bento, F.M. Comparison of bioremediation strategies for soil impacted with petrochemical oily sludge. *Int. Biodeterior. Biodegrad.* **2014**, *95*, 338–345. [CrossRef]
131. Chen, W.; Wang, F.; Zeng, L.; Li, Q. Bioremediation of petroleum-contaminated soil by semi-aerobic aged refuse biofilter: Optimization and mechanism. *J. Clean. Prod.* **2021**, *294*, 125354. [CrossRef]
132. Dillon, L. Race, waste, and space: Brownfield redevelopment and environmental justice at the Hunters Point Shipyard. *Antipode* **2014**, *46*, 1205–1221. [CrossRef]
133. Navy USDOT. *Finding of Suitability to Transfer for Parcel A (Revision 3), Hunters Point Shipyard*; Department of the Navy, Base Realignment and Closure: San Francisco, CA, USA, 2004.
134. U.S. EPA; Department of Defense; Department of Energy; Department of Interior I; National Aeronautics and Space Administration. *Tennessee Valley Authority; Coast Guard; Abstracts of Remediation Case Studies*; Federal Remediation Technologies Roundtable: Washington, DC, USA, 2006; Volume 10.
135. White, R. Deconstructing Mare Island: Reconnaissance in the Ruins. *Boom A J. Calif.* **2012**, *2*, 55–69. [CrossRef]
136. Battelle. *Remedial Investigation and Focused Feasibility Study (RI/FFS) Report for the Defense Reutilization and Marketing Office (DRMO), Former Mare Island Naval Shipyard*; Battelle, 505 King Avenue: Columbus, OH, USA, 2014.
137. Woods, E.E.S. *Designing an Integrated Waterfront: Responsive Redevelopment at the Philadelphia Navy Yard*; Massachusetts Institute of Technology: Cambridge, MA, USA, 2012.
138. U.S. EPA. *Philadelphia Navy Yard Site Spotlight*; U.S. EPA, Region III: Philadelphia, PA, USA, 2025.
139. Calisti, S.P. The Redevelopment of the Charlestown Navy Yard: A Model for the Military Base Closure Process. Ph.D. Thesis, University Of Florida, Gainesville, FL, USA, 1992.
140. U.S. EPA. *Environmental Summary Report for Charlestown Navy Yard Parcels 6 and 7*; U.S. EPA Region 1: Boston, MA, USA, 2010.
141. *Massachusetts Department of Environmental Protection (MassDEP)*; Phase II Site Assessment Report: Charlestown Navy Yard Building 105 Area. U.S. Army Corps of Engineers (USACE): Boston, MA, USA, 2012.
142. Andrén, S. Urban sustainable development from a place-based and a system-based approach: Case study Malmö. In Proceedings of the International Conference City Futures 09, Bangkok, Thailand, 7–9 March 2009.
143. Varvsstaden, A.B. *Varvstaden—An Example of Transformation and Upcycling of Local Resources*; Nordic Sustainable Construction: Malmö, Sweden, 2023.
144. Lehoux, A.P.; Petersen, K.; Leppänen, M.T.; Snowball, I.; Olsen, M. Status of contaminated marine sediments in four Nordic countries: Assessments, regulations, and remediation approaches. *J. Soils Sediments* **2020**, *20*, 2619–2629. [CrossRef]
145. Oslo Port, A. *Oslo Harbor Remediation Project Summary*; Oslo Havn KF: Oslo, Norway, 2015.
146. Pettersen, A.; Solerød, A.W.; Norwegian Geotechnical Institute (NGI); Port of Oslo. *The Oslo Harbour Remediation Project: Oslo Fjord Clean Up*; Oslo Havn KF: Oslo, Norway, 2007.
147. Ministry of Climate and Environment; Port of Oslo. *RenOslofjord: Final Report on Harbor Sediment Remediation*; Norwegian Environment Agency: Oslo, Norway, 2018.
148. U.S. EPA. Howards Bay Remediation, Great Lakes Areas of Concern. Available online: https://www.epa.gov/great-lakes-aocs/howards-bay-remediation (accessed on 19 May 2025).
149. U.S. EPA. *Howards Bay Area of Concern Cleanup Summary*; U.S. EPA Region 5: Chicago, IL, USA, 2021.

150. Wisconsin Department of Natural Resources. *Dredging Complete at Howards Bay*; WDNR Remediation and Redevelopment Program: Madison, WI, USA, 2022.
151. Arcadis; Wisconsin Department of Natural Resources. *Howards Bay Cleanup and Restoration Overview*; Arcadis and WDNR: Madison, WI USA, 2020.
152. City of Tongyeong. *Project Overview of Soil and Groundwater Remediation at Former Tongyeong Shipyard*; Internal Meeting Document: Tongyeong-si, Republic of Korea, 2023.
153. U.S. EPA. *Soil Washing Treatment: Engineering Bulletin*; EPA Office of Emergency and Remedial Response: Washington, DC, USA, 1990.
154. Lee, H.; Lee, Y.; Kim, J.; Kim, C. Field Application of Modified In Situ Soil Flushing in Combination with Air Sparging at a Military Site Polluted by Diesel and Gasoline in Korea. *Int. J. Environ. Res. Public Health* **2014**, *11*, 8806–8824. [CrossRef]
155. Haemers, T. *Comparison of Power Consumption for Thermal Conductive Heating vs Electrical Resistance in In Situ Thermal Desorption*; Haemers Technologies: Brussels, Belgium, 2018.
156. California Environmental Protection Agency – Department of Toxic Substances Control (DTSC). *Remediation of Chlorinated Volatile Organic Compounds in Vadose Zone Soil (PT&R Guidance)*; California Environmental Protection Agency, DTSC: Sacramento, CA, USA, 2010.
157. Wei, K.-H.; Ma, J.; Xi, B.-D.; Yu, M.-D.; Cui, J.; Chen, B.-L.; Li, Y.; Gu, Q.-B.; He, X.-S. Recent progress on in-situ chemical oxidation for the remediation of petroleum contaminated soil and groundwater. *J. Hazard. Mater.* **2022**, *432*, 128738. [CrossRef] [PubMed]
158. Andrade, L.N.; Araujo, S.F.; Matos, A.T.; Henriques, A.B.; Oliveira, L.C.; Souza, P.P.; Chagas, P.; Leão, M.M.D.; Amorim, C.C. Performance of different oxidants in the presence of oxisol: Remediation of groundwater contaminated by gasoline/ethanol blend. *Chem. Eng. J.* **2017**, *308*, 428–437. [CrossRef]
159. Piscitelli, D.; Zingaretti, D.; Verginelli, I.; Gavasci, R.; Baciocchi, R. The fate of MtBE during Fenton-like treatments through laboratory scale column tests. *J. Contam. Hydrol.* **2015**, *183*, 99–108. [CrossRef] [PubMed]
160. Tratnyek, P.G.; Johnson, R.L.; Lowry, G.V.; Brown, R.A. In Situ Chemical Reduction for Source Remediation. In *Chlorinated Solvent Source Zone Remediation*; Kueper, B.H., Stroo, H.F., Vogel, C.M., Ward, C.H., Eds.; Springer: New York, NY, USA, 2014; pp. 307–351.
161. Yuan, L.; Xu, X.; Li, H.; Wang, Q.; Wang, N.; Yu, H. The influence of macroelements on energy consumption during periodic power electrokinetic remediation of heavy metals contaminated black soil. *Electrochim. Acta* **2017**, *235*, 604–612. [CrossRef]
162. U.S. EPA. *Solidification/Stabilization Resource Guide*; EPA Office of Solid Waste and Emergency Response: Washington, DC, USA, 1999.
163. Wang, L.; Chen, G.; Du, X.; Li, M.; Zhang, Z.; Liang, H.; Gao, D. Bioremediation of PAHs-contaminated site in a full-scale biopiling system with immobilized enzymes: Removal efficiency and microbial communities. *Environ. Res.* **2024**, *262*, 119763. [CrossRef]
164. Pivetz, B.E. *Phytoremediation of Contaminated Soil and Ground Water at Hazardous Waste Sites*; US Environmental Protection Agency, Office of Research and Development: Washington DC, USA, 2001.
165. Lemming, G.; Hauschild, M.Z.; Bjerg, P.L. Life cycle assessment of soil and groundwater remediation technologies: Literature review. *Int. J. Life Cycle Assess.* **2010**, *15*, 115–127. [CrossRef]
166. Cadotte, M.; Deschênes, L.; Samson, R. Selection of a remediation scenario for a diesel-contaminated site using LCA. *Int. J. Life Cycle Assess.* **2007**, *12*, 239–251. [CrossRef]
167. Cai, Y.; Chen, H.; Zhao, W.; Sun, Y.; Wang, M.; Chen, Y.; Ma, J. Investigation of greenhouse gas emissions in soil remediation activities based on the integration of machine learning models and life cycle assessment. *J. Clean. Prod.* **2025**, *487*, 144630. [CrossRef]
168. Ashkanani, Z.; Mohtar, R.; Al-Enezi, S.; Smith, P.K.; Calabrese, S.; Ma, X.; Abdullah, M. AI-assisted systematic review on remediation of contaminated soils with PAHs and heavy metals. *J. Hazard. Mater.* **2024**, *468*, 133813. [CrossRef]
169. Khalid, S.; Shahid, M.; Niazi, N.K.; Murtaza, B.; Bibi, I.; Dumat, C. A comparison of technologies for remediation of heavy metal contaminated soils. *J. Geochem. Explor.* **2017**, *182*, 247–268. [CrossRef]
170. Falconi, M. *Soil Washing (SW) Report*; IMPEL: Brussels, Belgium, 2023; p. 120. ISBN 978-2-931225-30-1.
171. Song, Y.H.; Kim, G.Y.; Kim, D.Y.; Hwang, Y.W. Life Cycle Assessment of Crude Oil-Contaminated Soil Treated by Low-Temperature Thermal Desorption and Its Beneficial Reuse for Soil Amendment. *Sustainability* **2024**, *16*, 10900. [CrossRef]
172. Geoengineer.Org. Contaminated Soil Remediation through Thermal Desorption: Synthesis of Case Histories and Comparison of In-Situ and Ex-Situ Applications; Geoengineer.Org, 29 April 2021. Available online: https://www.geoengineer.org/education/web-class-projects/ce-176-environmental-geotechnics/assignments/contaminated-soil-remediation-through-thermal-desorption (accessed on 15 May 2025).
173. Vilén, A.; Laurell, P.; Vahala, R. Comparative life cycle assessment of activated carbon production from various raw materials. *J. Environ. Manag.* **2022**, *324*, 116356. [CrossRef] [PubMed]
174. U.S. EPA. *How to Evaluate Alternative Cleanup Technologies for Underground Storage Tank Sites: A Guide for Corrective Action Plan Reviewers*; EPA 510-B-17-003; EPA Office of Underground Storage Tanks: Washington, DC, USA, 2017.

175. Inogen Alliance. How Much Does It Cost for Site Remediation? *Inogen Alliance Website*. 23 June 2023. Available online: https://inogenalliance.com/ (accessed on 15 May 2025).
176. Deng, B.; Carter, R.A.; Cheng, Y.; Liu, Y.; Eddy, L.; Wyss, K.M.; Ucak-Astarlioglu, M.G.; Luong, D.X.; Gao, X.; JeBailey, K.; et al. High-temperature electrothermal remediation of multi-pollutants in soil. *Nat. Commun.* **2023**, *14*, 6371. [CrossRef]
177. Cameselle, C. Electrokinetic Remediation, Cost Estimation. In *Encyclopedia of Applied Electrochemistry*; Kreysa, G., Ota, K.-I., Savinell, R.F., Eds.; Springer: New York, NY, USA, 2014; pp. 723–725.
178. Fernández-Marchante, C.M.; Vieira dos Santos, E.; Souza, F.L.; Martínez-Huitle, C.A.; Rodríguez-Gómez, A.; Lobato, J.; Rodrigo, M.A. Environmental impact assessment of the electrokinetic adsorption barriers to remove different herbicides from agricultural soils. *Sci. Total Environ.* **2024**, *927*, 172287. [CrossRef]
179. Orellana, R.; Cumsille, A.; Piña-Gangas, P.; Rojas, C.; Arancibia, A.; Donghi, S.; Stuardo, C.; Cabrera, P.; Arancibia, G.; Cárdenas, F.; et al. Economic Evaluation of Bioremediation of Hydrocarbon-Contaminated Urban Soils in Chile. *Sustainability* **2022**, *14*, 11854. [CrossRef]
180. Vafa, H.J.; Pourbabaee, A.A.; Alikhani, H.A.; Yazdanfar, N.; Khanali, M. A comparative life cycle analysis of bioremediation approaches for old-aged petroleum pollution in hypersaline soil. *Chemosphere* **2025**, *373*, 144150. [CrossRef]
181. Aghili, S.; Golzary, A. Greening the earth, healing the soil: A comprehensive life cycle assessment of phytoremediation for heavy metal contamination. *Environ. Technol. Innov.* **2023**, *32*, 103241. [CrossRef]

processes

MDPI

Review

A Mini-Review on the Use of Chelating or Reducing Agents to Improve Fe(II)-Fe(III) Cycles in Persulfate/Fe(II) Systems

Lanbing Feng [1], Yun Zeng [1], Pan Wang [1], Na Duan [1], Haodong Ji [2] and Xiao Zhao [1,*]

1 College of Water Resources and Civil Engineering, China Agricultural University, Beijing 100083, China; woshifenglanbing@163.com (L.F.); zeng_unn99@163.com (Y.Z.); wangpan1102@hotmail.com (P.W.); duanna@cau.edu.cn (N.D.)

2 Eco-Environment and Resource Efficiency Research Laboratory, School of Environment and Energy, Peking University Shenzhen Graduate School, Shenzhen 518055, China; jihaodong@pku.edu.cn

* Correspondence: xiaozhao88@cau.edu.cn; Tel./Fax: +86-010-6273-6533

Abstract: Persulfates (PSs) have recently gained recognition as strong oxidants notably for soil and groundwater remediation. Ferrous iron (Fe(II)) has been extensively employed in activating PS. However, the PS/Fe(II) system still suffers from some inherent drawbacks, including the fast accumulation of ferric ion (Fe(III)) or precipitation of Fe(III), sluggish conversion from Fe(III) to Fe(II), and the unwanted scavenging of $SO_4^{\bullet-}$ by excess Fe(II). To overcome these issues, the following two types of reagents are usually used: chelating agents and reducing agents. There are 11 chelating agents mentioned in this paper (EDTA, CA, etc.) and 3 reducing agents (HA, thiosulfate and A-Boron). This review extracted data from studies to give a data-based overview on how the added agents can improve the PS/Fe(II) system under varied conditions. The mechanisms of improvement by chelating agents could be classified into the following three categories: to control excess Fe(II), reduce Fe(III) to Fe(II), and inhibit Fe precipitation. The data analysis suggested an optimal working agent/Fe(II) ratio of 0–2 with acidic and neutral pH. The mechanism of using a reducing agent for PS/Fe(II) system improvement is simply reducing Fe(III) to Fe(II), using hydroxylamine, thiosulfate, etc. The improvement was significant at a hydroxylamine/Fe (II) molar ratio between 0 and 10, yet overdosed hydroxylamine can scavenge the reactive radicals, resulting in negative effects. The future research trend was also proposed in this review.

Keywords: persulfate-based advanced oxidation processes; $SO_4^{\bullet-}$; Fe(II)/Fe(III) cycles; Fe(II)-persulfate

1. Introduction

Peroxymonosulfate (PMS) and peroxydisulfate (PDS) are two typical persulfates (PS) [1] that have recently gained recognition as emerging oxidants [2], notably for soil and groundwater remediation [3–9]. Over the past 15 years, PS has been considered an innovative advanced oxidation chemical, capable of generating reactive sulfate radical ($SO_4^{\bullet-}$) with a standard redox potential of 2.5 V to 3.1 V [10–14]. The activation of PS has also been extensively studied. Due to the high reactivity of $SO_4^{\bullet-}$, PS can oxidize lots of the organic pollutants [15–19]. Moreover, a PS-mediated sulfate radical with a longer half-life and wider working pH has shown significant potential in environmental remediation as an alternative to $OH^{\bullet}$ [20–22]. Ferrous iron (Fe(II)) has been broadly selected as the activator for producing $SO_4^{\bullet-}$ in PS (Equations (1) and (2), whereby $S_2O_8^{2-}$ is produced by PS and HSO_5^- is produced by PMS), owing to its environmentally friendly characteristics, cost-efficiency, and superior activation performance [23–26]. Fe(II) activation necessitates comparably low energy levels (14.8 kcal·mol^{-1}) [25,27,28]. However, the Fe(II) activation still suffers from some inherent drawbacks, including the swift accumulation of ferric ion (Fe(III)) [29,30], sluggish conversion from Fe(III) to Fe(II) [31–33], the formation of

Fe(III) precipitates at pH > 4 (Equation (3)) [34], and $SO_4^{\bullet-}$ scavenging by Fe(II) excess (Equation (4)) [35–37].

$$Fe^{2+} + S_2O_8^{2-} \rightarrow Fe^{3+} + SO_4^{\bullet-} + SO_4^{2-} \tag{1}$$

$$Fe^{2+} + HSO_5^{-} \rightarrow Fe^{3+} + SO_4^{\bullet-} + OH^{-} \tag{2}$$

$$Fe^{3+} + 3H_2O \rightarrow Fe(OH)_3 + 3H^{+} \tag{3}$$

$$SO_4^{\bullet-} + Fe^{2+} \rightarrow Fe^{3+} + SO_4^{2-} k = 4.9 \times 10^{9} M^{-1} s^{-1} \tag{4}$$

One of the main constraints of PS/Fe(II) system is the slow conversion efficiency from Fe(III) to Fe(II) [38,39]. Moreover, the presence of excessive Fe(II) or rapid conversion of Fe(II) to Fe(III) can hinder the ultimate oxidizing potential of the PS/Fe(II) system by diminishing sulfate radicals. [40–42]. Ideally, Fe(III) was expected to be converted from Fe(II) per Equation (4) and then be reduced by organic compounds or their intermediate products (Equation (5)) [43,44].

$$RH + Fe^{3+} \rightarrow Fe^{2+} + RH^{\bullet+} \tag{5}$$

As a result, Fe(II) can be produced to reactivate the PS and create a new Fe(II)-Fe(III) cycle. However, this process is highly dependent on the redox potential of Fe(III), which may not have the ability to oxidize resistant organics [44]. To circumvent these issues, two types of reagents are usually used, namely chelating agents and reducing agents. Chelating agents can be employed to stabilize the concentration of Fe(II) in solution [45]. Certain chelates have proven effective in enhancing PS activation by elevating Fe solubility and decelerating the reaction rate of Fe(II) [43,46]. Furthermore, some chelating agents can also function as reducing agents to reduce Fe(III) [28,47]. Liang et al. [48] reported the first investigation on the effects of chelating agents on the PS/Fe(II) system in 2004. In this work, citrate was used as a chelating agent to enhance the trichloroethylene (TCE) degradation by PS/Fe(II). Increasing the molar ratio of citrate/Fe(II) iron from 0.1 to 0.5 resulted in an enhanced degradation efficiency, rising from 82.25% to 99.91%. Using reductive agents is an obvious choice to reduce the produced Fe(III) to Fe(II) in PS/Fe(II) system. Some reducing agents, exhibiting slow reaction rates with reactive species, can aid in the conversion of Fe(III) [49]. C. Liang et al. first reported the use of reducing agents to enhance the PS/Fe(II) system in 2004 [50], demonstrating an increase in trichloroethylene (TCE) degradation efficiency from 47% to 92% with the introduction of 1 mL thiosulfate. To date, a total of 23 studies have been published addressing the enhancement of the PS/Fe(II) system using chelating and reducing agents. However, there remains a lack of a systematic review that consolidates the findings related to the application of these agents in PS/Fe(II) systems. Such a review would offer a thorough comprehension of the mechanisms behind the improved activation of PS, while also revealing knowledge gaps and guiding future research efforts.

Given the technological advancements in this field, this review aims to offer a cohesive understanding of the current state of the art in enhancing the Fe(II)-Fe(III) cycle in the PS/Fe(II) system. Initially, the agents used for improvement were classified into two groups, i.e., chelating and reducing agents. Moreover, the raw data from studies under varied experimental conditions were extracted and investigated to reveal the mechanisms behind. Further, the effects of chemical dosage (mainly varied agent/Fe ratios) and solution pH were systematically compared for both chelating and reducing agents. Finally, the review provided summary and future research prospects.

2. Data Collection and Analysis

We conducted a publication search on Web of Science by limiting the topic to "ferrous ion, persulfate, Fe(II)/Fe(III) cycles, persulfate-based AOPs, persulfate-based advanced oxidation processes and (chelating agents or reducing agents)" and total of 206 research papers were found. To ensure their relevance, each study was examined manually against our specific criteria. Articles on homogeneous experiments where chelating or reducing agents were used were selected and the publication years were set from 2000 to 2024. We excluded all irrelevant studies (mostly heterogeneous experiment) and a total of 23 studies were found directly pertinent to our research objectives. These studies primarily involved homogeneous experimental conditions, with key variables such as dosage and pH. The raw data from the 23 studies were then extracted (per Web Plot Digitizer), analyzed for summarization, and finally charted using Origin.

3. Use of Chelating Agents for Fe(II)-PS Improvement

To enhance the contaminant removal efficiency of the PS/Fe(II) system, chelating agents can be introduced to control the Fe(II) concentration. Since the first publication in 2004 [48] that used citric acid to chelate Fe(II) for PS/Fe(II) system improvement, a total of 11 chelating agents have been tested (Table 1). Ethylene diamine tetraacetic acid (EDTA), which can form stable chelates with a variety of metal ions, is widely used. It can also inhibit the catalytic effect of metal ions in certain reactions and reduce the reaction rate. However, it is not easily degraded in nature, and high concentrations of EDTA may be toxic to aquatic organisms. In addition, some EDTA metal chelates are not easy to separate, limiting the recovery of metals [43,44,51–53]. Citric acid (CA) can chelate with metal ions to prevent precipitation and oxidation reactions, and also buffer pH at a variety of application scenarios. However, high concentrations of CA are corrosive and less stable in certain circumstances [47,48,52,54]. Gallic acid (GA), a potent antioxidant that neutralizes free radicals, is commonly found in plants. However, it has low solubility, is broken down under light and high-temperature conditions, and it may cause toxicity or side effects when consumed in excess [28]. Diethylenetriaminepentaacetic acid (DTPA) can combine with a variety of metal ions to form stable chelates. However, like EDTA, it is not biodegradable, and toxic to some aquatic organisms at high concentrations [47]. Ethylenediaminetetraacetic acid disodium salt (EDTA-Na_2), with a strong chelating ability, good water solubility, and low toxicity, is widely used. However, it is also not easily degraded in the natural environment, with a certain toxicity and a high cost [18]. Oxalic acid (OA), a strong reducing agent, has a natural source and is widely used. However, it is toxic and corrosive to humans, and its emissions have an impact on water and soil, as well as a tendency to form crystals [51,55]. Ethylenediaminedisuccinic acid (EDDS) is a new, effective chelating agent with better biodegradability, good water solubility, and low toxicity. However, the cost is high and the chelating ability for some metal ions may not be as good as traditional chelating agents such as EDTA [51,55–57]. Citrate, derived from natural citric acid, is safer and regulates pH. However, its solubility in water is limited, while it may lose its activity under high temperature or light conditions [46–48,55,58,59]. Tartaric acid is safe from natural sources and has some stability and antioxidant capacity. However, it has limited solubility and high cost [55]. Pyrophosphates have a good chelating ability, a strong water solubility, and are relatively safe. However, excessive use will affect the environment and the cost is high [57]. These agents exhibit diverse physico-chemical properties, as summarized in Table 1. Most of them are organic, sharing similar mechanisms for pollutant removal.

Table 1. Summary of 11 types of chelating agents for Fe(II)-PS improvement.

Name	CAS Number	Molecular Formula	Structural Formula	Molecular Weight	Density (g/cm^3)	Flash Point (°C)	Melting Point (°C)	Boiling Point (°C)	Water Solubility (g/L)
Ethylene diamine tetraacetic acid (EDTA)	60-00-4	$C_{10}H_{16}N_2O_8$	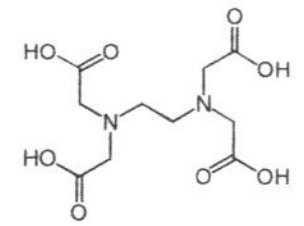	292.24	about 1.6	—	237–245 (dec.)	614.2	0.5 (25 °C)
Citric acid (CA)	77-92-9	$C_6H_8O_7$	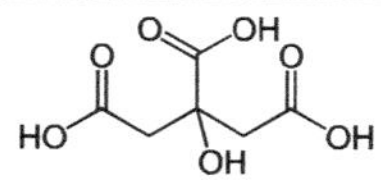	192.12	1.542	100	153~154.5	175 (decompose)	750 (20 °C)
Gallic acid (GA)	149-91-7	$C_7H_6O_5$	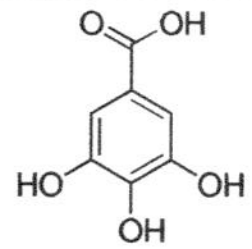	170.12	1.694	—	251 (dec.)	501.1	12 (cold water)
Sodium thiosulfate	7772-98-7	$Na_2S_2O_3$	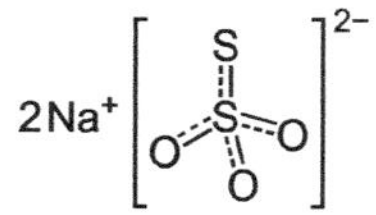	158.10	1.667	—	48	100	Soluble
Diethylenetriaminepentaacetic acid (DTPA)	67-43-6	$C_{14}H_{23}N_3O_{10}$	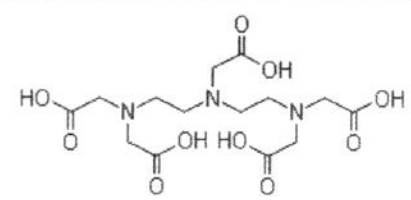	393.35	1.56	200	220	721.1 ± 60.0 (predicted)	5 (20 °C)
Ethylenediaminetetraacetic acid disodium salt (EDTA-Na$_2$)	139-33-3	$C_{10}H_{14}N_2Na_2O_8$	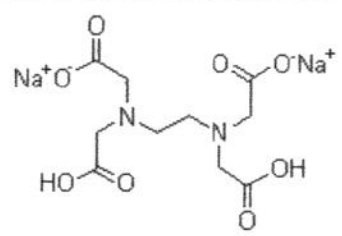	336.21	1.01	—	237~245	—	Soluble
Oxalic acid (OA)	144-62-7	$C_2H_2O_4$	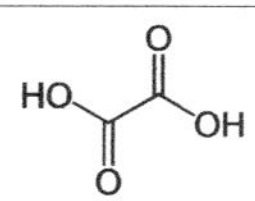	90.03	1.9	—	189~191	365.10	90 (20 °C)

Table 1. *Cont.*

Name	CAS Number	Molecular Formula	Structural Formula	Molecular Weight	Density (g/cm^3)	Flash Point (°C)	Melting Point (°C)	Boiling Point (°C)	Water Solubility (g/L)
Ethylenediaminedisuccinic acid (EDDS)	20846-91-7	$C_{10}H_{16}N_2O_8$	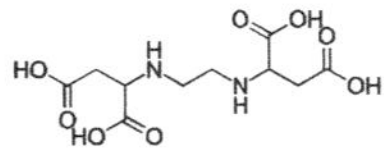	292.24	1.534	—	220~222	516.7 ± 50.0 (predicted)	150 (20 °C)
Citrate	126-44-3	$C_6H_5O_7$	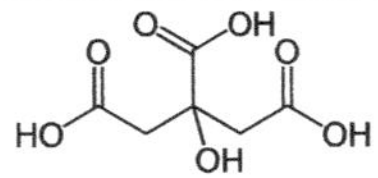	189.10	—	155.243 (cal.)	—	309.567 at 760 mmHg (cal.)	—
Tartaric acid (TA)	133-37-9	$C_4H_6O_6$	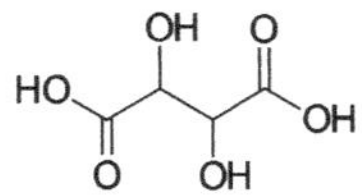	150.09	1.886	210	200~206	399.3	Soluble
Pyrophosphate	14000-31-8	O_7P_2	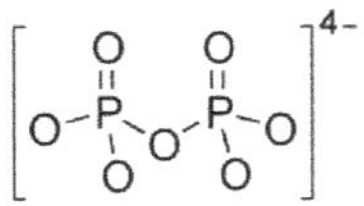	173.94	—	—	—	—	—

3.1. Mechanisms for Fe(II)-PS Improvement

A previous publication demonstrated that chelating agents can help to sustain the PS/Fe(II) system's efficiency across diverse pH ranges, regulating the quantity of Fe(II) available for reaction with PS [60]. In addition, chelating agents also alleviate the drawbacks of excessive Fe consumption and iron sludge accumulation. Furthermore, chelating agents have been found to promote the Fe(II)-Fe(III) redox cycle, which is crucial for the iron-based sulfate radical system [28,47].

The mechanisms of adding chelating agents for PS/Fe(II) system improvement can be classified into three categories (Figure 1). First, to control excess Fe(II). Adding a chelating agent (such as EDTA or CA) can chelate the excess Fe(II) to form Fe(II) complexes, preventing the reactions between $SO_4^{\bullet-}$ and Fe(II). Thereby, excessive Fe(II) can be used to activate PS per Equation (4) and reduce the consumption of $SO_4^{\bullet-}$ by excessive Fe(II) [43,44]; second, to decrease Fe(III) to Fe(II). Certain chelating agents can partially act as reducing agents to directly decrease Fe(III) to Fe(II), such as GA [28] and Na_2SO_3 [47]. This allows Fe(II) from a cycle of Fe(II)-Fe(III) to continue activating PS, leading to the improved removal efficiency of target pollutants; third, to inhibit Fe precipitation. The added chelating agent can directly chelate Fe(II) and Fe(III), preventing the hydrolytic precipitation of free Fe(II) and Fe(III) (Equations (3) and (6)), thus reducing the loss of Fe(II) and Fe(III) [56,58,61].

$$Fe^{2+} + 3H_2O \rightarrow Fe(OH)_3 \downarrow +3H^+ \tag{6}$$

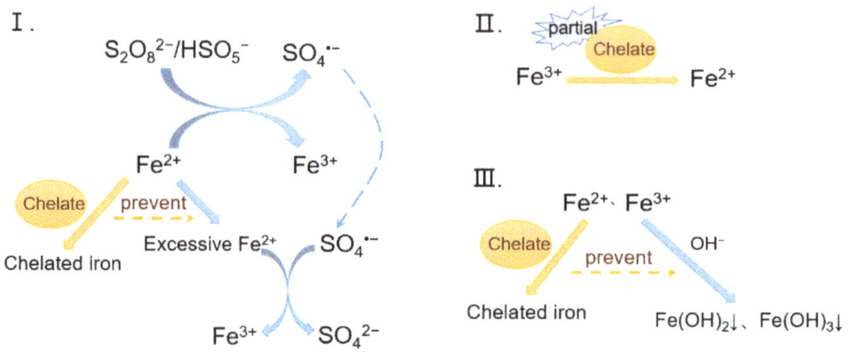

Figure 1. Mechanism of chelating agent for PS/Fe(II) system improvement. (**I**) Control of excess Fe(II); (**II**) Reduce Fe(III) to Fe(II); and (**III**) Inhibit Fe precipitation.

3.2. Effect of Molar Ratio of Chelate/Fe

Figure 2 illustrates the correlation between the removal efficiency (%) of target pollutants and the molar ratio of chelate/Fe. Generally, the scatter chart (Figure 2a) reveals an initial increase in removal efficiency, followed by a plateau or decrease as the chelate/Fe molar ratio increases. The peak values of the removal efficiency are observed at a molar ratio roughly between 0 and 2. The heatmap (Figure 2b), plotted using the raw data of removal efficiency, also confirmed this trend. As tabulated in Table 1, most of the chelating agents contain -COOH as the main functional group to chelate with Fe(II) ions, forming stable complexes. Taking EDTA as an example, ideally, one EDTA molecule (with four -COOH groups) can strongly coordinate with one Fe(II) ion, enveloping Fe(II) as a center metal ion, to form a stable Fe-EDTA complex. Apparently, the optimal molar ratios of chelate/Fe in Figure 2 did not absolutely follow the theoretical stoichiometry of complex forming by ligands and Fe ions.

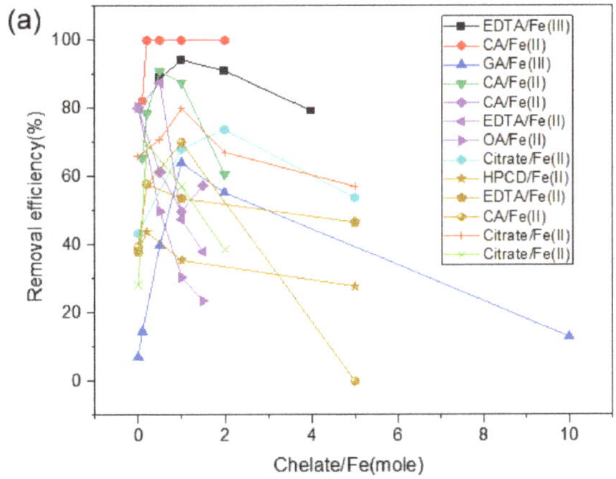

(b)

Chelate/Fe (mole)	0	0.1	0.2	0.5	1	1.5	2	4	5	10
CA/Fe(II)	80	73.767	67.534	61.3	49.78	57.39				
OA/Fe(II)	80.25	70.107	59.964	49.82	30.43	23.56				
EDTA/Fe(II)	80	82.514	85.027	87.54	47.54	37.97				
HPCD/Fe(II)	38.49	41.16	43.83	39.595	35.36	33.45	31.54	29.63	27.72	
Citrate/Fe(II)	28.06	48.84	69.62	63.385	57.15	47.93	38.71			
EDTA/Fe(II)	37.84	47.765	57.69	55.65	53.61	51.853	50.1	48.338	46.58	
CA/Fe(II)		82.25	99.91	99.91	99.91	99.1	99.91			
CA/Fe(II)		65.34	78.38	90.88	87.62	74.125	60.63			
GA/Fe(III)	7.15	14.57	27.24	39.91	63.98	59.635	55.29	41.173	27.057	12.94
EDTA/Fe(III)				88.82	94.2	92.6	91	79.07		
CA/Fe(II)	39.44	48.61	57.78	63.89	70	52.5	35	17.5		
Citrate/Fe(II)	65.96	67.586	69.213	70.84	79.96	73.515	67.07	62.07	57.07	
Citrate/Fe(II)	43.31	49.51	55.71	61.91	68.11	70.905	73.7	63.775	53.85	

Low Removal efficiency High

Figure 2. Target pollutants' removal efficiency (%) as a function of the molar ratio of chelate/Fe (Reference # is based on the References in the Supplementary Materials [28,43,44,46–48,57,61,62]) plotted as scatter charts (**a**) and a heatmap (**b**). Chelating agents: ethylene diamine tetraacetic acid (EDTA), citric acid (CA), gallic acid (GA), oxalic acid (OA), hydroxypropyl cyclodextrin (HPCD), citrate.

In the early stages of the reaction, the addition of chelating agents apparently improves the activation efficiency of Fe(II) on PS, thus enhancing the elimination of the target contaminants. However, overdosed chelating agents lead to the excessive chelation of Fe(II), and the high stability of the chelate reduces the stable formation of $SO_4^{\bullet -}$, resulting in a decline in the removal of the target pollutant. Moreover, surplus chelating agents can also complex with Fe(III), thereby interrupting the Fe(II)-Fe(III) cycle and initiating a decrease in contaminant removal. For example, when the pollutant iopamidol (IPM) was treated with the presence of chelator GA [28], the IPM removal efficiency was initially enhanced as the initial molar ratio of GA/Fe(II) increased from 0 to 1, and the removal efficiency peaked from 7.15% to 63.98%. Introducing an appropriate amount of GA greatly accelerated IPM degradation by preventing $SO_4^{\bullet -}$ consumption through complexation with Fe(II). GA also facilitated the Fe(II)-Fe(III) cycle by forming complexes with and reducing Fe(III), thereby efficiently boosting the activation of PS by Fe(II). Moreover, the Fe(III)/Fe(II)-GA complex exhibited greater stability than Fe(II) and Fe(III) ions, thus preventing the hydrolysis of Fe(II) and Fe(III). However, further increasing the GA/Fe(II) molar ratio from 1 to 10 resulted in a decline in removal efficiency to 12.94%. Excess GA can partially shield Fe(II) and Fe(III) ions, hindering their interaction with PS. Furthermore, excessive GA can act as a scavenger for free radicals. Therefore, the removal efficiency of IPM initially rose but subsequently declined as the molar ratio increased.

It is noteworthy that when the chelating agent CA was added in Fe(II)-PS system to treat TCE [48], the removal efficiency was 82.25% at a CA/Fe(II) molar ratio of 0.1. The removal efficiency reached a peak of 100% when the CA/Fe(II) ratio exceeded 0.2, highlighting the significant impact of CA on TCE removal. However, the addition of chelating agents CA and OA for trichloroacetic acid (TCA) treatment exhibited a contrasting trend [44], i.e., the TCA removal efficiency decreased with increasing molar ratio. This can be attributed to the competition for $SO_4^{\bullet-}$ between CA/OA and TCA. Additionally, the instability of the Fe(II)(OA) complex fails to prevent its reaction with Fe(II) and $SO_4^{\bullet-}$, leading to reduced TCA degradation. In summary, the improvement of chelating agents for the Fe(II)/PS system is varied at different chelate/Fe ratios, with a possible molar ratio roughly between 0 and 2. The optimal chelate/Fe molar ratios for pollutant removal did not fully follow the theoretical stoichiometry of complex reactions, yet overdosed chelating agents suppressed the improvement effect. For the same chelate, the improvement effectiveness may vary when different target pollutants are used. While the previous study mostly focused on the interaction between added chelating agents and Fe(II), the possible influence of chelating agents on target pollutants and radicals should also be considered.

3.3. *Effect of Initial pH on Chelating Agents*

Figure 3 demonstrates the removal efficiency (%) of the target pollutants versus the initial solution pH. A general decline in removal efficiency (%) is apparent as the initial pH rises. However, an exception to this trend was observed in the case using EDTA as chelating agent for TCE degradation [43], showing an opposite trend.

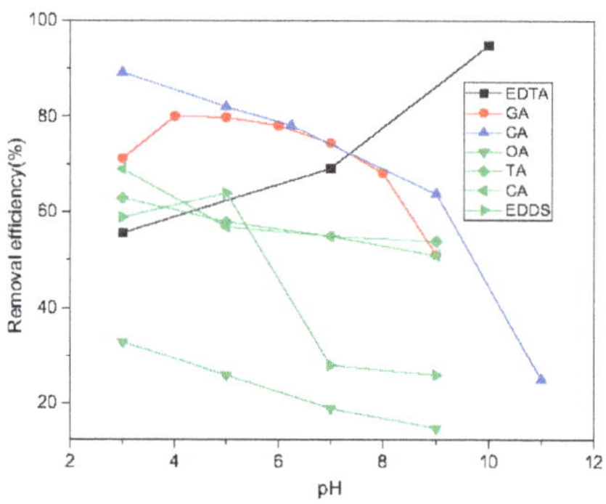

Figure 3. The correlation between target pollutant removal efficiency (%) and initial solution pH in the cases with the presence of varied chelating agents [28,43,55,58].

The solution pH is a key determinant in the removal of target contaminants, as it severely affects the forms of Fe, PS, and the target pollutants, and the interactions between any two of them in the triangle (Figure 4). For Fe species, at a pH higher than 8.5, the vast majority of Fe(II) are hydrolyzed and precipitated (Figure 5, calculated per Visual MinTeq 3.1), while the significant precipitation of Fe(III) occurs even at pH 4. On the other hand, solution pH also affects the forms of PS and their reactivity. PMS, with an operational pH range of 1 to 9.4, predominantly exists as the anion HSO_5^-, while it transforms to SO_5^{2-} at a pH higher than 9.4. This shift in speciation is critical as the presence of SO_5^{2-} can substantially inhibit the formation of the highly $SO_4^{\bullet-}$. Conversely, PDS, with a pKa2 of −3.5, remains in the form of $S_2O_8^{2-}$ across a wide pH range, indicating its stability and consistent behavior in various pH conditions [5]. Moreover, the solution pH can also impact the dynamics of radical species generation and transformation. In acidic conditions, hydrogen ions act as scavengers for both sulfate and hydroxyl radicals, reducing their availability for oxidation reactions. On the other hand, in alkaline conditions, sulfate radicals can react with hydroxide ions to form hydroxyl radicals, and this transformation is central to the base activation mechanism of PS. This shift from sulfate to hydroxyl radicals, however, comes with a possible decrease in oxidation potential. Solution pH also plays a pivotal role in determining the surface charge or species distribution of the target contaminants. For example, varied species with different surface charges of norfloxacin can exist at varied pH conditions. The introduction of chelating or reducing agents can further complicate the system (Figure 4), mainly by controlling the Fe(II)-Fe(III) cycle. Meanwhile, solution pH also poses a great impact on the introduced chelating or reducing agents. For example, in general, Fe^{2+}EDTA and Fe^{3+}EDTA are present in the low to neutral pH ranges, whereas at high pH ranges more hydroxyl complexes can form (Figure 6). Thus, the addition of chelating reagents, such as EDTA, can extend the stability or solubility of Fe(II) to a broader pH range [43].

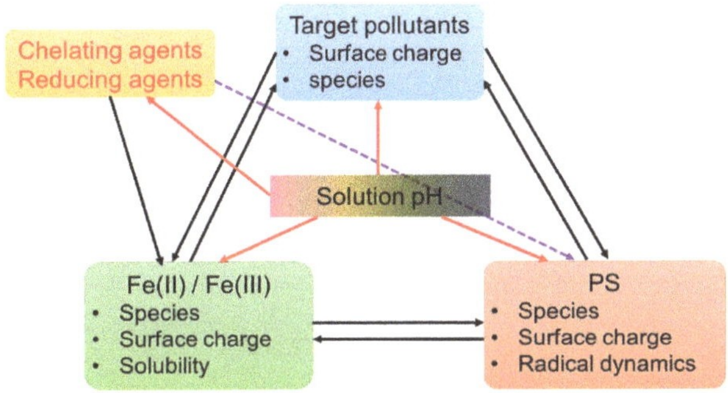

Figure 4. The mechanisms of how solution pH influence the PS/Fe(II) system for pollutant removal with the presence of chelating or reducing agents.

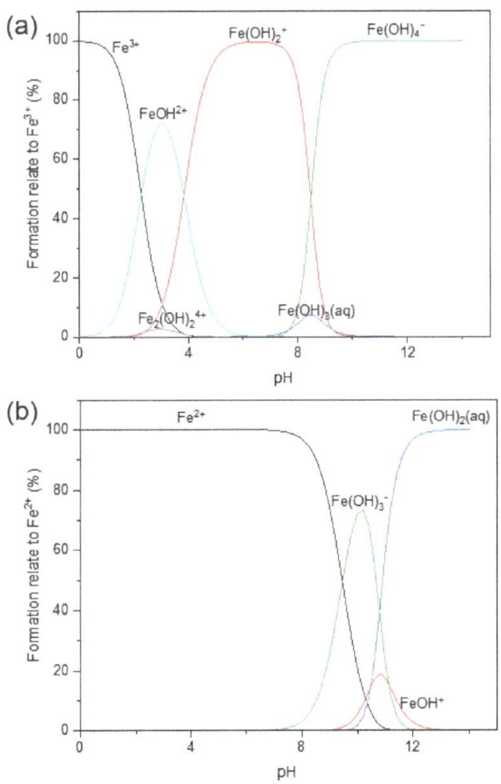

Figure 5. Distribution of (**a**) Fe(III) and (**b**) Fe(II) species at varied pH per theoretical calculation (calculated per Visual MINTEQ, ver. 3.1, Fe = 0.002 molal).

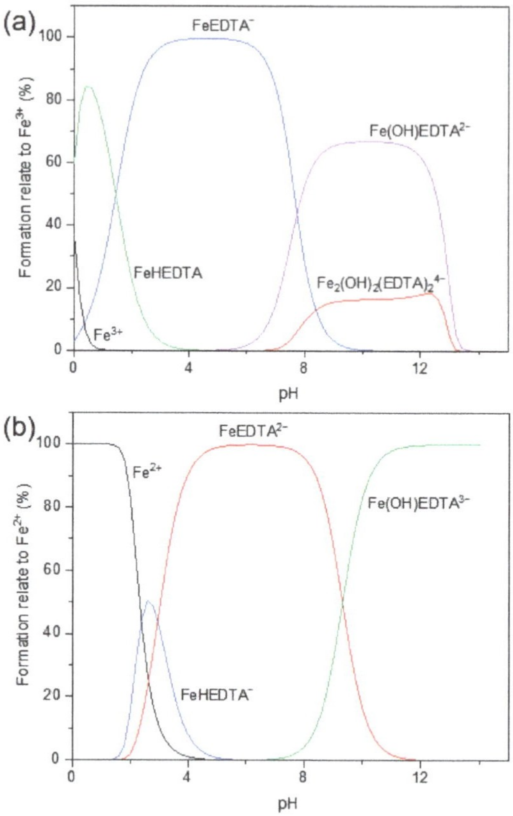

Figure 6. (**a**) EDTA/Fe(III) and (**b**) EDTA/Fe(II) complexes' speciation diagram with the presence of EDTA at an EDTA/Fe molar ratio of 1.

Most of the Fe ions will precipitate immediately at high pH conditions, but the addition of chelating agents can help maintain Fe(II) solubility and reduce the precipitation rate. This indicates that chelating agents could potentially alleviate the adverse effects of a high initial pH on the PS/Fe(II) system. Different chelating agents may function differently. Under alkaline conditions, citric acid (CA) can react with Fe ions, and Fe(III)(OH)CA and Fe(III)$(OH)_2$CA are the main species of Fe(III)-CA complexes [49]. However, they do not favor the conversion to Fe(II)-CA for oxidative activation, leading to a continuous decrease in pollutant removal efficiency as pH increases. The same impact was observed with the other two chelating agents, TA [55] and OA [55].

In the case of using GA in PS/Fe(II) system to treat IPM [28], the removal efficiency exhibited a pattern of an initial increasing and then decreasing trend as pH increased from 3 to 9. This was due to GA maintaining iron solubility effectively, thus extending the pH range of the reaction to 3.0–8.0. However, when the initial pH increased further to 9.0, the precipitation of both Fe(II) and Fe(III) overcomes the chelating and reducing power of GA complexation. In contrast, when EDTA was added [43], a greater removal efficiency of the target pollutant TCE was observed at even higher pH (pH 10). The EDTA/Fe complexes existed mainly in protonated form at pH 10 (e.g., Fe(III)$(OH)_2$EDTA or Fe(II)$(OH)_2$EDTA), with relatively low formation constants, and exhibited thermodynamic instability. Therefore, the quantity of Fe(II) capable of activating PS and removing target contaminants increases as pH increases, when EDTA is added.

In conclusion, the pollutant removal efficiency for the PS/Fe(II) system is intricately linked to the solution pH, which impacts greatly on the interactions among Fe cycles, PS, and target pollutants. Generally, as the solution pH increased, the improvement of removal efficiency from chelating agents was suppressed, implying that the agents mainly impact Fe cycles to enhance the PS/Fe(II) system.

4. Addition of Reducing Agents

A significant limitation of the PS/Fe(II) system is the buildup of Fe(III) or iron sludge, which can greatly impede the Fe cycle and subsequently decrease the pollutants' degradation efficiency. To mitigate this, reducing agents can also utilize for accelerate the conversion of Fe(III) to Fe(II) to improve the reaction efficiency. Hydroxylamine (HA), a widely recognized reducing agent, has been identified as a suitable reducing agent in previous articles for this purpose [63]. Moreover, the degradation products of HA with Fe(III) mainly consist of inorganic substances, including N_2, NO^{2-}, and NO^{3-} (Equation (7)) [64]. Thiosulfate ($S_2O_3{}^{2-}$), which promotes PS oxidation during polymerization, has likewise been investigated for promoting contaminant removal. The reaction scheme of the presumed persulfate with the copper–thiosulfate complex anion (i.e., $[Cu(S_2O_3)_2]^{3-}$) is expressed as Equation (8) [50].

$$NH_3OH^+ + Fe^{3+} \rightarrow Fe^{2+} + \text{nitrogenous products} \quad (7)$$

$$S_2O_8{}^{2-} + [Cu(S_2O_3)_2]^{3-} \rightarrow SO_4{}^{-\cdot} + SO_4{}^{2-} + 2S_2O_3{}^{2-} + Cu^{2+} \quad (8)$$

While for the reducing agents, sodium thiosulphate, an effective reducing agent, is environmentally friendly, soluble, and suitable for a wide range of application. However, it decomposes easily under light, and high doses of sodium thiosulphate may cause side effects and environmental issues [47]. Oxalic acid (OA) is a strong reducing agent usually found in plants. However, it can also pose toxic effects to humans [51,55].

4.1. Mechanism

The mechanisms of adding reducing agent for PS/Fe(II) system is straightforward. During the activation of PS using Fe(II), Fe(II) itself is oxidized to Fe(III) (Figure 7). The addition of a reductant can directly convert Fe(III) to Fe(II), allowing it to continue activating PS. This process realizes the Fe(II)-Fe(III) cycle, ultimately enhancing the overall reactivity of the PS/Fe(II) system and the removal efficiency of target pollutants [65].

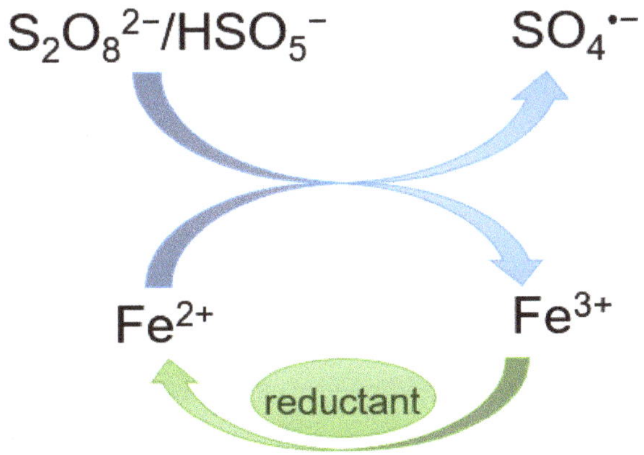

Figure 7. Mechanism diagram of reducing agents for PS/Fe(II) system improvement.

4.2. Effect of Molar Ratio of Reductant/Fe

Figure 8 shows the correlation between the removal efficiency of target pollutant (%) and the molar ratio of the reducing agent/Fe. So far, to the best of our knowledge, HA, thiosulfate and amorphous boron (A-Boron) [66] are all reducing agents used for the improvement of the PS/Fe(II) system. Like the trend found for chelating agents in Figure 2, these removal efficiency data primarily exhibit an initial rise followed by a decline as the molar ratio increases [67]. In most of the cases, the peak removal efficiencies roughly correspond to molar ratios between 0 and 10.

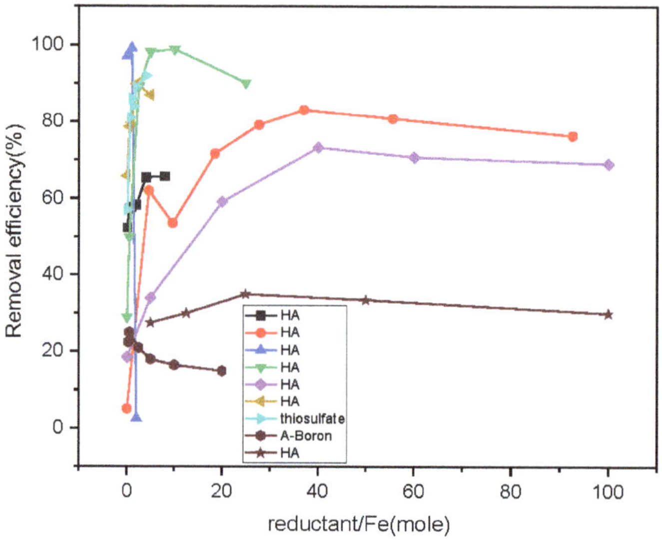

Figure 8. Correlation between target pollutant removal efficiency (%) and molar ratio of reductant/Fe [31,50,56,62,65–68].

Again, in the initial stage, an increase in the molar ratio of reductant/Fe leads to an enhanced removal of target pollutant. This is mainly due to the overwhelming effect of conversion of Fe(III) to Fe(II) (taking HA as an example, Equations (9) and (10)), which in turn activate PS, thereby increasing the removal efficiency.

$$Fe^{3+} + NH_2OH + \rightarrow NH_2O^{\bullet} + Fe^{2+} + H^+ \quad (9)$$

$$Fe^{3+} + NH_2O^{\bullet} + \rightarrow NHO + Fe^{2+} + H^+ \quad (10)$$

However, when the addition of reductant is beyond a certain value, the removal efficiency starts to decrease. Firstly, this may be due to the fact that when the concentration of $NH_2O^{\bullet}$ produced in Equation (11) reaches a certain level, it starts to decay (Equation (11)), leading to an inhibited reduction of Fe(III) in Equation (10) [62].

$$NH_2O^{\bullet} \rightarrow N_2 + H_2O \quad (11)$$

Additionally, HA is also capable of directly reducing PS via a non-radical pathway [56], leading to an inhibition in radical generation. Moreover, a high concentration of HA can also function as a scavenger for $SO_4^{\bullet-}$ and $OH^{\bullet}$, exhibiting rapid reaction rates, specifically $k < 5.0 \times 10^8\ M^{-1}\ s^{-1}$ for $OH^{\bullet}$ and $k = 1.5 \times 10^7\ M^{-1}\ s^{-1}$ for $SO_4^{\bullet-}$ [65]. In addition, upon the addition of HA, the chloride ions consume a portion of $SO_4^{\bullet-}$ to produce less oxidized dichloride radicals and hypochlorous acid [56].

A previous study investigated the impact of HA concentration on TCE degradation in the PS/Fe(II)/HA system [31]. Within 30 min, TCE degradation efficiency rose from 28.9% to 97.9% as HA concentration increased from 0 to 1.5 mM (i.e., reductant/Fe increased from 0 to 10). This efficiency improvement can be primarily attributed to the strong reduction of Fe(III) to Fe(II) by HA. The increase in Fe(II) concentration in the solution was also observed to confirm this conversion. However, a subsequent increase in HA dosage resulted in a decrease in TCE removal. In this contest, HA mainly presented mainly in the form of NH_3OH^+, along with the introduction of more Cl^-, and it scavenges reactive oxygen species ($SO_4^{\bullet -}$ and $OH^\bullet$) produced in the PS/Fe(II)/HA system, leading to reduced TCE removal.

4.3. Effect of Initial pH on Reducing Agents

Analogous to chelating agents, the solution pH governs the distribution of HA, Fe(II), and Fe(III) species, along with the self-decomposition kinetics of PS. Figure 9 illustrates the removal efficiency of target pollutants in relation to the initial solution pH. The data present an upward trend in removal efficiency with an increase in pH, reaching a plateau at pH 3–5, followed by a sharp decrease at pH above 6.

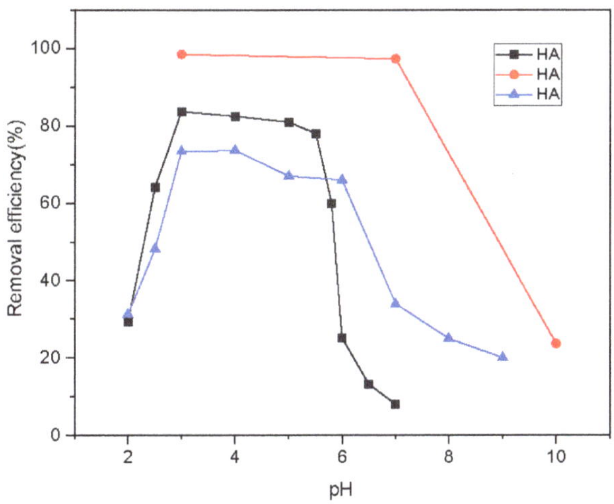

Figure 9. Correlation between target pollutant removal efficiency (%) and initial solution pH in the cases with the presence of varied reducing agents [56,67,68].

A prior study on the HA/Fe(II)/PMS process revealed a marked increase in benzylamine degradation, from 29.23% to a peak of 83.73%, as the initial pH rose from 2.0 to 3.0 [68]. This might be attributed to the formation of $Fe(OH)_2$ under higher pH conditions, which was reported to be more active than Fe(II) ions in the Fenton process. However, within a pH range of 5.5 to 7.0, the removal efficiency of benzylamine began to decrease significantly. This was because Fe(III) ions start to precipitate as iron hydroxide at pH 3.0 and above. In addition, NH_2OH gradually became the main form of HA when the pH increased to 6.0. Consequently, most of the generated active oxidant will be consumed by HA through a side reaction with a high-rate constant, resulting in a decreased removal efficiency.

In summary, solution pH again showed great impacts on Fe cycles, PS, target pollutants, and added reducing agents. Generally, the improvement of removal efficiency from reducing agents decreased at higher pH, especially above pH 6. This could be attributed to the fact that at elevated pH, the Fe(II) regeneration was suppressed and the consumption of PS by reducing agents was more significant.

5. Conclusions and Prospects

In the study, we summarized the publications detailing the use of chelating agents and reducing agents to improve the PS/Fe(II) systems for pollutant degradation. This review aimed to provide a better understanding of the underlying reaction mechanisms and the influence of different experimental conditions based on data extracted from the existing literature. The main conclusions were as follows:

(i) The mechanisms for improvement of the PS/Fe(II) system by chelating agents could be classified into the following three categories: to control excess Fe(II), to reduce Fe(III) to Fe(II), and to inhibit Fe precipitation;
(ii) The target pollutant degradation was most effective when the molar ratio of chelating agent to Fe was maintained between 0 and 2;
(iii) The basic solution was not favored for PS/Fe(II) system improvement by chelating agents, due to hydrolysis and precipitation of the vast majority of Fe(II) and Fe(III) at high pH;
(iv) The mechanism of using a reducing agent for PS/Fe(II) system improvement is straightforward, i.e., the introduced reducing agent can improve the conversion of Fe(III) to Fe(II);
(v) So far, HA has been widely used as reducing agent in several studies for PS/Fe(II) system improvement. In most of the cases, as the dose of HA increased, the removal efficiency of the target pollutant first increased to peak values (at molar ratio of HA/Fe(II) between 0 and 10) then decreased. This decline was attributed to the scavenges of reactive radicals by excessive HA and side reactions;
(vi) Upon adding a reductant, the removal efficiency of target pollutants tended to rise as the pH increased, reaching a plateau at pH 3–5, followed by a sharp decrease at pH 6 and higher. The inhibition at basic solution could be due to the precipitation of Fe and side reactions of PS consumption by reducing agents.

As this field continues to evolve, further research is needed, detailed as follows:

a. Are there any other more powerful chelating agents or reducing agents to enhance the overall performance of the PS/Fe(II) system?
b. The present study did not concentrate on the changes in the processes and products of chelating or reducing agents in persulfate-based advanced oxidation processes.
c. Future research should focus on comparing the removal of same target pollutants by chelating agents and reducing agents;
d. A cost assessment about the cost-effectiveness of PS/Fe(II) system improvement is highly expected, especially for the decision makers in the industry;
e. So far, the studies in this field have focused more on the radical pathways, while the possible mechanisms about non-radical pathways are still unknown;
f. A life-cycle assessment for using chelating or reducing agents to enhance the PS/Fe(II) system or the long-term investigation is needed to reveal the possible negative effects of the introduced agents.

Supplementary Materials: The following supporting information can be downloaded at: https://www.mdpi.com/article/10.3390/pr12112361/s1, Table S1: Correlation between target pollutant removal efficiency (%) and molar ratio of Chelate/Fe; Table S2: Correlation between target pollutant removal efficiency (%) and molar ratio of reductant/Fe; Table S3: Correlation between target pollutant removal efficiency (%) and initial solution pH in the cases with the presence of varied chelating agents and reducing agent.

Funding: This work was supported by the National Key R&D Program of China (No. 2022YFD1700700), National Natural Science Foundation of China (Nos. 52279051 and U21A20162), and Chinese Universities Scientific Fund (No. 2024TC027).

Conflicts of Interest: The authors declare no conflict of interest.

Abbreviations

The following abbreviations are used in this manuscript:

PS	Persulfates
PMS	Peroxymonosulfate
PDS	Peroxydisulfate
EDTA	Ethylene diamine tetraacetic acid
CA	Citric acid
GA	Gallic acid
OA	Oxalic acid
EDDS	Ethylenediaminedisuccinic acid
TA	Tartaric acid
IPM	Iopamidol
TCE	Trichloroethylene
TCA	Trichloroacetic acid
HA	Hydroxylamine

References

1. Wang, P.; Zhang, H.; Wu, Z.; Zhao, X.; Sun, Y.; Duan, N.; Liu, Z.; Liu, W. A data-based review on norfloxacin degradation by persulfate-based advanced oxidation processes: Systematic evaluation and mechanisms. *Chin. Chem. Lett.* **2023**, *34*, 108722. [CrossRef]
2. Long, X.; Luo, J.; Zhong, Z.; Zhu, Y.; Zhang, C.; Wan, J.; Zhou, H.; Zhang, B.; Xia, D. Performance and mechanism of carbamazepine removal by $FeS^{-}S_2O_8{}^{2-}$ process: Experimental investigation and DFT calculations. *Front. Environ. Sci. Eng.* **2023**, *17*, 113. [CrossRef]
3. Xiao, S.; Cheng, M.; Zhong, H.; Liu, Z.; Liu, Y.; Yang, X.; Liang, Q. Iron-mediated activation of persulfate and peroxymonosulfate in both homogeneous and heterogeneous ways: A review. *Chem. Eng. J.* **2020**, *384*, 123265. [CrossRef]
4. Karim, A.V.; Jiao, Y.; Zhou, M.; Nidheesh, P.V. Iron-based persulfate activation process for environmental decontamination in water and soil. *Chemosphere* **2021**, *265*, 129057. [CrossRef] [PubMed]
5. Yang, N.; Cui, J.; Zhang, L.; Xiao, W.; Alshawabkeh, A.N.; Mao, X. Iron electrolysis-assisted peroxymonosulfate chemical oxidation for the remediation of chlorophenol-contaminated groundwater. *J. Chem. Technol. Biotechnol.* **2016**, *91*, 938–947. [CrossRef]
6. Lyu, C.; He, D.; Chang, Y.; Zhang, Q.; Wen, F.; Wang, X. Cobalt oxyhydroxide as an efficient heterogeneous catalyst of peroxymonosulfate activation for oil-contaminated soil remediation. *Sci. Total Environ.* **2019**, *680*, 61–69. [CrossRef]
7. Liu, S.; Gu, C.; Zhang, J.; Luo, C.; Rong, X.; Yue, G.; Liu, H.; Wen, J.; Ma, J. Degradation of 1,2,3-trichloropropane by unactivated persulfate and the implications for groundwater remediation. *Sci. Total Environ.* **2023**, *865*, 161201. [CrossRef]
8. Ma, M.; Chen, L.; Zhao, J.; Liu, W.; Ji, H. Efficient activation of peroxymonosulfate by hollow cobalt hydroxide for degradation of ibuprofen and theoretical study. *Chin. Chem. Lett.* **2019**, *30*, 2191–2195. [CrossRef]
9. Yang, Z.; Zhang, Z.; Zuo, Y.; Zhang, J.; Zhang, P. Comparison of exogenous degrader-enhanced bioremediation with low-dose persulfate oxidation for polycyclic aromatic hydrocarbon removal in alkaline soil: Efficiency and influence on ecological health. *Front. Environ. Sci. Eng.* **2023**, *17*, 133. [CrossRef]
10. Gao, Y.Q.; Gao, N.Y.; Wang, W.; Kang, S.F.; Xu, J.H.; Xiang, H.M.; Yin, D.Q. Ultrasound-assisted heterogeneous activation of persulfate by nano zero-valent iron (nZVI) for the propranolol degradation in water. *Ultrason. Sonochem.* **2018**, *49*, 33–40. [CrossRef]
11. Pan, F.; Ji, H.; Du, P.; Huang, T.; Wang, C.; Liu, W. Insights into catalytic activation of peroxymonosulfate for carbamazepine degradation by MnO_2 nanoparticles in-situ anchored titanate nanotubes: Mechanism, ecotoxicity and DFT study. *J. Hazard. Mater.* **2021**, *402*, 123779. [CrossRef] [PubMed]
12. Liu, M.; Chen, H.; Xiao, P.; Ji, H. Sulfite activation by Jahn-Teller-driven oxygen vacancies Cu-Mn composite oxide for chlortetracycline degradation. *J. Hazard. Mater.* **2024**, *461*, 132658. [CrossRef] [PubMed]
13. Xu, J.; Wang, D.; Hu, D.; Zhang, Z.; Chen, J.; Wang, Y.; Zhang, Y. Magnetic Co-doped 1D/2D structured γ-Fe_2O_3/MoS_2 effectively activated peroxymonosulfate for efficient abatement of bisphenol A via both radical and non-radical pathways. *Front. Environ. Sci. Eng.* **2024**, *18*, 37. [CrossRef]
14. Wu, R.; Wang, S. Integration of microbial reductive dehalogenation with persulfate activation and oxidation (Bio-RD-PAO) for complete attenuation of organohalides. *Front. Environ. Sci. Eng.* **2022**, *16*, 22. [CrossRef]
15. Luo, J.; Gao, Y.; Song, T.; Chen, Y. Activation of peroxymonosulfate by biochar and biochar-based materials for degrading refractory organics in water: A review. *Water Sci. Technol.* **2021**, *83*, 2327–2344. [CrossRef]
16. Nfodzo, P.; Choi, H. Triclosan decomposition by sulfate radicals: Effects of oxidant and metal doses. *Chem. Eng. J.* **2011**, *174*, 629–634. [CrossRef]
17. Gao, Y.; Luo, J.; Song, T.; Yu, X. Research progress on nano-Fe^0/PS system for degradation of refractory organics in aqueous solution. *J. Environ. Chem. Eng.* **2021**, *9*, 105345. [CrossRef]

18. Xiao, R.; Luo, Z.; Wei, Z.; Luo, S.; Spinney, R.; Yang, W.; Dionysiou, D.D. Activation of peroxymonosulfate/persulfate by nanomaterials for sulfate radical-based advanced oxidation technologies. *Curr. Opin. Chem. Eng.* **2018**, *19*, 51–58. [CrossRef]
19. Gao, Y.; Yang, F.; Jian, H.; Zhen, K.; Zhang, P.; Tang, X.; Fu, Z.; Xu, W.; Wang, C.; Sun, H. Pyrene degradation in an aqueous system using ferrous citrate complex activated persulfate over a wide pH rangin. *J. Environ. Chem. Eng.* **2021**, *9*, 106733. [CrossRef]
20. Hasija, V.; Raizada, P.; Thakur, V.K.; Ahamad, T.; Alshehri, S.M.; Thakur, S.; Nguyen, V.-H.; Van Le, Q.; Singh, P. An overview on photocatalytic sulfate radical formation via doped graphitic carbon nitride for water remediation. *Curr. Opin. Chem. Eng.* **2022**, *37*, 100841. [CrossRef]
21. Shen, Y.; Zhao, S.; Lu, Y.; Yang, J.; Wang, J.; Zhang, S. Effective degradation of VOCs from wood by Fe_2^+ chelate activated dual oxidant (H_2O_2-PS). *Chemosphere* **2022**, *291*, 132882. [CrossRef] [PubMed]
22. Zhang, L.; Yang, F.; Cang, Z.; Zhang, J.; Chen, Z.; Yu, C.; Wang, Q. Insight into the degradation of sulfamethoxazole in novel persulfate activation system based on steel slag-based geopolymer. *J. Environ. Chem. Eng.* **2023**, *11*, 110095. [CrossRef]
23. Zhu, S.; Zhou, S.; Yu, Y.; Gao, N.; Dong, B. Degradation of Orange II by electrochemical activation of persulfate: Performance and mechanism. *Desalination Water Treat.* **2017**, *89*, 189–196. [CrossRef]
24. Yuan, Y.; Tao, H.; Fan, J.; Ma, L. Degradation of p-chloroaniline by persulfand activated with ferrous sulfide ore particles. *Chem. Eng. J.* **2015**, *268*, 38–46. [CrossRef]
25. Wang, Q.; Shao, Y.; Gao, N.; Chu, W.; Deng, J.; Shen, X.; Lu, X.; Zhu, Y.; Wei, X. Degradation of alachlor with zero-valent iron activating persulfate oxidation. *J. Taiwan Inst. Chem. Eng.* **2016**, *63*, 379–385. [CrossRef]
26. Zeng, X.; Zhu, J.; Zhang, G.; Wu, Z.; Lu, J.; Ji, H. Molecular-level understanding on complexation-adsorption-degradation during the simultaneous removal of aqueous binary pollutants by magnetic composite aerogels. *Chem. Eng. J.* **2023**, *468*, 143536. [CrossRef]
27. Cao, X.; Song, H.; Li, X.-X.; Qiao, Q.-A.; Zhao, Y.; Wang, Y. Which is the real oxidant in competitive ligand self-hydroxylation and substrate oxidation—A biomimetic iron-hydroperoxo species or an oxo-iron-hydroxy one? *Dalton Trans.* **2022**, *51*, 7571–7580. [CrossRef]
28. Dong, H.; Qiang, Z.; Hu, J.; Sans, C. Accelerated degradation of iopamidol in iron activated persulfate systems: Roles of complexing agents. *Chem. Eng. J.* **2017**, *316*, 288–295. [CrossRef]
29. Li, Y.; Feng, Y.; Yang, B.; Yang, Z.; Shih, K. Activation of peroxymonosulfate by molybdenum disulfide-mediated traces of Fe(III) for sulfadiazine degradation. *Chemosphere* **2021**, *283*, 131212. [CrossRef]
30. Yang, L.; Hai, C.; Hao, X.; Meng, S.; Liu, Y.; Xiong, Z.; Guo, Y.; Zhang, H.; Zhou, P.; Lai, B. Zero-valent tungsten boosted Fenton-like oxidation (Fe(III)/peroxydisulfate) towards long-lasting oxidation of carbamazepine: Performance and mechanism. *Sep. Purif. Technol.* **2023**, *316*, 123780. [CrossRef]
31. Wu, X.; Gu, X.; Lu, S.; Qiu, Z.; Sui, Q.; Zang, X.; Miao, Z.; Xu, M. Strong enhancement of trichloroethylene degradation in ferrous ion activated persulfate system by promoting ferric and ferrous ion cycles with hydroxylamine. *Sep. Purif. Technol.* **2015**, *147*, 186–193. [CrossRef]
32. Khandarkhaeva, M.; Batoeva, A.; Sizykh, M.; Aseev, D.; Garkusheva, N. Photo-Fenton-like degradation of bisphenol A by persulfate and solar irradiation. *J. Environ. Manag.* **2019**, *249*, 109348. [CrossRef] [PubMed]
33. Yang, Z.; Ren, X.; Ding, S.; Chen, R.; Tian, M. Preparation of 1 T-WS2 under different conditions and its enhancement of Fe(III)/Fe(II) cycle, synergistic catalysis of PMS activation and degradation of organic pollutants. *J. Environ. Chem. Eng.* **2023**, *11*, 111444. [CrossRef]
34. Li, C.; Zhu, X.; Wang, Z.; Zhu, D.; Yang, S.; Liu, H.; Li, B.; Tian, S. Insight into utilisation of ilex pur-purea waste extract as an eco-friendly promoter in Fe activated persulfate oxidation for remediation of organic contaminated soil. *Chem. Eng. J.* **2022**, *431*, 134222. [CrossRef]
35. Peluffo, M.; Pardo, F.; Santos, A.; Romero, A. Use of different kinds of persulfate activation with iron for the remediation of a PAH-contaminated soil. *Sci. Total Environ.* **2016**, *563*, 649–656. [CrossRef]
36. Park, S.-M.; Lee, S.-W.; Jeon, P.-Y.; Baek, K. Iron Anode-Mediated Activation of Persulfate. *Water Air Soil Pollut.* **2016**, *227*, 462. [CrossRef]
37. Song, M.; Nguyen, Q.B.; Kim, C.; Hwang, I. Sustained activation of persulfate by slow release of Fe(II) from silica-coated nanosized zero-valent iron for in situ chemical oxidation. *Water Res.* **2023**, *246*, 120715. [CrossRef]
38. Ding, Y.; Zhang, M.; Zhou, S.; Xie, L.; Li, A.; Wang, P. Degradation of dimethyl phthalate through Fe(II)/peroxymonosulphate heightened by fulvic acid: Efficiency and possible mechanism. *Environ. Technol.* **2023**, *44*, 1850–1862. [CrossRef]
39. Yuan, L.; Liu, H.; Lu, Y.; Lu, Y.; Wang, D. Enhancing the dewaterability of waste activated sludge by the combined ascorbic acid and zero-valent iron/persulfate system. *Chemosphere* **2022**, *303*, 135104. [CrossRef]
40. Sheng, B.; Yang, F.; Wang, Y.; Wang, Z.; Li, Q.; Guo, Y.; Lou, Z.; Liu, J. Pivotal roles of MoS2 in boosting catalytic degradation of aqueous organic pollutants by Fe(II)/PMS. *Chem. Eng. J.* **2019**, *375*, 121989. [CrossRef]
41. Qi, C.; Wen, Y.; Zhao, Y.; Dai, Y.; Li, Y.; Xu, C.; Yang, S.; He, H. Enhanced degradation of organic contaminants by Fe(III)/peroxymonosulfate process with l-cysteine. *Chin. Chem. Lett.* **2022**, *33*, 2125–2128. [CrossRef]
42. Hou, K.; Pi, Z.; Yao, F.; Wu, B.; He, L.; Li, X.; Wang, D.; Dong, H.; Yang, Q. A critical review on the mechanisms of persulfate activation by iron-based materials: Clarifying some ambiguity and controversies. *Chem. Eng. J.* **2021**, *407*, 127078. [CrossRef]
43. Liang, C.; Liang, C.P.; Chen, C.C. pH dependence of persulfate activation by EDTA/Fe(III) for degradation of trichloroethylene. *J. Contam. Hydrol.* **2009**, *106*, 173–182. [CrossRef]

44. Zhang, K.; Zhou, X.; Zhang, T.; Yu, L.; Qian, Z.; Liao, W.; Li, C. Degradation of the earthy and musty odorant 2,4,6-tricholoroanisole by persulfate activated with iron of different valences. *Environ. Sci. Pollut. Res.* **2018**, *25*, 3435–3445. [CrossRef]
45. Miao, Z.; Gu, X.; Lu, S.; Dionysiou, D.D.; Al-Abed, S.R.; Zang, X.; Wu, X.; Qiu, Z.; Sui, Q.; Danish, M. Mechanism of PCE oxidation by percarbonate in a chelated Fe(II)-based catalyzed system. *Chem. Eng. J.* **2015**, *275*, 53–62. [CrossRef]
46. Liang, C.; Huang, C.F.; Chen, Y.J. Potential for activated persulfate degradation of BTEX contamination. *Water Res.* **2008**, *42*, 4091–4100. [CrossRef]
47. Zhou, L.; Zheng, W.; Ji, Y.; Zhang, J.; Zeng, C.; Zhang, Y.; Wang, Q.; Yang, X. Ferrous-activated persulfate oxidation of arsenic(III) and diuron in aquatic system. *J. Hazard. Mater.* **2013**, *263*, 422–430. [CrossRef]
48. Liang, C.; Bruell, C.J.; Marley, M.C.; Sperry, K.L. Persulfate oxidation for in situ remediation of TCE. II. Activated by chelated ferrous ion. *Chemosphere* **2004**, *55*, 1225–1233. [CrossRef]
49. Li, Z.Y.; Liu, Y.L.; He, P.N.; Zhang, X.; Wang, L.; Gu, H.T.; Zhang, H.G.; Ma, J. Further understanding the role of hydroxylamine in transformation of reactive species in Fe(II)/peroxydisulfate system. *Chem. Eng. J.* **2021**, *418*, 129464. [CrossRef]
50. Liang, C.; Clifford Bruell, J.; Michael Marley, C.; Kenneth Sperry, L. Persulfate oxidation for in situ remediation of TCE. I. Activated by ferrous ion with and without a persulfate–thiosulfate redox couple. *Chemosphere* **2004**, *55*, 1213–1223. [CrossRef]
51. Cai, S.; Hu, X.; Lu, D.; Zhang, L.; Jiang, C.; Cai, T. Ferrous-activated persulfate oxidation of triclosan in soil and groundwater: The roles of natural mineral and organic matter. *Sci. Total Environ.* **2021**, *762*, 143092. [CrossRef]
52. Ji, Y.; Ferronato, C.; Salvador, A.; Yang, X.; Chovelon, J.M. Degradation of ciprofloxacin and sulfamethoxazole by ferrous-activated persulfate: Implications for remediation of groundwater contaminated by antibiotics. *Sci. Total Environ.* **2014**, *472*, 800–808. [CrossRef]
53. Adewuyi, Y.G.; Khan, M.A. Nitric oxide removal by combined persulfate and ferrous-EDTA reaction systems. *Chem. Eng. J.* **2015**, *281*, 575–587. [CrossRef]
54. Wu, X.; Gu, X.; Lu, S.; Xu, M.; Zang, X.; Miao, Z.; Qiu, Z.; Sui, Q. Degradation of trichloroethylene in aqueous solution by persulfate activated with citric acid chelated ferrous ion. *Chem. Eng. J.* **2014**, *255*, 585–592. [CrossRef]
55. Han, D.; Wan, J.; Ma, Y.; Wang, Y.; Li, Y.; Li, D.; Guan, Z. New insights into the role of organic chelating agents in Fe(II) activated persulfate processes. *Chem. Eng. J.* **2015**, *269*, 425–433. [CrossRef]
56. Han, D.; Wan, J.; Ma, Y.; Wang, Y.; Huang, M.; Chen, Y.; Li, D.; Guan, Z.; Li, Y. Enhanced decolorization of Orange G in a Fe(II)-EDDS activated persulfate process by accelerating the regeneration of ferrous iron with hydroxylamine. *Chem. Eng. J.* **2014**, *256*, 316–323. [CrossRef]
57. Rastogi, A.; Al-Abed, S.R.; Dionysiou, D.D. Effect of inorganic, synthetic and naturally occurring chelating agents on Fe(II) mediated advanced oxidation of chlorophenols. *Water Res.* **2009**, *43*, 684–694. [CrossRef]
58. Yu, S.; Gu, X.; Lu, S.; Xue, Y.; Zhang, X.; Xu, M.; Qiu, Z.; Sui, Q. Degradation of phenanthrene in aqueous solution by a persulfate/percarbonate system activated with CA chelated-Fe(II). *Chem. Eng. J.* **2018**, *333*, 122–131. [CrossRef]
59. Anotai, J.; Bunmahotama, W.; Lu, M.C. Oxidation of aniline with sulfate radicals in the presence of citric acid. *Environ. Eng. Sci.* **2011**, *28*, 207–215. [CrossRef]
60. Zheng, F.; Gao, B.; Sun, Y.; Shi, X.; Xu, H.; Wu, J.; Gao, Y. Removal of tetrachloroethylene from homogeneous and heterogeneous porous media: Combined effects of surfactant solubilization and oxidant degradation. *Chem. Eng. J.* **2016**, *283*, 595–603. [CrossRef]
61. Rastogi, A.; Al-Abed, S.R.; Dionysiou, D.D. Sulfate radical-based ferrous–peroxymonosulfate oxidative system for PCBs degradation in aqueous and sediment systems. *Appl. Catal. B-Environ.* **2009**, *85*, 171–179. [CrossRef]
62. Tan, C.; Gao, N.; Chu, W.; Li, C.; Templeton, R.M. Degradation of diuron by persulfate activated with ferrous ion. *Sep. Purif. Technol.* **2012**, *95*, 44–48. [CrossRef]
63. Zhang, H.; Liu, X.; Lin, C.; Li, X.; Zhou, Z.; Fan, G.; Ma, J. Peroxymonosulfate activation by hydroxylamine-drinking water treatment residuals for the degradation of atrazine. *Chemosphere* **2019**, *224*, 689–697. [CrossRef] [PubMed]
64. Li, D.; Zheng, T.; Yu, J.; He, H.; Shi, W.; Ma, J. Enhancement of the electro-Fenton degradation of organic contaminant by accelerating $Fe_3{}^+/Fe_2{}^+$ cycle using hydroxylamine. *J. Ind. Eng. Chem.* **2022**, *105*, 405–413. [CrossRef]
65. Peng, H.; Zhang, W.; Liu, L.; Lin, K. Degradation performance and mechanism of decabromodiphenyl ether (BDE209) by ferrous-activated persulfate in spiked soil. *Chem. Eng. J.* **2017**, *307*, 750–755. [CrossRef]
66. Meng, S.; Zhou, P.; Sun, Y.; Zhang, P.; Zhou, C.; Xiong, Z.; Zhang, H.; Liang, J.; Lai, B. Reducing agents enhanced Fenton-like oxidation (Fe(III)/Peroxydisulfate): Substrate specific reactivity of reactive oxygen species. *Water Res.* **2022**, *218*, 118412. [CrossRef]
67. Liu, G.; Li, X.; Han, B.; Chen, L.; Zhu, L.; Campos, L.C. Efficient degradation of sulfamethoxazole by the Fe(II)/HSO5− process enhanced by hydroxylamine: Efficiency and mechanism. *J. Hazard. Mater.* **2017**, *322*, 461–468. [CrossRef]
68. Zou, J.; Ma, J.; Chen, L.; Li, X.; Guan, Y.; Xie, P.; Pan, C. Rapid acceleration of ferrous iron/peroxymonosulfate oxidation of organic pollutants by promoting Fe(III)/Fe(II) cycle with hydroxylamine. *Environ. Sci. Technol.* **2013**, *47*, 11685–11691. [CrossRef]

MDPI AG
Grosspeteranlage 5
4052 Basel
Switzerland
Tel.: +41 61 683 77 34

Processes Editorial Office
E-mail: processes@mdpi.com
www.mdpi.com/journal/processes

www.ingramcontent.com/pod-product-compliance
Lightning Source LLC
Chambersburg PA
CBHW061610220926
51850CB00005B/470
9783725855797